Artificial Intelligence

R. Kruse E. Schwecke J. Heinsohn

Uncertainty and Vagueness in Knowledge Based Systems

Numerical Methods

With 59 Figures

Springer-Verlag
Berlin Heidelberg New York
London Paris Tokyo
Hong Kong Barcelona
Budapest

Rudolf Kruse
Erhard Schwecke
Department of Computer Science
Technical University of Braunschweig
Bültenweg 74/75
W-3300 Braunschweig
FRG

Jochen Heinsohn
German Research Center
for Artificial Intelligence (DFKI)
Stuhlsatzenhausweg 3
W-6600 Saarbrücken 11
FRG

ISBN 3-540-54165-9 Springer-Verlag Berlin Heidelberg New York
ISBN 0-387-54165-9 Springer-Verlag New York Berlin Heidelberg

Typesetting: Camera ready by author
Printed in the United States of America

45/3140-543210 – Printed on acid-free paper

Preface

The primary aim of this monograph is to provide a formal framework for the representation and management of uncertainty and vagueness in the field of artificial intelligence. It puts particular emphasis on a thorough analysis of these phenomena and on the development of sound mathematical modeling approaches. Beyond this theoretical basis the scope of the book includes also implementational aspects and a valuation of existing models and systems.

The fundamental ambition of this book is to show that vagueness and uncertainty can be handled adequately by using measure-theoretic methods. The presentation of applicable knowledge representation formalisms and reasoning algorithms substantiates the claim that efficiency requirements do not necessarily require renunciation of an uncompromising mathematical modeling. These results are used to evaluate systems based on probabilistic methods as well as on non-standard concepts such as certainty factors, fuzzy sets or belief functions.

The book is intended to be self-contained and addresses researchers and practioneers in the field of knowledge based systems. It is in particular suitable as a textbook for graduate-level students in AI, operations research and applied probability. A solid mathematical background is necessary for reading this book. Essential parts of the material have been the subject of courses given by the first author for students of computer science and mathematics held since 1984 at the University in Braunschweig.

In large parts this research was funded under research contract T/RF0033/I0020/C1315, administered by the Fraunhofer-Gesellschaft. The industrial working group "Evaluation of expert systems" initiated by Hartmut Wolff gave us helpful stimulation concerning applicational aspects. Valuable support for the prototype implementations was provided by the staff members of the AI-department of Dornier GmbH. The authors are also indebted to Philips Research Laboratories, especially Peter Blume, for help and comments on an early version of this book, and also to the German Research Center for Artificial Intelligence (DFKI), especially Wolfgang Wahlster, for support.

The book benefited from the advice of many researchers. We would like to thank especially Piero P. Bonissone, Didier Dubois, David Heckerman, Serafin Moral, Judea Pearl, Henri Prade, Enrique Ruspini, Prakash Shenoy, David Spiegelhalter, Philippe Smets, and Lotfi Zadeh as well as the members of the ESPRIT Basic Research Action 3085, Defeasible Reasoning and Uncertainty Management Systems.

For careful proofreading and asking critical questions we thank the doctoral candidates Jörg Gebhardt, Frank Klawonn and Detlef Nauck, for typesetting the manuscript our students Heiner Bunjes, Yan Chu, Niklas Matthiess, Andreas Mellen, Roland Stellmark, and Thomas Wahlbuhl.

We are grateful to Springer-Verlag for their steady support and the competent and constructive remarks of anonymous referees.

Finally we thank our families for their patience.

Braunschweig, April 1991 R. Kruse, E. Schwecke, J. Heinsohn

Table of Contents

Chapter 1

General Considerations of Uncertainty and Vagueness

This chapter gives a brief introduction to the field of artificial intelligence and the basic epistemic aspects of modeling in its most general meaning. Furthermore a survey of the concept and organization of the book is provided.

1.1 Artificial Intelligence

At the initial stage of automated data processing, the practical application of computing devices was restricted to those problems that require the handling of great amounts of data or the performance of complex calculations. Examples are the 1890 census in the USA where the Hollerith punched-card system was used, or the World War II COLOSSUS project, where with the cooperation of Alan Turing a vacuum tube-calculator was constructed to decipher (German) military codes. In the following decades the dramatically increasing performance of computers allowed the development of problem oriented programming languages, but nevertheless, the typical fields of application of data processing were still databases, simulation systems or numerical methods.

In the early 1970s, first attempts at the adaptation of human problem solving strategies were made in order to enable the application of computers in fields where exact theories are missing but where only the long term experience of human "experts" is available. The aim of knowledge based approaches is to encode knowledge and to perform symbol manipulations which correspond to reasoning processes of human beings. Although the euphoria in the beginning turned out to be inappropriate, today knowledge based techniques are successfully applied in the area of image and speech processing as well as in the field of robotics, and "artificial intelligence" is nowadays an important and accepted branch of knowledge, not only of computer sciences.

An important subject within this research area is that of expert systems. These "intelligent" information retrieval systems aim to answer inquiries of users concerning some restricted domain based on expert knowledge provided in a knowledge acquisition process. Note that "artificial intelligence" does *not* mean making computers think.

The development of expert systems leads principally to two important conceptual problems. Firstly the knowledge has to be adequately represented, which means a model for the codification of knowledge has to be formulated, in order to equip symbols with suitable semantics. Secondly the reasoning process itself has to be modeled by operations on these symbols. At first glance the calculus of formal logic seems to be appropriate, but logic distinguishes just between true and false propositions whereas humans are capable of coping efficiently and successfully with highly subjective partial beliefs and of drawing reasonable conclusions from distorted, hazy information of restricted reliability. Thus the development of knowledge representation formalisms requires us to take into account that we have to deal with *uncertainty* and *vagueness*, which are the two most important forms of partial *ignorance*. The purpose of this book is to consider measure theoretic methods rather than logical approaches. Also the field of non-monotonic reasoning is not contained since this would require another monograph. Some of the links to other research fields are however shown in Chap. 15.

As mentioned above, the pivots of this book are uncertainty and vagueness; in Sect. 2.3 they are extensively considered from a conceptual point of view, while the rest of the book deals with the problem of a formal representation of those phenomena.

1.2 Modeling Ignorance

The purpose of this book is to provide and to analyze methods which allow us to handle *ignorance* within knowledge based systems. Ignorance arises from a restricted reliability of technical devices, from partial knowledge, from insufficiencies of observations or from other causes. In the sequel we distinguish between two different types of ignorance: *uncertainty* and *vagueness*. The notion of vagueness refers to the ambiguity (in its colloquial meaning) remaining due to the hazy content of some datum concerning a questionable facet of the actual problem. Vagueness arises whenever a datum, although its *meaning* is not in doubt, lacks the desired precision. Examples may be the result of some measuring process which yields that some electrical potential is *in the range* from 10V to 12V or a driver's report that it took *approximately* 2 hours to get from town A to town B, whereas the exact driving time is not known. Uncertainty, on the other hand, corresponds to a human being's valuation of some datum, reflecting his or her faith or doubt in its source. This concept covers those cases in which the actual state of affairs or process is not completely determined but where we have to rely on some human expert's subjective preferences among the different possibilities. The toss of a coin may serve as an example. There is no vagueness since heads or tails are the only possible results but nevertheless there is uncertainty due to the fact that the actual outcome is open. Another example is a physician's uncertainty concerning the type of cancer a patient suffers from.

As one of our basic assumptions we consider uncertainty and vagueness to superpose each other independently, so an integrated treatment arises largely from the consideration of the solitary phenomena. Although other approaches are possible, the scope of our book is restricted to the analysis of *numerical models*, since they turn out to be very useful with respect to their operational character. Early models for the description of ignorance in range of knowledge based systems, for example the certainty factor calculus [Shortliffe 1976], were developed in order to cover the representation of uncertain and vague knowledge as a whole. The special merits of those models are their restricted requirements with respect to computational effort as well as to the expert in the context of knowledge acquisition. Their main deficiency is that they are often *heuristic*, which means they evade validation (or falsification), since due to the lack of exact semantics an objective measurement of the quantities considered is not possible.

Before we turn to the essential intent of this book, namely to provide methods for an integrated treatment of uncertainty and vagueness in the range of knowledge based systems, some epistemological remarks are indispensible. Since the application of mathematical techniques always relies on "models" of the actual problems, it should be clarified what "modeling" means, what can be achieved with models, and where the limits are. Of course it would be beyond the scope of this book to go into the philosophical details of modeling theory. Our aim is just to give brief review of those aspects which are important for our further considerations.

From our point of view, the basic intention of any model is to reflect properties of the real world, i.e. to enable the prediction of a system's behavior in the real world. Thus a model, and also a model for handling uncertainty and vagueness, should be *objective*, i.e. it should be possible to valuate a model's quality by comparising the model's predictions with what really happens. Every objective model has the potential to be found false, since it separates things considered to be impossible from those considered to be possible. A model that allows everything to happen is of no use. So as soon as an "impossible" thing is observed the model is falsified. Note that from an epistomological point of view a model can *never* be verified, and the only reasonable argument for its validity is that all efforts to falsify it have failed.

Of course any model is designed to cover just a restricted aspect of the real world, e.g. Newton's classical gravitational theory is restricted to those processes where the speed of any movement is far below the speed of light. Thus it is clear that there are borderlands beyond which a model's predictions do not fit with the empirical findings. Unfortunately there is no way to determine *within* the model whether the results of its application fit with reality – this has to be done by empirical procedures. The guidelines for the choice of a model should be *efficiency* and *sufficiency*, where sufficiency means that the precision of the predictions we obtain meets our requirements, and efficiency refers to their practical applicability. A model should be never created just for its own sake, but in order to obtain operational descriptions of what we observe in reality, i.e. natural laws. So the choice of a model always means a compromise between efficiency and precision, and making a good choice requires a lot of empirical effort or a lot of experience. Finally, in this context it should be stressed that it is of particular importance to keep apart the structures considered in a model and those of the real world. All those entities our models refer to are just constructions which enable and facilitate our talking and thinking about perceptions.

In the handling of uncertainty, the use of probabilistic models has a long tradition. Probabilistic models allow predictions when the number of influences affecting the questionable process is vast, but the predictions are very weak, in contrast to those we obtain, e.g. from Newton's theory of gravity. If we consider the results of continued tosses of a fair coin, the typical probabilistic model tells us that (on a long term basis) the relative frequency of "heads" is 50%, but it tells us very little about an actual toss. Nevertheless, also probabilistic models are often refutable: if in our (fair) coin tossing experiment we observe, in a long run, the relative frequency of "heads" to be 99%, then this points to the fact that the model is not valid. Especially the occurence of zero-probability events often gives rise to the suspicion that the model used is inadequate. Note that also in the handling of uncertainty, real world properties and modeling techniques have to be thoroughly distinguished. For example, *randomness* is a useful concept within some models but does not necessarily correspond to a structure in the real world. Some researchers hold that "there is no probability at all but only probabilistic models" [de Finetti 1970], [Matheron 1987].

In the area of knowledge representation we often have to rely on subjective probabilities, reflecting degrees of rational belief rather than objectively measureable quantities. A crucial point of this approach and all subjective theories which consider "truth" as a property of some person's state of mind is their lack of objectivity, since they can easily evade any criticism and therefore are almost irrefutable. Nevertheless they are of value as they allow the long run experience of experts to be utilized.

Two other subjective theories we consider in our book are the theory of fuzzy sets and the theory of evidence. Fuzzy sets [Zadeh 1965] were originally intended to model vagueness, but they turned out to be also useful for the description of uncertainty. Whenever the theory of fuzzy sets is applied in this way we talk of "approximate reasoning". In contrast to probability theory, the theory of fuzzy sets relies on the attachment of membership degrees, but it lacks an exact semantics of these quantities, which are indispensible for performing "objective" measurements.

The theory of evidence is based on belief functions and Dempster's rule of combination. Whereas in [Shafer 1976] a very intuitive idea of "belief" causes problems with the balance of model inputs and real world parameters to arise [Pearl 1986], the probabilistic interpretation of "evidence masses" we propose leads to sufficient models. From our point of view an adequate model for uncertainty should allow us to quantify trust, thus numbers on a ratio scale should be used in order to enable us to talk about relative frequencies or ratios. So in this book we restrict ourselves to a description of uncertainty by probabilistic models, including mass distributions which we conceive to be probabilistic in nature.

The various schools of probability have applied probabilistic models to handle uncertainty but have left out of consideration such other kinds of ignorance as vagueness, distortion, etc. [Smithson 1989]. Of course it is possible to use probabilistic approaches also in these cases, but several objections have been made to the employment of "pure" probability theory concerning the efficiency as well as the sufficiency of probabilistic models, especially if the treatment of vagueness is required. Vagueness, in our opinion, corresponds to the intuitive idea of "gradedness", where different descriptions are obtained within different contexts. Thus we apply the calculus of L-sets, which are essentially layered sets. As the first approximation the reader may imagine each level as one shade of grey, for each of which a corresponding crisp picture of the entity in question is extracted from a hazy picture.

Note that in principle each of the theories we mentioned above can be applied in order to model vagueness as well as uncertainty, but in practical applications it turns out that the success of the respective models is very different. Nevertheless, choosing an adequate model is still a difficult task. From our point of view, for the representation of vagueness the L-set calculus should be applied, whereas for the description of uncertainty we prefer probabilistic approaches. When uncertainty and vagueness have to be treated simultaneously then we have to deal with the attachment of (probability) masses to L-sets. In [Kruse and Meyer 1987] randomness and vagueness are considered in the context of statistical evaluations.

1.3 The Scope of the Book

As we have said already, the content of this book revolves around the notions of vagueness and uncertainty. So the concept of the book is as follows. *Firstly* the formal representation of these phenomena is analyzed from a rigorous mathematical point of view. The result of these efforts is a general model that allows us to deal with the problems of knowledge representation and reasoning. *Secondly* we describe the experiences we have gained during the prototypical implementation of three special cases of this general model. *Thirdly* we compare our ideas with existing systems and evaluate the models behind these systems using the results of the earlier parts of the book.

The essential problems arising from the theoretical treatment of ignorance in the context of knowledge representation and reasoning are:

- The choice of an adequate "frame of discernment" consisting of atomic "states of the world". Its elements are subject to knowledge about facts, rules, etc. It has to be assured that the "frame of discernment" sufficiently covers the actual domain and allows the desired resolution.
- A suitable "modularization" of the domain. In this context it is important that the structure of the knowledge induces the modularization. Perceiving it the other way round leads to serious problems. The existence of pieces of knowledge concerning only partial aspects of the domain, for example rules involving only a subset of variables, indicates that such modularization is possible. If a decomposition into subdomains is performed, then their mutual relationships have to be explicitly determined and represented. Note that up to now neither uncertainty nor vagueness have appeared on stage.
- The adequate numerical representation of uncertain and vague pieces of information, to enable the expert to express his or her (partial) knowledge and ignorance.
- The choice of an inference mechanism, in order to integrate the expert's knowledge and the consulting user's observations. The respective mechanism should reflect what the expert considers a priori to be "admissible" states.
- Making reasonable decisions based on the results of the inference process.

The approach to the treatment of uncertainty and vagueness we present in this book implies solutions to these problems. These solutions, which are mainly choices among possibilities, reflect our conceptual view of modeling ignorance as well as our ideas concerning the development of mathematically tractable descriptions of the concepts in question.

We use frames of discernment with a product space structure, spread by the quantities and characteristics considered to be of relevance. The subsequent inclusion of additional characteristics turns out to be a further refinement of the description.

The qualitative dependency structure of the domain is represented by hypergraphs (strictly speaking by hyper*trees*). The pieces of information can be expressed within subspaces of the original product space, each of them corresponding to some hyperedge. Note that the hyperedges are determined by the scope of the different pieces of knowledge — *not* the other way round.

Uncertainty in expert and user information we represent by probabilities, sets of probabilities, and weights or evidence masses, whereas sets, L-sets or fuzzy-sets cover vagueness. We cope with the coincidence of uncertainty and vagueness by (generalized) mass distributions.

Second-order uncertainties are considered only so far as they are of importance for the interpretation of the different experts' knowledge. The several items of information have to be synthesized according the structure of the domain.

Besides the knowledge, the expert has to specify explicitly an adequate inference mechanism. For this purpose we provide updating schemes, which rely in the (pure) probabilistic case on conditioning or in more complex cases on mechanisms such as specialization. From our point of view, inference is in essence a refutation of possibilities, in which new pieces of information cause probabilities (or mass distributions) considered a priori admissible to be rejected.

Finally we have to evaluate and interpret the result we obtain from the inference procedure. This is done by making use of decision theoretical methods. In this context the idea of a most likely hyothesis, given a present state of knowledge, plays an important role.

Of course the final aim of our efforts is the development of concrete systems for application purposes. For reasons of complexity the implementation is not straightforward, thus we emphasize that the underlying product space should *not* be directly represented, but that based on a *modularization* of the domain efficient structures can be derived. Whenever possible parallel algorithms should be applied for the inference procedures.

The content of the book is organized as follows. In the first six chapters we present the different mathematical theories we use in the sequel with emphasis on formal properties in order to provide the theoretical basis for the rest of the book. The subject of the following chapters is the application of the mathematical calculi to model uncertainty and vagueness in the context of knowledge based systems. Besides a pure knowledge representation, which we can imagine as the static aspect of the reasoning problem, we have to take into account the combination and integration of knowledge. This requires us to rely on special structural, qualitative (meta)knowledge beyond the numerical information we obtain directly. The notion of metaknowledge covers our ideas about the physical properties which determine the behavior of what we call the domain. These ideas we obtain from experience, from (possibly contradictory) advice of specialists and practitioneers. The process of updating actual valuations in the light of additional information reflects the dynamics of reasoning.

We present three approaches to the treatment of vagueness, uncertainty, and special cases of their superposition: probabilistic reasoning, fuzzy reasoning and evidential reasoning. The probabilistic reasoning approach is directly adapted from the literature whereas in the case of fuzzy reasoning and evidential reasoning we propose as alternative approaches L-sets and weighted sets, respectively. A reader who is only interested in probabilistic models should read Chaps. 1, 2, 4, 7, 8, and 12. Fuzzy models are described in Chaps. 1, 2, 3, 7, 8, 10, and 11, evidential models in Chaps. 2, 5, 6, 7, 8, 13, and 14. Heuristic models are treated in Chap. 9, whereas in Chap. 15 relations to other methods for handling uncertainty are considered.

Chapter 2

Introduction

This chapter provides the set-theoretical notations we use throughout the whole book. In particular, notations for (finite) product spaces are thoroughly introduced. Moreover a simple example is presented, which will be referred to whenever new concepts have to be illustrated. Finally we return to the notions of uncertainty and vagueness, in order to clarify our *intuitive* understanding of these phenomena.

2.1 Basic Notations

This section introduces the basic notations we use in the sequel. Since our mathematical model requires the extensive handling of product spaces, we first provide the definitions of projection and cylindrical extension and some well known theorems concerning these operations.

Let Ω, a non-empty set, be the formal basis of our further exertions. In the context of uncertainty representation this set Ω is often called the *universe of discourse* or *frame of discernment*. In the sequel we mainly consider *finite sets*. In this case the number of elements in Ω, its cardinality, is abbreviated by card(Ω) or $|\Omega|$. As usual the symbols $\cup$, $\cap$, and $\overline{}$ denote set-theoretic union, intersection, and complement, respectively. $A \subseteq B$ means that A is a subset of B; the empty set $\emptyset$ is a subset of every $B \subseteq \Omega$. For the sets of the natural and real numbers we use the symbols $\mathbb{N}$ and $\mathbb{R}$, respectively. 2^Ω denotes the power set of Ω, i.e. $2^\Omega = \{A \mid A \subseteq \Omega\}$ is the set of all subsets of Ω.

Let f be a function from Θ to Ω. By the *image* of $A \subseteq \Theta$ under f we mean the set of elements $\omega \in \Omega$ such that $\omega = f(\theta)$ for some $\theta \in A$. We denote this image by $f(A)$. The *inverse image* of $B \subseteq \Omega$ under f is the set of those $\theta \in \Theta$ for which $f(\theta) \in B$. We write

$$f^{-1}(B) = \{\theta \in \Theta \mid f(\theta) \in B\}.$$

In order to compare different universes of discourse we introduce the notion of a *refinement*.

(2.1) Definition. *A set Ω' is called a refinement of Ω if there is a mapping $\widehat{\Pi} : 2^{\Omega} \to 2^{\Omega'}$ such that*

(i) $\widehat{\Pi}(\{\omega\}) \neq \emptyset$ *for all* $\omega \in \Omega$,

(ii) $\widehat{\Pi}(\{\omega\}) \cap \widehat{\Pi}(\{\omega'\}) = \emptyset$, *if* $\omega \neq \omega'$,

(iii) $\bigcup\Big\{\widehat{\Pi}(\{\omega\}) \;\Big|\; \omega \in \Omega\Big\} = \Omega'$, *and*

(iv) $\widehat{\Pi}(A) = \bigcup\Big\{\widehat{\Pi}(\{\omega\}) \;\Big|\; \omega \in A\Big\}$.

$\widehat{\Pi}$ is called a *refinement mapping.* If such a mapping exists, then the sets Ω and Ω' are compatible, and the refined space Ω' is able to carry more information than Ω. In this context Ω is said to be a *coarsening* of Ω'.

The next problem we have to solve is to decide for every $\omega \in \Omega$ whether information concerning some set $A' \subseteq \Omega'$ may be of relevance for the valuation of ω or not. For this purpose we define the mapping Π.

(2.2) Definition. *Let Ω' be a refinement of Ω where $\widehat{\Pi} : 2^{\Omega} \to 2^{\Omega'}$ is the corresponding refinement mapping. The mapping*

$$\Pi : 2^{\Omega'} \to 2^{\Omega}, \quad \Pi(A') \stackrel{\mathrm{d}}{=} \{\omega \in \Omega \mid \widehat{\Pi}(\{\omega\}) \cap A' \neq \emptyset\}$$

is called the outer reduction induced by $\widehat{\Pi}$.

$\Pi(A')$, the outer reduction of $A' \subseteq \Omega'$, contains those $\omega \in \Omega$ which have one or more corresponding elements $\omega' \in \widehat{\Pi}(\{\omega\})$ within A'. Note that Π is essentially a projection that attaches to each element $\omega' \in \Omega'$ that element ω_0 with $\omega' \in \widehat{\Pi}(\{\omega_0\})$.

A typical example of a frame of discernment is the set $\Omega = \{$not_at_sea, at_sea$\}$. If the granularity of this set turns out to be too coarse we may switch to a refined set $\Omega' = \{$open_sea, 12-mile-zone, 3-mile-zone, canal, refueling_dock, loading_dock$\}$. The sets Ω and Ω' are compatible, and we obtain the refinement mapping $\widehat{\Pi} : 2^{\Omega} \to 2^{\Omega'}$ where

$$\widehat{\Pi}(\{\text{at\_sea}\}) = \{\text{open\_sea, 12-mile-zone, 3-mile-zone}\}$$
$$\widehat{\Pi}(\{\text{not\_at\_sea}\}) = \{\text{canal, refueling\_dock, loading\_dock}\}.$$

On the other hand we derive for the projection mapping for example

$$\Pi(\{\text{open\_sea}\}) = \{\text{at\_sea}\}$$
$$\Pi(\{\text{canal}\}) = \{\text{not\_at\_sea}\}.$$

As mentioned earlier the mathematical theory is based on the consideration of product spaces, therefore we briefly repeat some well known notions and properties. A finite product space is defined as the Cartesian product of finite sets $\Omega^{(1)}, \ldots, \Omega^{(m)}$. The elements of the product space therefore are m-tuples $(\omega^{(1)}, \ldots, \omega^{(m)})$, where $\omega^{(i)} \in \Omega^{(i)}$, $i = 1, \ldots, m$. A family $\mathcal{U}$ of sets $\Omega^{(i)}$, $i = 1, \ldots, m$, $m \in \mathbb{N}$, we declare to be a *universe*. $M = \{1, \ldots, m\}$ is called the *index set* of $\mathcal{U}$.

Let $\mathcal{U}$ be a universe with index set M. For every index subset $N \subseteq M$ we define the product space

$$\Omega^N \stackrel{\mathrm{d}}{=} \begin{cases} \prod_{k \in N} \Omega^{(k)} & \text{if } N \neq \emptyset, \\ \{\varepsilon\} & \text{otherwise.} \end{cases} \tag{2.1}$$

$\omega^\emptyset$ denotes the empty tuple ε which is the only element of $\Omega^\emptyset$.

Let $N = \{i_1, \ldots, i_n\} \subseteq M$. Then each vector $\omega^N \in \Omega^N$ can be written as $(\omega^{(i_1)}, \ldots, \omega^{(i_n)})$.

(2.3) Definition. *Let $\mathcal{U}$ be a universe with index set M. If S, T, and C are index subsets of M such that $T = S \cup C$, $S \cap C = \emptyset$, then we define the pointwise projection by setting*
(i) $\pi_S^T : \Omega^T \to \Omega^S$, $\pi_S^T(\omega^T) \stackrel{\mathrm{d}}{=} y^S$, where $y^{(i)} = \omega^{(i)}$ for all $i \in S$, $S \neq \emptyset$ and
(ii) $\pi_\emptyset^T : \Omega^T \to \Omega^\emptyset$, $\pi_\emptyset^T(\omega^T) \stackrel{\mathrm{d}}{=} y^\emptyset = \varepsilon$.
In the sequel we use the notation $\omega^T = (\omega^S, \omega^C)$ if

$$\pi_S^T(\omega^T) = \omega^S \quad \textit{and} \quad \pi_C^T(\omega^T) = \omega^C$$

hold.

The concept of pointwise projection coincides in the case of the usual three-dimensional space with geometric projection. For example, let $\mathbb{R}^{\{1,2,3\}} = \mathbb{R}^3 = \mathbb{R} \times \mathbb{R} \times \mathbb{R}$ denote the three-dimensional space of real numbers. In this case the projection $\pi_{\{2\}}^{\{1,2,3\}}$ is just the mapping that assigns to each 3-tuple $(\omega, y, z) \in \mathbb{R}^3$ the value y.

If we want to project subsets of Ω^T to Ω^S, we can easily extend the notation of pointwise projections.

(2.4) Definition. *Let $\mathcal{U}$ be a universe with index set M. If S, T, and C are index subsets of M such that $T = S \cup C$, $S \cap C = \emptyset$, then*
(i) the mapping

$$\Pi_S^T : 2^{(\Omega^T)} \to 2^{(\Omega^S)},$$

$$\Pi_S^T(A) \stackrel{\mathrm{d}}{=} \{\omega^S \in \Omega^S \mid \exists \omega^T \in A : \pi_S^T(\omega^T) = \omega^S\} \tag{2.2}$$

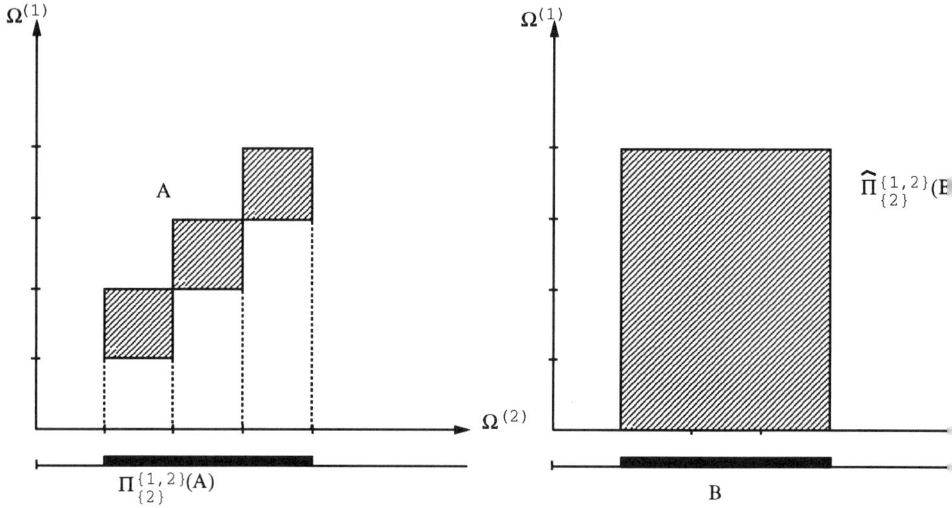

Fig. 2.1. Projection and cylindrical extension

is called the projection of Ω^T onto Ω^S, whereas

(ii) the mapping

$$\widehat{\Pi}_S^T : 2^{(\Omega^S)} \to 2^{(\Omega^T)},$$
$$\widehat{\Pi}_S^T(B) \stackrel{\mathrm{d}}{=} \left\{\omega^T \in \Omega^T \;\middle|\; \pi_S^T(\omega^T) \in B\right\} \tag{2.3}$$

is called the cylindrical extension of Ω^S onto Ω^T.

For the case $S = \emptyset$ we obtain

$$\Pi_\emptyset^T(A) = \begin{cases} \{\omega^\emptyset\} & \text{if } A \neq \emptyset, \\ \emptyset & \text{otherwise} \end{cases}$$
$$\widehat{\Pi}_\emptyset^T(\emptyset) = \emptyset, \text{ and } \widehat{\Pi}_\emptyset^T(\{\omega^\emptyset\}) = \Omega^T.$$

The set Ω^T is a refinement of Ω^S where the refinement mapping $\widehat{\Pi}$ is identical to the cylindrical extension $\widehat{\Pi}_S^T$, and the outer reduction Π coincides with the projection Π_S^T.

Figure 2.1 illustrates projection and cylindrical extension by an example concerning some two-dimensional space $\Omega^{(1)} \times \Omega^{(2)}$.

Let S, T, and U be index subsets of M such that $S \subseteq U \subseteq T$ holds. Then we can compose projection and cylindrical extension in several ways. We derive

- $\Pi_S^U(\Pi_U^T(A)) = \Pi_S^T(A)$ for all $A \subseteq \Omega^T$,
- $\widehat{\Pi}_U^T(\widehat{\Pi}_S^U(B)) = \widehat{\Pi}_S^T(B)$ for all $B \subseteq \Omega^S$,
- $\Pi_S^T(\widehat{\Pi}_S^T(B)) = B$ for all $B \subseteq \Omega^S$,
- $\widehat{\Pi}_S^T(\Pi_S^T(A)) \supseteq A$ for all $A \subseteq \Omega^T$.
  (The equality is in general not valid, as we can easily see from Fig. 2.1.)

In the following we restrict ourselves to the consideration of product spaces $\Omega^{(1)} \times \ldots \times \Omega^{(m)}$, as defined in this section.

2.2 A Simple Example

In this book we use a multivariate approach, i.e. we do not ask if statements are true or false, but we consider characteristics and their values within some domains. Instead of the proposition "The ship is in the loading dock", the characteristic "Location of the ship" together with its possible values, e.g. *open sea, 12-mile zone, 3-mile zone, canal, refueling dock, loading dock* is the subject of our investigations.[1] In many cases this is the appropriate way to model the "part of the world" under consideration. Formally speaking we have to deal with a finite set of characteristics $X^{(1)}, \ldots, X^{(m)}$ and their domains $\Omega^{(1)}, \ldots, \Omega^{(m)}$. The product space

$$\Omega = \prod_{i=1}^{m} \Omega^{(i)} = \left\{ (\omega^{(1)}, \ldots, \omega^{(m)}) \mid \omega^{(i)} \in \Omega^{(i)}, i = 1, \ldots, m \right\}. \tag{2.4}$$

contains m-tuples, each of them representing one "possible world". However, most of the "possible worlds" need not represent a consistent state of the world. To illustrate our approach in this paragraph we provide a simple example in which all relevant statements are absolutely precise and uncertainty appears only in its most basic form, so that the reader is able to focus attention on the essential ideas and get an intuitive understanding of the problems. Throughout the book we will return to this example whenever new concepts have to be clarified and explained.

The subject of our example is the oil contamination of waters by trading vessels. Suppose we know that the locations of interest are *open sea* (z3), *12-mile zone* (z2), *3-mile zone* (z1), *canal (ca)*, *refueling dock* (rd) and *loading dock*

[1] The idea of using the identification of ship locations as an example was inspired by [Lawrence et al. 1986].

(ld). In our example we assume the six locations or areas, respectively, to be disjoint and exhaustive. So the set of possible locations is

$$L = \{\mathrm{z3}, \mathrm{z2}, \mathrm{z1}, \mathrm{ca}, \mathrm{rd}, \mathrm{ld}\}.$$

Assuming that the authorities in charge are *fire department* (fd), *harbor control* (hc), *water police* (wp) and *national preserved area police* (np) and that they can pronounce *admonitions* (ad), make out an *account* (ac) and inflict a *fine* (fi) or *imprisonment* (im), the space of interest is

$$\Omega = L \times A \times M$$

where $A = \{\mathrm{fd}, \mathrm{hc}, \mathrm{wp}, \mathrm{np}\}$ and $M = \{\mathrm{ad}, \mathrm{ac}, \mathrm{fi}, \mathrm{im}\}$. The tuple $(\mathrm{ld}, \mathrm{hc}, \mathrm{fi}) \in \Omega$, for example, means that the harbor control has imposed a fine for an oil contamination on the loading dock.

In this example the product space Ω is what later on will be called the *evidence space*. Now we have to investigate how to represent information obtained from different sources in terms of "structures" defined on this product space. Suppose for simplicity that there is an expert θ_1 who is able to give detailed information on the spheres of responsibility, an expert θ_2 on competences, and an observer who knows the facts about the actual oil contamination under consideration. The information given by θ_1 is the following set of propositions:

(P1) Exclusively the *national preserved area police* is responsible for the *open sea* and the *12-mile zone.*

(P2) The range of responsibility of the *harbor control* is restricted to the *refueling dock* and the *loading dock.*

(P3) The *water police* and the *national preserved area police* are not responsible for the docks.

(P4) The *national preserved area police* is not responsible for the *canal* and the *docks.*

(P5) The *harbor control* is the only authority responsible for the *refueling dock.*

(P6) The *fire department* has no equipment for actions at sea.

Each proposition gives one piece of information on the subspace $L \times A$. Let us consider the first proposition in detail. For our expert θ_1 the tuples (z2, np) and (z3, np) are possibly true, whereas the tuples (z3, wp) or (z3, hc), for instance, are definitely impossible. z1 is not contained in the premise of the rule, so the expert gives no information on z1 by P1. In this way we obtain the following table.

P1	z3	z2	z1	ca	rd	ld
fd	∘	∘	•	•	•	•
hc	∘	∘	•	•	•	•
wp	∘	∘	•	•	•	•
np	•	•	•	•	•	•

An entry ∘ means that the corresponding 2-tuple is definitely impossible, whereas an entry • indicates that the corresponding 2-tuple is either possibly true or that no information is given. Each of the six propositions (P<i>) corresponds to such a table. We can summarize this set of information by superposition of the six tables, where a ∘ dominates an arbitrary number of •'s. It is easy to see that this yields

θ_1	z3	z2	z1	ca	rd	ld
fd	∘	∘	∘	•	∘	•
hc	∘	∘	∘	∘	•	•
wp	∘	∘	•	•	∘	∘
np	•	•	•	∘	∘	∘

Let us consider next the expert θ_2 who is able to provide information on the competences. Assume that the result is the following table concerning the subspace $M \times A$.

θ_2	fd	hc	wp	np
ad	∘	∘	∘	•
ac	•	•	•	∘
fi	•	∘	•	∘
im	∘	∘	•	∘

The observer may provide the following statements:

(O1) The water police was not involved.

(O2) The accident took place either in the canal or at the refueling dock.

We call such pieces of information *evidences*, since they represent direct knowledge about facts. They can be modeled as follows:

O1	
fd	•
hc	•
wp	∘
np	•

O2	
z3	∘
z2	∘
z1	∘
ca	•
rd	•
ld	∘

In order to get a formal representation for the information of the experts and the observer we can identify each of the above tables with a set containing those elements or tuples that are *not impossible*. So from the statements P<i>, $i = 1, \ldots, 6$, we obtain relations $R_i^{A\times L} \subseteq A \times L$, and from O1, O2 we obtain the sets $E^A \subseteq A$ and $E^L \subseteq L$. The superposition of tables obviously corresponds to the intersection of the respective sets, i.e. we derive

$$R^{A\times L} = R_1^{A\times L} \cap \ldots \cap R_6^{A\times L}$$

from the total information provided by θ_1. So finally we have the two relations $R^{A\times L}$ and $R^{M\times A}$ as well as two sets E^A and E^L. Their combination requires a unified representation, so we have to extend them to Ω, i.e. we have to calculate the cylindrical extensions. This leads to

$$\begin{aligned} R_{\theta_1} &= \widehat{\Pi}_{L\times A}^{L\times A\times M}(R^{A\times L}), \\ R_{\theta_2} &= \widehat{\Pi}_{A\times M}^{L\times A\times M}(R^{M\times A}), \\ E &= \widehat{\Pi}_{A}^{L\times A\times M}(E^{A}) \cap \widehat{\Pi}_{L}^{L\times A\times M}(E^{L}). \end{aligned} \tag{2.5}$$

We obtain three subsets of $L \times A \times M$, firstly R_{θ_1} and R_{θ_2} summarizing the experts' knowledge and secondly E summarizing the user's observations. What remains after the transformation of the data into a compatible representation is their evaluation.

From a mathematical point of view we construct two mappings

$$\Gamma_i : \{\theta_i\} \to 2^{L\times A\times M}; \qquad \theta_i \mapsto R_{\theta_i}, \quad i = 1, 2 \tag{2.6}$$

and the set E. The calculation of the results is carried out in two steps. The first step is to combine the expert knowledge. This is done by the construction of some "joint expert" integrating the knowledge of θ_1 and θ_2, we define

$$R \stackrel{\mathrm{d}}{=} R_{\theta_1} \cap R_{\theta_2} \tag{2.7}$$

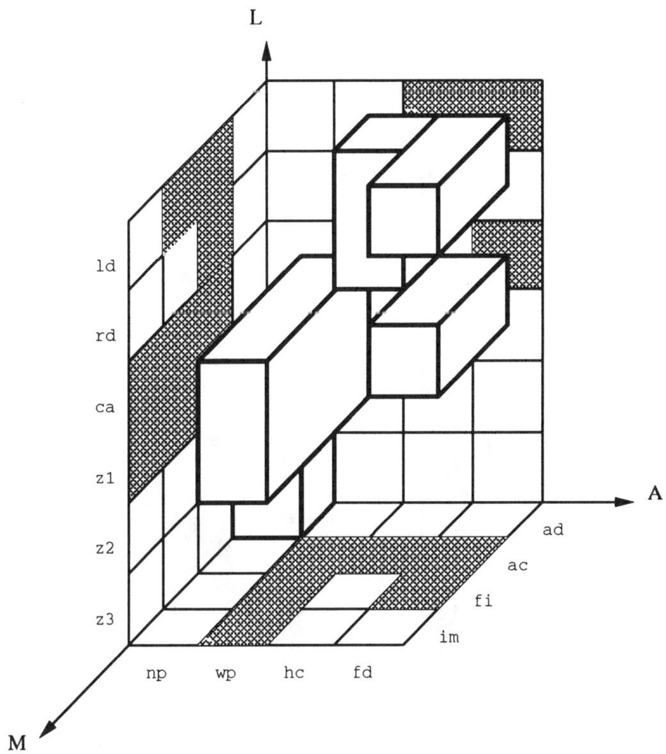

Fig. 2.2 (a). $R = \widehat{\Pi}_{L\times A}^{L\times A\times M}(R^{L\times A}) \cap \widehat{\Pi}_{A\times M}^{L\times A\times M}(R^{A\times M})$

since both, θ_1 and θ_2, are assumed to be reliable. In the second step we have to combine the knowledge with the observations, and again the appropriate operation is intersection; finally we obtain

$$A = R \cap E. \tag{2.8}$$

Figure 2.2 (a)–(c) illustrates the single steps of the above considerations. To answer, for example, a question concerning the reaction of the authorities to the actual accident we have to calculate the projection of A onto the set M. In our example we obtain $\{\text{ac}, \text{fi}\}$, which means an account or a fine is given.

Of course, our example oversimplifies the situation, since in practice the propositions and observations are not always exact. A typical formulation would be "The accident occured *approximately 10 miles* away from the coast". Besides

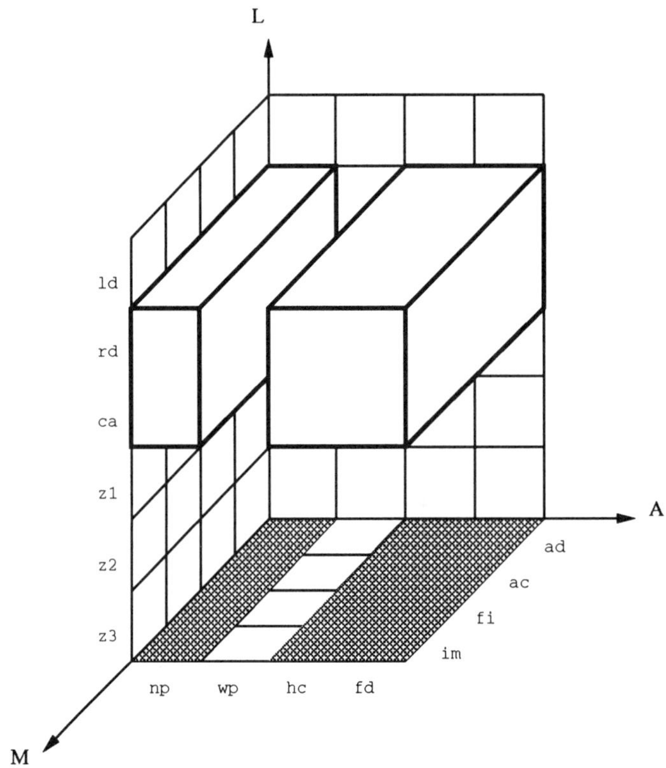

Fig. 2.2 (b). $E = \widehat{\Pi}_A^{L\times A\times M}(E^A) \cap \widehat{\Pi}_L^{L\times A\times M}(E^L)$

such imprecision, or more generally, *vagueness* arising from the formulation of some statement, often its validity is questionable. That means, in many cases the purport of some vague statement is affected with *uncertainty*. For example, the proposition "fire departments have no equipment for actions at sea" may be not simply true or false but hold with a chance of 95%. In the next section we will have a closer look at such phenomena.

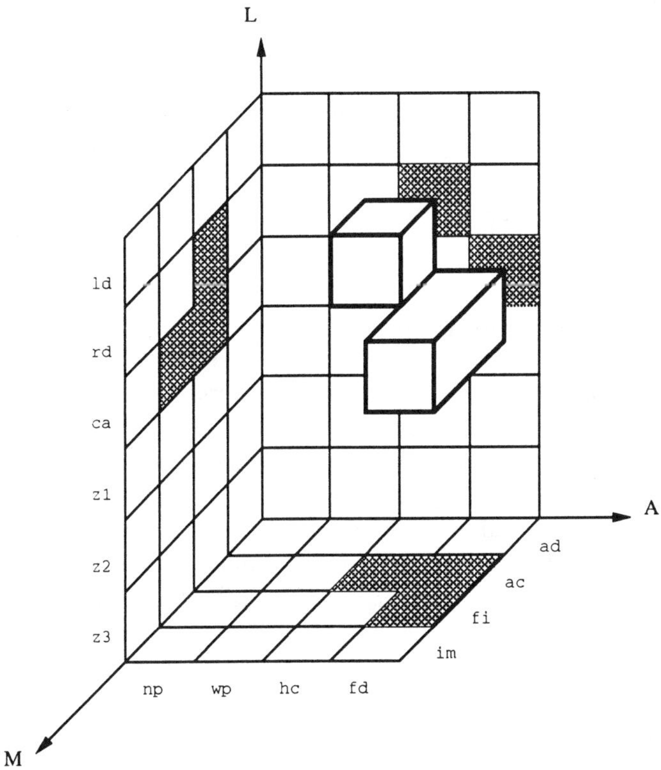

Fig. 2.2 (c). $A = R \cap E, \quad \Pi_M^{L\times A\times M}(A) = \{\mathbf{ac}, \mathbf{fi}\}$

2.3 Vagueness and Uncertainty

As we pointed out in Chap. 1, the purpose of this book is to provide suitable theoretical tools for the treatment of vague and uncertain information about facts and causal relations. Although all our efforts rely on the formulation of suitable models the reader should keep in mind that models, in essence, are nothing but grids which enable us to arrange the vast number of our impressions by means of notions we *create*. The *mathematical* models considered in the sequel allow us to represent perceptions by formal statements and to interpret formal statements against the background of what we perceive to be the "real world". As in the methodology of physical research we first focus on idealized problems in order to obtain sufficient descriptions for delimitable aspects of the mechanisms in question. Later on we will try to extend, combine, and modify the results of our investigation into idealized phenomena. Assuming that they independently superpose each other we aspire to derive methods suitable for application purposes. However, the final criterion for the quality of any model is the usefulness of its results.

The two basic concepts we examine more closely in this book are vagueness and uncertainty in representing the idealized phenomena we mentioned above. In the sequel we give a characterization of these concepts and a short introduction to the formal approaches which are most appropriate from our point of view.

2.3.1 Modeling Vague Data

The final aim of the reasoning processes considered in this book is to determine the true but for the present unknown state of the world. Observations allow us to restrict the set Ω of possible states of the world; these observations may be precise but in general they will contain some inherent ambiguity. If only projections of the world are accessible (for example two-dimensional pictures) then sets of possibilities will remain compatible with the observations. In the worst case the only thing we can observe is a "silhouette on milky glass", where several grey levels invite speculations. Such hazy images consisting of various shades of grey are what we have in mind if we use the term *vague data.* Conceptually we conceive any vague datum to be the vague projection of some crisp observation.

Suppose that in our example concerning the locations of ships, where

$$\Omega = \{\mathrm{z3}, \mathrm{z2}, \mathrm{z1}, \mathrm{ca}, \mathrm{rd}, \mathrm{ld}\},$$

a radar device provides vague outputs, consisting of the three grey levels black, grey and white. Assume that the locations $\{\mathrm{ca}, \mathrm{rd}, \mathrm{ld}\}$ appear in black, while $\{\mathrm{z1}\}$ appears in grey and $\{\mathrm{z2}, \mathrm{z3}\}$ appear in white. (Note that in contrast to the nomenclature we introduced above, where we had in mind shadows on a milky glass, on a radar screen white color usually indicates "there might be something".)

We have to distinguish between models for describing a *vague datum* and the *uncertainty* about the location of the ship based on the vague datum. The information contained in the vague radar device image should be encoded by a function $\mu : \Omega \to \{\mathrm{black}, \mathrm{grey}, \mathrm{white}\}$. In general the range of such a function depends on the application. We often have metric scales, but also ordinal scales or other scales play an important role for modeling measurement values.

This kind of information can be modeled in various ways. We could, for example, rely on heuristic methods in order to cover ignorance as a whole, or we could apply the theory of probability by using likelihood functions. However, we learn from our experience that such techniques do not always fit with the above interpretation of vagueness. It turns out that a set-theoretic approach is promising, even if a single set in general is not sufficient and the specification of multiple crisp descriptions is required.

In our example, using the different grey levels we obtain a scale of acceptability, which allows us to evaluate whether some region of the image covers the actual state of the world. So the basis of the representation is a set L of distinguishable degrees of acceptability.

L should contain besides "white" at least one element that corresponds to the lightest shade of grey actually appearing and denotes the basic level of absolute reliability. The description at this level we call the *sure region* $\mu_{\text{sure}} \subseteq \Omega$; it has to be assured that μ_{sure} covers the true value, i.e. μ_{sure} is the set of those elements of Ω that remain possible in the light of the actual information. So the use of μ_{sure} as a conservative approximation is reasonable. Adding a third "grey" level allows us to obtain a second *speculative* description μ_{spec}, the speculative region, arising from an optimistic rather than conservative valuation of the available evidence. So of course we have $\mu_{\text{spec}} \subseteq \mu_{\text{sure}}$. Indeed on the speculative level we obtain a stronger restriction on the set Ω of possible states of the world, but the price we have to pay is the possibility of overshooting the mark, i.e. of restricting Ω to a strict subset of A so that the final truth might be not among the elements of μ_{spec}. In general we have to consider the case of not only one but two or more grey levels arranged in a natural order, where a high level corresponds to a strong restriction and thus a high danger of "losing the truth". The number of levels depends on the expert's ability to distinguish degrees of speculation. Thus we obtain $\mu_{\text{sure}} = \{\text{z1}, \text{z2}, \text{z3}\}$, $\mu_{\text{spec}} = \{\text{z2}, \text{z3}\}$.

More generally, we consider a totally ordered finite set $(L, \leq)$ of acceptability degrees. The expert is allowed to specify for each acceptability degree $l \in L$ a corresponding region μ_l. This leads to the mathematical calculus of layered sets which provides a suitable formal framework for handling those kinds of vagueness we have to deal with. Note that in the case of $L = \{\text{white}, \text{black}\}$ the layered-set calculus coincides with the usual set calculus.

Layered sets can be equivalently characterized by functions

$$\mu : \Omega \to L, \tag{2.9}$$

where $\mu(\omega)$, $\omega \in \Omega$, is the biggest value l of L such that $\omega \in \mu_l$.

Note that in the case of ordinary sets the functions μ coincide with the well known indicator functions, where $L = \{\text{white}, \text{black}\}$ corresponds to the set $\{0, 1\}$. For our above example we obtain

$$\mu(\omega) = \begin{cases} \text{black} & \text{if } \omega \in \{\text{ca}, \text{rd}, \text{ld}\}, \\ \text{grey} & \text{if } \omega \in \{\text{z1}\}, \\ \text{white} & \text{if } \omega \in \{\text{z2}, \text{z3}\}. \end{cases}$$

In practical applications the representation of layered sets by these functions, called L-sets, is very useful since it allows us to perform efficiently the (generalized) set operations.

Vague data arise in different situations. Their origin cannot be described by a single mechanism. Therefore throughout this book we will often use the term "vagueness" in a rather narrow meaning in order to derive well defined semantics of vague data. While "imprecision" covers those cases in which the true but unknown value $\omega_0 \in \Omega$ is within a subset A of Ω we assume vagueness to arise from the fact that the identification problem is considered from a number of different points of view, called *contexts*. Within each of those contexts we obtain an imprecise description of ω_0, so we get a class of subsets instead of a single subset. In Chap. 3 we will show that L-sets can be seen as condensed representations of such mathematical structure.

2.3.2 Modeling Partial Belief

As mentioned before, the (possibly vague) statements we have to deal with are often not simply true or false but their validity becomes a matter of degree. Such lack of confidence in general arises from a restriction on the reliability of the source which provides the questionable information. So if we consider uncertainty, the subject of our modelling efforts is a human being's judgement concerning the credibility of data. As in the case of vagueness (in principle) we are free to chose an appropriate mathematical calculus. Here it turns out that the decision maker should be enabled to *quantify* his or her degree of belief, which is determined either by objective measurement or by subjective valuation. So one adequate formal approach is the theory of probability.

The basic idea of probabilistic models is the idea of drawing balls out of urns. We are interested in the frequency of certain results (e.g. how often a red ball is drawn). In practice we observe relative frequencies in order to obtain "probabilities", i.e. numbers assigned to events evaluating the likelihood of their occurrence. This "frequentistic" approach is useful in situations where large amounts of data are available. In such cases a suitable model should be based on the theory of probability.

The starting point is Θ, a finite set of distinct possible outcomes of some random experiment. In this context Θ is called the *sample space*. An event is a collection of possible outcomes, i.e. a subset of Θ. So, for any sample space Θ, a probability P is assumed to be a real-valued function mapping the set of possible events into the unit interval and satisfying the Kolmogorov axioms *(i)* $P(A) \geq 0$ for all events $A \subseteq \Theta$,

(ii) $P(\Theta) = 1$,

(iii) additivity condition:

If $A, B \subseteq \Theta$ are mutually exclusive, then

$$P(A \cup B) = P(A) + P(B). \tag{2.10}$$

These three axioms ensure that the assignment of probability numbers to the different events is consistent with our intuitive idea. The additivity condition is sufficient as we restrict ourselves to considering only finite sample spaces. If Θ is infinite we have to use the more general third Kolmogorov axiom

*(iii)** If $A_i \subseteq \Theta$, $i = 1, 2, \ldots$ are mutually exclusive, then

$$P(\bigcup_{i=1}^{\infty} A_i) = \sum_{i=1}^{\infty} P(A_i).$$

Some properties of the theory of probability can be soundly interpreted within the context we are dealing with. $P(\Theta) = 1$ means for example that with probability one Θ is a complete representation of the actual problem; this is a kind of "closed world assumption", since there is no possibility of representing that Θ does not cover the actual problem. In the range of the representation of uncertainty there are two particular situations: firstly, the total absence of uncertainty which, in a probabilistic approach, is coded by some distribution P_0 with $P_0(\{\theta_0\}) = 1$; secondly, the state of total ignorance, which reveals one of the main deficiencies of the theory of probability in the range of uncertainty handling as it cannot be expressed canonically by a single probability distribution (although often the uniform distribution is applied for this purpose). Many cases, e.g. if an expert is only able to specify probability ranges or if a probability distribution remains underspecified at the end of the knowledge acquisition process, the treatment of sets of probability distributions is inevitable. We will discuss how to manage this problem in detail later on.

The methods of probability theory can also be used to model a human decision maker's estimates concerning future occurrences of certain events. With such *subjective* interpretation, probabilities become degrees of rational belief. From the "Bayesian" point of view all probabilities are conditional with respect to the information available at the time of assessment. In this context Bayes' Theorem is of particular importance, thus its name was adapted. A touchstone to clarify the difference between the classical and the Bayesian school is the question whether the probability of some hypothesis to be true may change in the light of data. While with a subjectivistic interpretation such change reflects just increasing information, in the classical view probabilities are model parameters which cannot be changed *within* the model. The assessment of subjective probabilities can be based on the valuation of (virtual) bets. Probability in this interpretation has the nature of an individual assessment and loses its objective character.

We may conceive not the elements of Θ to be the subject of primary interest but their observations of some data space Ω, represented by a mapping Γ that attaches to each sensor $\theta \in \Theta$ its "output" which is an element of Ω, a subset of Ω, or, in the most general case, an L-set of Ω. So a probability P on Θ induces via Γ a structure on Ω which represents the partial beliefs about the actual state of the world ω_0. If the sensors provide set-valued observations at random, i.e. if Γ maps Θ to 2^Ω, the pair (P, Γ) is described as a "random set". Whenever a human decision maker quantifies his or her subjective valuation of imprecise entities we can think of movable *evidence masses* or weights directly attached to the subsets of Ω. The expert has to partition one unit of "belief", attributing bigger amounts to the more reliable pieces of knowledge. This interpretation of the numbers will be subsumed under the label *credibility*.

Let us consider our example where Ω contains 6 elements

$$\Omega = \{\text{z3}, \text{z2}, \text{z1}, \text{ca}, \text{ld}, \text{rd}\}.$$

The propositional statement *in port* corresponds to the event

$$\{\text{ca}, \text{rd}, \text{ld}\}.$$

Such a set may represent the maximum level of differentiation for the expert. So in this case he specifies a *mass distribution* m which is a mapping from 2^Ω to the unit interval where Ω in this context is named the *frame of discernment*. The function $m : 2^\Omega \to [0,1]$ is assumed to satisfy the following axioms:

$$\begin{aligned} &(i)\ \ m(\emptyset) = 0, \\ &(ii)\ \sum_{A:A\subseteq\Omega} m(A) = 1. \end{aligned} \tag{2.11}$$

Subsets $A \subseteq \Omega$ with $m(A)>0$ are called *focal elements* of the mass distribution[2] m. The quantity $m(A)$ is understood to be the measure of belief committed *exactly* to A. To obtain the *total* amount of belief that is bound to A, i.e. the *credibility* of A, we have to add the quantities $m(B)$ where $B \subseteq A$. If we are interested in the maximum amount of belief movable to A we have to sum up those $m(B)$ with $B \cap A \neq \emptyset$. This leads to the following definitions:

If m is a mass distribution the real-valued function

$$\text{Bel}_m : 2^\Omega \to [0,1]; \quad \text{Bel}_m(A) \stackrel{\text{d}}{=} \sum_{B:B\subseteq A} m(B) \tag{2.12}$$

[2] In [Shafer 1976] mass distributions are denoted as *basic probability assignments* (bpa).

is called a *belief function* and the real-valued function

$$\mathrm{Pl}_m : 2^{\Omega} \to [0,1]; \quad \mathrm{Pl}_m(A) \stackrel{\mathrm{d}}{=} \sum_{B: B \cap A \neq \emptyset} m(B) \tag{2.13}$$

is called a *plausibility function.*

In any case we have $\mathrm{Bel}(\Omega) = 1$, which is also some kind of closed world assumption; no amount of belief is allowed to leave Ω. There are two extreme types of mass distribution. $m_0 : 2^{\Omega} \to [0,1]$ defined by $m_0(\Omega) = 1$, $m_0(A) = 0$ for all $A \neq \Omega$ leads, besides to $\mathrm{Bel}(\Omega) = \mathrm{Pl}(\Omega) = 1$, to $\mathrm{Bel}(A) = 0$, $\mathrm{Pl}(A) = 1$ for all $A \neq \Omega$ and represents *total ignorance.* In order to model an ordinary probability we use a function $m_1 : 2^{\Omega} \to [0,1]$ with $m_1(\{\omega\}) = p_\omega$ and $m_1(A) = 0$ for all sets A with $\mathrm{card}(A) > 1$. The belief function induced by such a mass distribution is called a Bayesian belief function, since it obeys the Kolmogorov axioms. In this case the evidence mass is completely distributed to the atomic events. Finally we have to consider the situation of exact knowledge about the "state of the world"which is represented by $m_2 : 2^{\Omega} \to [0,1]$, $m_2(\{\omega_0\}) = 1$, $m_2(A) = 0$ for all $A \neq \{\omega_0\}$.

The crucial point is the assignment of the masses $m(A)$, thus we prefer to consider mass distributions to be induced structures, determined by a "weighted" set acting on some inaccessible sensor space [Kruse, Schwecke, and Klawonn 1991].

Having once constructed a mass distribution it is possible to obtain reasonable results if we consider "belief" to be a partially movable evidence mass. In this model, $m(A)$ is the amount of mass which can move to every subset of A but is committed to A as out of ignorance we do not know its distribution among the subsets. To get a vivid impression of this idea let us return to our example. Consider a statement like "The ship is *in port* with a degree of certainty of 0.6, further evidence is not availiable". This would be modeled by a mass distribution $m : 2^{\Omega} \to [0,1]$, $m(\textit{in port}) = 0.6$, $m(\Omega) = 0.4$, and $m(A) = 0$ otherwise. The important feature is the assignment $m(\Omega) = 0.4$, since it represents the inability to attach that amount of mass to any subset of Ω. Assigning for example $m(\Omega \cap \overline{\textit{in port}}) = 0.4$ would exceed the information given by the expert's statement.

Besides Bayesian belief functions there is another important special case of belief functions which is of interest in the context of uncertain reasoning. If the focal elements $A_1, \ldots, A_n$ of a mass distribution m are nested, i.e. if they can be arranged such that $A_1 \subseteq A_2 \subseteq \ldots \subseteq A_n$, then the induced belief function is called a *consonant belief function* since the underlying evidence points "consonantly in a single direction".

To clarify the special properties of consonant belief functions let us first consider the simplest case of a consonant belief function, that is, a belief function with only one focal element $A \subseteq \Omega$ where $m(A) = 1$. This implies

$$\mathrm{Bel}_m(B) = \begin{cases} 1 & \text{if } A \subseteq B, \\ 0 & \text{otherwise,} \end{cases} \quad \text{and} \quad \mathrm{Pl}_m(B) = \begin{cases} 1 & \text{if } A \cap B \neq \emptyset, \\ 0 & \text{otherwise.} \end{cases}$$

Recall that we assumed Ω to be the set of possible states of the world containing the actual but unknown state ω_0. In the light of the information coded by the above mass distribution each element of A is a candidate for being ω_0 whereas we definitely know $\omega_0 \notin \overline{A}$. So this information can be equivalently represented by a function $\rho : \Omega \to \{0,1\}$; $\rho(\omega) = \mathrm{Pl}(\{\omega\})$, attaching 1 to the possible and 0 to the impossible elements, which is, on the formal level, the indicator function of some set R containing the consistent elements of Ω. Such a "possibility distribution" was considered in the example of Sect. 2.2, though not explicitly. Based on this function ρ we can evaluate the *possibility* of some set $B \subseteq \Omega$, i.e. the truth of the statement "possibly $\omega_0 \in B$". Obviously it is true if $\max\{\rho(\omega) \mid \omega \in B\} = 1$. The *necessity* of some set $B \subseteq \Omega$, i.e. the truth of the statement "necessarily $\omega_0 \in B$", is strongly related to the possibility of $\overline{B}$. It is true if $\max\{\rho(\omega) \mid \omega \in \overline{B}\} = 0$. The relationship of possibility and necessity is intuitively clear: "necessarily $\omega_0 \in B$" being true requires "possibly $\omega_0 \in \overline{B}$" being false.

If we treat consonant belief functions in general, then possibility (and so necessity) becomes a matter of degree. Instead of using the function $\rho : \Omega \to \{0,1\}$ we consider a real valued *contour function* $\pi : \Omega \to [0,1]$, $\pi(\omega) = \mathrm{Pl}(\{\omega\})$ and define the corresponding *possibility measure* by setting

$$\mathrm{poss} : 2^{\Omega} \to [0,1], \quad \mathrm{poss}(B) = \max\{\pi(\omega) \mid \omega \in B\}.$$

The dual measure is called *necessity*, and is defined by

$$\mathrm{nec} : 2^{\Omega} \to [0,1], \quad \mathrm{nec}(B) = 1 - \mathrm{poss}(\overline{B}). \tag{2.14}$$

Note that poss and nec are plausibility and belief functions, respectively, as we will show in Chap. 5. Possibility measures satisfy the three conditions

(i) $\mathrm{poss}(\emptyset) = 0$

(ii) $\mathrm{poss}(\Omega) = 1$

(iii) $\mathrm{poss}(A \cup B) = \max\{\mathrm{poss}(A), \mathrm{poss}(B)\}$ for all $A, B \subseteq \Omega$ (2.15)

which are sufficient to characterize this class of functions completely. These properties, especially the third one, are helpful to clarify the interpretation of possibility measures. They do not require us to think in terms of sets but allow us to concentrate on elements. The possibility of some set is determined by its "most possible" element.

From $\mathrm{nec}(\Omega) = 1 - \mathrm{poss}(\emptyset) = 1$ follows that we have a closed world assumption, i.e. the statement "necessarily $\omega_0 \in \Omega$" is required to be true. Within the framework of possibility measures the state of total ignorance is represented by $\text{poss}(B) = 1$, $\text{nec}(B) = 0$ for all $B \neq \emptyset$, $B \neq \Omega$ (corresponding to the vacuous belief function which of course is a consonant belief function). The state of exact knowledge leads to $\text{poss}(\{\omega\}) = \text{nec}(\{\omega\}) = 0$ for all $\omega \neq \omega_0$ and $\text{poss}(\{\omega_0\}) = \text{nec}(\{\omega_0\}) = 1$ (corresponding to a mass distribution with $\{\omega_0\}$ as its only focal element).

Let us finally consider our example concerning ship locations. The starting point may be the contour function defined by

$$\begin{aligned} \pi(\text{z3}) &= \pi(\text{z2}) = 0 \\ \pi(\text{z1}) &= \pi(\text{ld}) = 0.3 \\ \pi(\text{ca}) &= 0.6 \\ \pi(\text{rd}) &= 1. \end{aligned}$$

This implies $\text{poss}(\{\text{z3}, \text{z2}\}) = 0$ and $\text{nec}(\{\text{z1}, \text{ca}, \text{rd}, \text{ld}\}) = 1$, i.e. it is known to be impossible that the ship's location is in $\{\text{z3}, \text{z2}\}$ and consequently we have $\omega_0 \in \{\text{z1}, \text{ca}, \text{rd}, \text{ld}\}$. $\text{poss}(\{\text{ca}, \text{rd}\}) = 1$, $\text{nec}(\{\text{ca}, \text{rd}\}) = 0.7$ reads "the ship's location is possibly but not with certainty in $\{\text{ca}, \text{rd}\}$".

Chapter 3

Vague Data

In this chapter we present the theory of L-sets as a tool for handling vagueness. A thorough analysis of the vague data provides us with well defined semantics. It turns out that vagueness has to be conceived as imprecision with respect to different contexts. Due to this interpretation L-sets have to be distinguished from fuzzy sets although there are various formal similarities. Finally we address the evaluation of vague data by the use of the concepts of possibility and necessity, and show how to apply these techniques in practice.

3.1 Basic Concepts

As indicated in the introduction we have to analyze statements we encounter in practice with respect to two aspects: firstly the exact meaning of some datum has to be determined, and secondly its validity has to be evaluated. The respective concepts we denoted by vagueness and uncertainty. Throughout this chapter let us assume, for simplicity, that we perfectly trust any source of information, so that we can focus on vagueness.

The essential nature of "information" is that it allows one to restrict some set of disjoint and exhaustive alternatives. By *restriction* we mean the ability to separate things that are compatible with some perception from those that are incompatible. Throughout the whole book the underlying intuitive idea of "vague" data is that of observations, made in order to find out some unknown but *uniquely* determined "true" value. Note that in contrast to those cases we consider in this book, the original entities may also be subsets of Ω instead of point values, for example if a ship's location has to be determined with respect to a very fine frame of discernment where due to its size the ship projects into several regions. [Kruse and Gebhardt 1990].

Let us return to our overall example. Using this example we will clarify the hierarchy of vagueness consisting of the three stages: precision, imprecision and vagueness in general. In the best case the situation is characterized by the total absence of vagueness. Of course, the property of precision depends on the definition of the frame of discernment used to represent the actual domain.

In our example the set of alternatives is $\Omega = \{\text{z3}, \ldots, \text{rd}\}$. Indeed, the most pleasant case is to know "The ship is in z2", but such exact knowledge in practice very rarely appears. It is more realistic to assume that we obtain, for example, information like "The location of the ship is either z1 or z2". This knowledge remains ambiguous about the exact location of the ship. A natural way to describe this imprecise information on the formal level is to consider the set $\{\text{z1}, \text{z2}\}$ which is a subset of Ω. The concept of imprecision covers that of interval analysis which is commonly applied in physics to handle the inaccuracies of measuring instruments. Vague data in its most general form appears, for example, if some body is observed through milky glass. In this simple case we can represent a vague image by a mapping $\eta : \Omega \to L$, where L is the set of (physically existing) grey levels, and to each point ω its grey-value is assigned. This is a quite natural representation of a vague image. In practical applications L may cover physical scales such as energy, intensity, etc.

L is assumed to be an ordered set, where the ordering relation is denoted by $\sqsubseteq$. The scale is *not* required to be a metric scale, but an ordinal scale turns out to be sufficient.

(3.1) Definition. *Each function $\eta : \Omega \to L$ is called an L-set of Ω. $\mathcal{F}_L(\Omega)$ denotes the set of all L-sets of Ω*

Given a vague datum in terms of an L-set $\eta : \Omega \to L$ the question of interest always concerns the actual location of the original entity $\omega_0 \in \Omega$. So the aim of our modeling efforts is to represent the information contained by the vague datum η with respect to this question.

(3.2) Example. Let $\Omega = \{\text{z3}, \text{z2}, \text{z1}, \text{ca}, \text{ld}, \text{rd}\}$, and let

$$\eta_1 : \Omega \to L_1 = \{\text{black}, \text{white}\},$$

where

$$\eta_1(\omega) = \begin{cases} \text{white} & \text{if } \omega \in \{\text{z3}, \text{z2}, \text{z1}\} \\ \text{black} & \text{otherwise,} \end{cases}$$

be a radar image concerning some ship's location. Note that in contrast to the black-and-white pictures we usually bear in mind, in the case of radar images white regions on the screen indicate "there might be something". Based on this image, we know that $A = \{\text{z3}, \text{z2}, \text{z1}\}$ is a *minimal* set that *necessarily* covers ω_0.

Now suppose that the radar device produces three grey levels, i.e. we obtain a vague datum of the form $\eta_2 : \Omega \to L_2 = \{\text{black}, \text{grey}, \text{white}\}$, where

$$\eta_2(\omega) = \begin{cases} \text{white} & \text{if } \omega \in \{\text{z3}, \text{z2}\} \\ \text{grey} & \text{if } \omega \in \{\text{ld}\} \\ \text{black} & \text{otherwise.} \end{cases}$$

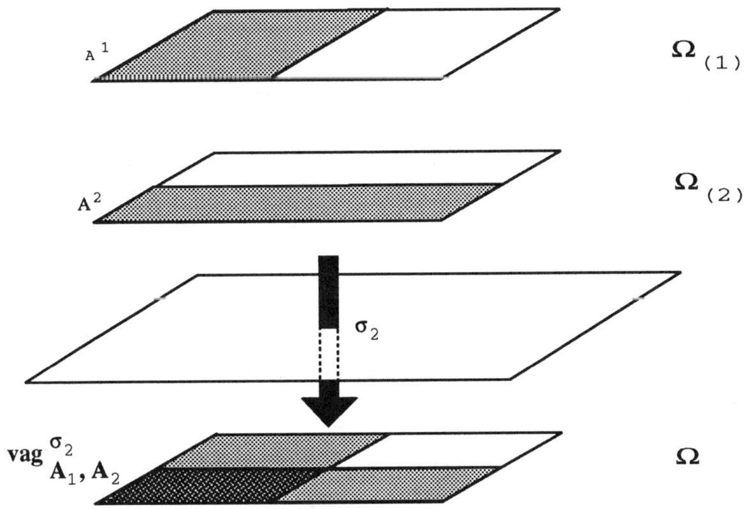

Fig. 3.1. The superposition of shaded patterns

Obviously the statement "The ship is necessarily at sea" is not true, since the location ld appears at the grey level, i.e. it is possible to some degree that the ship is in the loading dock. But on the other hand it would not be reasonable to consider the above statement simply to be false. This example shows that modeling vague data requires an extension of the ordinary set calculus, but before we can turn to the formulation of models we have to clarify our intuitive idea of vagueness. ◇

3.2 On the Origin of Vague Data

Grey levels often arise physically from the superposition of multiple layers of shaded patterns, where the grey gradation corresponds to the number of levels. In the simplest case we can distinguish just between black and white, but in general there is a variety of hues. Figure 3.1 shows the superposition of three patterns, consisting of four grey tones including white.

There is a large variety of sources of vagueness. In this book we cannot consider all sources but restrict ourselves to describe one rather narrow approach that typically arises in knowledge based systems. In the sequel we consider a vague datum to represent the superposition of *restrictions*, each of which reflects what is known about ω_0 in some well defined *context*. In the presence of n different contexts therefore we obtain n restrictions $A^1, \ldots, A^n$ for the unknown original ω_0. The sets $A^i, i = 1, \ldots, n$, we imagine to refer to n "copies" $\Omega_{(1)}, \ldots, \Omega_{(n)}$ of Ω, respectively, so on the formal level we have to deal with the set $\Omega_n \stackrel{\text{d}}{=} \bigcup_{\omega:\omega\in\Omega} \{\omega^1, \ldots, \omega^n\}$ which is a refinement of Ω. The corresponding refinement mapping is $\hat{\sigma}_n : 2^\Omega \to 2^{\Omega_n}$, where

$$\hat{\sigma}_n(\{\omega\}) = \{\omega^1, \ldots, \omega^n\},$$

and for the projection we obtain

$$\sigma_n : 2^{\Omega_n} \to 2^\Omega; \quad \sigma_n(\{\omega^i\}) = \{\omega\} \text{ for } i = 1, \ldots, n.$$

Define $\Omega_{(i)} \stackrel{\text{d}}{=} \{\omega^i \mid \omega \in \Omega\}$. If subsets $A^1, \ldots, A^n$ of $\Omega_{(1)}, \ldots, \Omega_{(n)}$, respectively, are specified, then these sets induce a vague datum which we conceive to be a condensed representation of various restrictions within different contexts. At the perception level we obtain an image consisting of up to n grey tones. On the formal level the superposition of layered restrictions can be easily described by a mapping

$$\text{vag}^{\sigma_n}_{A^1,\ldots,A^n} : \Omega \to \{0, \ldots, n\}; \quad \text{vag}^{\sigma_n}_{A^1,\ldots,A^n}(\omega) \stackrel{\text{d}}{=} \text{card}\big(\{i \mid \omega^i \in A^i\}\big). \quad (3.1)$$

The grey level of ω is determined by the number of contexts in which ω is considered to be admissible. Obviously the value 0 corresponds to white. In the general case a vague datum on Ω is induced by a subset A of Ω_n.

(3.3) Definition. *Each subset $A \subseteq \Omega_n, n \in \mathbb{N}$, induces via the projection mapping σ_n the grey level image*

$$\text{vag}^{\sigma_n}_A : \Omega \to \{0, \ldots, n\}; \quad \text{vag}^{\sigma_n}_A(\omega) \stackrel{\text{d}}{=} \text{card}\Big(\hat{\sigma}_n(\{\omega\}) \cap A\Big)$$

on Ω.

Note that counting contexts does *not* necessarily mean rating importance of contexts. $\text{vag}^{\sigma_n}_A$ is just used as a condensed representation of a class of subsets. This approach can also be applied in product spaces. The mapping vag, the ordinary projection and the cylindrical extension can be canonically combined. Let Ω^M be a product space as defined in Sect. 2.1 and let $\Omega^S, \Omega^T, S \subseteq T$, $T \subseteq M$, be subspaces of Ω^M. Each set $A \subseteq (\Omega^T)_n$ we can conceive to consist of subsets $A^i \subseteq (\Omega^T)_{(i)}$, $i = 1, \ldots, n$, i.e. $A = A^1 \cup \ldots \cup A^n$. The projection of A to $(\Omega^S)_n$ is determined by the union of the projections of the sets A^i to $(\Omega^S)_i$. We obtain

$$\Pi_S^T(A) \stackrel{\mathrm{d}}{=} \bigcup_{i=1}^{n} \Pi_S^T(A^i).$$

Intuitively, the projection of A emerges from the projections within the different contexts. Thus we derive

$$\begin{aligned}
\mathrm{vag}_{\Pi_S^T(A)}^{\sigma_n^S}(\omega^S) &= \mathrm{card}\Big(\hat{\sigma}_n^S(\{\omega^S\}) \cap \Pi_S^T(A)\Big) \\
&= \mathrm{card}\left(\left\{i \;\middle|\; \exists(\omega^T) \in A : \Pi_S^T\Big(\{(\omega^T)^i\}\Big) = \{(\omega^S)^i\}\right\}\right) \\
&= \max\left\{\mathrm{card}\Big(\hat{\sigma}_n^T(\{\omega^T\}) \cap A\Big) \;\middle|\; \Pi_S^T(\{\omega^T\}) = \{\omega^S\}\right\} \\
&= \max\left\{\mathrm{vag}_A^{\sigma_n^T}(\omega^T) \;\middle|\; \Pi_S^T(\{\omega^T\}) = \omega^S\right\} \qquad (3.2)
\end{aligned}$$

where $\sigma_n^S\big((\omega^S)^i\big) = \omega^S$ and $\sigma_n^T\big((\omega^T)^i\big) = \omega^T$ for all i. This means that the vague image in Ω^S is completely determined by the vague image in Ω^T, and no recourse to the original set is required. The same holds for the cylindrical extension which is defined in strict analogy. We obtain

$$\mathrm{vag}_{\hat{\Pi}_S^T(B)}^{\sigma_n^T}(\omega^T) = \mathrm{vag}_B^{\sigma_n^S}\left(\Pi_S^T(\omega^T)\right) \text{ for all } B \subseteq \Omega^S. \qquad (3.3)$$

3.3 Uncertainty Handling by Means of Layered Contexts

In practice the origin of some vague datum is often not known, so we have to rely just on the grey level image. If we know that a one-to-one correspondence between the levels $l \in L$ and the number of layers in an original space as considered in Chap. 2 exists, then the analysis is facilitated. Thus to each $l \in L$ we can assign a number $j(l)$ which is interpreted as the number of superposed layers, i.e. we can define a mapping $j : L \to \{0, \ldots, n\}$. Note that the restrictions in various contexts might coincide. In this case the number of layers would exceed the number of grey levels that actually appear, i.e. it would *not* be that for each $i \in \{0, \ldots, n\}$, there is some $l \in L$ such that $i = j(l)$. For the moment let us assume that any vague datum $\eta : \Omega \to L$ corresponds to a mapping $\mu : \Omega \to \{0, \ldots, n\}$ and that there is an (unknown) subset A of Ω_n with the property $\mathrm{vag}_A^{\sigma_n} = \mu$.

3.3.1 Possibility and Necessity

Recall that our basic aim is to determine the actual state of the world $\omega_0 \in \Omega$. So based on some vague datum $\eta : \Omega \to L$ we have to evaluate whether a given set $A \subseteq \Omega$ covers the element ω_0 or not. More precisely, with respect to $\mu : \Omega \to \{0, \ldots, n\}$ we have to answer the questions

"Is it possible that ω_0 is within a given set $A \subseteq \Omega$"

and

"Is it necessary that ω_0 is within a given set $A \subseteq \Omega$".

On the intuitive level it is clear that dark areas of Ω are more likely to cover ω_0 than light areas. In order to formalize this idea we introduce the notions of graded possibility and necessity.

(3.4) Definition. *Let $\mu : \Omega \to \{0, \ldots, n\}$. $\omega \in \Omega$ is called j-possible with respect to μ, if and only if $\mu(\omega) = j$. A subset $A \neq \emptyset$ of Ω is called j-possible, if $A \cap \{\omega \in \Omega \mid \mu(\omega) \geq j\} \neq \emptyset$, but $A \cap \{\omega \in \Omega \mid \mu(\omega) \geq j+1\} = \emptyset$. $\emptyset$ is 0-possible by definition. (For the sake of vivid nomenclature, instead of 0-possible we use the term impossible.)*

From a mathematical point of view we obtain a set-function

$$\mathrm{Poss}_\mu : 2^\Omega \to \{0, \ldots, n\},$$

which assigns to each subset $A \subseteq \Omega$ its degree of possibility. It is easy to see that $\mathrm{Poss}_\mu(A) = \max\{\mu(\omega) \mid \omega \in A\}$. Since $\mathrm{Poss}_\mu(A) = j$ means that given μ, the degree of possibility for the unknown state ω_0 to be within A is j. In the literature we often encounter notations like $\mathrm{Poss}(\omega_0 \in A|\mu) = j$. Poss_μ has the following properties

(i) $$\mathrm{Poss}_\mu(\emptyset) = 0$$

(ii) $$\mathrm{Poss}_\mu(\Omega) \leq n$$

(iii) $$\mathrm{Poss}_\mu(A \cup B) = \max\big(\mathrm{Poss}_\mu(A), \mathrm{Poss}_\mu(B)\big)$$

(iv) $$\mathrm{Poss}_\mu(A \cap B) \leq \min\big(\mathrm{Poss}_\mu(A), \mathrm{Poss}_\mu(B)\big)$$

where the equality in (iv) is in general not valid.

(3.5) Example. For the vague datum

$$\eta_2 : \Omega \to L_2; \quad \eta_2(\omega) = \begin{cases} \text{white} & \text{if } \omega \in \{\text{z3}, \text{z2}\} \\ \text{grey} & \text{if } \omega \in \{\text{ld}\} \\ \text{black} & \text{otherwise.} \end{cases}$$

considered in Sect. 3.1 we obtain

$$\mu_2 : \Omega \to \{0,1,2\}; \quad \mu_2(\omega) = \begin{cases} 2 & \text{if } \omega \in \{\text{z3}, \text{z2}\} \\ 1 & \text{if } \omega \in \{\text{ld}\} \\ 0 & \text{otherwise.} \end{cases}$$

where we use the identification $j(\text{white}) = 2$, $j(\text{grey}) = 1$, $j(\text{black}) = 0$. Doing so, we assume, that there is a subset A of Ω_2 which induces μ_2. Obviously z2^1, z3^1 as well as z2^2, z3^2 are contained in A, while we do not know whether $\text{ld}^1 \in A$ or $\text{ld}^2 \in A$. So z2 and z3 are 2-possible, ld is 1-possible and the other elements of Ω are impossible with respect to μ_2. ◇

Note that from the above considerations it is obvious that all concepts can be easily extended to the case of set-valued instead of point-valued original entities.

Similarly to a possibility measure we can define a necessity measure.

(3.6) Definition. *Let $\mu : \Omega \to \{0,\ldots,n\}$. A subset $A \subseteq \Omega$ is called j-necessary, $j \neq 0$, with respect to μ, if*

$$\{\omega \in \Omega \mid \mu(\omega) \geq j\} \subseteq A, \quad \textit{but} \quad \{\omega \in \Omega \mid \mu(\omega) \geq j-1\} \not\subseteq A.$$

A is defined to be 0-necessary (not necessary) if there is no $j \neq 0$ such that A is j-necessary.

The function $\text{Nec}_\mu : 2^\Omega \to \{0,\ldots,n\}$ assigning to each subset of Ω its degree of necessity has the following properties

(i) $\text{Nec}_\mu(\emptyset) = 0$

(ii) $\text{Nec}_\mu(\Omega) \leq n$

(iii) $\text{Nec}_\mu(A \cap B) = \min(\text{Nec}_\mu(A), \text{Nec}_\mu(B))$

(iv) $\text{Nec}_\mu(A \cup B) \geq \max(\text{Nec}_\mu(A), \text{Nec}_\mu(B))$

where the equality in (iv) is in general not valid.

The definition of j-possibility and j-necessity with respect to a function $\mu : 2^\Omega \to \{0,\ldots,n\}$ mainly relies on sets of the form $\mu_j \stackrel{\text{d}}{=} \{\omega \in \Omega \mid \mu(\omega) \geq j\}$ which are called *level sets*. We have

$$\text{(i)} \quad \mu_0 = \Omega, \text{ and}$$
$$\text{(ii)} \quad j < j' \Longrightarrow \mu_j \supseteq \mu_{j'}. \tag{3.4}$$

The function μ is completely defined by the total number of level sets since

$$\mu(\omega) = \max\Big\{\min(j, \mathbb{1}_{\mu_j}(\omega)) \;\Big|\; j = 0,\ldots,n\Big\}, \tag{3.5}$$

where $1\!\!1_A(\omega) = n$ if $\omega \in A$ and $1\!\!1_A(\omega) = 0$ otherwise. This equation is referred to as the *representation theorem* [Zadeh 1971a].

The set μ_j is the minimal j-necessary set for $j \geq 1$, so we can determine a vague datum by using this property "the other way round", i.e. by specifying the minimal j-necessary sets $A_j, j = 0, \ldots, n$. These sets should satisfy the property $j < j' \Longrightarrow A_j \supseteq A_{j'}$. Assuming $A_0 = \Omega$ we obtain the corresponding function μ by

$$\mu(\omega) \stackrel{\mathrm{d}}{=} \max\Big\{\min(j, 1\!\!1_{A_j}(\omega)) \;\Big|\; j = 1, \ldots, n\Big\}.$$

In the simplest case of interest we have levels 0, 1 and 2, therefore a minimal 1-necessary (sure) region and a minimal 2-necessary (speculative) region has to be specified. This example leads to the notion of a flou set.

(3.7) Definition. *Let f be the 2-tuple (E_1, E_2), where $E_1 \subseteq E_2 \subseteq \Omega$. E_1 is called the speculative region, E_2 the sure region. The set of all 2-tuples with the above property is denoted by* $\mathrm{Fl}(\Omega)$.

(3.8) Example. The vague data considered in Sect. 3.1 can be described by flou sets $f_1 = (\{\text{z2, z3}\}, \{\text{z2, z3}\})$ and $f_2 = (\{\text{z2, z3, ld}\}, \{\text{z2, z3}\})$, respectively. These sets are assumed to be vague images of unknown subsets. The corresponding possibility and necessity measures consequently take their values within $\{0, 1, 2\}$. ◇

3.3.2 Operations with Vague Data

In the case of imprecise information with respect to some frame of discernment the definition of *operations* is straightforward. If A and B certainly contain the actual state of the world ω_0, then of course it is an element of their intersection $A \cap B$. If ω_0 is either in A or in B, then it is certainly in $A \cup B$. If we know $\omega_0 \in A$ then for each superset B of A we also know $\omega_0 \in B$.

Our aim is the derivation of those operations for vague data which we conceive to be caused by operations performed in some original space Ω_n. Let us start with the treatment of inclusion. If A, B are subsets of Ω_n such that $A \subseteq B$ then we obtain

$$\mathrm{vag}_A^{\sigma_n}(\omega) \leq \mathrm{vag}_B^{\sigma_n}(\omega)$$

for all $\omega \in \Omega$. Thus we define the specialization relation as follows:

(3.9) Definition. *Let $\mu : \Omega \to \{0, \ldots, n\}$ and $\mu' : \Omega \to \{0, \ldots, n\}$. μ is called a specialization of μ' (written $\mu \sqsubseteq \mu'$), if and only if $\mu(\omega) \leq \mu'(\omega)$ holds for all $\omega \in \Omega$.*

Intuitively, $\mu \sqsubseteq \mu'$ means that the vague datum represented by μ' is "as least as" grey as the one respresented by μ. Let A and A' be subsets of Ω_n, inducing μ and μ', respectively, i.e. $\text{vag}_A^{\sigma_n} = \mu$ and $\text{vag}_{A'}^{\sigma_n} = \mu'$. Note that $\mu \sqsubseteq \mu'$ does *not* imply $A \subseteq A'$. Obviously the notion of specialization is not a characterization of subset relationships in Ω_n but it allows us to compare the degrees of possibility derived for the elements of Ω with respect to μ and μ'.

In addition to the definition there is a simple, equivalent characterization of the specialization relation. We have $\mu \sqsubseteq \mu'$ if and only if there are subsets $A, A', E \subseteq \Omega_n$ with $\text{vag}_A^{\sigma_n} = \mu$, $\text{vag}_{A'}^{\sigma_n} = \mu'$ and $A = A' \cap E$. From this point of view a specialization is the result of a *monotonic* data revision.

For *handling* vague data, various operators like union, intersectio projection, etc., acting on spaces $2^{\Omega_n} \times \ldots \times 2^{\Omega_n}$ are needed. The intersection of two pieces of (vague) information may serve as an example. For $A, B \subseteq \Omega_n$, $A \cap B$ corresponds to the (conjunctive) "A as well as B" combination of A and B. Unfortunately in practice we are not able to observe directly the sets A and B but only the vague images $\text{vag}_A^{\sigma_n}, \text{vag}_B^{\sigma_n}$ and $\text{vag}_{A\cap B}^{\sigma_n}$. Thus a technique is required that allows us to determine $\text{vag}_{A\cap B}^{\sigma_n}$ (or a sufficient approximation) just from $\text{vag}_A^{\sigma_n}$ and $\text{vag}_B^{\sigma_n}$.

(3.10) Definition. *Let T be a mapping from $(2^{\Omega_n})^t$ to $2^{\Omega'_n}$. Let $\mathcal{F}_n(\Omega)$ denote the set of all mappings $\mu : \Omega \to \{0, \ldots, n\}$ and $\mathcal{F}_n(\Omega')$ the set of all mappings $\mu' : \Omega' \to \{0, \ldots, n\}$. We call a mapping $\varphi_T : (\mathcal{F}_n(\Omega))^t \to \mathcal{F}_n(\Omega')$ correctness preserving with respect to T if and only if for all $\mu_1, \ldots, \mu_t \in \mathcal{F}_n(\Omega)$, and all $A_1, \ldots, A_t \subseteq \Omega_n$ with $\text{vag}_{A_i}^{\sigma_n} = \mu_i$, $i = 1, \ldots, t$, we have*

$$\text{vag}_{T(A_1,\ldots,A_t)}^{\sigma_n} \sqsubseteq \varphi_T(\mu_1, \ldots, \mu_t).$$

Note that the mapping $1\!\!1_T^{(n)} : \left(\mathcal{F}_n(\Omega)\right)^t \to \mathcal{F}_n(\Omega')$ defined by

$$1\!\!1_T^n(\omega_1, \ldots, \omega_t) = n$$

is of course correctness preserving for each mapping $T : (2^{\Omega_n})^t \to 2^{\Omega'}$, but useless in practice.

(3.11) Example. Let $e = (E_1, E_2)$ and $f = (F_1, F_2)$ be elements of $\mathcal{F}_2(\Omega)$. The operator $\cap : \mathcal{F}_2(\Omega) \times \mathcal{F}_2(\Omega)$ defined by $e \cap f = (E_1, E_2) \cap (F_1, F_2) = (E_1 \cap F_1, E_2 \cap F_2)$ is correctness preserving with respect to the intersection $\cap : 2^{\Omega_2} \times 2^{\Omega_2} \to 2^{\Omega_2}$; $(A, A') \mapsto A \cap A'$ since we have

$$0 \leq \mathrm{vag}^{\sigma_2}_{A\cap A'}(\omega) = \mathrm{card}\Big(\hat{\sigma}_2(\{\omega\}) \cap A \cap A'\Big)$$
$$\leq \min\Big\{\mathrm{card}\Big(\hat{\sigma}_2(\{\omega\}) \cap A\Big), \mathrm{card}\Big(\hat{\sigma}_2(\{\omega\}) \cap A'\Big)\Big\}$$
$$= \min\{\mathrm{vag}^{\sigma_2}_{A}(\omega), \mathrm{vag}^{\sigma_2}_{A'}(\omega)\}$$

for all $\omega \in \Omega$ and $A, A' \in \Omega_2$. If we know that the original sets A, A' are subsets of Ω_2 then the "best", i.e. most specific, correctness preserving operation with respect to the intersection is $\varphi_\cap$ defined by

$$\varphi_\cap(e,f)(\omega) = \max\{\mathrm{vag}^{\sigma_2}_{A\cap A'}(\omega) \mid \mathrm{vag}^{\sigma_2}_{A}(\omega) = e, \mathrm{vag}^{\sigma_2}_{A'}(\omega) = f, A, A' \subseteq \Omega_2\}.$$

Note that the operator $\cup$ defined by

$$e \cup f = (E_1, E_2) \cup (F_1, F_2) = (E_1 \cup F_1, E_2 \cup F_2)$$

is *not* correctness preserving with respect to the union in Ω_2. ◇

Similar methods can be applied to other operations, even if the refined space is Ω_n instead of Ω_2. Roughly speaking, we used an "extension principle" in order to extend T to an operator φ_T acting on vague data. The interpretation of the results causes no difficulties, since possibility and necessity degrees can be easily determined.

In the range of knowledge representation, special cases are of particular importance. Often we know from some absolutely reliable source that $\mathrm{poss}_\mu(B) = 0$, i.e. $\omega_0 \in B$ is impossible. Such a piece of information can be integrated by intersecting μ and $1\!\!1_{\overline{B}}$. We obtain

$$\varphi_{\cap(1\!\!1_{\overline{B}},\mu)}(\omega) = \begin{cases} \mu(\omega) & \text{if } \omega \in \overline{B} \\ 0 & \text{otherwise.} \end{cases}$$

The corresponding possibility measure yields that B is impossible with respect to $\varphi_{\cap(1\!\!1_{\overline{B}},\mu)}$. In the general case such a piece of information concerns Ω_n. In this case we derive a specialization η of μ, i.e. $\mathrm{Poss}_\eta(A) \leq \mathrm{Poss}_\mu(A)$ for all $A \subseteq \Omega$.

3.3.3 On the Interpretation of Vague Data

In a more general case which in practice mostly appears, not only are the originals which induce the vague data unknown, but even the refined space Ω_n lies in the dark. So we have to rely just on functions mapping Ω into L, which contains the different grey levels. Thus the first step in the analysis of vague data should be to determine the nature of the scale formed by L. If physical indices are considered we have to deal with a metric scale, i.e. the quotient of two different degrees has a meaning. Ordinal scales arise, for example, if marks are given to rate quality; i.e. if just a total order on L is available. In practice it is mostly sufficient to rely on finite metric scales, where the quotient of two elements is a rational number. The precision suggested by real numbers is often an illusion.

So in the sequel we assume that L is a nonempty, *finite* set with a total order $\leq$. We furthermore assume that there are two different elements l_{min} and l_{max} with $l_{min} \leq l \leq l_{max}$ for all $l \in L$. The value l_{min} indicates impossibility, i.e. if $\mu(\omega) = l_{min}$ then ω is surely not the unknown actual state ω_0. If such a value does not appear, it can be added without any complications.

The second step is to explain where the vague data come from. Recall that the refined space Ω_n is inaccessible, so the number of contexts involved is unknown. Thus a suitable refinement of Ω has to be constructed. For this purpose a one-to-one transformation of L to a subset of the natural numbers is performed.

If L is a metric scale (where we assume the quotient of two elements of L to be a rational number), then we can use each transformation of the set

$$J_L = \left\{ j : L \to \mathbb{N} \cup \{0\} \;\middle|\; j(l_{min}) = 0 \text{ and } \frac{l}{l'} = \frac{j(l)}{j(l')} \text{ for all } l, l' \in L, l' \neq l_{min} \right\}.$$

J_L is not empty, the transformations are unique up to multiplication with a constant. If L is just an ordinal scale, then we have to use all the transformations of

$$J_L = \Big\{ j : L \to \mathbb{N} \cup \{0\} \;\Big|\; j(l_{min}) = 0 < j(l) < j(l') \\ \text{for all } l, l' \in L \text{ with } l_{min} < l < l' \Big\}.$$

Using them, each L-set η can be considered to be induced by a set within some space $\Omega_{j(l_{max})}$, and the set of admissible originals of η with respects to J_L is

$$\left\{ E \subseteq \Omega_{j(l_{max})} \;\middle|\; j \in J_L ; \mathrm{vag}_E^{\sigma_{j(l_{max})}}(\omega) = j\big(\eta(\omega)\big) \text{ for all } \omega \in \Omega \right\}.$$

In the sequel we assume that a suitable set J_L of transformations is specified. Based on η, degrees of possibility and necessity can be determined for subsets of Ω.

(3.12) Definition. *(a) A is called l-possible with respect to η, if and only if*

(i) $A \cap \{\omega \in \Omega \mid \eta(\omega) \geq l\} \neq \emptyset$ *and*

(ii) $A \cap \{\omega \in \Omega \mid \eta(\omega) \geq l'\} = \emptyset$ *for all* $l' > l$

(b) A is called l-necessary with respect to η, if and only if

(i) $A \supseteq \{\omega \in \Omega \mid \eta(\omega) \geq l\}$ *and*

(ii) $A \not\supseteq \{\omega \in \Omega \mid \eta(\omega) \geq l'\}$ *for all* $l' > l$.

These definitions are compatible with those given in Definition 3.4: we have that A is l-possible with respect to η, if and only if

$$\forall j \in J_L, \forall E \subseteq \Omega_{j(l_{max})}: \Big[\text{vag}_E^{\sigma_j(l_{max})}(\omega) = j(\eta(\omega)) \text{ for all } \omega \in \Omega$$
$$\Longrightarrow A \text{ is } j(l)\text{-possible with respect to } \text{vag}_E^{\sigma_j(l_{max})}\Big].$$

An analogous equivalence holds for necessity.

If $\eta : \Omega \to L$ is an L-set, then it is completely determined by its level sets $\mu_{(l)} \stackrel{\text{d}}{=} \{\omega \in \Omega \mid \mu(\omega) \geq l\}$, $l \in L$. If, on the other hand, for each level $l \in L$ a subset E_l of Ω is specified, where from $l < l'$ follows $E_l \supseteq E_l'$ and where $E_{l_{min}} = \Omega$ holds, then we obtain an L-set by the equation

$$\mu(\omega) \stackrel{\text{d}}{=} \max\Big\{\min\big(l, 1\!\!\mathrm{I}_{E_l}(\omega)\big)\Big\}, \quad l \in L, \omega \in \Omega. \tag{3.6}$$

Thus we define the "horizontal view" of an L-set.

(3.13) Definition. *Let L be a totally ordered finite set. Let f be a family $f = (E_l)_{l \in L}$ such that*

(i) $E_{\min L} = \Omega$

(ii) $E_l \subseteq E_{l'}$ *for all* $l, l' \in L$ *with* $l \geq l'$.

The set of all these families is denoted by $\text{Fl}_L(\Omega)$.

Obviously the classes $\text{Fl}_L(\Omega)$ and $\mathcal{F}_L(\Omega)$ contain corresponding elements. This leads in a canonical way to a concept of specialization.

(3.14) Definition. *Let* $\eta, \eta' \in \mathcal{F}_L(\Omega)$. *$\eta$ is called a specialization of η' (written $\eta \sqsubseteq \eta'$) if and only if $\eta(\omega) \leq \eta'(\omega)$ for all $\omega \in \Omega$.*

The transfer of the specialization relation to the elements of $\text{Fl}_L(\Omega)$ causes no difficulties, but if arbitrary operations on $\mathcal{F}_L(\Omega)$ are considered, the situation gets more complicated, as the relevant refinement Ω_n is unknown. A straightforward generalization of Definition 3.10 is the following one.

(3.15) Definition. *Let* $T = \{T_n\}_{n \in IN}$ *be a family of mappings, where* $T_n : 2^{(\Omega^{(1)})_n} \times \ldots \times 2^{(\Omega^{(t)})_n} \to 2^{(\Omega')_n}$. *The mapping*

$$\varphi_T : \mathcal{F}_L(\Omega^{(1)}) \times \ldots \times \mathcal{F}_L(\Omega^{(n)}) \to \mathcal{F}_L(\Omega')$$

is called correctness preserving with respect to T if and only if for all $\eta_i \in \mathcal{F}_L(\Omega^{(i)}), i = 1, \ldots, t$, *for each transformation* $j \in J_L$ *and for all subsets*

$A_i \subseteq (\Omega^{(i)})_{j(l_{max})}$, *where* $\mathrm{vag}_{A_i}^{\sigma_{j(l_{max})}}(\omega') = j(\eta_i(\omega'))$ *for each* $\omega' \in \Omega'$ *and* $i = 1, \ldots, t$, *we have*

$$j(\varphi_T(\eta_1, \ldots, \eta_t)(\omega')) \geq \mathrm{vag}_{T(A_1,\ldots,A_t)}^{\sigma_{j(l_{max})}}(\omega') \text{ for all } \omega' \in \Omega'.$$

It is easy to see that once more the pointwise intersection $\cap : \mathcal{F}_L(\Omega) \times \mathcal{F}_L(\Omega) \to \mathcal{F}_L(\Omega)$ defined by $(\eta' \cap \eta)(\omega) = \min(\eta'(\omega), \eta(\omega))$ for all $\omega \in \Omega$ is correctness preserving with respect to the ordinary intersection of sets, or more formally presented, $\cap$ is correctness preserving with respect to the family $\{T_n\}_{n \in I\!N}$, where $T_n : 2^{\Omega_n} \times 2^{\Omega_n} \to 2^{\Omega_n}; T_n(A, B) = A \cap B$ for all $n \in I\!N$. In this sense the pointwise union defined by $(\eta' \cup \eta)(\omega) = \max(\eta'(\omega), \eta(\omega))$, $\omega \in \Omega$, is *not* correctness preserving with respect to the ordinary union.

Let Ω^M be a product space and let $\Omega^T, \Omega^S, S \subseteq T \subseteq M$, be subspaces of Ω^M. Π_S^T, $\hat{\Pi}_S^T$ may denote projection and cylindrical extension as defined in Sect. 2.1. The cylindrical extension of $e = (E_l)_{l \in L}, E_l \subseteq \Omega^S$ is defined by

$$\hat{\Pi}_S^T(e) = \left(\hat{\Pi}_S^T(E_l)\right)_{l \in L},$$

whereas the projection of $f = (F_l)_{l \in L}, F_l \subseteq \Omega^S$, is defined by

$$\Pi_S^T(f) = \left(\Pi_S^T(F_l)\right)_{l \in L}.$$

For the corresponding L-sets μ_e and μ_f we obtain

$$\hat{\Pi}_S^T(\mu_e)(\omega^T) = \mu_e\left(\Pi_S^T(\omega^T)\right)$$
$$\Pi_S^T(\mu_f)(\omega^S) = \max\{\mu_f(\omega^T) \mid \omega^T \in \hat{\Pi}_S^T(\omega^S)\}$$

Also projection and cylindrical extension are correctness preserving with respect to the corresponding ordinary operations. In the case of projection we obtain for example

$$T_n : 2^{(\Omega^T)_n} \to 2^{(\Omega^S)_n}; \quad T_n(A) = T_n(A^1 \cup \ldots \cup A^n) = \Pi_S^T(A^1) \cup \ldots \cup \Pi_S^T(A^n),$$

and in the equation of Definition 3.15 we have equality. If we obtain vague data $\mu' : \Omega \to L'$, $\mu'' : \Omega \to L''$, then with our interpretation it should be possible to find a common refinement L of L' and L'', such that μ' as well as μ'' can be embedded and equivalently represented as L-sets.

3.4 The General Case

Now let us finally consider the case where we do not have sufficient information about the origin of the vague datum under consideration. The problem is how to express the subjective, partial beliefs about the identity of some unknown state ω_0. Of course, for this purpose the theory of probability can be applied, but from the preceding consideration it should be clear that if for example no metric scale is available (as in the case of subjectively determined preferences) the use of possibility/necessity measures offers a reasonable alternative. The procedure we recommend goes as follows:

- As a first step the observer has to specify a suitable scale L. Mostly L will be a finite set, since humans are in general not able to distinguish between more than a finite number of possibility degrees. L should contain at least the value impossible (l_{min}) and one additional degree; further elements of L represent levels of *speculation*.
- Secondly the observer is asked to specify the set $E_{l_{min}}$ which is minimal but covers the unknown state ω_0 with certainty. If nothing is known, i.e. in the case of total ignorance, we have $E_{l_{min}} = \Omega$; if the observation is precise, we obtain $E_{l_{min}} = \{\omega_0\}$.
- Finally for each speculative level $l \in L, l > l_{min}$ the observer is asked to provide a more optimistic but still conservative restriction for ω_0. In the case of imprecise data we have $E_l = E_{l_{min}}$ for all $l \in L$.

If it turns out that $\omega_0 \notin E_l, l > l_{min}$, this does not invalidate an L-set model. It just reflects the fact that the observer "overshot" the mark when he or she provided optimistic restrictions. If, on the other hand, we learn $\omega_0 \notin E_{l_{\text{min}}}$, this means a radical error which falsifies the model. The sets E_l constructed by this procedure are the minimal l-necessary sets with respect to the actual datum. We can also consider within this model the notions of possibility and necessity according to Definition 3.12 as well as a specialization concept (Definition 3.14). Problems occur if we want to integrate several pieces of information in cases where the sources and the relations between sources are not completely known. In this case different scales L and different aggregation methods may be appropriate. Often it is reasonable to consider a common refinement of the respective scales and to use the L-set-intersection as a conservative upper approximation of the unknown correct aggregation. In Chap. 11 this approach is considered in detail.

3.5 Concluding Remarks

From our point of view, *imprecision* means that the value $\omega_0 \in \Omega$ is unknown (in its narrow sense) but restricted by some set $A \subseteq \Omega$ which covers ω_0 with certainty. *Vagueness* arises essentially from the fact that the problem of determining ω_0 is considered in several different contexts, each of which yields an imprecise description. In the case of n contexts we obtain a mapping $\mu \in \mathcal{F}_n(\Omega)$ which we conceive to be a condensed representation of sets $A^1, \ldots, A^n$ corresponding to the n contexts. We conceive an L-set to be induced by unknown imprecise descriptions in an unknown number of contexts. The most important result of this chapter is that the standard set operations (intersection, projection and cylindrical extension) can be extended to L-sets in a correctness preserving way. It should be emphasized that the numbers used in this approach are not weights, but are just used as a condensed representation of the available information [Gebhardt and Kruse 1991].

Although the semantics is of course different, the interpretation of vague data in Sect. 3.2 and 3.3 relies on ideas similar to the definition of a one-point coverage function in the context of random sets (see [Goodman and Nguyen 1985], [Wang and Sanchez 1986], or [Dubois and Prade 1989]). In general the source of the vagueness is *not* known, so the integration of several pieces of information should be performed very carefully. Our interpretation of vague data is in essence an extension of interval analysis [Moore 1966, 1979] traditionally used in physics. The approach applied throughout this book for the representation of subjective knowledge was also used for the development of the software tool SOLD that allows evaluation of vague data by statistical methods [Kruse and Meyer 1987]. We assume vague data to be the result of a special projection operation applied to an ordinary set in a refined space. Based on such pieces of information we obtain a representation of our partial belief concerning the identity of some unknown but uniquely determined state of the world ω_0. By possibility and necessity measures, subsets of Ω are evaluated with respect to the question whether they possibly or necessarily contain ω_0 [Zadeh 1978], [Dubois and Prade 1988], [Yager 1980].

The notion of vagueness is by no means unique within the vast number of publications in this field of research. The classical paper on vagueness is [Black 1937]. Black distinguishes between ambiguity, generality and vagueness, and uses generality as a synonym for non-specificity and vagueness for fuzziness. These notions are also used in the field of linguistic philosophy. The notions used in the AI community are strongly influenced by L.A. Zadeh, who founded fuzzy set theory in 1963 [Zadeh 1963].

The concept of fuzziness, which is the subject of numerous publications and was introduced in the 1970s, is different from the approach we have presented. Fuzzy sets aim at the representation of lexical imprecision in natural language such as in the expression "The ship is *far away* from the coast", where the class of locations *far away* from the coast has no sharp boundaries, but there are borderlands. So ordinary sets are not sufficient to model the subjective, vague concept "far away". A natural and simple way to achieve the intended graded membership is to generalize the well known notion of a characteristic function $I_A : \Omega \to \{0,1\}$, i.e. to consider membership functions mapping Ω to the unit interval. The concept *far away* may serve as an example. We might obtain $\mu_{\text{far\_away}} : \Omega \to [0,1]$;

$$\mu_{\text{far\_away}}(\omega) = \begin{cases} 1 & \text{if } \omega = \text{z3} \\ 0.5 & \text{if } \omega = \text{z2} \\ 0 & \text{otherwise,} \end{cases}$$

where $\mu_{\text{far\_away}}(\text{z3}) = 1$ indicates that z3 completely fits with the (subjective) concept far away, while $\mu_{\text{far-away}}(\text{z1})$ means that z3 contradicts the intuitive idea of a location far away from the coast.

Thus a fuzzy set is defined to be a mapping $\mu : \Omega \to [0,1]$. Fuzziness addresses the intuitive idea of partial truth. If we consider some ship located in z2, then the statement "the ship is far away from the coast" is true just to a degree of 0.7. To clarify the intended meaning one could write

$$\text{truth(the ship is far away from the coast} \mid \text{the ship is in z2)} = 0.5.$$

Further details of the application of fuzzy sets are presented in Chap. 10.

Fuzzy sets try to capture *vague concepts*. We on the contrary consider only *vague data*. Based on such information it is not possible to assert that the statement "the ship is at z2" is true or false; the only thing we can derive is a valuation of the possibility and the necessity of the ship location to be in some region $A \subseteq \Omega$. Besides its use for representing lexical imprecision in natural language the idea of fuzziness can also be canonically applied to pattern recognition [Bellman 1966] and approximate reasoning (see Chap. 11). The most successful application was found in the field of process control (fuzzy controlling, see [Sugeno 1985]).

Chapter 4

Probability Theory

In this chapter we present probability theory as the basic tool for handling uncertainty, i.e. partial beliefs. In order to obtain a mathematically sound representation we always keep in mind a "relative frequency" interpretation or alternatively a degree of confirmation of probabilities. The main shortcoming of many textbooks is that the treatment of imprecise probabilities, which is important for obtaining suitable knowledge representation facilities, is left out. Our approach to this problem relies on the consideration of classes of probabilities.

4.1 Basic Concepts

Mathematical probability theory offers axioms and methods for manipulating numerical probabilities. This axiomatic theory is generally accepted. Unfortunately there are several different operational definitions of probability [Fine 1973]. To clarify the differences let us consider the situation where we want to determine the unknown location of a ship under consideration. Assume we know that the ship is at exactly one of the mutually exclusive locations within $\Omega = \{z1, z2, z3, ld, rd, ca\}$. Then using a classical definition of probability we have to determine first a number of cases which are equally possible. Using the *principle of indifference*, i.e. the alternatives are judged to be "equiprobable", if we have no reason to prefer one over the other, the result for the probability, that the ship is at sea, is the quantity

$$P(\text{at sea}) = \frac{|\{z1, z2, z3\}|}{|\Omega|} = \frac{3}{6}$$

Performing this calculation for all interesting locations we obtain a set function $P : 2^{\Omega} \to [0, 1]$ satisfying

(i) $P(A) = \frac{|A|}{|\Omega|} \geq 0$ for all $A \subseteq \Omega$,

(ii) $P(\Omega) = 1$,

(iii) $P(A \cup B) = P(A) + P(B)$, if $A, B \subseteq \Omega \quad A \cap B = \emptyset$, (4.1)

as one can easily verify. The principle of indifference as well as other metarules such as invariance [Jeffrey 1948] or maximum entropy principles [Jaynes 1968] should be applied very carefully, as the ambiguities inherent in an appeal to these principles can lead easily to contradictory results. So in textbooks of probability the purely *axiomatic* approach based on measure theory is preferred.

4.1.1 Axiomatic Probability Theory

Let Ω denote a finite collection of mutually exclusive statements about the world. Then by $\mathcal{E}$ we denote the set of all events. Some events are of particular importance. On the one hand the empty set $\emptyset$, a subset of every set by definition, is called the *impossible event*, since the outcome of a random selection can never be an element of $\emptyset$. On the other hand the set Ω itself always contains the actual outcome, therefore it is called the *certain event*. If A and B are events, then so are the union of A and B, the intersection of A and B and the complements of A and B, respectively. The event $A \cup B$ for example occurs if and only if A occurs *or* B occurs; the occurrences of $A \cap B$, $\overline{A}$, and $\overline{B}$ are defined analogously.

In the sequel we assume $\mathcal{E}$ to be the power set of Ω unless stated otherwise. The pair $(\Omega, \mathcal{E})$ is called the *sample space*. If it is clear what set of events we refer to, the explicit notation is omitted and Ω itself is referred to as the sample space. We define a function $P : \mathcal{E} \to [0,1]$ to be a probability if it could be induced in the way described above, i.e. if it satisfies the conditions (i) to (iii) which are well known as the Kolmogorov axioms.

(4.1) Definition. *Let $(\Omega, \mathcal{E})$ be a sample space* [1]*, where Ω is a finite set and $\mathcal{E} = 2^{\Omega}$. A real-valued set function $P : \mathcal{E} \to [0,1]$ defined on $(\Omega, \mathcal{E})$ is called a probability (function), if it satisfies the following conditions:*

(i) $P(A) \geq 0$ *for all* $A \subseteq \Omega$,

(ii) $P(\Omega) = 1$,

(iii) For $A, B \subseteq \Omega$ *from* $A \cap B \neq \emptyset$ *follows* $P(A \cup B) = P(A) + P(B)$.

[1] The *canonical* space $\mathcal{E}$ for events is a σ-field, i.e. a non-empty class of subsets of Ω that is closed under the formation of countable unions, countable intersections, and complements and contains the sets $\emptyset$ and Ω. In the general case where Ω is not necessarily a finite set and $\mathcal{E}$ is a σ-field, condition *(iii)* has to be replaced by the countable additivity condition

*(iii)** If $A_1, A_2, \ldots$ is a sequence of mutually exclusive events in $\mathcal{E}$, i.e. $A_i \cap A_k = \emptyset$ for $i \neq k$, then $P(\bigcup_{i=1}^{\infty} A_i) = \sum_{i=1}^{\infty} P(A_i)$.

Each probability satisfying the above definition has various properties; we summarize some of them in the following list:

(i) $P(\emptyset) = 0$,

(ii) $P(A) \leq 1$ for each $A \subseteq \Omega$,

(iii) $P(\overline{A}) = 1 - P(A)$ for each $A \subseteq \Omega$,

(iv) $A \subseteq B \Rightarrow P(A) \leq P(B)$ for each $A, B \subseteq \Omega$, (monotonicity)

(v) $P(A \cup B) \leq P(A) + P(B)$ for each $A, B \subseteq \Omega$, (subadditivity)

(vi) $A \subseteq B \Rightarrow P(B \cap \overline{A}) = P(B) - P(A)$ for each $A, B \subseteq \Omega$, (subtractivity)

(vii) $P(A \cup B) = P(A) + P(B) - P(A \cap B)$ for each $A, B \subseteq \Omega$,

(viii) $P(\bigcup_{i=1}^{n} A_i) = \sum_{I:I\subseteq\{1,\ldots,n\},I\neq\emptyset} (-1)^{|I|+1} P(\bigcap_{i\in I} A_i)$
for all $A_1, \ldots, A_n$, $n \in \mathbb{N}$. (4.2)

If it is known for certain in the context that an event E holds, than we can integrate this information by a conditioning process.

(4.2) Definition. *If P is a probability on the sample space Ω and $E \subseteq \Omega$ is an event with $P(E) > 0$, then for every event $A \subseteq \Omega$ the quantity*

$$P(A|E) \stackrel{\mathrm{d}}{=} \frac{P(A \cap E)}{P(E)}$$

is called the conditional probability of A given E. The conditional probability remains undefined if $P(E) = 0$.

It is easy to show that for a fixed event E with $P(E) > 0$ the set function $P(\,.\,|E) : 2^{\Omega} \to [0,1]$ defined by

$$P(\,.\,|E)(A) = P(A|E) \qquad \text{for all } A \subseteq \Omega$$

is also a probability on Ω. Let P be a probability and let E_1, E_2 be events with $P(E_1 \cap E_2) > 0$. For the probabilities

$$P_1 \stackrel{\mathrm{d}}{=} P(\,.\,|E_1), \quad P_2 \stackrel{\mathrm{d}}{=} P(\,.\,|E_2)$$

we have the property

$$P_1(\,.\,|E_2) = P_2(\,.\,|E_1) = P(\,.\,|E_1 \cap E_2).$$

From the definition of the conditional probability of A we obtain

$$P(A \cap E) = P(E) \cdot P(A|E). \tag{4.3}$$

This way of calculating the probability of an intersection can be generalized to a sequence of n events. If $A_1, \ldots, A_n \subseteq \Omega$ are events such that $P(\bigcap_{j=1}^{n-1} A_j) > 0$, then

$$P(\bigcap_{j=1}^{n} A_j) = P(A_1) \cdot P(A_2|A_1) \cdot P(A_3|A_1 \cap A_2) \cdot \ldots \cdot P(A_n| \bigcap_{j=1}^{n-1} A_j) \quad (4.4)$$

holds. This formula is known as the *general multiplication rule* or *chaining rule.*

Another rule, called the *total probability rule* (see e.g. [Feller 1966]), states the following: Let $H_1, \ldots, H_n$ be a partition of Ω, i.e. $H_i \cap H_j = \emptyset$, if $i \neq j$ and $H_1 \cup \ldots \cup H_n = \Omega$. If $P(H_i) > 0$ for $i = 1, \ldots, n$, then

$$P(A) = \sum_{i=1}^{n} P(H_i) \cdot P(A|H_i) \quad (4.5)$$

holds for each event A. If $P(A \mid B) = P(A)$ hold, we say that A and B are independent, and in this case we obtain

$$P(A \cap B) = P(A|B) \cdot P(B) = P(A) \cdot P(B).$$

This formula can be used to give a formal definition of the concept of *independence.*

(4.3) Definition. *Two events A and B are said to be (statistically) independent, if and only if*

$$P(A \cap B) = P(A) \cdot P(B).$$

If $P(A) = 0$ holds, then A is obviously independent of every B, and

$$P(A) > 0$$

implies that A is independent of $\emptyset$ and Ω. If A and B are independent, then so are A and $\overline{B}$, $\overline{A}$ and $\overline{B}$, as well as $\overline{A}$ and B.

It has to be emphasized that independence of events is not to be confused with mutual exclusion: If A and B are independent and $P(A) > 0$, $P(B) > 0$, then A and B cannot be mutually exclusive. The only test of independence with respect to a given probability in the theory is to check the multiplicative property of the definition. But in the range of application, independence is not a property which has to be tested; usually it is to be assumed because of the analyst's intuitive feeling, so the lack of equivalent and easily testable characterizations of independence is not grievous.

If we have a family of events, then it is possible to extend the concept of independence.

(4.4) Definition. *A collection of events $\mathcal{A} = \{A_1, A_2, A_3, \ldots\}$ is said to be*

(a) pairwise independent, if and only if for every pair of distinct events $A, B \in \mathcal{A}$, the equation

$$P(A \cap B) = P(A) \cdot P(B)$$

is valid,

(b) completely independent, if and only if for every subcollection $\{A_{i_1}, \ldots A_{i_k}\}$ of $\mathcal{A}$ the following equation holds:

$$P(A_{i_1} \cap A_{i_2} \cap \ldots \cap A_{i_k}) = \prod_{j=1}^{k} P(A_{i_j}).$$

Complete independence, of course, implies pairwise independence, but the converse statement is *not* true.

In applications the notion of "absolute independence" is often not appropriate. More reasonable is the use of the concept of conditional independence: If $P(A \cap B \mid C) = P(A \mid C) \cdot P(B \mid C)$, we say that A and B are conditionally independent given C. In that case the equality $P(A \mid B, C) = P(A \mid B \cap C) = P(A \mid C)$ holds.

4.1.2 On the Interpretation of a Probability

Beside the classical interpretation, many different semantics of probability have been presented. As R. Carnap [Carnap 1971] pointed out, there are two main but totally different interpretations, which are often confused as we use the term probability for both:

- probability_1 = degree of confirmation[2]
- probability_2 = relative frequency in a long run

A typical example for the use of probability_2 is the following statement

"The probabilty of casting a 'one' with this die is $\frac{1}{6}$"

Such statements can be tested by statistical investigations using sufficiently long series of throws. Therefore it is argued that probabilities of type 2 have objective reality apart from an observer. There are several strong hints coming from physics that some probabilities (e.g. electron diffraction) should be in fact conceived as *objective*.

---

[2] Confirmation is used here as a measure of absolute belief, not as a measure of the change of belief or corroboration [Good 1960].

In the context of building expert systems the concept of probablity$_1$ is more relevant. For example the probability of the failure of a structure like a power plant presents greater difficulties of interpretation than casting a "one" with a die, since the failures are often unrepeatable; there are only experiments deemed to be governed by the same probabilities. In this application, the uncertainty simply represents our degree of confirmation relative to a particular situation. These probabilities are *subjective*, and depend on an observer. De Finetti [DeFinetti 1972] has derived an axiomatic system of subjective probabilities in a betting context; he defends the extreme position that *all* probabilities are subjective.

4.1.3 Practical Aspects

The determination of probabilities in terms of the space Ω mostly reflects knowledge about the structure of a problem, for example symmetries, physical properties, etc. In many cases there is wide agreement about the adequacy of specific mass distributions for a certain type of problem. Coin flipping may serve as an example. By convention a probability of 0.5 is assigned to both outcomes, heads and tails for each coin. Doing so we make use of the "insufficient reason principle", which states that if there is no basis for judging one member of an exhaustive set of mutually exclusive events to be more likely than another, they should be treated as equally probable. The principle involves the danger of obtaining counterintuitive results and leads to various paradoxes, thus it should not be applied in order to obtain descriptions of real world phenomena.

The most common practical procedure is in fact to collect a large number of observations and then to consider relative frequencies. Such an approach is justified by the success of empirical sciences and relies on the assumption of *stationarity* (the future will be like the past) and *identicality* (the future event which is subject to our prediction is like the events that form the basis of that prediction).

This procedure is partly justified by De Finetti's Representation Theorem [DeFinetti 1937b]. In fact many of the statements in knowledge based systems fall within the area of frequency theories, especially if inference is based on statistical material. As an example, suppose that an observer has the knowledge

(*) "The relative frequency to be at sea among all ships of Hamburg is $\frac{1}{6}$."
Consider the hypotheses

(**) "Doris is at sea"

where Doris is defined as Ship No. 52 of Hamburg.

The correct representation of a probability in the sense of confirmation has to be

"The probability of (**) with respect to (*) is $\frac{1}{6}$".

Circumscriptions in this case such as "The probability that Doris is at sea is $\frac{1}{6}$" or "Since the relative frequency to be at sea among all ships of Hamburg is $\frac{1}{6}$, the probability that Doris is at sea is $\frac{1}{6}$" are incorrect and misleading, see [Carnap 1971].

In the context of expert systems it is required to elicit probability judgements directly from people. Here some experts' assignments are highly subjective and reflect mainly these experts' states of knowledge and belief. This shows that uncertainty assessment is a branch of psychophysics [Stevens 1951]. The methods involved are based on direct numerical estimation [Goodman 1972], pairwise comparisons or ranking, etc. [Stillwell, Seaver, and Schwartz 1981].

4.2 Probabilities on Different Sample Spaces

This section concerns the properties of different but related sample spaces, covering also the special case of product spaces. The concept of "refinement" allows the description and analysis of such properties, especially the relationship of probabilities induced on refined spaces.

(4.5) Example. Consider the two sample spaces

$$\Omega = \{\text{radar-devices}, \text{sonar-devices}\}$$

and

$$\Omega' = \{\text{r1}, \text{r2}, \text{r3}, \text{s1}, \text{s2}, \text{s3}\},$$

where r1, r2, r3 and s1, s2, s3 are the particular radar and sonar devices, respectively. Obviously Ω' is a refinement of Ω, and we have

$$\widehat{\Pi}\big(\{\text{radar-devices}\}\big) = \{\text{r1}, \text{r2}, \text{r3}\}$$

and

$$\widehat{\Pi}\big(\{\text{sonar-devices}\}\big) = \{\text{s1}, \text{s2}, \text{s3}\}.$$

Starting with a probability P' on Ω' the probability P on Ω defined by

$$P\big(\{\text{radar-devices}\}\big) = P'\big(\{\text{r1}, \text{r2}, \text{r3}\}\big)$$

and

$$P\big(\{\text{sonar-devices}\}\big) = P'\big(\{\text{s1}, \text{s2}, \text{s3}\}\big)$$

is obviously compatible with P'. On the other hand, if a probability P on Ω is given, then each probability P' on Ω' satisfying the above conditions is compatible with P. In this case P' is not uniquely determined. ◇

The above example shows that in the course of "projecting" a probability P' on Ω' the probability mass of those elements $\omega' \in \widehat{\Pi}(\{\omega\})$, $\omega \in \Omega$, should be concentrated on ω. Thus the probability

$$\Pi[P'] : 2^{\Omega} \to [0,1]; \qquad \Pi[P'](\{\omega\}) \stackrel{\mathrm{d}}{=} \sum_{\omega' : \omega' \in \widehat{\Pi}(\{\omega\})} P'(\{\omega'\}) \tag{4.6}$$

is called the projection of P' to Ω. Whenever we have two probabilities P and P' on Ω and Ω', respectively, where Ω' is a refinement of Ω with $\widehat{\Pi}$ as the corresponding refinement mapping, then if $P = \Pi[P']$ holds we call P' a *refinement* of P.

Let $(\Omega^{(i)}, \mathcal{E}_i)$, $i \in M = \{1, \ldots, m\}$ be a finite set of sample spaces. The description of interactions between different spaces requires the notion of the product sample space $(\Omega^M, \mathcal{E})$. If for all[3] $i \in M$ we have $\mathcal{E}_i = 2^{\Omega^{(i)}}$, then we obtain

$$\begin{aligned} \Omega^M &= \prod_{i=1}^{m} \Omega^{(i)} \\ &= \{(\omega^{(1)}, \ldots, \omega^{(m)}) \mid \omega^{(i)} \in \Omega^{(i)}, i = 1, \ldots, m\} \text{ and} \\ \mathcal{E} &= 2^{\Omega^M}. \end{aligned} \tag{4.7}$$

In the next part of this section we will clarify the dependencies between the probability P^M on a product space Ω^M and probabilities P^N on subspaces Ω^N, $N \subseteq M$. In the sequel we restrict our considerations to *finite* sets[4] $\Omega^{(i)}$, $i \in M$, so that also Ω^M is finite. The simplest subspace is $\Omega^{\emptyset}$, containing the empty tuple ε as its only element. This implies $P^{\emptyset}(\{\varepsilon\}) = 1$. If a probability P^M on Ω^M is given, then it induces on every subspace Ω^N a marginal probability $P^N(B) \stackrel{\mathrm{d}}{=} P^M(\widehat{\Pi}_N^M(B))$ for all $B \subseteq \Omega^N$.

(4.6) Example. Let $M = \{1,2,3\}$ and $\Omega^{(1)} = \{\text{radar}, \text{sonar}\}$, $\Omega^{(2)} = \{\text{dist}, \text{dir}\}$, $\Omega^{(3)} = \{\text{active}, \text{passive}\}$, where "dist" and "dir" denote components for the determination of distance and direction, and where "active" and "passive" denote active and passive components. The product space $\Omega^M = \Omega^{\{1,2,3\}} = \Omega^{(1)} \times \Omega^{(2)} \times \Omega^{(3)}$ contains 8 elements as depicted in Fig. 4.1. An example of a two-dimensional subspace is

$$\Omega^{\{2,3\}} = \{(\text{dist}, \text{active}), (\text{dist}, \text{passive}), (\text{dir}, \text{active}), (\text{dir}, \text{passive})\}$$

---

3 If $\mathcal{E}_i \neq 2^{\Omega^{(i)}}$ for some $i \in M$, then $\mathcal{E}$ is defined as the smallest σ-field $\mathcal{E}'$ on Ω^M, such that each cylinder $A_1 \times A_2 \times \ldots \times A_m$, $A_i \in \mathcal{E}_i$, $i = 1, \ldots, m$, is contained in $\mathcal{E}'$.

4 In the case of finite sets the appropriate σ-fields are the power sets, so no measurability problems occur.

The projection of the set

$$A = \{(\text{radar}, \text{dist}, \text{active}), (\text{radar}, \text{dir}, \text{active}), (\text{sonar}, \text{dist}, \text{passive})\}$$

on $\Omega^{\{1,2\}}$ yields

$$\Pi^{\{1,2,3\}}_{\{1,2\}}(A) = \{(\text{radar}, \text{dist}), (\text{radar}, \text{dir}), (\text{sonar}, \text{dist})\}.$$

$(\omega^{(1)}, \omega^{(2)}, \omega^{(3)}) \in \Omega^{\{1,23\}}$	$P(\{\omega^{(1)}, \omega^{(2)}, \omega^{(3)}\})$
(radar, dist, active)	0.25
(radar, dist, passive)	0.2
(radar, dir, active)	0
(radar, dir, passive)	0.05
(sonar, dist, active)	0.2
(sonar, dist, passive)	0.1
(sonar, dir, active)	0.1
(sonar, dir, passive)	0.1

Fig. 4.1. The product space $\Omega^{\{1,2,3\}}$

With the numerical values provided in Fig. 4.1 for the marginal probabilities on $\Omega^{(1)}$ and $\Omega^{\{2,3\}}$ we obtain

$\omega^{(1)} \in \Omega^{(1)}$	$P(\{\omega^{(1)}\})$
radar	0.5
sonar	0.5

$(\omega^{(2)}, \omega^{(3)}) \in \Omega^{\{2,3\}}$	$P\Big(\{(\omega^{(2)}, \omega^{(3)})\}\Big)$
(dist, active)	0.45
(dist, passive)	0.3
(dir, active)	0.1
(dir, passive)	0.15

◇

A probability P^M on Ω^M together with its marginal probabilities P^N on Ω^N, $N \subseteq M$, forms a consistent family of probabilities, which means we have

$$P^S(B) = P^T\big(\widehat{\Pi}^T_S(B)\big) \tag{4.8}$$

for each $S, T \subseteq M$, where $S \subseteq T$, and for each $B \subseteq \Omega^S$. In the sequel we consider the question of the conditions under which it is possible to construct a consistent family of probabilities. The main purpose of the following exertions is to provide the theoretical basis for problems we encounter in Chap. 12, which deals with probabilistic knowledge representation in product spaces.

(4.6) Example. (continued) Suppose

$$\begin{aligned} P^{\{1\}}(\{\text{radar}\}) &= P^{\{1\}}(\{\text{sonar}\}) = 0.5, \\ P^{\{1\}}(\{\text{dist}\}) &= P^{\{1\}}(\{\text{dir}\}) = 0.5. \end{aligned}$$

In order to obtain a consistent family of probabilities we have to require

$$\begin{aligned} P^{\{1,2\}}\Big(\{(\text{radar},\text{dist}),(\text{radar},\text{dir})\}\Big) &= P^{\{1,2\}}\Big(\{(\text{sonar},\text{dist}),(\text{sonar},\text{dir})\}\Big) \\ &= P^{\{1,2\}}\Big(\{(\text{radar},\text{dist}),(\text{sonar},\text{dist})\}\Big) \\ &= P^{\{1,2\}}\Big(\{(\text{radar},\text{dir}),(\text{sonar},\text{dir})\}\Big) \\ &= 0.5. \end{aligned}$$

For $P^{\{1,2\}}$ we are free to choose one of the probabilities within the set

$$\Big\{Q \;\Big|\; Q\Big(\{(\text{radar},\text{dist})\}\Big) = Q\Big(\{(\text{sonar},\text{dir})\}\Big), \\ Q\Big(\{(\text{radar},\text{dir})\}\Big) + Q\Big(\{(\text{sonar},\text{dist})\}\Big) = 0.5\Big\}.$$

The probability P defined by

$$\begin{aligned} P\Big(\{(\text{radar},\text{dist})\}\Big) &= P\Big(\{(\text{radar},\text{dir})\}\Big) \\ &= P\Big(\{(\text{sonar},\text{dist})\}\Big) = P\Big(\{(\text{sonar},\text{dir})\}\Big) = 0.25 \end{aligned}$$

is of special interest, since the cylindrical extensions of events in $\Omega^{\{1\}}$ are independent of the cylindrical extensions of events in $\Omega^{\{2\}}$. We have, for example,

$$\begin{aligned} & P\Big(\widehat{\Pi}_{\{1\}}^{\{1,2\}}(\{\text{radar}\})\Big) \cdot P\Big(\widehat{\Pi}_{\{2\}}^{\{1,2\}}(\{\text{dist}\})\Big) \\ =& P\Big(\{(\text{radar},\text{dist}),(\text{radar},\text{dir})\}\Big) \cdot P\Big(\{(\text{sonar},\text{dist}),(\text{radar},\text{dist})\}\Big) \\ =& 0.5 \cdot 0.5 = 0.25 = P\Big(\{(\text{radar},\text{dist})\}\Big) \\ =& P\Big(\widehat{\Pi}_{\{1\}}^{\{1,2\}}(\{\text{radar}\}) \cap \widehat{\Pi}_{\{2\}}^{\{1,2\}}(\{\text{dist}\})\Big). \end{aligned}$$

P is the only probability that satisfies this independence condition. ◇

If a consistent family of probabilities $\{P^N \mid N \subseteq M\}$ is given, then the probability $P^N(A)$ of some subset A of Ω^N always equals the probability $P^M(\widehat{\Pi}_N^M(A))$ of its cylindrical extension. The notation $P(A|B)$, where $A \subseteq \Omega^S$ and $B \subseteq \Omega^T$, refers to the cylindrical extensions of A and B. In terms of a formal definition,

$$P(A|B) \stackrel{\mathrm{d}}{=} P^M(\widehat{\Pi}_S^M(A)|\widehat{\Pi}_T^M(B)). \tag{4.9}$$

Furthermore we can express the (statistical) independence of two sets $A \subseteq \Omega^S$ and $B \subseteq \Omega^T$. A, B are independent if and only if

$$P^M(\widehat{\Pi}_S^M(A) \cap \widehat{\Pi}_T^M(B)) = P^M(\widehat{\Pi}_S^M(A)) \cdot P^M(\widehat{\Pi}_T^M(B)) \tag{4.10}$$

or equivalently

$$P^{S\cup T}(\widehat{\Pi}_S^{S\cup T}(A) \cap \widehat{\Pi}_T^{S\cup T}(B)) = P^S(A) \cdot P^T(B). \tag{4.11}$$

In the sequel we consider two methods for the construction of a consistent family of probabilities. Both of them ensure that the result is uniquely determined. The first method is based on independence assumptions, the second one makes use of the general multiplication rule.

(i) Let $S_1, \dots, S_n$ be a partition of M, that is, we have

$$S_1 \cup \ldots \cup S_n = M \text{ and } S_i \cap S_j = \emptyset \text{ for } i \neq j.$$

If for each S_i a probability P^{S_i} on Ω^{S_i} is given, then we obtain a consistent family of probabilities as follows. We determine the probability P^M on Ω^M by setting

$$P^M(A) = P^{S_1}(\Pi_{S_1}^M(A)) \cdot \ldots \cdot P^{S_n}(\Pi_{S_n}^M(A)).$$

The probabilities P^N, $N \subseteq M$, are defined to be the marginal probabilities of P^M. For a set $\emptyset \neq B \subseteq \Omega^{S_i}$, $i \in \{1, \dots, n\}$ we obtain

$$\begin{aligned}
&P^M(\widehat{\Pi}_{S_i}^M(B)) \\
&= P^{S_1}\left(\Pi_{S_1}^M(\widehat{\Pi}_{S_i}^M(B))\right) \cdot \ldots \cdot P^{S_n}\left(\Pi_{S_n}^M(\widehat{\Pi}_{S_i}^M(B))\right) \\
&= P^{S_i}\left(\Pi_{S_i}^M(\widehat{\Pi}_{S_i}^M(B))\right) \cdot \prod_{\substack{j=1 \\ j\neq i}}^{n} P^{S_j}\left(\Pi_{S_j}^M(\widehat{\Pi}_{S_i}^M(B))\right) \\
&= P^{S_i}(B) \cdot \prod_{\substack{j=1 \\ j\neq i}}^{n} P^{S_j}(\Omega^{S_j}) \\
&= P^{S_i}(B).
\end{aligned}$$

As a consequence of this construction two sets $A \subseteq \Omega^{S_i}$ and $B \subseteq \Omega^{S_j}$, where S_i and S_j are two different elements of the partition, are always independent.

(ii) Again, let $S_1, \ldots, S_n$ be a partition of M. We assume that for each $i \in \{2, \ldots, n\}$ and for each A and B the conditional probability $P(B|A)$ is known, where A is a subset of $\Omega^{S_1 \cup \ldots \cup S_{i-1}}$ and B is a subset of Ω^{S_i}. Furthermore we assume the (unconditional) probabilities for all subsets of Ω^{S_1} to be given. Then the general multiplication rule yields that there is only one family of consistent probabilities satisfying the above constraints.

(4.7) Example. (continued) If $Q^{\{1\}}$ is a probability on $\Omega^{\{1\}}$, if

$$Q^{\{2\}|\{1\}}(\,.\mid\{\text{radar}\}),\quad Q^{\{2\}|\{1\}}(\,.\mid\{\text{sonar}\})$$

are probabilities on $\Omega^{\{2\}}$, and if

$$Q^{\{3\}|\{1,2\}}\Big(\,.\;\Big|\;\{(\text{radar},\text{dist})\}\Big),\quad Q^{\{3\}|\{1,2\}}\Big(\,.\;\Big|\;\{(\text{radar},\text{dir})\}\Big),$$

$$Q^{\{3\}|\{1,2\}}\Big(\,.\;\Big|\;\{(\text{sonar},\text{dist})\}\Big),\quad Q^{\{3\}|\{1,2\}}\Big(\,.\;\Big|\;\{(\text{sonar},\text{dir})\}\Big)$$

are probabilities on $\Omega^{\{3\}}$, then there is exactly one probability $P^{\{1,2,3\}}$ on $\Omega^{\{1,2,3\}}$ such that

$$\begin{aligned} P^{\{1,2,3\}}(\{x\}\times\Omega^{(2)}\times\Omega^{(3)}) &= Q^{\{1\}}(\{x\}),\\ P^{\{1,2,3\}}(\Omega^{(1)}\times\{y\}\times\Omega^{(3)}\mid\{x\}\times\Omega^{(2)}\times\Omega^{(3)}) &= Q^{\{2\}|\{1\}}(\{y\}|\{x\}),\\ P^{\{1,2,3\}}(\Omega^{(1)}\times\Omega^{(2)}\times\{z\}\mid\{x\}\times\{y\}\times\Omega^{(3)}) &= Q^{\{3\}|\{1,2\}}(\{z\}|\{x,y\}) \end{aligned}$$

hold for all $x \in \{\text{radar},\text{sonar}\}$, $y \in \{\text{dist},\text{dir}\}$, $z \in \{\text{active},\text{passive}\}$. Using the multiplication rule we obtain, for example,

$$\begin{aligned} &P^{\{1,2,3\}}\Big(\{(\text{sonar},\text{dist},\text{passive})\}\Big)\\ &= P^{\{1,2,3\}}(\{\text{sonar}\}\times\Omega^{(2)}\times\Omega^{(3)})\\ &\cdot P^{\{1,2,3\}}(\Omega^{(1)}\times\{\text{dist}\}\times\Omega^{(3)}\mid\{\text{sonar}\}\times\Omega^{(2)}\times\Omega^{(3)})\\ &\cdot P^{\{1,2,3\}}(\Omega^{(1)}\times\Omega^{(2)}\times\{\text{passive}\}\mid\{\text{sonar}\}\times\{\text{dist}\}\times\Omega^{(3)})\\ &= Q_{\{1\}}(\{\text{sonar}\})\cdot Q_{\{2\}|\{1\}}(\{\text{dist}\}\mid\{\text{sonar}\})\\ &\qquad\cdot Q_{\{3\}|\{1,2\}}\Big(\{\text{passive}\}\mid(\text{dist},\text{sonar})\}\Big). \end{aligned}$$

◇

Now let us, for the first time, make an attempt to model the generation of observations by sensors. For the moment our aim is to clarify the problem of choosing an adequate probabilistic model.

(4.8) Example. Suppose we have to classify some ship under consideration with respect to the set $\Omega = \{\text{tv}, \text{ps}\}$ containing the two types tv (trading vessel) and ps (passenger ship). Suppose furthermore that a set $\Omega = \{\omega_1, \ldots, \omega_4\}$ of sensors is available, each of them observing the ship and providing its information. Assume we know the following things about the nature of the sensors:

- Sensor ω_1 tells the truth,
- Sensor ω_2 tells the opposite of the truth,
- Sensor ω_3's answer is always ps,
- Sensor ω_4's answer is always tv.

One of the sensors is selected at random, where the probability for selecting ω_i is p_i for $i = 1, \ldots, 4$. Then the chosen sensor has to transmit its information. Given that we obtain the output "The ship is of type ps" how can our belief that the ship is actually of type ps be quantified? Let us first formalize the situation. We have a mapping

$$X : \Omega \times \Omega \to \Omega$$

that assigns to each sensor and each type the answer of this sensor according to our "meta-knowledge", encoded in the following table

	ps	tv
ω_1	ps	tv
ω_2	tv	ps
ω_3	ps	ps
ω_4	tv	tv

The product space $\Omega \times \Omega \times \Omega$ corresponding to the set of (sensor, ship type, sensor output)-triples contains 16 elements, where 8 elements turn out to be impossible. We obtain

$$\Omega' = \{(\omega_1, \text{ps}, \text{ps}), (\omega_1, \text{tv}, \text{tv}), (\omega_2, \text{ps}, \text{tv}), (\omega_2, \text{tv}, \text{ps}),$$

$$(\omega_3, \text{ps}, \text{ps}), (\omega_3, \text{tv}, \text{ps}), (\omega_4, \text{ps}, \text{tv}), (\omega_4, \text{tv}, \text{tv})\}.$$

Following the Bayesian analysis we have to require that there is *one* underlying probability on Ω'. From the description of the problem it is clear that we have no probability assignment p_{ps} for the statement "The ship's type is ps". So we should use a "minimum specificity" argument and take the value $p_{\text{ps}} = 0.5$. Otherwise we would try to assess a meta-probability on the values of p_{ps}, which is difficult to obtain in this case. Furthermore we assume that the selection of a sensor and the selection of a ship type are (stochastically) independent. We obtain

$$P\{(\omega_1, \text{ps}, \text{ps})\} = P\{(\omega_1, \text{tv}, \text{tv})\} = \ldots = P\{(\omega_4, \text{tv}, \text{tv})\} = \frac{1}{8}.$$

The information "The sensor's message is ps" corresponds to the set

$$\{(\omega_1, \text{ps}, \text{ps}), (\omega_2, \text{tv}, \text{ps}), (\omega_3, \text{ps}, \text{ps}), (\omega_3, \text{tv}, \text{ps})\}$$

whereas the information "The ship is actually of type ps" can be described by the set

$$\{(\omega_1, \text{ps}, \text{ps}), (\omega_2, \text{ps}, \text{tv}), (\omega_3, \text{ps}, \text{ps}), (\omega_4, \text{ps}, \text{tv})\}.$$

So we have

$$\begin{aligned} &P(\text{ship type is ps}|\text{sensor message is ps}) \\ &\quad = \frac{P\Big(\{(\omega_1, \text{ps}, \text{ps}), (\omega_3, \text{ps}, \text{ps})\}\Big)}{P\Big(\{(\omega_1, \text{ps}, \text{ps}), (\omega_2, \text{tv}, \text{ps}), (\omega_3, \text{ps}, \text{ps}), (\omega_3, \text{tv}, \text{ps})\}\Big)} \\ &\quad = 0.5 \end{aligned}$$

(to be continued). ◇

4.3 Bayesian Inference

In Bayesian methods, propositions are given numerical values quantifying the (personal) degree of confirmation under some body of knowledge. For the mathematical formalism, probability theory is chosen, where $P(A \mid K)$ specifies the degree of confirmation in A under the assumption that the body of knowledge K is given. It is always assumed that the underlying space Ω is a collection of mutually exclusive statements about the world, exactly one of which is true, and that the probability measure $P(\ . \mid K)$ assigns a number to each possible subset A of Ω according to the degree of confirmation that this subset contains the true, but unknown statement. This implies that if new evidence is utilized, the probability value will generally change. The only exception in this variability is when the probability is zero or one. Note that in this interpretation, probabilities are inherently "subjective".

Since all quantities are *conditional* probabilities, for convenience we leave out the symbol K if the background information K does *not* change. But in general, in Bayesian theory the formula (4.5) should read

$$P(A \mid K) = \sum_{i=1}^{n} P(H_i \mid K) \cdot P(A \mid H_i \cap K).$$

A simple consequence of this rule is the famous Bayes' Theorem, which is given here in shorthand notation.

(4.9) Theorem. *Let $H_1, \ldots, H_n$ be a partition of Ω such that $P(H_i) > 0$ for all $i = 1, \ldots, n$, and let A be an event with $P(A) > 0$. Then*

$$P(H_j|A) = \frac{P(H_j) \cdot P(A|H_j)}{\sum_{i=1}^{n} P(H_i) \cdot P(A|H_i)}.$$

Whenever probabilities are used with the Bayes interpretation the members of the partition mentioned in Theorem 4.9 are called "hypotheses". Let H_j be the set corresponding to the j-th hypothesis. Assume that the probabilities $P(H_j)$, $j = 1, \ldots, n$, called the *prior probabilities*, can be assigned. If we interpret $P(A \mid H)$ as the likelihood that A will materialize if H is true, then this information leads according to the Bayesian point of view to a "reassessment" of the prior probabilities $P(H_j)$; they are replaced by the conditional probabilities $P(H_j|A)$, which in this context are called *posterior probabilities.*

In the most simple case the partition mentioned in Theorem 4.9 consists of two sets H, $\overline{H}$. Given some datum A, from Bayes Theorem we obtain

$$P(H|A) = \frac{P(H) \cdot P(A|H)}{P(A)} \quad \text{and} \quad P(\overline{H}|A) = \frac{P(\overline{H}) \cdot P(A|\overline{H})}{P(A)} \tag{4.12}$$

Dividing those equations we obtain the so-called *odds-likelihood ratio form* of Bayes Theorem, i.e.

$$O(H|E) = \frac{P(H|E)}{P(\overline{H}|E)} = \frac{P(H)}{P(\overline{H})} \cdot \frac{P(E|H)}{P(E|\overline{H})} = O(H) \cdot \lambda(E|H) . \tag{4.13}$$

$O(H) = \frac{P(H)}{P(\overline{H})}$ and $O(H|E)$ are called prior odds and posterior odds, respectively, while $\lambda(E|H) \stackrel{\mathrm{d}}{=} \frac{P(E|H)}{P(E|\overline{H})}$ is usually called the *likelihood ratio* (a notion well known from classical statistics). It is very important to realize that these formulas are not only used for calculating special probabilities, but that these rules are viewed as a *normative* method for updating degrees of confirmation in reponse to evidence E.

Odds are commonly used in the field of gambling theory. Moreover they are useful representations of uncertainty and therefore often used within probability assessment procedures [Smith 1961].

The final aim of any probabilistic analysis is to obtain support in finding decisions. To elucidate how decisions can be derived let us consider an idealized and very simple decision problem. Assume you are invited to bet on the result of randomly selecting some element $\omega_0 \in \Omega$. The bets are:

(a) You bet on $\omega_0 \in A$ at odds $\frac{x}{y}$. (i.e. stake x, payoff $x+y$ if you win)

(b) You bet on $\omega_0 \in \overline{A}$ at odds $\frac{y}{x}$. (i.e. stake y, payoff $x+y$ if you win)

and you are forced to choose one of them. As a rational being you will compare the stake to the "expected payoff", i.e. you will consider the differences

$$P(A) \cdot (x+y) - x \text{ and } P(\overline{A}) \cdot (x+y) - y$$

as you expect to win, in the long run, in $P(A) \cdot 100\%$ or $P(\overline{A}) \cdot 100\%$ of all cases, respectively. There is no danger of being cheated, since

$$\big(P(A) \cdot (x+y) - x < 0\big) \text{ and } \big(P(\overline{A}) \cdot (x+y) - y < 0\big)$$

is not possible as

$$\frac{x}{x+y} < P(A) \quad \text{and} \quad P(\overline{A}) < \frac{y}{x+y}$$

is obviously contradictory. So a rational individual should accept either bet (a) or bet (b). If $P(A) = \frac{x}{x+y}$ then the individual will hesitate to select either bet since in both cases the expected payoff equals the stake.

(4.10) Example. In a roulette game the bank offers, for example, the bet $\omega_0 = \boxed{13}$ at odds $\frac{1}{35}$ while the probability of $\boxed{13}$ occurring is $\frac{1}{37}$ (be aware of the $\boxed{0}$), so we have

$$\frac{1}{37}(1+35) - 1 < 0.$$

Thus a rational individual should accept the complementary bet $\omega_0 \neq \boxed{13}$ at odds $\frac{35}{1}$ where we obtain

$$\frac{36}{37}(1+35) - 35 > 0.$$

(This is what the croupier does.) ◇

Let us illustrate the basic ideas of Bayesian inference by the following.

(4.11) Example. Consider again our example where a radar device ω provides information about the location of a ship. It is assumed that the true location is exactly one of the elements of $\Omega = \{$z1,z2,z3,rd,ld,ca$\}$. Results of radar devices are typically imprecise, and a possible result is "the ship is at sea". So each non-empty subset of Ω is a possible evidence.

The application of the Bayesian framework needs information about the reliability or sensitivity of the device which could be based on technical properties as well as statistical evaluation of former observations. The aim is to quantify a measure of confirmation for statements

"if the ship is actually at position w"

then ω gives A as a result"

by conditional probabilities. Formally we have to consider the space of hypotheses $H = \Omega$ and the space of possible evidences, here $E = 2^{\Omega} \cap \overline{\{\emptyset\}}$. The background knowledge K is assumed to be always the same. All information has to be expressed in the product space $H \times E$, and the conditional probabilities corresponding to the rule are $P(H \times \{A\} \mid \{w\} \times E)$. For convenience we always use the abbreviated notion $P(\{A\} \mid \{w\})$. If the available knowledge on Ω based on the background can be described by a prior probability P on Ω, and if the information given by the radar device is $A \in \mathcal{E}$, then we obtain as the posterior probability

$$P(\{\omega\} \mid \{A\}) = \frac{P(\{\omega\}) \cdot P(\{A\} \mid \{\omega\})}{P(\{A\})}, \text{ for all } \omega \in \Omega,$$

where $P(\{A\})$ only plays the role of a "normalizing constant". In case of n radar devices the situation can be described analogously by considering the hypothesis space $H = \Omega$ and the evidence space E^n. This would of course require enormous calculations, so one tries to reduce the complexity by introducing independence assumptions. If it is reasonable to assume that the result of one radar device depends only on the true location of the ship and not on the results of the other devices, the conditional independence assumption

$$P(A_1, \ldots, A_n \mid \omega) = \prod_{i=1}^{n} P(A_i \mid \omega)$$

for all $A_i \in E$, and all $\omega \in \Omega$ leads to the result

$$P(\omega \mid A_1, \ldots, A_n) = \alpha \cdot P(\omega) \cdot \prod_{i=1}^{n} P(A_i \mid \omega),$$

where $\alpha = \frac{1}{P(A_1,\ldots,A_n)}$ is again just a normalizing constant. Note that we have used the shorthand notation $P(A_1,\ldots,A_n \mid \omega)$ for the probability

$$P(\{A_1\} \times \ldots \{A_n\} \times \Omega \mid E^n \times \{\omega\}).$$

◇

Bayesian updating has the interesting property of recursivity. Suppose that $E = \{e_1,\ldots,e_n\}$ and $F = \{f_1,\ldots,f_n\}$ are evidence spaces and that $H = \{h_1,\ldots,h_n\}$ is a hypothesis such that $P(f \mid e,h) = P(f \mid h)$ holds. Then we obtain

$$O(h \mid e,f) = O(h \mid e) \cdot \lambda(f \mid h). \tag{4.14}$$

That is, the updating process can be performed by first integrating the evidence e and afterwards the knowledge contained in f by using this odds-likelihood ratio calculation.

One basic problem with the Bayesian approach is that the available information is often not sufficient to determine one particular (conditional) probability. The normal way to solve this difficulty is to supply the missing constraints by using "conditional independence assumptions" or "maximum entropy assumptions".

4.4 Classes of Probabilities

The Bayesian framework offers a consistent framework for handling uncertainty. In the last section we considered the case where one expert quantifies his belief by a probability measure. If several experts provide their personal beliefs then conceptually there are two alternatives. The first approach is to combine the probability distributions in order to find a simple, common probability. This approach is analyzed in detail in the paper [Genest and Zidek 1986]. A second alternative is to consider the information provided by the experts as a set of probabilities. A "practical" Bayesian could see sets of probabilities as a way to integrate a sensitivity analysis during the assessment process into the probability calculus.

One can also use the results obtained with this approach to describe some aspects of interval-valued probabilities ([Kyburg 1987] and [Good 1950]). If a single probability does not seem to be sufficient for representing adequately some individual's partial beliefs then a straightforward way to model ignorance concerning a probability $P(A)$ is to consider a *probability interval*, which is minimal with respect to the individual's state of knowledge and covers the "true" probability. Such an approach leads to various difficulties. If we consider the complementary events A and $\overline{A}$, then by no means every pair (x, y), where

x is taken from the probability interval of A and y from the probability interval of $\overline{A}$, satisfies the condition $x+y=1$. Therefore we consider a set $\mathcal{L}$ containing "permissible" [Levi 1974] probability functions in order to represent the beliefs of an individual, where each element of $\mathcal{L}$ is consistent with the individual's knowledge about some problem.

If the set $\mathcal{L}$ is assumed to be convex, i.e. $P, P' \in \mathcal{L}$ implies $a \cdot P + (1-a) \cdot P' \in \mathcal{L}$ for all $a \in (0,1)$, we obtain (possibly open) probability intervals for the events $A \subseteq \Omega$, but in contrast to the direct assignment of probability intervals endowed with a well-defined formal background. Note that in the sequel we do *not* restrict ourselves to convex sets $\mathcal{L}$ but consider arbitrary classes of probabilities.

Since a possibly infinite set $\mathcal{L}$ of probabilities is difficult to deal with, our aim is to derive finite methods for its description and handling. The main difficulty with such approaches is the inability always to get a single decision, see Sect. 4.5.

(4.12) Definition. *Let $\mathcal{L}$ be a non-empty class of probability measures. Then $\mathcal{L}$ defines the set functions*

$$\begin{aligned}
(P_{\mathcal{L}})_* &: 2^{\Omega} \to [0,1], \qquad A \mapsto \inf\{P(A) \mid P \in \mathcal{L}\},\\
(P_{\mathcal{L}})^* &: 2^{\Omega} \to [0,1], \qquad A \mapsto \sup\{P(A) \mid P \in \mathcal{L}\}.
\end{aligned}$$

$(P_{\mathcal{L}})_*(A)$ gives a lower bound whereas $(P_{\mathcal{L}})^*(A)$ gives an upper bound for the probabilities $P(A)$ where $P \in \mathcal{L}$. We have the following properties.

(4.13) Theorem. *Given a non-empty class $\mathcal{L}$ of probability measures the set functions $(P_{\mathcal{L}})_*$ and $(P_{\mathcal{L}})^*$ satisfy the following properties:*

(i) $(P_{\mathcal{L}})_*(\emptyset) = (P_{\mathcal{L}})^*(\emptyset) = 0$ *and* $(P_\ell)_*(\Omega) = (P_\ell)^*(\Omega) = 1$,

(ii) $0 \le (P_{\mathcal{L}})_*(A) \le (P_{\mathcal{L}})^*(A) \le 1$ *for each* $A \subseteq \Omega$,

(iii) $(P_{\mathcal{L}})^*(A) = 1 - (P_{\mathcal{L}})_*(\overline{A})$ *for each* $A \subseteq \Omega$, *and*

(iv) For all $A, B \subseteq \Omega$ *such that* $A \cap B = \emptyset$ *holds*

$$\begin{aligned}
(P_{\mathcal{L}})_*(A) + (P_{\mathcal{L}})_*(B) &\le (P_{\mathcal{L}})_*(A \cup B)\\
&\le (P_{\mathcal{L}})_*(A) + (P_{\mathcal{L}})^*(B)\\
&\le (P_{\mathcal{L}})^*(A \cup B) \le (P_{\mathcal{L}})^*(A) + (P_{\mathcal{L}})^*(B)
\end{aligned}$$

Proof.

(iii) For all $A \subseteq \Omega$ we have

$$\begin{aligned}(P_{\mathcal{L}})_*(\overline{A}) &+ (P_{\mathcal{L}})^*(A) \\ &= \inf\{1 - P(A) \mid P \in \mathcal{L}\} + \sup\{P(A) \mid P \in \mathcal{L}\} \\ &= -\sup\{P(A) \mid P \in \mathcal{L}\} + 1 + \sup\{P(A) \mid P \in \mathcal{L}\} \\ &= 1.\end{aligned}$$

(iv) Let $A, B \subseteq \Omega$ be such that $A \cap B = \emptyset$. From

$$(P_{\mathcal{L}})^*(A \cup B) = \inf\{P(A \cup B) \mid P \in \mathcal{L}\}$$

we know that there exists a sequence $(P_i)_{i=1}^{\infty}$ of probability measures in $\mathcal{L}$ such that

$$\lim_{i \to \infty} P_i(A \cup B) = (P_{\mathcal{L}})^*(A \cup B).$$

On the other hand we have

$$\begin{aligned}\lim_{i \to \infty} P_i(A \cup B) &= \lim_{i \to \infty} \big(P_i(A) + P_i(B)\big) \\ &\geq \liminf_{i \to \infty} \big(P_i(A)\big) + \liminf_{i \to \infty} \big(P_i(B)\big) \\ &\geq (P_{\mathcal{L}})_*(A) + (P_{\mathcal{L}})_*(B).\end{aligned}$$

The remaining inequalities are proved similarly. □

Note that in general the condition

$$\forall A, B \subseteq \Omega : (P_{\mathcal{L}})_*(A \cap B) + (P_{\mathcal{L}})_*(A \cup B) \geq (P_{\mathcal{L}})_*(A) + (P_{\mathcal{L}})_*(B) \quad (4.15)$$

is not satisfied. Let us consider next a probability P which is unknown except that for each event A in Ω for $P(A)$ there are bounds $R_1(A)$ and $R_2(A)$:

$$\forall A \subseteq \Omega : R_1(A) \leq P(A) \leq R_2(A). \quad (4.16)$$

In this case our aim is to determine the class of all probabilities compatible with R_1 and R_2.

(4.14) Definition. *A class $\mathcal{L}$ of probability measures P on $\Omega = \{\omega_1, \ldots, \omega_n\}$ is said to be of type 1, if there exist functions R_1 and R_2 from 2^{Ω} to $[0,1]$ such that*

$$\mathcal{L} = \{P \mid \forall A \subseteq \Omega : R_1(A) \leq P(A) \leq R_2(A).\}$$

If arbitrary functions R_1, R_2 are chosen, then a set $\{P \mid R_1 \leq P \leq R_2\}$ may be empty. On the other hand we possibly have

$$(P_{\mathcal{L}})_*(A) = \inf\{P(A) \mid P \in \mathcal{L}\} > R_1(A)$$

or

$$(P_{\mathcal{L}})^*(A) = \sup\{P(A) \mid P \in \mathcal{L}\} < R_2(A)$$

for some $A \subseteq \Omega$.

In order to clarify the properties of the set functions $(P_{\mathcal{L}})_*$, $(P_{\mathcal{L}})^*$ where $\mathcal{L}$ is a class of type 1 let us consider a finite space Ω, i.e. $\Omega = \{\omega_1, \ldots, \omega_n\}$. In this case each probability measure P on Ω can be identified by the point $\big(P(\{\omega_1\}), \ldots, P(\{\omega_n\})\big)$ in the n-dimensional space $I\!R^n$. From the conditions $P(\{\omega_1\}) + \ldots + P(\{\omega_n\}) = 1$ and $0 \leq P(\{\omega_i\}) \leq 1$ for $i = 1, \ldots, n$ we know that the points are located on a simplex. If the class $\mathcal{L}$ is of type 1 then $P \in \mathcal{L}$ if and only if $(P_{\mathcal{L}})_*(A) \leq P(A) \leq (P_{\mathcal{L}})^*(A)$ for all $A \subseteq \Omega$. Therefore a probability P is in $\mathcal{L}$ if and only if its representation $\big(P(\{\omega_1\}), \ldots, P(\{\omega_n\})\big)$ is within the set

$$\Big\{(r_1, \ldots r_n) \in I\!R^n \mid \exists Q : \big(\forall A \subseteq \Omega : (P_{\mathcal{L}})_*(A) \leq Q(A) \leq (P_{\mathcal{L}})^*(A) \text{ and } r_i = Q(\{\omega_i\}), i = 1, \ldots, n\big)\Big\}$$

$$= \bigcap_{A: A \subseteq \Omega} \Big\{(r_1, \ldots, r_n) \in I\!R^n \mid \exists Q : \forall i \in \{1, \ldots, n\} : r_i = Q(\{\omega_i\}) \text{ and } (P_{\mathcal{L}})_*(A) \leq \sum_{\omega: \omega \in A} Q(\{\omega\})\Big\}$$

$$\cap \bigcap_{A: A \subseteq \Omega} \Big\{(r_1, \ldots r_n) \in I\!R^n \mid \exists Q : \forall i \in \{1, \ldots, n\} : r_i = Q(\{\omega_i\}) \text{ and } (P_{\mathcal{L}})^*(A) \geq \sum_{\omega: \omega \in A} Q(\{\omega\})\Big\}$$

This shows that we have to consider the finite intersection of regions which are bordered by a hyperplane with a simplex.

From an application point of view, sets of probabilities are of course useful for an expert to express his imprecise knowledge about some underlying "true" probability distribution. A class of type 1 corresponds to the case where the expert is able to give lower and upper bounds for the probabilities of some events. It is obvious that the bounds have to be chosen very carefully, since otherwise there might remain no probability that fits all restrictions.

(4.15) Definition. *Let $\mathcal{L}$ be a class of type 1. We say that $\mathcal{L}$ is a class of type 2 if and only if for every positive integer n and every collection $A_1, \ldots, A_n$ of subsets of Ω the inequality*

$$(P_{\mathcal{L}})_*(A_1 \cup \ldots \cup A_n) \geq \sum_{I:\emptyset \neq I \subseteq \{1,\ldots,n\}} (-1)^{|I|+1} \cdot (P_{\mathcal{L}})_*(\bigcap_{i \in I} A_i)$$

holds.

The corresponding upper probability $(P_{\mathcal{L}})^*$ of a class of type 2 has the dual property

$$(P_{\mathcal{L}})^*(A_1 \cap \ldots \cap A_n) \leq \sum_{I:\emptyset \neq I \subseteq \{1,\ldots,n\}} (-1)^{|I|+1} \cdot (P_{\mathcal{L}})^*(\bigcup_{i \in I} A_i) \quad (4.17)$$

for all $n \in \mathbb{N}$ and $A_1, \ldots, A_n \subseteq \Omega$.

(4.16) Example. Consider the class $\mathcal{L}$ of probability measures P on $\Omega = \{\omega_1, \omega_2, \omega_3\}$ which fit the restrictions

$$\frac{1}{2} \leq P(\{\omega_1, \omega_2\}) \leq 1,$$
$$\frac{1}{2} \leq P(\{\omega_2, \omega_3\}) \leq 1, \text{ and}$$
$$\frac{1}{2} \leq P(\{\omega_1, \omega_3\}) \leq 1.$$

Then the characterizing region is the volume shown in Fig. 4.2. From this figure we can directly find the lower and upper probabilities induced by $\mathcal{L}$. We have, for example,

$$(P_{\mathcal{L}})_*(\{\omega_1\}) = (P_{\mathcal{L}})_*(\{\omega_2\}) = (P_{\mathcal{L}})_*(\{\omega_3\}) = 0,$$

$$(P_{\mathcal{L}})_*(\{\omega_1, \omega_2\}) = (P_{\mathcal{L}})_*(\{\omega_1, \omega_3\}) = (P_{\mathcal{L}})_*(\{\omega_2, \omega_3\}) = \frac{1}{2}.$$

By setting $A_1 = \{\omega_1, \omega_2\}$, $A_2 = \{\omega_2, \omega_3\}$, $A_3 = \{\omega_1, \omega_3\}$ we obtain $(P_{\mathcal{L}})_*(A_1 \cup A_2 \cup A_3) = 1$, but

$$\begin{aligned}
&(P_{\mathcal{L}})_*(A_1) + (P_{\mathcal{L}})_*(A_2) + (P_{\mathcal{L}})_*(A_3) \\
&-(P_{\mathcal{L}})_*(A_1 \cap A_2) - (P_{\mathcal{L}})_*(A_2 \cap A_3) - (P_{\mathcal{L}})_*(A_1 \cap A_3) \\
&+(P_{\mathcal{L}})_*(A_1 \cap A_2 \cap A_3) \\
&= \frac{3}{2}.
\end{aligned}$$

This shows that L is of type 1 but not of type 2. ◇

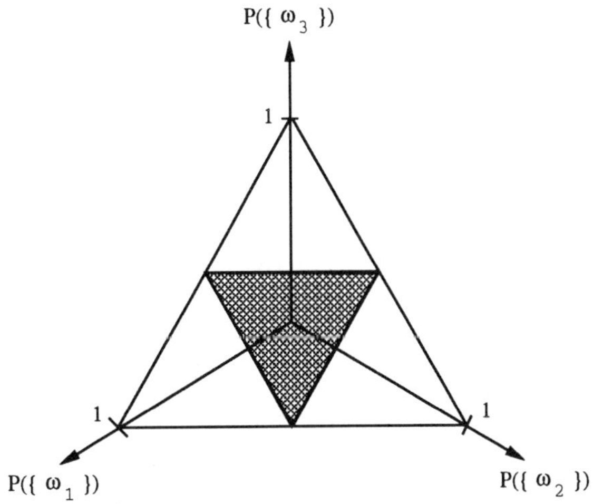

Fig. 4.2. Characterizing region of class $\mathcal{L}$

Let $R_1 : 2^{\Omega} \to [0,1]$ be a set function satisfying the conditions

$$R_1(\emptyset) = 0,$$
$$R_1(\Omega) = 1,$$
$$R_1(\bigcup_{i=1}^{n} A_i) \geq \sum_{I:\emptyset \neq I \subseteq \{1,\ldots,n\}} (-1)^{\mathrm{card}(I)+1} R_1(\bigcap_{i \in I} A_i)$$

for all $n \in \mathbb{N}$ and all $A_1, \ldots, A_n \subseteq \Omega$. Define $R_2 : 2^{\Omega} \to [0,1]$ by setting $R_2(A) \stackrel{\mathrm{d}}{=} 1 - R_1(\overline{A})$. Then the set

$$\mathcal{L} = \{P \mid \forall A \subseteq \Omega : R_1(A) \leq P(A) \leq R_2(A)\}$$

is of type 2. In order to determine this set of probabilities, we use the Moebius transformation[5] to define a set function $m : 2^{\Omega} \to [0,1]$

$$m(A) = \sum_{B:B \subseteq A} (-1)^{|A \cap \overline{B}|} R_1(B). \tag{4.18}$$

[5] This transformation is performed in Theorem 6.6 concerning belief functions. We obtain that $m(\emptyset) = 0$ and $\sum_{A:A \subseteq \Omega} m(A) = 1$ hold.

Let $\Omega = \{\omega_1, \ldots, \omega_n\}$, $S_{d_1,d_2,\ldots d_n} = \{\omega_i \in \Omega \mid d_i = 1\}$, and $P_{d_1,d_2,\ldots d_n} = m(S_{d_1,d_2,\ldots d_n})$, where $d_1, d_2, \ldots d_n \in \{0,1\}$. So $S_{1,0,\ldots,0,1}$ denotes for example the set $\{\omega_1, \omega_n\}$, whereas $S_{0,\ldots,0}$ denotes the empty set. The probabilities $P_{d_1,\ldots,d_n}$ we assume to be divided into portions $P^{(i)}_{d_1,\ldots,d_n} \geq 0$, so that

$$P_{d_1,d_2,\ldots,d_n} = \sum_{i=1}^{n} P^{(i)}_{d_1,d_2,\ldots,d_n}, \tag{4.19}$$

where $P^{(i)}_{d_1,\ldots,d_n} = 0$ unless $d_i = 1$. $P^{(i)}_{d_1,\ldots,d_n}$ is that piece of the probability mass $P_{d_1,\ldots,d_n}$ attached to $\omega_i \in S_{d_1,\ldots,d_n}$. With this interpretation it becomes obvious that $P^{(i)}_{d_1,\ldots,d_n} > 0$ requires $d_i = 1$. Gathering the different pieces some element $\omega_i \in \Omega$ gets from the probabilities $P^{(i)}_{d_1,\ldots,d_n}$ we obtain

$$P\{\omega_i\} = \sum_{d_1,d_2,\ldots,d_n \in \{0,1\}} P^{(i)}_{d_1,d_2,\ldots,d_n}. \tag{4.20}$$

Each division according to (4.20) uniquely determines one probability on Ω, i.e. one element of the class $\mathcal{L}$.

From an application point of view it is of course different to specify directly functions R_1 and R_2 satisfying the above conditions (in [Weichselberger and Pöhlmann 1990] some critical examples are presented), but the next theorem shows where sets of type 2 often appear.

(4.17) Theorem. *Let P be a probability on Ω, and let Ω' be a refinement of Ω with refinement mapping $\widehat{\Pi}$. Then the set $\mathcal{L}$ of all probabilities on Ω', which are refinements of P, is a class of type 2.*

Proof. For simplicity the proof relies on random sets, which will be considered in Chap. 5. Consider the random set $\Gamma : \Omega \to 2^{\Omega'}$, $\Gamma(\omega) = \widehat{\Pi}(\{\omega\})$. P' is a refinement of P

$$\Longleftrightarrow \forall \omega \in \Omega : P(\{\omega\}) = P'\Big(\widehat{\Pi}(\{\omega\})\Big)$$

$$\Longleftrightarrow \forall A' \in \Omega' : \sum_{\omega \in \Omega : \widehat{\Pi}(\{\omega\}) \subseteq A'} P(\{\omega\}) \leq P'(A') \leq \sum_{\omega \in \Omega : \widehat{\Pi}(\{\omega\}) \cap A' \neq \emptyset} P(\{\omega\})$$

$$\Longleftrightarrow \forall A' \in \Omega' : (P_\Gamma)_*(A') \leq P'(A') \leq (P_\Gamma)^*(A').$$

It follows that $(P_\Gamma)_* = (P_{\mathcal{L}})_*$ and $(P_\Gamma)^* = (P_{\mathcal{L}})^*$. □

The next topic of this section concerns the concept of conditioning. If we consider a class $\mathcal{L}$ of probabilities, then we have to compute for each $P \in \mathcal{L}$ the conditional probability $P(\,.\,|B)$, where $B \subseteq \Omega$ is the conditioning event. This yields

$$(P_{\mathcal{L}})_*(A|B) = \inf\{P(A|B) \mid P \in \mathcal{L} \text{ and } P(B) > 0\}, \quad A \subseteq \Omega, \quad (4.21)$$

and

$$(P_{\mathcal{L}})^*(A|B) = \sup\{P(A|B) \mid P \in \mathcal{L} \text{ and } P(B) > 0\}, \quad A \subseteq \Omega, \quad (4.22)$$

which (in case of existence) are called the lower and upper conditional probabilities, respectively.

If $\mathcal{L}$ is a class of type 1, and if $(P_{\mathcal{L}})_*$ is 2 monotone (4.26), then we obtain [Moral 1989]

$$(P_{\mathcal{L}})_*(A|B) = \frac{(P_{\mathcal{L}})_*(A \cap B)}{(P_{\mathcal{L}})_*(A \cap B) + (P_{\mathcal{L}})^*(B \cap \overline{A})}, \quad (4.23)$$

$$(P_{\mathcal{L}})^*(A|B) = \frac{(P_{\mathcal{L}})^*(A \cap B)}{(P_{\mathcal{L}})^*(A \cap B) + (P_{\mathcal{L}})_*(B \cap \overline{A})}. \quad (4.24)$$

In practical applications we often have to calculate sets of probabilities which satisfy various independence assumptions.

(4.18) Example. Let $\Omega^{(1)} = \{\omega_1, \omega_2, \omega_3\}$ and $\Omega^{(2)} = \{\omega_4, \omega_5, \omega_6\}$. The best way to express independencies is to consider the product space $\Omega^{\{1,2\}} = \Omega_1 \times \Omega_2$. Let P be a probability on $\Omega^{\{1,2\}}$ and $P^{(1)} \stackrel{\mathrm{d}}{=} \Pi^{\{1,2\}}_{\{1\}}(P)$ and $P^{(2)} \stackrel{\mathrm{d}}{=} \Pi^{\{1,2\}}_{\{2\}}(P)$ the marginal probabilities of P, so that for example

$$P^{(1)}(\{\omega_2\}) = P\Big(\{(\omega_2, \omega_4), (\omega_2, \omega_5), (\omega_2, \omega_6)\}\Big),$$

$$P^{(2)}(\{\omega_5\}) = P\Big(\{(\omega_1, \omega_5), (\omega_2, \omega_5), (\omega_3, \omega_5)\}\Big).$$

$\{\omega_2\}$ and $\{\omega_5\}$ are said to be independent, if and only if

$$\Big(P^{(1)}(\{\omega_2\})\Big) \cdot \Big(P^{(2)}(\{\omega_5\})\Big) = P\Big(\{(\omega_2, \omega_5)\}\Big).$$

If on the other hand two probabilities P_1 on Ω_1 and P_2 on Ω_2 are given, then we obtain a class $\mathcal{L}$ of probabilities on $\Omega^{\{1,2\}}$ compatible with P_1 and P_2 by requiring

$$P \in \mathcal{L} \text{ if and only if } P^{(1)} = P_1 \text{ and } P^{(2)} = P_2.$$

Further independence assumptions lead to a reduction of this set $\mathcal{L}$. If we require that ω_2 and ω_5 have to be independent, we obtain a set $\mathcal{L}'$ of compatible probabilities, where

$$P \in \mathcal{L}' \text{ if and only if } P^{(1)} = P_1 \text{ and } P^{(2)} = P_2$$
$$\text{and } P_1(\{\omega_2\}) \cdot P_2(\{\omega_5\}) = P(\{\omega_2, \omega_5\}).$$

◇

Analogously, if we know that the conditional probability $P(A|B)$ is equal to c, then the set

$$\{P \mid P(A|B) = c\}$$

is compatible with this information. If there are several different conditions, then we have to intersect the sets of probabilities. In the best case there remains only one probability compatible with all concepts, in the most general case none – which means that the knowledge is contradictory and has to be considered again. One possible way to manage the problem of handling sets of probabilities is to use representatives. The set of all specializations of some probability R

$$\{P \mid P \sqsubseteq R\}$$

can of course be represented by R. The same holds for the set of all conditionings of R, i.e. the set

$$\Big\{P \,\Big|\, \exists A : (R(A) > 0 \text{ and } \forall B : P(B) = R(B|A))\Big\}.$$

(4.8) Example. (revisited) In a second analysis we again assume that the selection of the ships is (stochastically) independent of the selection of the sensors, but that the probability p_{ps} for the event "The ship type is actually ps" is unknown. In that case we have

$$\begin{aligned}
P\{(\omega_1, \text{ps}, \text{ps})\} &= 0.25 \cdot p_{\text{ps}} \\
P\{(\omega_1, \text{tv}, \text{tv})\} &= 0.25 \cdot (1 - p_{\text{ps}}) \\
P\{(\omega_2, \text{ps}, \text{tv})\} &= 0.25 \cdot p_{\text{ps}} \\
P\{(\omega_2, \text{tv}, \text{ps})\} &= 0.25 \cdot (1 - p_{\text{ps}}) \\
P\{(\omega_3, \text{ps}, \text{ps})\} &= 0.25 \cdot p_{\text{ps}} \\
P\{(\omega_3, \text{tv}, \text{ps})\} &= 0.25 \cdot (1 - p_{\text{ps}}) \\
P\{(\omega_4, \text{ps}, \text{tv})\} &= 0.25 \cdot p_{\text{ps}} \\
P\{(\omega_4, \text{tv}, \text{tv})\} &= 0.25 \cdot (1 - p_{\text{ps}}).
\end{aligned}$$

That is, we obtain a *set* $\mathcal{L}$ of possibly underlying probabilities. The calculation of the lower and upper probabilities yields

$$\begin{aligned}
&(P_{\mathcal{L}})_*(\text{ship type is ps}|\text{sensor message is ps}) \\
&\quad = \inf\Big\{\frac{0.25 \cdot p_{\text{ps}} + 0.25 \cdot p_{\text{ps}}}{0.25 \cdot p_{\text{ps}} + 0.25 \cdot (1 - p_{\text{ps}}) + 0.25} \mid p_{\text{ps}} \in [0,1]\Big\} \\
&\quad = \inf\big\{p_{\text{ps}} \mid p_{\text{ps}} \in [0,1]\big\} = 0 \\
&(P_{\mathcal{L}})^*(\text{ship type is ps}|\text{sensor message is ps}) \\
&\quad = \sup\big\{p_{\text{ps}} \mid p_{\text{ps}} \in [0,1]\big\} = 1
\end{aligned}$$

which represents total ignorance. In this approach we assume that there is in fact an underlying probability p_{ps} for the statement "The ship's type is ps". (to be revisited) ◇

4.5 Decision Making Aspects

As we pointed out in Sect. 1.2, models allow us to describe environments, tasks, or people's state of mind, but the final aim of reasoning processes is in general not just description. The aim is support in finding decisions or choosing actions. A physician, for example, is not only asked for a diagnosis but based on his diagnosis he has to choose a therapy. In this context the decision for or against some medicament relies on an evaluation of its advantages and disadvantages with respect to the result of the examination, but not directly on the evaluation of the actual symptoms. So if the aim is not merely the adaptation of some naive expert's decision behavior, the decision problem has to be conceptually distinguished from the descriptive task. Thus a suitable model should, beyond description, permit us to derive *prescriptions* for actions or conditions some environment is required to meet.

In order to model decision making define $D = \{d_1, \ldots, d_k\}$ to be the set of possible decisions. Besides D we have the set $C = \{c_1, \ldots, c_N\}$ of consequences, forming the scale for measuring the benefits of the different decisions. The actual consequence $c \in C$ depends on the "state of the world" $\omega_0 \in \Omega$ as well as on the chosen action $d \in D$, where $\Omega = \{\omega_1, \ldots, \omega_m\}$ is the frame of discernment. The relationship between the true "state of the world", the decision and its consequence is given by the mapping

$$T : \Omega \times D \to C.$$

Since we aspire to compare the benefits of different decisions, the existence of a total order $\leq$ on C which reflects our preferences is required. In this context $c \leq c'$ denotes that the consequence c' is preferred to c or at least equivalent. Equivalence with respect to relation $\leq$ means that $c \leq c'$ as well as $c' \leq c$ is valid. If $c' \leq c$ holds whereas $c \leq c'$ does not hold, then c is preferred.

Whenever our interest is not only to compare consequences with respect to their benefit but to quantify the profit they cause, we consider the elements of C to be associated with real numbers, which we imagine to be that amount of money we obtain if c is the actual consequence. Adapting a notion from game theory we call this number the payoff. On the formal level a function

$$U : C \to \mathbb{R},$$

which we call the *utility measure*, determines the payoff, i.e. the utility of the different decisions.

(4.19) Example. Recall the betting problem we considered in Sect. 4.3. The problem was to choose between two bets

(a) You bet on $\omega_0 \in A$ at odds $\frac{x}{y}$.

(b) You bet on $\omega_0 \in \overline{A}$ at odds $\frac{y}{x}$.

In terms of the notions we introduced above this reads as follows:

$$D = \{a, b\}, \quad C = \{\text{win\_a}, \text{win\_b}, \text{lose\_a}, \text{lose\_b}\}$$

$$T(\omega, d) = \begin{cases} \text{win\_a} & \text{if } \omega_0 \in A \text{ and } d = a \\ \text{win\_b} & \text{if } \omega_0 \in \overline{A} \text{ and } d = b \\ \text{lose\_a} & \text{if } \omega_0 \in \overline{A} \text{ and } d = a \\ \text{lose\_b} & \text{if } \omega_0 \in A \text{ and } d = b \end{cases}$$

$$U(c) = \begin{cases} y & \text{if } c = \text{win\_a} \\ x & \text{if } c = \text{win\_b} \\ -x & \text{if } c = \text{lose\_a} \\ -y & \text{if } c = \text{lose\_b} \end{cases}$$

The following table summarizes these relations:

result	a	b	probability
A	y	$-y$	$P(A)$
$\overline{A}$	$-x$	x	$P(\overline{A})$
expected utility	$P(A) \cdot y - P(\overline{A}) \cdot x$	$P(\overline{A}) \cdot x - P(A) \cdot y$	

We have

$$U(a) = P(A) \cdot y - \big(1 - P(A)\big) \cdot x = P(A) \cdot (x + y) - x$$

$$U(b) = P(\overline{A}) \cdot x - \big(1 - P(\overline{A})\big) \cdot y = P(\overline{A}) \cdot (x + y) - y.$$

As we pointed out in Sect. 4.3 we have $U(a) \cdot U(b) \leq 0$, thus applying a *maximum expected utility criterion* reduces to choosing the bet $d \in \{a, b\}$ with $U(d) > 0$. ◇

Recall that our aim is decision making in the presence of uncertainty, which means we do not know the exact identity of the true "state of the world" ω_0, but have to rely on probabilities or, in the most general case, on mass distributions on Ω. The next example shows how to handle purely probabilistic "knowledge" about Ω.

(4.20) Example. Suppose we have to determine the location of some ship with respect to the frame of discernment $\Omega = \{at\text{-}sea, not\text{-}at\text{-}sea\}$. Our knowledge may be encoded by a probability Q on Ω, where $Q(\{at\text{-}sea\}) = 0.8$, $Q(\{not\text{-}at\text{-}sea\}) = 0.2$. Since our aim is just to find the ship, the possible decisions are

- to assume that the ship is at-sea (d1),
- to assume that the ship is not-at-sea (d2), or
- to resign ourselves to the fact that the ship position is unknown (d3).

Thus we have $D = \{d1, d2, d3\}$. Depending on the decision the search for the ship will be either successful or not, i.e. $C = \{\text{success}, \text{no success}\}$. The benefit of the two consequences may be quantified be the utility measure $U : C \to I\!R$, where $U(\text{success}) = 100$, $U(\text{no success}) = 0$. The following table summarizes the relations, consequences and utilities:

state	d_1	d_2	d_3	state probability
at-sea	100	0	0	0.8
not-at-sea	0	100	0	0.2
expected utility	80	20	0	

The last row contains the expected utility $U(d)$ of the decisions defined by

$$U(d) = \sum_{i=1}^{n} U\big(T(\omega_i, d)\big) \cdot Q(\{\omega_i\}), \quad \text{where } \Omega = \{\omega_1, \ldots, \omega_n\}, \tag{4.25}$$

for each $d \in D$. Obviously in our example d_1 would be a good choice, since it promises the highest benefit. In more general terms our strategy is to rely on the *maximum expected utility criterion.* ◇

Up to now we restricted ourselves to considering the consequences of some decision maker's choice among the possible locations. Doing so we assumed the decision maker to be a passive observer of the situation before his decision. In practice an intervention is often possible which facilitates the decision. Such intervention in general causes costs which have to be taken into account when we choose whether to perform it or not.

(4.20) Example. (continued) Recall that our aim is to determine the location of some ship. Assume that by an intervention, say a radar bearing which costs \$t, it is possible to obtain a result such as:

"If the ship is at-sea, then there is a 90% probability that the bearing will confirm it. If the ship is not at-sea there is an 80% probability of a correct result".

The result of the radar bearing is "positive" if the ship is located at sea and "negative" otherwise. The question we have to answer is whether it is reasonable to perform the bearing and, if so, what cost is acceptable, or not to perform it. ◇

In the field of decision theory, diagrams are used to answer such questions. Decision diagrams consist of decision nodes (graphically represented by squares) and chance nodes (represented by circles). Branches emanating from decision nodes correspond to options available to the decision maker, whereas the branches emanating from chance nodes end in the possible outcome of the corresponding random experiment. Each of these branches is labelled with a conditional probability. At each leaf of the diagram we find the utility associated with the path from the root node to that leaf. The symbol $|_f$ indicates that an amount \$t has to be paid to enter this region of the diagram.

(4.21) Example. In our example we obtain the diagram depicted in Fig. 4.3. In order to analyse the above graph we have to calculate the probabilities for the branches at the chance nodes. We start from the known probabilities, i.e.

$$\begin{aligned} P\big(\{\text{at-sea}\}\big) &= 0.8 \\ P\big(\{\text{not-at-sea}\}\big) &= 0.2 \\ P\big(\{\text{positive}\} \mid \{\text{at-sea}\}\big) &= 0.9 \\ P\big(\{\text{positive}\} \mid \{\text{not-at-sea}\}\big) &= 0.2 \\ P\big(\{\text{negative}\} \mid \{\text{at-sea}\}\big) &= 0.1 \\ P\big(\{\text{negative}\} \mid \{\text{not-at-sea}\}\big) &= 0.8. \end{aligned}$$

By Bayes' Theorem (Theorem 4.3) we obtain

$$\begin{aligned} P\Big(\big\{(\text{positive}, \text{at-sea})\big\}\Big) &= 0.72 \\ P\Big(\big\{(\text{positive}, \text{not-at-sea})\big\}\Big) &= 0.04 \\ P\Big(\big\{(\text{negative}, \text{at-sea})\big\}\Big) &= 0.08 \\ P\Big(\big\{(\text{negative}, \text{not-at-sea})\big\}\Big) &= 0.16. \end{aligned}$$

This yields

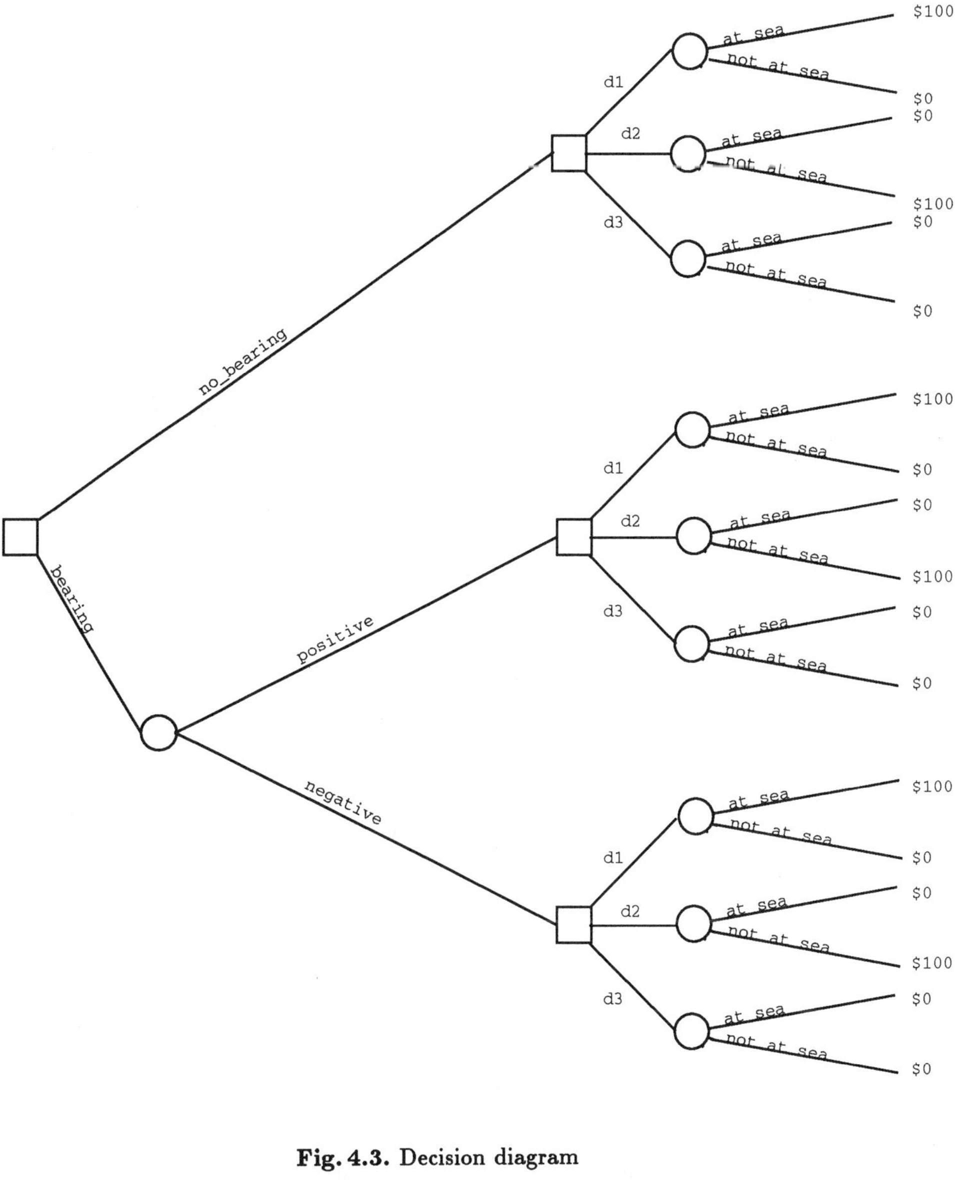

Fig. 4.3. Decision diagram

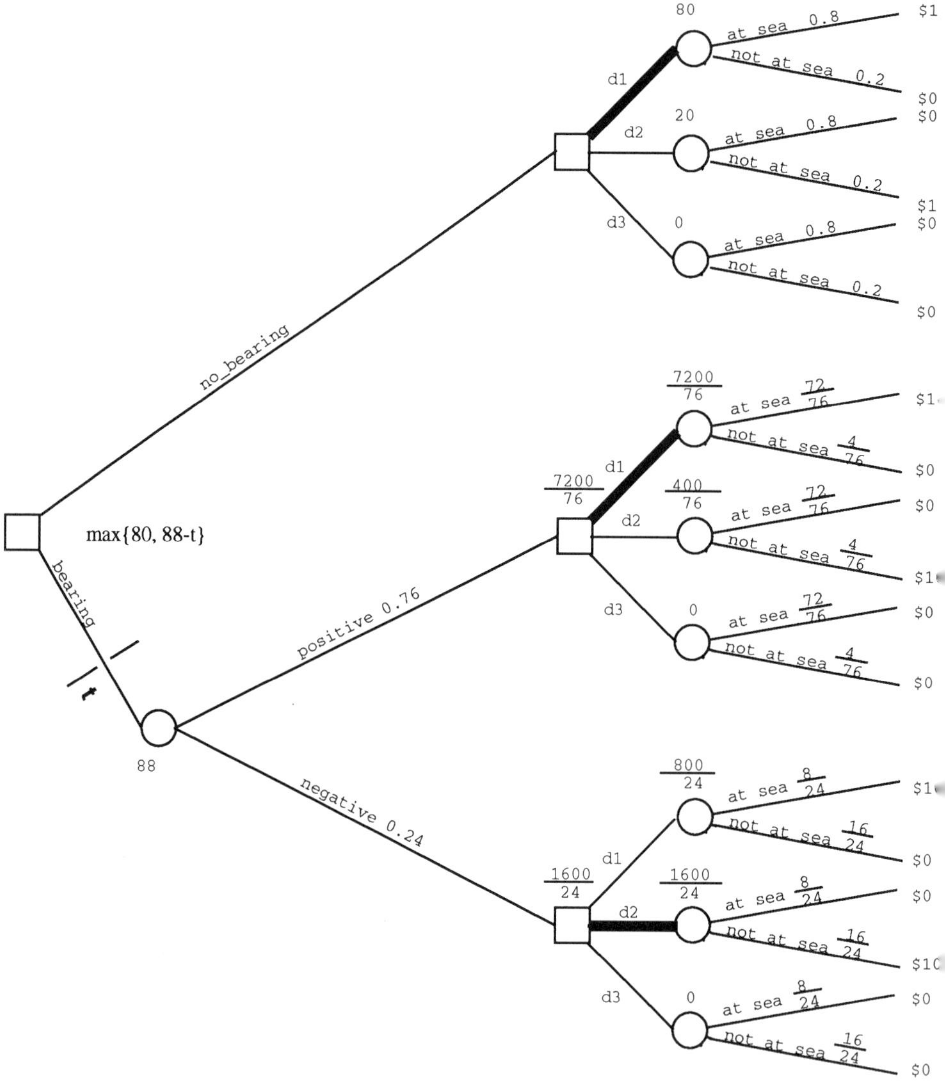

Fig. 4.4. Decision diagram with marked branches

$$P(\{\text{positive}\}) = 0.76, \qquad P(\{\text{negative}\}) = 0.24,$$

$$P(\{\text{at-sea}\} \mid \{\text{positive}\}) = \frac{72}{76},$$

$$P(\{\text{at-sea}\} \mid \{\text{negative}\}) = \frac{8}{24},$$

$$P(\{\text{not-at-sea}\} \mid \{\text{positive}\}) = \frac{4}{76},$$

$$P(\{\text{not-at-sea}\} \mid \{\text{negative}\}) = \frac{16}{24}.$$

These are the probabilities we used to complete our decision diagram. Based on these probabilities we can perform a so-called "roll-back analysis" providing the expected utility for each decision node. Since the maximization of the expected utility is our decision criterion we choose the maximum-expected-utility decision (and in the diagram we mark the corresponding branch as depicted in Fig. 4.4). Starting at the leaves we are able to determine the optimal decisions step by step. The completed decision diagram contains all relevant quantities. The result of our analysis is that a radar bearing is reasonable only if the costs do not exceed \$8, otherwise it should not be performed. The marked branches represent the optimal decisions. ◇

Decision making in the simplest case means choosing one of the events in $\Omega = \{\omega_1, \ldots, \omega_m\}$ to be considered as the result of some random experiment according to a probability Q on Ω, where the decision $d_i \in D = \{d_1 \ldots, d_m\}$ represents the choice of ω_i. Let $C = \{0,1\}$ be the set of consequences. We obtain

$$T : D \times \Omega \to C; \qquad T(d_i, \omega_j) = \begin{cases} 1, & \text{if } i = j \\ 0, & \text{if } i \neq j. \end{cases}$$

With $U : C \to I\!R$; $U(c) = c$, the maximum-expected-utility criterion obviously favors the decision d_i where $Q(\{\omega_i\})$ is maximal. Unfortunately in the context of knowledge representation the situation often gets more complicated, if instead of a single probability a set $\mathcal{P}$ of probabilities on Ω is used.

(4.22) Example. Consider once more $\Omega = \{\text{at-sea}, \text{not-at-sea}\}$. The set $\mathcal{P}$ of probabilities on Ω may be defined by

$$\mathcal{P} = \Big\{ Q_t : 2^\Omega \to [0,1] \,\Big|\, Q_t(\{\text{at-sea}\}) = t, \\ Q_t(\{\text{not-at-sea}\}) = 1 - t, t \in [0.4, 0.6] \Big\}$$

The function $T : \Omega \times D \to C$ is listed in the following table:

	d_1	d_2	state probability
at-sea	10	20	t
not-at-sea	20	10	$1-t$
expected utility	$20-10t$	$10+10t$	

(Note that the decisions as well as the associated utilities differ from those in Example 4.21). For $t = 0.4$ we obtain $U(d_1) = 16 \geq U(d_2) = 14$ whereas $t = 0.6$ yields $U(d_1) = 14 \leq U(d_2) = 16$. So it depends on the parameter t which decision is optimal. ◇

As the above example shows, in the general case it depends on the choice of a probability $Q \in \mathcal{P}$, which decision turns out to be optimal. The respective utilities are given by

$$U(d|Q) = \sum_{i=1}^{m} U\big(T(\omega_i, d)\big) \cdot Q\big(\{\omega_i\}\big),$$

where $d \in D$ and $Q \in \mathcal{P}$. If there is a decision $d_0 \in D$ such that for all $Q \in \mathcal{P}$ $U(d_0|Q) \geq \max\{U(d|Q) \mid d \in D\}$ then d_0 is the "unique" best decision. In general such a decision does not exist, but it is possible to compare selected decisions. For this purpose we define the notion of *dominance*. d dominates d' if and only if $U(d|Q) \geq U(d'|Q)$ for all $Q \in \mathcal{P}$.

(4.22) Example. (continued) In our example there are no dominances, but adding a third decision as listed in the following table this new decision is dominated by d_1 as well as by d_2.

state	d_1	d_2	d_3	state probability
at-sea	10	20	10	t
not-at-sea	20	10	10	$1-t$
expected utility	$20-10t$	$10+10t$	10	$t \in \{0.4, 0.6, 0.8\}$

Since in this case we have neither that d_1 dominates d_2 nor that d_2 dominates d_1 a unique "best" decision does not exists.

If preferences among the elements $\mathcal{P}$ are specified in the form of a probability P on $\mathcal{P}$, then we can apply an analysis similar to the approach of "second order probabilities". The straightforward way is to consider weighted utilities, i.e. to define

$$U(d) = \sum_{Q \in \mathcal{P}} \left(\sum_{i=1}^{m} U\big(T(d, \omega_i)\big) \cdot Q\big(\{\omega_i\}\big) \right) \cdot P(Q)$$

and then to maximize the expected utility by the choice of d. In the above example we obtain for the uniform distribution on $\mathcal{P}$

$$U(d_1) = (16 + 14 + 12)\frac{1}{3} = \frac{42}{3} = 14$$

$$U(d_2) = (14 + 16 + 18)\frac{1}{3} = \frac{48}{3} = 16$$

$$U(d_3) = 10.$$

Obviously d_2 is the best decision and should be chosen. ◇

4.6 Aggregating Probability Distributions

In the following we address the problem of aggregating a number of expert opinions which have been expressed in the form of probability distributions. If a meta-expert is able to assign weights to the n given subjective measures $P_1, \ldots, P_n$ then the linear opinion pool

$$T(P_1, \ldots, P_n) = \sum_{i=1}^{n} w_i P_i$$

with non-negative weights w_i such that $\sum_{i=0}^{n} w_i = 1$ is appropriate [Store 1961], [Wagner 1982] and [Genest 1984].

(4.23) Example. Let $\Omega = \{z3, z2, z1\}$ be the set of possible ship locations. Suppose we have three different information sources, each providing a single probability P_i on Ω, and the weight of source i is described by the function $Q : \{1, 2, 3\} \rightarrow [0, 1]$.

$P_i(\{\omega\})$	P_1	P_2	P_3
z1	0.4	0.3	0.2
z2	0.4	0.4	0.4
z3	0.2	0.3	0.4

i	$Q(\{i\})$
1	0.1
2	0.8
3	0.1

Then the aggregated probability P is given by

$$P(\{z1\}) = 0.4 \cdot 0.1 + 0.3 \cdot 0.8 + 0.2 \cdot 0.1 = 0.3$$

$$P(\{z2\}) = 0.4 \quad P(\{z3\}) = 0.3$$

if we interpret the weight w_i as the probability with which the source i is selected then an approach which was originally proposed by DeFinetti [DeFinetti 1972] is much more informative. Consider the random variable

$$X_A : \{1, 2, 3\} \to [0, 1]; j \mapsto P_j(A), \quad \text{where } A \subseteq \Omega,$$

that assigns to each source j the probability P_j of an event A of Ω. Using this random variable we can consider the probability on the interval $[0, 1]$ which is induced by Q to obtain a reasonable description of the situation in question. (see Fig. 4.5)

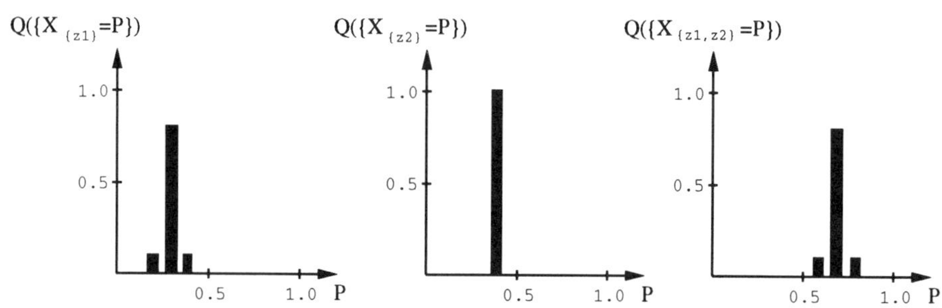

Fig. 4.5. Second order probabilities

If we learn that the true result is within the set {z1, z2}, then we have to modify the random variables. We obtain

$$X_A(\,.\,\mid\{\text{z1},\text{z2}\}) : \{1, 2, 3\} \to [0, 1]; \qquad j \mapsto P_j\big(A \mid \{\text{z1},\text{z2}\}\big)$$

If a *single* value reasonably reflecting the "probability of A" is requested, then the expected value of X_A may be used, i.e. $E(X_A) = \sum_{j=1}^{3} Q\big(\{i\}\big) \cdot P_j(A)$.

◇

In the literature alternative approaches have been considered. In the context of Bayesian theory one can introduce a SUPRA Bayesian [Winkler 1968], [Morris 1974], but there plenty of other pooling recipes [Genest and Zidek 1986], [French 1985] or consensus rules [Berenstein et al. 1986].

4.7 Concluding Remarks

As Pearl [Pearl 1988] pointed out, probabilities are useful in knowledge based system since they can be used as "a faithful guardian of common sense". The main reason for this fact seems to be that probabilities behave like relative frequencies. Axiomatic justifications are given in [Cox 1946], [Ramsey 1931] and [DeFinetti 1937], whereas in [Shafer 1986] these arguments are critically reconsidered. We think that using a single probability offers in general the most efficient but not always an adequate method for uncertainty representation. If a decision has to be taken, then one should use probabilities ([Lindley 1982] and [DeGroot 1970]).

Probability in its formal sense seems to have had its beginnings in games of chance in the 16th and 17th centuries. This theory was related to the mathematical theory of combinatorics, where the probability of an event is defined to be the quotient of the number of those cases in which some particular event occurs and the number of "possible" cases. This so-called Laplace definition relies on the classical interpretation of probability and is appropriate in finite spaces of equally likely events.

In the 19th century D. Poisson tried to define probability as a limit of a long-run relative frequency in order to handle more complex models. These approaches turned out to be unsatisfactory in this century. The modern axiomatic foundation of mathematical probability is embedded in measure theory [Halmos 1950]. Our Definition 4.1 recalls the Kolmogorov axioms [Kolmogorov 1950]. In this formalization the long-run relative frequency properties are reflected by the so-called "law of large numbers". The notion of conditioning is not a derived concept but has to be defined, so its properties are mainly determined by the aims underlying the definition (Definition 4.3). This view of probability is well established; we recommend the books [Feller 1960, 1966] and [Rény 1970] as introductory monographs.

An alternative view of probability is to consider a probability as a "personal degree of confidence" that an individual attaches to an uncertain event or, in other words, as the description of the "epistemic state of an individual". There are strong arguments for the use of probability theory to model personal belief due to Dutch Book arguments [Ramsey 1931]. The Bayesian approach [Bayes 1763] is due to B. de Finetti [DeFinetti 1937] and L.J. Savage [Savage 1954]. Besides these original publications there are several commendable advanced books, e.g. [Raiffa and Schlaifer 1968], [Kenney and Raiffa 1976], [Kahneman 1982] and [Winterfeld and Edwards 1986]. The ideas underlying subjective probabilities are described in [Kyburg and Smokler 1964]. The axioms of subjective probability have been extensively studied in [Fishburn 1968] and the problem of aggregation of probability, which is especially important in the context of this book, is reviewed in [Genest and Zidek 1986]. For the updating of a subjective probability the rule of conditioning is used although there are other proposals, see e.g. [Jeffrey 1965].

Other possible interpretations of probabilistic models and schools of probability in the context of uncertainty handling, such as logical probabilities and nonquantitative probabilities, are presented in [Smithson 1988] and compared mathematically in [Fine 1973]. In [Heckerman 1988] axioms for measures of "change of belief" in the spirit of Cox axioms are given. It turns out that these "belief updates" are monotonic transformations of a likelihood ratio.

The problem of incompletely specified distributions is often solved by using assumptions of conditional independence or meta principles such as the insufficient reasoning principle or the maximum entropy principle. In [Lemmer 1983] a generalized Bayesian updating scheme is proposed. On method of using classes of probabilities as considered in Definitions 4.14 and 4.15 can be found in [Dempster 1967], [Kyburg 1974], and [Gärdenfors 1979]. These classes can be used in situations where we know only that the probability $P(A)$ lies in the range between two limits $P_*(A)$ and $P^*(A)$ called the upper and lower probability of A, respectively. We obtain a convex set of probabilities. The notion of conditioning used for classes of type 1 is described in [de Campos, Lamata, and Moral 1989] and goes back to [Good 1950]. In [Weichselberger and Pöhlmann 1987] sets of probabilities and probability intervals are considered in detail.

The problem of handling this special type of set of probabilities is strongly connected to (and often facilitated by) the direct use of upper and lower probabilities. In [Huber 1981] the respective theorems are considered. Lower and upper probabilities are fuzzy measures [Sugeno 1974]. Set functions have been employed in many different fields such as the theory of capacities [Choquet 1954]. In this context a set function g is called 2-monotone, if

$$g(A \cup B) \geq g(A) + g(B) - g(A \cap B)$$

holds for all $A, B \subseteq \Omega$, whereas a set function g satisfying the equation

$$g\left(\bigcup_{i=1}^{n} A_i\right) \geq \sum_{i=1}^{n} g(A_i) - \sum_{1<i<j<n} g(A_i \cap A_j) \pm \ldots \pm (-1)^{n+1} \cdot g(A_1 \cap \ldots \cap A_n) \tag{4.26}$$

for all $n \in \mathbb{N}$ and every collection of subsets of Ω, is called monotone of order infinity.

Such set functions are also called belief functions, although the semantics are totally different from the probabilistic interpretation we have used throughout this chapter. It should be pointed out that in practice we do not always have a convex set of probabilities, so this treatment is rather rough.

Since from our point of view D.V. Lindley is right when he points out [Lindley 1982] that whenever decision making is required uncertainty has to be quantified by probabilities, our aim is to obtain a probabilistic decision making environment. This enables us to follow the approach presented in [Raiffa 1968], which we recommend as introductory reading. There is an extensive literature on statistical decision making. This theory was founded by J. von Neumann, I. Morgenstern [Neumann and Morgenstern 1947] and A. Wald [Wald 1950]. The case of multiple underlying probabilities is treated following ideas of A.P. Dempster and A. Kong [Dempster and Kong 1987]. A recommendable monograph in this area is [Berger 1980]. Philosophical aspects of decision making are considered in [Gärdenfors et al. 1988].

Chapter 5

Random Sets

In this chapter we address the problem of aggregating a number of expert opinions which have been expressed by vague data. In our approach a meta-expert attaches nonnegative weights of importance to each of his or her experts $\theta_i \in \Theta\ i = 1, \ldots, n$. Therefore we have to discuss the problem of finding suitable weights, integrating pieces of certain knowledge and giving methods for decision making. To clarify our basic idea let us consider a set Ξ (which is for the moment assumed to be finite) containing n "atomic agents" each of which is equally important. If we imagine these atomic agents voting on some motion, then the weight of some "party" $A \subseteq \Xi$ is given by $|A|$. So selecting agents $\xi \in \Xi$ randomly would yield, in a long run, members of A in $(\frac{1}{n}|A| \cdot 100)\%$ of all cases. Thus we consider $P(A) = \frac{1}{n}|A|$ to be the probability of A. Unfortunately, in general not the space Ξ but only a projection Θ is accessible, where ϕ may denote the point-wise projection mapping, i.e. $\phi : \Xi \to \Theta$. On the intuitive level Θ represents an assembly of "floor leaders". Thus the weight of each $\theta \in \Theta$ is given by the "probability" of $\phi^{-}(\theta) = \{\xi \in \Xi \mid \phi(\xi) = \theta\}$, which means it is determined by the number of supporters. So we can interpret the importance weights here as a probability (with a classical or frequentistic interpretation).

One could argue that the notion of a *weighted set* would be more appropriate in this chapter.

5.1 Random Variables

In the above we considered the uncertainty structure on some space Θ. Now we assume that the elements of Θ are sensors or agents, providing data concerning the evidence space Ω. So depending on the type of data we have to consider mappings of the form

$$\begin{array}{ll} \Gamma : \Theta \to \Omega, & \text{(precise data)} \\ \Gamma : \Theta \to 2^{\Omega} \text{ or} & \text{(imprecise data)} \\ \Gamma : \Theta \to \mathcal{F}_L(\Omega) & \text{(vague data)} \end{array}$$

By such an observation mapping the uncertainty structure on Θ (i.e. a probability $P : 2^{\Theta} \to [0,1]$) induces an uncertainty structure on Ω. In the case of precise data this leads to the well known notion of a *random variable.*

(5.1) Definition. *Let $(\Theta, \mathcal{A}, P)$ be a probability space, and let $(\Omega, \mathcal{E})$ be a frame of discernment. A mapping $\Gamma : \Theta \to \Omega$ is called a random variable if the inverse images under Γ of all events in $\mathcal{E}$ are events in $\mathcal{A}$, that is, if*

$$\Gamma^{-1}(B) = \{\theta \in \Theta \mid \Gamma(\theta) \in B\} \in \mathcal{A} \textit{ for all } B \in \mathcal{E}.$$

If $\mathcal{A}$ is equal to the power set of Θ, as is usually the case in this book, then each mapping $\Gamma : \Theta \to \Omega$ is a random variable [1]. Let P be a probability on Θ. Then each random variable Γ induces a probability Q on Ω by means of the correspondence

$$Q(B) = P(\Gamma^{-1}(B)) = P(\{\theta \in \Theta \mid \Gamma(\theta) \in B\}). \tag{5.1}$$

(5.2) Example. Consider a set $\Theta = \{\theta_1, \ldots, \theta_n\}$ of sensors, which are able to determine exactly the location of some ship under consideration. We can describe this situation by a mapping

$$\Gamma : \Theta \to \Omega,$$

$\Omega = \{\text{z3}, \ldots, \text{ld}\}$, which assigns to each sensor $\theta \in \Theta$ the ship's location $\Gamma(\theta) \in \Omega$. If we are interested in the question whether the ship is *in port* or not, we can determine the set $\Gamma^{-1}(A) = \{\theta \in \Theta \mid \Gamma(\theta) \in A\}$, where $A = \{\text{ca}, \text{rd}, \text{ld}\}$ is the subset of Ω corresponding to the propositional statement *in port.* $P(\Gamma^{-1}(A))$ quantifies the importance of those sensors announcing the ship to be *in port.* So the quotient $\frac{P(\Gamma^{-1}(A))}{P(\Gamma^{-1}(\Omega))}$ meets our intuitive requirements of a measure for the evidence that the ship actually is *in port.* Mathematically we have used a probability P on Θ, and a random variable $\Gamma : \Theta \to \Omega$. Thus we determine the induced probability $Q(A) = P(\Gamma^{-1}(A))$ which is obviously a reasonable measure of the "importance" of the sensors modeled by the probability distribution P.

Now suppose an absolutely reliable source of information establishes that the ship under consideration is *at sea.* Again we are interested in a measure for the evidence that the ship is located within some area $C \subseteq \Omega$. In this case we can only trust those sensors indicating the ship is *at sea.* Let $B = \{\text{z3}, \text{z2}, \text{z1}\}$

---

1 A special role for applications is played by the frame of discernment $(\mathbb{R}, \mathcal{E})$ of real numbers. In the statistical literature a function is denoted as a random variable, if and only if (i) $\Omega = \mathbb{R}$ and (ii) $\mathcal{E}$ is equal to the Borel σ-field [Rohatgi 1976].

be the set representing the proposition *at sea*. An appropriate measure for this evidence is

$$Q(C|B) = \frac{P(\Gamma^{-1}(C \cap B))}{P(\Gamma^{-1}(B))} = \frac{Q(C \cap B)}{Q(B)}.$$

Mathematically we have considered the *conditional probability*

$$Q(\,\cdot\,|B) : 2^{\Omega} \to [0,1]; \quad Q(C|B) = \frac{Q(C \cap B)}{Q(B)}$$

as pointed out in Sect. 4.1. ◇

If we consider Γ to map Θ to 2^{Ω} where only sets $A \in 2^{\Omega}$ with $\mathrm{card}(A) \leq 1$ actually appear, then we obtain the probability

$$\Gamma[P] : 2^{\Omega} \to [0,1], \quad \Gamma[P](B) = \frac{P(\Gamma^{-1}(B))}{P(\Gamma^{-1}(\Omega))}, \tag{5.2}$$

where $\Gamma^{-1}(B) = \{\theta \in \Theta \mid \Gamma(\theta) \subseteq B, \Gamma(\theta) \neq \emptyset\}$. (Note that there might be elements $\theta \in \Theta$ with $\Gamma(\theta) = \emptyset$.) From this point of view we can model the impact of the information $\omega_0 \in E$ by switching from Γ to $\Gamma_E : \Theta \to 2^{\Omega}$, where

$$\Gamma_E(\theta) \stackrel{\mathrm{d}}{=} \begin{cases} \Gamma(\theta) & \text{if } \Gamma(\theta) \subseteq E \\ \emptyset & \text{otherwise,} \end{cases}$$

for all $\theta \in \Theta$. Doing so we derive

$$\Gamma[P](B|E) = \Gamma_E[P](B) = \frac{P(\Gamma_E^{-1}(B))}{P(\Gamma_E^{-1}(\Omega))} = \frac{P(\Gamma^{-1}(B \cap E))}{P(\Gamma^{-1}(E))} = \frac{\Gamma[P](B \cap E)}{\Gamma[P](E)}$$

which coincides with Definition 4.3. Obviously $\Gamma[P](\,.\,|E)$ restricts Ω, the domain of interest, to a subset $E \subseteq \Omega$, thus $\Gamma[P](\,.\,|E)$ carries more information than $\Gamma[P]$; in some sense it is more *specific*. The increasing specificity is caused by the fact that $\Gamma(\theta) = \emptyset$ implies $\Gamma_E(\theta) = \emptyset$ for all $\theta \in \Theta$ but not vice versa. Thus we define an observation mapping $S : \Theta \to 2^{\Omega}$ to be a *specialization* of another observation mapping $T : \Theta \to 2^{\Omega}$, if $T(\theta) = \emptyset$ implies $S(\theta) = \emptyset$, $\theta \in \Theta$.

(5.3) Definition. *Let $S, T : \Theta \to 2^{\Omega}$ be two observation mappings such that for all $\theta \in \Theta$ we have $\mathrm{card}(S(\theta)) \leq 1$ and $\mathrm{card}(T(\theta)) \leq 1$. S is called a specialization of T (written $S \sqsubseteq T$) if and only if $T(\theta) = \emptyset \Rightarrow S(\theta) = \emptyset$ for all $\theta \in \Theta$.*

Now assume that Ω, the frame of discernment which is accessible for us, is just a coarsening of some original space Ω' which is subject to the observations of the sensors. Let $\widehat{\Pi} : 2^{\Omega} \to 2^{\Omega'}$ be the respective refinement mapping. If $\Gamma' : \Theta \to 2^{\Omega'}$ (Note that we still assume $\mathrm{card}(\Gamma'(\theta)) \leq 1$ for all $\theta \in \Theta$) is the original observation mapping we perceive only its projection defined by

$$\Pi(\Gamma') : \Theta \to 2^{\Omega'}; \quad \Pi(\Gamma')(\theta) \stackrel{\mathrm{d}}{=} \Pi\big(\Gamma'(\theta)\big).$$

In this case Γ' is denoted to be a *refinement* of $\Pi(\Gamma')$. Let Q' be a probability on Ω' induced by Γ' and let Q be the probability on Ω induced by $\Gamma = \Pi(\Gamma')$. Obviously we have $\Pi[Q'] = Q$, i.e. Q' is a refinement of Q.

The next theorem provides an intuitive impression of the concept of specializations. We obtain that it is caused by a projected conditioning of a refined observation mapping.

(5.4) Theorem. *Let $S, T : \Theta \to 2^{\Omega}$, such that for all $\theta \in \Theta$ we have $\mathrm{card}\big(S(\theta)\big) \leq 1$, $\mathrm{card}\big(T(\theta)\big) \leq 1$. S is a specialization of T if and only if there are refinements S', T' of S, T, respectively, and if there is a set $E' \subseteq \Omega'$ such that $S' = T'_{E'}$.*

In practical applications the sensor space Θ often remains in the dark, and the only thing we perceive are the induced probabilities on Ω. Therefore we define the specialization relation for probabilities as follows.

(5.5) Definition. *Let Q_1, Q_2 be probabilities defined on Ω. Q_1 is a specialization of Q_2 (written $Q_1 \sqsubseteq Q_2$) if and only if there is a probability space $(\Theta, 2^{\Theta}, P)$ and two observation mappings $S, T : \Theta \to 2^{\Omega}$ where $\mathrm{card}\big(S(\theta)\big) \leq 1$ and $\mathrm{card}\big(T(\theta)\big) \leq 1$ for all $\theta \in \Theta$, where $S \sqsubseteq T$ and where $Q_1 = S[P]$, $Q_2 = T[P]$, for all $A \subseteq \Omega$.*

(5.6) Theorem. *Let Q_1, Q_2 be two probabilities defined on Ω. The following two statements are equivalent.*

(i) *There are refinements Q'_1, Q'_2 of Q_1, Q_2 defined on a refinement Ω' of Ω and there is an event $E' \subseteq \Omega'$ such that $Q'_1 = Q'_2(\,.\,|E')$.*

(ii) $Q_1 \sqsubseteq Q_2$.

The proofs of the above theorems are straightforward.

(5.7) Theorem. *Let Q_1 and Q_2 be two probabilities on $\Omega = \{\omega_1, \ldots, \omega_n\}$. The following three statements are equivalent*

(i) $Q_1 \sqsubseteq Q_2$

(ii) *From $Q_2(E) = 0$ follows $Q_1(E) = 0$ for all $E \subseteq \Omega$.*

(iii) *There are numbers $a_1, \ldots, a_n$ such that $0 \leq a_i \leq Q_2\big(\{\omega_i\}\big)$ and $Q_1\big(\{\omega_i\}\big) = \frac{a_i}{\sum_{j=1}^{n} a_j}$ holds for $i = 1, \ldots, n$.*

Proof. (i) $\Rightarrow$ (ii) From $Q_2(E) = 0$ follows $Q'_2(\widehat{\Pi}(E)) = 0$, and we obtain

$$\begin{aligned} 0 \le Q_1(E) &= Q'_1(\widehat{\Pi}(E)) \\ &= Q'_2(\widehat{\Pi}(E)|A') \\ &= \frac{Q'_2(\widehat{\Pi}(E) \cap A')}{Q'_2(A')} \\ &\le \frac{Q'_2(\widehat{\Pi}(E))}{Q'_2(A')} = 0. \end{aligned}$$

(ii) $\Rightarrow$ (iii) Define $A \stackrel{\mathrm{d}}{=} \{\omega_i \mid Q_2(\{\omega_i\}) > 0\}$, $b \stackrel{\mathrm{d}}{=} \min_{\omega_i \in A}\left\{Q_2(\{\omega_i\})\right\}$ and $a_i \stackrel{\mathrm{d}}{=} b \cdot Q_1(\{\omega_i\})$ for $i = 1, \ldots, n$. This yields $0 \le a_i \le b \le Q_2(\{\omega_i\})$ for $\omega_i \in A$ and $a_i = 0$ for $\omega_i \notin A$, since

$$0 \le a_i \le Q_2(\{\omega_i\}) = 0$$

holds for $\omega_i \notin A$. It follows that

$$\frac{a_i}{\sum_{i=j}^{n} a_j} = \frac{b \cdot Q_1(\{\omega_i\})}{\sum_{i=j}^{n} b \cdot Q_1(\{\omega_j\})} = Q_1(\{\omega_i\})$$

for $i = 1, \ldots, n$.

(iii) $\Rightarrow$ (i) Consider the space

$$\Omega' = \{\omega_1^1, \omega_1^2, \omega_2^1, \omega_2^2, \ldots, \omega_n^1, \omega_n^2\}.$$

Ω' is a refinement of Ω with the refinement mapping

$$\widehat{\Pi} : 2^{\Omega} \to 2^{\Omega'}; \quad \widehat{\Pi}\{\omega_i\} = \{\omega_i^1, \omega_i^2\}, \quad i = 1, \ldots, n.$$

The refinement Q'_2 of Q_2 we define by

$$Q'_2\{\omega_i^1\} = a_i, \quad Q'_2\{\omega_i^2\} = Q'_2\{\omega_i\} - a_i, \quad i = 1, \ldots, n.$$

Conditioning with the set $A' = \{\omega_1^1, \omega_2^1, \ldots, \omega_n^1\}$ yields

$$\begin{aligned} Q'_2(\{\omega_i^j\}|A') &= \begin{cases} \frac{a_i}{\sum_{j=1}^{n} a_j}, & \text{if } j = 1 \\ 0, & \text{if } j = 2 \end{cases} \\ &= Q'_1(\{\omega_i^j\}) \end{aligned}$$

for $i = 1, \ldots, n$, $j = 1, 2$. Therefore $Q'_2(B'|A') = Q'_1(B')$ is valid for all $B' \subseteq \Omega'$. □

As mentioned before, our intuitive interpretation of the relation $\sqsubseteq$ is that if $Q_1 \sqsubseteq Q_2$ holds then the probability Q_1 carries at least as much information as the probability Q_2, i.e. the information encoded by Q_1 is more specific than that encoded by Q_2 in the sense that whatever can be concluded from Q_2 can also be concluded from Q_1. If besides $Q_1 \sqsubseteq Q_2$ also $Q_2 \sqsubseteq Q_1$ is valid, then the probabilities Q_1 and Q_2 are equivalent in the qualitative sense captured by the concept of specialization. Formalizing our idea of equivalent probabilities we define

$$Q_1 \equiv Q_2 \overset{\text{d}}{\Longleftrightarrow} Q_1 \sqsubseteq Q_2 \text{ and } Q_2 \sqsubseteq Q_1. \tag{5.3}$$

The basic structural properties of the relation $\sqsubseteq$ can easily be verified; we have

$$\begin{aligned} Q_1 \sqsubseteq Q_1 &\quad \text{(reflexivity)}, \\ (Q_1 \sqsubseteq Q_2 \wedge Q_2 \sqsubseteq Q_3) \Rightarrow Q_1 \sqsubseteq Q_3 &\quad \text{(transitivity)}. \end{aligned}$$

So the quotient set of the set of all probabilities with respect to the equivalence relation $\equiv$ together with the relation $\sqsubseteq$ forms a partially ordered set. Therefore the relation $\sqsubseteq$ allows us to compare two probabilities with regard to the amount of information they bear.

(5.8) Example. Let $\Omega = \{\text{z1}, \text{z2}, \text{z3}\}$. The three probabilities Q_1, Q_2 and Q_3 on Ω are listed in the following table:

	Q_1	Q_2	Q_3
z1	0.4	0.5	0.3
z2	0.6	0.5	0.3
z3	0	0	0.4

We have $Q_1 \equiv Q_2$, $Q_1 \sqsubseteq Q_3$, $Q_2 \sqsubseteq Q_3$, not$(Q_1 \equiv Q_3)$, not$(Q_2 \equiv Q_3)$, which means Q_1 and Q_2 carry more information than Q_3 (in the qualitative sense we mentioned above). This result is caused by the different valuation of $\{\text{z3}\}$. The probabilities Q_1 and Q_2 mean we consider the event $\{\omega_3\}$ to be impossible, whereas Q_3 assigns a positive probability. ◇

5.2 The Notion of a Random Set

Now let us turn to the case of imprecise data, i.e. we have to deal with a set-valued observation mapping of the form $\Gamma : \Theta \to 2^{\Omega}$.

(5.9) Example. Let us consider a set $\Theta = \{\theta_1, \ldots, \theta_n\}$ of sensors, each of which is able to determine the location of some ship under consideration within the frame of discernment

$$\Omega = \{\text{z3}, \text{z2}, \text{z1}, \text{ca}, \text{ld}, \text{rd}\}.$$

We assume the results of the sensors to be propositions like "The ship is in z3 or in z2". These propositions have to be represented by subsets of Ω. So the observation of the ship can be characterized by a multivalued mapping $\Gamma : \Theta \to 2^\Omega$, assigning to each sensor $\theta \in \Theta$ its observation $\Gamma(\theta) \subseteq \Omega$. Each proposition corresponds to a subset A of Ω. Consider, for example, *in-port* which is represented by the set

$$A = \{\text{ca}, \text{ld}, \text{rd}\}.$$

A determines two sets of sensors, first $\Gamma_*(A) \subseteq \Theta$, containing those sensors which consider the proposition to be *necessarily* true, and second $\Gamma^*(A)$, containing those sensors which consider it to be *possibly* true. On the formal level we derive

$$\begin{aligned} \Gamma_*(A) &= \{\theta \in \Theta \mid \Gamma(\theta) \neq \emptyset, \Gamma(\theta) \subseteq A\}, \\ \Gamma^*(A) &= \{\theta \in \Theta \mid \Gamma(\theta) \cap A \neq \emptyset\}. \end{aligned} \tag{5.2}$$

Obviously we have $\Gamma_*(A) \subseteq \Gamma^*(A)$. The assumption that the ship's location is within Ω with certainty implies that the information $\Gamma(\theta) = \emptyset$ is inconsistent. Such information has to be neglected, and indicates θ to be "out of order". (Without a "closed world assumption" $\Gamma(\theta) = \emptyset$ may be a reasonable result, caused by the fact the ship is "outside" of Ω, i.e. that Ω is not a sufficient frame of discernment.) In our example the quotient $\frac{P(\Gamma_*(A))}{P(\Gamma_*(\Omega))}$ in some sense measures the degree to which the available evidence substantiates that the ship *is necessarily in-port* whereas $\frac{P(\Gamma^*(A))}{P(\Gamma^*(\Omega))}$ measures the degree to which it substantiates that the ship *is possibly in-port* (assuming that the sensors are likewise reliable). So these two quotients can be interpreted as lower and upper bounds of the degree to which the evidence supports the hypothesis that the ship is *actually in-port.* ◇

(5.10) Definition. *Let P be a probability defined on the sample space Θ, and let $\Gamma : \Theta \to 2^\Omega$ be a set-valued mapping. In this case the pair (P, Γ) is called a random set*[2].

---

[2] We assume our sample space Θ to be endowed with the σ-field 2^Θ. Therefore each subset of Θ is an event and no measurability problems occur. The general case is discussed in , e.g., [Matheron 1975].

5.2.1 Weighted Sets versus Random Sets

The concept of random sets, although known for a long time in connection with set-valued functions and geometric probabilities, was recently formalized in [Matheron 1975] and [Kendall 1974].

In random set theory the realisations of an experiment are geometrical objects, where the objects are precisely described by subsets of a given space Ω (for example $\mathbb{R}^3$). There is no imprecision or vagueness in the description of the data. In order to not confuse the sampling mechanism with the data space, Matheron used a probability space $(\Theta, 2^\Theta, P)$ for describing the randomness, and the multivalued mapping $\Gamma : \Theta \to 2^\Omega \cap \overline{\{\emptyset\}}$ assigns to each outcome of the random experiment the corresponding geometrical object $\Gamma(\Theta) \subseteq \Omega$.

P induces via Γ a probability on $(2^\Omega, 2^{2^\Omega})$, i.e. the *elementary events are sets.* In our approach considered here the data are also sets, but they are considered to be imprecise. In this case of course it makes sense to change data in the light of new evidence, which is nonsense in ordinary random set theory. A similar view can be used for the foundation of statistics with imprecise data [Kruse and Meyer 1987]. We use in the following the same notations as in random set theory, although one could argue that the notation *weighted set* instead of random set would be more appropriate in our context.

(5.11) Definition. *Let (P, Γ) be a random set. Define*

$$\Gamma_*(A) = \{\theta \in \Theta \mid \Gamma(\theta) \neq \emptyset \text{ and } \Gamma(\theta) \subseteq A\},$$
$$\Gamma^*(A) = \{\theta \in \Theta \mid \Gamma(\theta) \cap A \neq \emptyset\}$$

for all $A \subseteq \Omega$. If $P(\Gamma^*(\Omega)) > 0$, then

$$(P_\Gamma)^*(A) \stackrel{\mathrm{d}}{=} \frac{P(\Gamma^*(A))}{P(\Gamma^*(\Omega))}$$

is called the upper probability of A, and

$$(P_\Gamma)_*(A) \stackrel{\mathrm{d}}{=} \frac{P(\Gamma_*(A)}{P(\Gamma_*(\Omega))}$$

the lower probability of A with respect to (P, Γ).

From the definition of Γ^* and Γ_* we obviously get

$$\Gamma_*(\Omega) = \Gamma^*(\Omega) = \{\theta \in \Theta \mid \Gamma(\theta) \neq \emptyset\}.$$

To avoid an unnecessary complexity in the notation, we use, if there is no risk of confusion, P_* and P^* instead of $(P_\Gamma)^*$ and $(P_\Gamma)_*$, respectively.

(5.12) Example. Suppose there are four sensors, i.e. $\Theta = \{\theta_1, \theta_2, \theta_3, \theta_4\}$, each of which is able to determine the location of some ship. The probability P,

$$P(\{\theta_1\}) = 0.4, \quad P(\{\theta_2\}) = P(\{\theta_3\}) = P(\{\theta_4\}) = 0.2,$$

interpreted as an uncertainty structure on Θ, reflects the credibility of the sensors. If all sensors are identical with respect to their priority or technical reliability, we might choose the uniform distribution on Θ, i.e.

$$P'(\{\theta\}) = 0.25 \quad \text{for } \theta = \theta_1, \theta_2, \theta_3, \theta_4.$$

The mapping $\Gamma : \Theta \to 2^{\Omega}$ describes the sensors observations. Let

$$\Gamma(\theta_1) = \{\mathrm{z1}\}, \quad \Gamma(\theta_2) = \{\mathrm{z1}, \mathrm{z2}\}, \quad \Gamma(\theta_3) = \{\mathrm{z2}, \mathrm{z3}\}, \quad \Gamma(\theta_4) = \emptyset.$$

We obtain

$$P(\Gamma^*(\Omega)) = P(\Gamma_*(\Omega)) = P(\{\theta_1, \theta_2, \theta_3\}) = 0.8,$$

since sensor θ_4 has to be neglected. For each $A \subseteq \Omega$ we can calculate the quantities $P_*(A)$ and $P^*(A)$, forming an "interval of confidence". The range of this interval depends on the quality of the sensors. If the sensors provided only exact information, i.e. if the sets $\Gamma(\theta)$, $\theta \in \Theta$ contained only one element, then we would obtain an ordinary probability on Ω. Table 5.1 summarizes the results from our above example.

A	$\Gamma_*(A)$	$\Gamma^*(A)$	$P_*(A)$	$P^*(A)$
$\emptyset$	$\emptyset$	$\emptyset$	0	0
$\{\mathrm{z1}\}$	$\{\theta_1\}$	$\{\theta_1, \theta_2\}$	0.5	0.75
$\{\mathrm{z2}\}$	$\emptyset$	$\{\theta_2, \theta_3\}$	0	0.5
$\{\mathrm{z3}\}$	$\emptyset$	$\{\theta_3\}$	0	0.25
$\{\mathrm{z1}, \mathrm{z2}\}$	$\{\theta_1, \theta_2\}$	$\{\theta_1, \theta_2, \theta_3\}$	0.75	1
$\{\mathrm{z1}, \mathrm{z3}\}$	$\{\theta_1\}$	$\{\theta_1, \theta_2, \theta_3\}$	0.5	1
$\{\mathrm{z2}, \mathrm{z3}\}$	$\{\theta_3\}$	$\{\theta_2, \theta_3\}$	0.25	0.5
$\{\mathrm{z1}, \mathrm{z2}, \mathrm{z3}\}$	$\{\theta_1, \theta_2, \theta_3\}$	$\{\theta_1, \theta_2, \theta_3\}$	1	1

Table 5.1.

◇

The functions $P^* : 2^\Omega \to [0,1]$ and $P_* : 2^\Omega \to [0,1]$ are *not* probabilities; we have for example

$$0.75 = P_*(\{z1\}) + P_*(\{z2, z3\}) \neq P_*(\{z1, z2, z3\}) = 1,$$
$$1.25 = P^*(\{z1\}) + P^*(\{z2, z3\}) \neq P^*(\{z1, z2, z3\}) = 1.$$

Obviously the additivity condition of the Kolmogorov axioms is not satisfied.

(5.13) Theorem. *For each pair of upper and lower probability functions P^*, P_* we have the following properties:*

(i) P_* *and* P^* *are functions from* 2^Ω *into* $[0,1]$,
(ii) $P_*(A) \leq P^*(A)$ *for all* $A \subseteq \Omega$,
(iii) $P_*(\emptyset) = P^*(\emptyset) = 0,\ P_*(\Omega) = P^*(\Omega) = 1$,
(iv) $A \subseteq B \Rightarrow P_*(A) \leq P_*(B)$ *and* $P^*(A) \leq P^*(B)$ *for all* $A, B \subseteq \Omega$,
(v) $P^*(A) = 1 - P_*(\overline{A})$ *for all* $A \subseteq \Omega$,
(vi) $P_*(A \cup B) \geq P_*(A) + P_*(B) - P_*(A \cap B)$ *for all* $A, B \subseteq \Omega$,
(vii) $P^*(A \cup B) \leq P^*(A) + P^*(B) - P^*(A \cap B)$ *for all* $A, B \subseteq \Omega$.

Proof. We restrict ourselves to proving the properties (v) to (vii). Let A be a subset of Ω. We have

$$\begin{aligned}
\Gamma_*(\overline{A}) &= \{\theta \in \Theta \mid \Gamma(\theta) \neq \emptyset \text{ and } \Gamma(\theta) \subseteq \overline{A}\} \\
&= \{\theta \in \Theta \mid \Gamma(\theta) \neq \emptyset \text{ and } \Gamma(\theta) \cap A = \emptyset\} \\
&= \{\theta \in \Theta \mid \Gamma(\theta) \neq \emptyset\} \cap \overline{\{\theta \in \Theta \mid \Gamma(\theta) \neq \emptyset \text{ and } \Gamma(\theta) \cap A \neq \emptyset\}} \\
&= \Gamma_*(\Omega) \cap \overline{\Gamma^*(A)}.
\end{aligned}$$

From $\Gamma^*(A) \subseteq \Gamma_*(\Omega)$ we obtain

$$P(\Gamma_*(\overline{A})) = P(\Gamma_*(\Omega)) - P(\Gamma^*(A)).$$

This yields

$$\begin{aligned}
P^*(A) &= \frac{P(\Gamma^*(A))}{P(\Gamma^*(\Omega))} \\
&= \frac{P(\Gamma_*(\Omega)) - P(\Gamma_*(\overline{A}))}{P(\Gamma_*(\Omega))} = 1 - P_*(\overline{A}).
\end{aligned}$$

To prove (vi) let us consider two arbitrary subsets A, B of Ω. The set

$$\{\theta \mid \Gamma(\theta) \neq \emptyset \text{ and } \Gamma(\theta) \subseteq A \cup B\}$$

can be split up into the three disjoint subsets

$$\{\theta \mid \Gamma(\theta) \neq \emptyset \text{ and } \Gamma(\theta) \subseteq A \text{ and } \Gamma(\theta) \cap \overline{A \cap B} \neq \emptyset\},$$

$$\{\theta \mid \Gamma(\theta) \neq \emptyset \text{ and } \Gamma(\theta) \subseteq B\},$$

$$\{\theta \mid \Gamma(\theta) \neq \emptyset \text{ and } \Gamma(\theta) \subseteq A \cup B$$
$$\text{and } \Gamma(\theta) \cap A \cap \overline{B} \neq \emptyset \text{ and } \Gamma(\theta) \cap B \cap \overline{A} \neq \emptyset\}.$$

since

$$\{\theta \mid \Gamma(\theta) \neq \emptyset \text{ and } \Gamma(\theta) \subseteq A\} \cap \overline{\{\theta \mid \Gamma(\theta) \neq \emptyset \text{ and } \Gamma(\theta) \subseteq A \cap B\}}$$

and

$$\{\theta \mid \Gamma(\theta) \neq \emptyset \text{ and } \Gamma(\theta) \subseteq B\}$$

are disjoint subsets of

$$\{\theta \mid \Gamma(\theta) \neq \emptyset \text{ and } \Gamma(\theta) \subseteq A \cup B\},$$

we can conclude

$$P_*(A \cup B) \geq P_*(A) + P_*(B) - P_*(A \cap B).$$

This property holds for arbitrary A, B. Considering the case

$$P_*(\overline{A} \cup \overline{B}) \geq P_*(\overline{A}) + P_*(\overline{B}) - P_*(\overline{A} \cap \overline{B}),$$

which is equivalent to

$$P_*(\overline{A \cap B}) \geq P_*(\overline{A}) + P_*(\overline{B}) - P_*(\overline{A \cup B}).$$

From (v) we obtain the inequality (vii). □

Property (vi) is called strong superadditivity. With the same methods we can find the even more general formula

$$P_*(\bigcup_{i=1}^{n} A_i) \geq \sum_{\emptyset \neq I : I \subseteq \{1,\ldots,n\}} (-1)^{|I|+1} \cdot P_*(\bigcap_{i \in I} A_i), \tag{5.3}$$

which holds for arbitrary events $A_1, \ldots, A_n$. An analogous inequality holds for P^*.

(4.9) Example. (revisited) If a probability p_{ps} does *not* exist, i.e. there is no random selection of the ship's type, then we cannot allocate probabilities to the singletons, but we have to allocate probabilities to sets. We obtain

$$Q\{(\theta_1, \text{ps}, \text{ps}), (\theta_1, \text{tv}, \text{tv})\} = \frac{1}{4}$$

$$Q\{(\theta_2, \text{ps}, \text{tv}), (\theta_2, \text{tv}, \text{ps})\} = \frac{1}{4}$$

$$Q\{(\theta_3, \text{ps}, \text{ps}), (\theta_3, \text{tv}, \text{ps})\} = \frac{1}{4}$$

$$Q\{(\theta_4, \text{ps}, \text{tv}), (\theta_4, \text{tv}, \text{tv})\} = \frac{1}{4}$$

Q is not an ordinary probability, since it is not possible to find probabilities for the singletons, e.g. to $\{(\theta_1, \text{ps}, \text{ps})\}$. We can describe this situation by choosing a probability space (Θ, P), $P(\theta_i) = \frac{1}{4}$, $i = 1, \ldots, 4$, which analyses the random choice of the sensors and a mapping $\Gamma : \Theta \to 2^{\Omega}$ that assigns to each sensor its information set. We obtain a random set Γ given by

$$\theta_1 \to \{(\theta_1, \text{ps}, \text{ps}), (\theta_1, \text{tv}, \text{tv})\}$$

$$\theta_2 \to \{(\theta_2, \text{ps}, \text{tv}), (\theta_2, \text{tv}, \text{ps})\}$$

$$\theta_3 \to \{(\theta_3, \text{ps}, \text{ps}), (\theta_3, \text{tv}, \text{ps})\}$$

$$\theta_4 \to \{(\theta_4, \text{ps}, \text{tv}), (\theta_4, \text{tv}, \text{tv})\}.$$

In the context of random set the information "sensor's answer is ps" can be integrated by geometric conditioning to obtain (for other concepts see Chap. 6.1)

$$(P_\Gamma)_*(\text{ship type is ps} \mid \text{sensor message is ps})$$
$$= \frac{P(\{\theta_i \mid \Gamma(\theta_i) \subseteq \{(\theta_1, \text{ps}, \text{ps}), (\theta_3, \text{ps}, \text{ps})\}\})}{P(\{\theta_i \mid \Gamma(\theta_i) \subseteq \{(\theta_1, \text{ps}, \text{ps}), (\theta_2, \text{tv}, \text{ps}), (\theta_3, \text{ps}, \text{ps}), (\theta_3, \text{tv}, \text{ps})\}\})} = 0$$

$$(P_\Gamma)^*(\text{ship type is ps} \mid \text{sensor message is ps})$$
$$= \frac{\begin{array}{c}P(\{\theta_i \mid \Gamma(\theta_i) \cap \{(\theta_1, \text{ps}, \text{ps}), (\theta_3, \text{ps}, \text{ps})\} \neq \emptyset, \\ \theta_i \subseteq \{(\theta_1, \text{ps}, \text{ps}), (\theta_2, \text{tv}, \text{ps}), (\theta_3, \text{ps}, \text{ps}), (\theta_3, \text{tv}, \text{ps})\}\})\end{array}}{P(\{\theta_i \mid \Gamma(\theta_i) \subseteq \{(\theta_1, \text{ps}, \text{ps}), (\theta_2, \text{tv}, \text{ps}), (\theta_3, \text{ps}, \text{ps}), (\theta_3, \text{tv}, \text{ps})\}\})} = 1$$

◇

(5.12) Example. (continued) Now suppose we obtain additional information indicating that the ship's location is z1 or z2 with certainty. In this case the sensors θ_3 and θ_4 gave incorrect results, thus they have to be neglected. This can be modeled either by switching from P to $P' \stackrel{\mathrm{d}}{=} P\Big(. \Big| \Gamma_*(\{\mathrm{z1},\mathrm{z2}\})\Big)$ or by using

$$\Gamma' : \Theta \to 2^{\Omega}; \quad \Gamma'(\theta) \stackrel{\mathrm{d}}{=} \begin{cases} \Gamma(\theta) & \text{if } \Gamma(\theta) \subseteq \{\mathrm{z1},\mathrm{z2}\} \\ \emptyset & \text{otherwise,} \end{cases}$$

instead of Γ. We derive

$$\begin{aligned} (P_\Gamma)_*(A|\{\mathrm{z1},\mathrm{z2}\}) &= (P_{\Gamma'})_*(A) \\ &= \frac{Q\Big(\Big\{\theta \in \Theta \Big| \Gamma(\theta) \subseteq A \text{ and } \Gamma(\theta) \subseteq \{\mathrm{z1},\mathrm{z2}\}\Big\}\Big)}{Q\Big(\Big\{\theta \in \Theta \Big| \Gamma(\theta) \subseteq \{\mathrm{z1},\mathrm{z2}\}\Big\}\Big)} \\ &= \frac{(P_\Gamma)_*(A \cap \{\mathrm{z1},\mathrm{z2}\})}{(P_\Gamma)_*(\{\mathrm{z1},\mathrm{z2}\})}. \end{aligned}$$

◇

5.2.2 Updating Random Sets

In order to model the impact of additional information we prefer to change the observation mapping, since this is the more general approach.

(5.14) Definition. *Let $\Gamma : \Theta \to 2^{\Omega}$ be an observation mapping and let E be a nonempty subset of Ω such that $\Gamma_*(E) \neq \emptyset$. The mapping*

$$\Gamma_E : \Theta \to 2^{\Omega}; \quad \Gamma_E(\theta) \stackrel{\mathrm{d}}{=} \begin{cases} \Gamma(\theta) & \text{if } \Gamma(\theta) \subseteq E \\ \emptyset & \text{otherwise,} \end{cases}$$

is called the conditional observation mapping with respect to E.

(5.15) Definition. *Let (P,Γ) be a random set. Define the conditional lower and upper probabilities of A under the condition E by setting*

$$(P_\Gamma)_*(A|E) \stackrel{\mathrm{d}}{=} (P_{\Gamma_E})_*(A),$$

$$(P_\Gamma)^*(A|E) \stackrel{\mathrm{d}}{=} (P_{\Gamma_E})^*(A).$$

From this definition we obtain

$$\begin{aligned}
(P_\Gamma)_*(A|E) &= \frac{P(\{\theta \in \Theta \mid \Gamma_E(\theta) \neq \emptyset \text{ and } \Gamma_E(\theta) \subseteq A\})}{P(\{\theta \in \Theta \mid \Gamma_E(\theta) \neq \emptyset\})} \\
&= \frac{P(\{\theta \in \Theta \mid \Gamma(\theta) \neq \emptyset \text{ and } \Gamma(\theta) \subseteq A \cap E\})}{P(\{\theta \in \Theta \mid \Gamma(\theta) \neq \emptyset \text{ and } \Gamma(\theta) \subseteq E\})} \\
&= \frac{(P_\Gamma)_*(A \cap E)}{(P_\Gamma)_*(E)}, \qquad (5.4) \\
(P_\Gamma)^*(A|E) &= \frac{P(\{\theta \in \Theta \mid \Gamma_E(\theta) \cap A \neq \emptyset\})}{P(\{\theta \in \Theta \mid \Gamma_E(\theta) \neq \emptyset\})} \\
&= \frac{P(\{\theta \in \Theta \mid \Gamma(\theta) \cap A \neq \emptyset \text{ and } \Gamma(\theta) \subseteq E\})}{P(\{\theta \in \Theta \mid \Gamma(\theta) \neq \emptyset \text{ and } \Gamma(\theta) \subseteq E\})} \\
&= \frac{(P_\Gamma)^*(A \cup \overline{E}) - (P_\Gamma)^*(\overline{E})}{1 - (P_\Gamma)^*(\overline{E})}. \qquad (5.5)
\end{aligned}$$

In the case of existence the functions $(P_\Gamma)_*(\,.\,|B)$, $(P_\Gamma)^*(\,.\,|B)$ are lower, upper probabilities themselves, respectively. In the literature the above definition is known to as the *geometric rule of conditioning.*

Dempster proposed to use instead of Γ_E the mapping

$$\Gamma^E : \Theta \to 2^\Omega; \quad \Gamma^E \stackrel{\text{d}}{=} \Gamma(\theta) \cap E$$

in order to model the conditioning process. From a probabilistic point of view, doing so means to perform a *data revision* rather than conditioning. So it is very important not to confuse these totally different concepts.

(5.16) Definition. *Let $\Gamma : \Theta \to 2^\Omega$ be an observation mapping. The mapping*

$$\Gamma^E : \Theta \to 2^\Omega; \quad \Gamma^E(A) \stackrel{\text{d}}{=} \Gamma(A) \cap E$$

is called the revision of Γ with respect to E.

The revised upper and lower probability exist, if

$$P\big(\{\theta \in \Theta \mid \Gamma(\theta) \cap E \neq \emptyset\}\big) > 0.$$

From this definition we obtain

$$(P_{\Gamma^E})_*(A) = \frac{(P_\Gamma)_*(A \cup \overline{E}) - (P_\Gamma)_*(\overline{E})}{(1 - (P_\Gamma)_*(\overline{E}))}, \qquad (5.6)$$

$$(P_{\Gamma^E})^*(A) = \frac{(P_\Gamma)^*(A \cap E)}{(P_\Gamma)^*(E)}, \quad A \subseteq \Omega. \qquad (5.7)$$

Again we implicitly used a closed world assumption since those sensors with $\Gamma(\theta) = \emptyset$ have been neglected. Assuming an open world requires that these sensors are also taken into account since there may be additional states of the world not contained in the frame of discernment Ω. So we have to consider mass distributions on at least $\widetilde{\Omega} = \Omega \cup \{\alpha\}$ (compare the final remarks of Sect. 4.1).

Once more it has to be emphasized that the reader should carefully avoid any confusion of random sets and sets of probabilities although there are some relationships on the formal level. Indeed if Γ is a random set, then

$$\mathcal{L} = \{P \mid (P_\Gamma)_* \leq P \leq (P_\Gamma)^*\}$$

is of type 2 and we have $(P_\Gamma)_* = (P_\mathcal{L})_*$, $(P_\Gamma)^* = (P_\mathcal{L})^*$, but as soon as we consider the conditioning concepts it becomes obvious that the respective underlying semantics are different. So one should keep in mind that a random set and the corresponding set of probabilities do *not* encode equivalent information.

Let $\Gamma : \Theta \to 2^\Omega$ be an observation mapping for which the property $\text{card}\big(\Gamma(\theta)\big) \leq 1$ holds for each $\theta \in \Theta$. Then Γ obviously corresponds to the random variable $X_\Gamma : \Theta \to \Omega$, $X_\Gamma(\theta) = \omega$, where $\omega \in \Gamma(\theta)$. In this case we have

$$\Gamma_E(\theta) = \Gamma^E(\theta) = \begin{cases} \{\omega\} & \text{if } \omega \in E, \\ \emptyset & \text{otherwise,} \end{cases}$$

and we obtain

$$\begin{aligned}(P_{\Gamma_E})_*(A) = (P_{\Gamma_E})^*(A) &= (P_{\Gamma^E})_*(A) = (P_{\Gamma^E})^*(A) \\ &= \frac{P(\{\theta \in \Theta \mid X_\Gamma(\theta) \in A \cap E\})}{P(\{\theta \in \Theta \mid X_\Gamma(\theta) \in E\})}\end{aligned}$$

for all $A \subseteq \Omega$ (if the conditionals exist), so the concept of a random set is a canonical extension of that of a random variable.

In this section we firstly have to clarify the relationship between refinements and random sets. Let Ω' be a refinement of Ω with $\widehat{\Pi} : 2^\Omega \to 2^{\Omega'}$ as the corresponding refinement mapping. The information carried by some random set (P, Γ), where P is a probability on Θ and $\Gamma : \Theta \to 2^\Omega$ is the observation mapping, can be equivalently represented by the random set $\big(P, \widehat{\Pi}(\Gamma)\big)$, where $\widehat{\Pi}(\Gamma)$, defined by

$$\widehat{\Pi}(\Gamma) : \Theta \to 2^{\Omega'}, \quad \widehat{\Pi}(\Gamma)(\theta) \stackrel{\text{d}}{=} \widehat{\Pi}\big(\Gamma(\theta)\big), \tag{5.8}$$

is called the *vacuous extension* of Γ.

The reverse process is projection. Starting with a random set (P, Γ'), where $\Gamma' : \Theta \to 2^{\Omega'}$ is the observation mapping, we define the projected random set $\big(P, \Pi(\Gamma')\big)$ by setting

$$\Pi(\Gamma') : \Theta \to 2^{\Omega}, \quad \Pi(\Gamma')(\theta) \stackrel{\text{d}}{=} \Pi\big(\Gamma'(\theta)\big), \tag{5.9}$$

which we call the *outer reduction* of Γ. No further restriction to subsets of $\Pi\big(\Gamma'(\theta)\big)$ is reasonable, since for every $\omega \in \Pi\big(\Gamma'(\theta)\big)$ there is at least one $\omega' \in \widehat{\Pi}\big(\{\omega\}\big)$ in $\Gamma'(\theta)$ that may turn out to be the final "truth". $\Gamma'(\theta)$ might have been assigned just in order to cover this state of the world. In contrast to the vacuous extension projection generally means a loss of information.

The concept of specialization we introduced in Sect. 5.1 can also be applied to random sets.

(5.17) Definition. *Let $S, T : \Theta \to 2^{\Omega}$ be two observation mappings. S is called a specialization of T (written $S \sqsubseteq T$) if and only if $S(\theta) \subseteq T(\theta)$ holds for all $\theta \in \Theta$.*

The concept of specialization can be equivalently characterized by revised refinements. Although we prefer our original approach we provide the following theorem which verifies this relationship. A random set $\Gamma' : \Theta \to 2^{\Omega'}$ is called a refinement of $\Gamma : \Theta \to 2^{\Omega}$ if there is a refinement mapping $\widehat{\Pi} : 2^{\Omega} \to 2^{\Omega'}$ satisfying the conditions of Definition 1.1, and if $\Gamma(\theta) = \Pi\big(\Gamma'(\theta)\big)$ holds for all $\theta \in \Theta$.

(5.18) Theorem. *Let $S, T : \Theta \to 2^{\Omega}$ be two observation mappings. S is a specialization of T if and only if there are refinements S', T' of S, T, respectively, and if there is a set $E' \subseteq \Omega'$ such that $S' = T'^{E'}$ holds.*

Proof. "$\Leftarrow$" $S'(\theta) \subseteq T'(\theta)$ holds for all $\theta \in \Theta$, and therefore $S(\theta) = \Pi\big(S'(\theta)\big) \subseteq \Pi\big(T'(\theta)\big) = T(\theta)$ holds for all $\theta \in \Theta$.

"$\Rightarrow$" Define $\Omega' = \Theta \times \Omega$, $\widehat{\Pi}\big(\{\omega\}\big) = \Theta \times \{\omega\}$ for $\omega \in \Omega$, $E' = \bigcup_{\theta \in \Theta} \big(\{\theta\} \times S(\theta)\big)$, $T'(\theta) = \widehat{\Pi}\big(T(\theta)\big)$ and $S'(\theta) = T'(\theta) \cap E'$ for all $\theta \in \Theta$. We have $\Pi\big(T'(\theta)\big) = \Pi\big(\widehat{\Pi}(\theta)\big) = T(\theta)$ for all $\theta \in \Theta$. So T' is a refinement of T. Furthermore we have the equation

$$\begin{aligned}
\Pi\big(S'(\theta)\big) &= \Pi\big(T'(\theta) \cap E'\big) \\
&= \Pi\Big(\widehat{\Pi}\big(T(\theta)\big) \cap \bigcup_{\theta \in \Theta} \big(\{\theta\} \times S(\theta)\big)\Big) \\
&= \Pi\Big(\big\{(\theta,\omega) \mid \omega \in T(\theta), \theta \in \Theta\big\} \cap \big\{(\theta,\omega) \mid \omega \in S(\theta)\big\}\Big) \\
&= \Pi\Big(\big\{(\theta,\omega) \mid \omega \in S(\theta)\big\}\Big) \\
&= S(\theta),
\end{aligned}$$

i.e. S' is a refinement of S. □

(5.19) Definition. *Let $S, T : \Theta \to 2^{\Omega}$ be two observation mappings. S is called a strict specialization of T (written $S \le T$) if and only if $S(\theta) = T(\theta)$ or $S(\theta) = \emptyset$ holds for all $\theta \in \Theta$.*

The following theorem shows that the strict specialization relation can be equivalently characterized as a projected conditioning. (As the proof is obvious it is omitted.)

(5.20) Theorem. *Let $S, T : \Theta \to 2^{\Omega}$ be two observation mappings. S is a strict specialization if and only if there are refinements S', T' of S, T, respectively, and if there is a set $E' \subseteq \Omega'$ such that $E' = T'_{E'}$.*

Often the subject of an expert's valuation are vague, ill-defined entities which cannot be described adequately by crisp sets. Thus in the most general case we have to deal with observation mappings of the form

$$\Gamma : \Theta \to \mathcal{F}_L(\Omega).$$

If such vague observation mappings are used, we talk of *generalized random sets*. In a similar way as in equation (5.2) we define

$$\Gamma_*(\mu) = \{\theta \in \Theta \mid \Gamma(\theta) \neq \mathbb{1}_\emptyset \text{ and } \Gamma(\theta) \subseteq \mu\},$$
$$\Gamma^*(\mu) = \{\theta \in \Theta \mid \Gamma(\theta) \cap \mu \neq \mathbb{1}_\emptyset\}.$$

Obviously we have $\Gamma_*(\mu) \subseteq \Gamma^*(\mu)$ for all $\mu \in \mathcal{F}_L(\Omega)$. In strict analogy to Definition 5.11 we obtain

$$(P_\Gamma)^*(\mu) = \frac{P(\Gamma^*(\mu))}{P(\Gamma^*(\mathbb{1}_\Omega))} \quad \text{and} \quad (P_\Gamma)_*(\mu) = \frac{P(\Gamma_*(\mu))}{P(\Gamma_*(\mathbb{1}_\Omega))}$$

as the respective generalized upper and lower probabilities. Conditioning and revision with respect to the L-set $\eta \in \mathcal{F}_L(\Omega)$ is also done by switching from Γ to

$$\Gamma_\eta : \Theta \to \mathcal{F}_L(\Omega); \quad \Gamma_\eta(\theta) = \begin{cases} \Gamma(\theta) & \text{if } \Gamma(\theta) \subseteq \eta, \\ \mathbb{1}_\emptyset & \text{otherwise,} \end{cases}$$

and

$$\Gamma^\eta : \Theta \to \mathcal{F}_L(\Omega); \quad \Gamma^\eta(\theta) = \Gamma(\theta) \cap \eta,$$

respectively. Note that in the case of generalized random sets the equations (5.4) to (5.7) are *not* valid, since there is no complement $\overline{\eta}$.

An important role in this book is given to *consonant random sets.* Let $\Gamma : \Theta \to 2^{\Omega}$ be a random set. Γ is called consonant if

(i) there are subsets $A_1, \ldots, A_n \subseteq \Omega$ with $A_1 \subseteq A_2 \subseteq \ldots \subseteq A_n$, and

(ii) if $B \neq A_i$ for all $i = 1, \ldots, n$ then $Q\Big(\{\theta \in \Theta \mid \Gamma(\theta) = B\}\Big) = 0.$

Obviously a random set is consonant if the possible outcomes are nested sets (with probability one). It is easy to check that for each consonant random set Γ we have

$$(P_\Gamma)^*(A \cup B) = \max\{(P_\Gamma)^*(A), (P_\Gamma)^*(B)\}, \tag{5.10}$$

$$(P_\Gamma)_*(A \cap B) = \min\{(P_\Gamma)_*(A), (P_\Gamma)_*(B)\}, \tag{5.11}$$

$$(P_\Gamma)_*(A|B) = \frac{\min\{(P_\Gamma)_*(A), (P_\Gamma)_*(B)\}}{(P_\Gamma)_*(B)}, \tag{5.12}$$

$$(P_\Gamma)^*(A|B) = \frac{\max\{(P_\Gamma)^*(A), (P_\Gamma)^*(B)\} - (P_\Gamma)^*(\overline{B})}{1 - (P_\Gamma)^*(\overline{B})}, \tag{5.13}$$

for all $A, B \subseteq \Omega$ in the case of existence.

If Ω is a product space, the above relations are of special interest. Let $\Omega^{(i)}$, $i \in M = \{1, \ldots, m\}$, be a set of frames of discernment. Recall that Ω^N denotes the subspace $\Omega^{(i_1)} \times \ldots \times \Omega^{(i_n)}$, where $N = \{i_1, \ldots, i_n\} \subseteq M$, and that Π_S^T and $\widehat{\Pi}_S^T$ denote projection and cylindrical extension, respectively. A space Ω^T is a refinement of Ω^S with $\widehat{\Pi}_S^T$ as the refinement mapping, if and only if $S \subseteq T$. Given the random set (P, Γ'), $\Gamma' : \Theta \to 2^{\Omega^T}$ the tuple $\big(P, \Pi_S^T(\Gamma')\big)$, where $S \subseteq T$, is called a marginal random set. On the other hand if (P, Γ), $\Gamma : \Theta \to 2^{\Omega^S}$, is given we denote $\big(P, \widehat{\Pi}_S^T(\Gamma')\big)$ as a vacuous extension.

(5.21) Example. Let $\Theta = \{\theta_1, \theta_2, \theta_3, \theta_4\}$ and $P : 2^{\Theta} \to [0, 1]$,

$$P(\{\theta_i\}) = \begin{cases} 0.5, & \text{if } i = 1, \\ 0.2, & \text{if } i = 2, \\ 0.1, & \text{if } i = 3, \\ 0.2, & \text{if } i = 4. \end{cases}$$

Once more let $\Omega^{(1)} = \{\mathrm{z3}, \mathrm{z2}, \mathrm{z1}, \mathrm{ca}, \mathrm{rd}, \mathrm{ld}\}$. Suppose we want to analyze besides the location also the type of ship. The possible types may be trading vessel (tv), cruiser (cr), inland ship (is) and yacht (ya), therefore we have $\Omega^{(2)} = \{\mathrm{tv}, \mathrm{cr}, \mathrm{is}, \mathrm{ya}\}$. In this case the common information about a ship has to be represented in terms of the space

$$\Omega^{(1)} \times \Omega^{(2)} = \{(\mathrm{z3}, \mathrm{tv}), \ldots, (\mathrm{ld}, \mathrm{ya})\},$$

where the first and second component of a tuple indicate the location and type of the ship, respectively. Our uncertain knowledge about this domain is encoded by a random set (P, Γ'), $\Gamma' : \Theta \to 2^{\Omega^{(1)} \times \Omega^{(2)}}$. Suppose the information concerning the location is encoded by the observation mapping

$$\Gamma : \Theta \to 2^{\Omega^{(1)}}; \quad \Gamma(\theta) = \begin{cases} \{\mathrm{z2}, \mathrm{z1}\}, & \text{if } \theta \in \{\theta_1\}, \\ \{\mathrm{z2}, \mathrm{z3}\}, & \text{if } \theta \in \{\theta_2, \theta_3, \theta_4\}. \end{cases}$$

If no further information is available we can use the vacuous extension of Γ to $\Omega^{(1)} \times \Omega^{(2)}$ for the transport of the information to the common space $\Omega^{(1)} \times \Omega^{(2)}$:

$$\widehat{\Pi}(\Gamma) : \Theta \to 2^{\Omega^{(1)} \times \Omega^{(2)}}; \quad \widehat{\Pi}(\Gamma)(\theta) = \begin{cases} \{\mathrm{z1}, \mathrm{z2}\} \times \Omega^{(2)}, & \text{if } \theta \in \{\theta_1\}, \\ \{\mathrm{z2}, \mathrm{z3}\} \times \Omega^{(2)}, & \text{if } \theta \in \{\theta_2, \theta_3, \theta_4\}. \end{cases}$$

The projection of an observation mapping $\widehat{\Pi}(\Gamma)$ onto $\Omega^{(1)}$ yields again Γ.

Conversely let $\Gamma' : \Theta \to \Omega^{(1)} \times \Omega^{(2)}$ be an observation mapping defined by

$$\Gamma'(\theta_i) = \begin{cases} \{(\mathrm{tv}, \mathrm{cr}), (\mathrm{z1}, \mathrm{cr}), (\mathrm{z2}, \mathrm{cr})\}, & \text{if } i = 1, \\ \{(\mathrm{z3}, \mathrm{tv}), (\mathrm{z3}, \mathrm{cr})\}, & \text{if } i = 2, \\ \{(\mathrm{z3}, \mathrm{ya})\}, & \text{if } i = 3, \\ \Omega^{(1)} \times \Omega^{(2)}, & \text{if } i = 4. \end{cases}$$

Then for the marginal observation mapping $\Pi(\Gamma') : \Theta \to 2^{\Omega^{(2)}}$ we obtain

$$\Pi(\Gamma')(\theta_i) = \begin{cases} \{\mathrm{tv}, \mathrm{cr}\}, & \text{if } i \in \{1, 2\}, \\ \{\mathrm{ya}\}, & \text{if } i = 3, \\ \Omega^{(2)}, & \text{if } i = 4. \end{cases}$$

◇

In the context of knowledge representation the most interesting observation mappings are those which yield cylinders, which means mappings

$$\Gamma : \Theta \to \mathcal{Q}(\Omega^N),$$

$M \supseteq N = \{i_1, \ldots, i_n\}$, where

$$\mathcal{Q}(\Omega^N) \stackrel{\mathrm{d}}{=} \{A^{(i_1)} \times \ldots \times A^{(i_n)} \mid A^{(i_j)} \subseteq \Omega^{(i_j)}, j = 1, \ldots, n\}.$$

5.3 Decision Making in the Context of Vague Data

Now let us focus our interest on the case of imprecise data, i.e. the situation where we have an observation mapping $\Gamma : \Theta \to 2^{\Omega}$ and a probability P on Θ. The possible decisions may be $d_1, \dots, d_m$, where choosing d_i means taking $\omega_i \in \Omega = \{\omega_1, \dots, \omega_m\}$ as the actual state of the world. Again our aim is to find the best "decision", thus the generalization of utility measures is required. For this purpose we conceive each imprecise datum $A \subseteq \Omega$ to be the vague image of an (unknown) value $\omega \in A$. Then each random set turns out to be the vague image of an (unknown) random variable $U : \Theta \to \Omega$ which we call the *original* of Γ. For all $\theta \in \Theta$, $U(\theta) \in \Gamma(\theta)$ holds. Knowing only Γ, there are $\prod_{i=1}^{n} |\Gamma(\theta_i)|$ different random variables, called the *selectors* of Γ, satisfying the above condition and thus candidates for the unknown original.

Each selector $S : \Theta \to \Omega$ of Γ induces a probability $S[P]$ on Ω where

$$S[P](\{\omega\}) = \sum_{\theta \in \Theta : S(\theta) = \omega} P(\{\theta\}) \tag{5.14}$$

for all $\omega \in \Omega$. On the intuitive level choosing a selector S of Γ means that some meta-expert advises the experts $\theta \in \Theta$ to select *one* element $\omega \in \Gamma(\theta)$ as the original. The decision for some element of Ω to be the actual state of the world then relies on $S[P]$ and is obtained by the maximum expected utility criterion.

Let $\mathcal{S}$ denote the set of all selectors of the random set Γ. The set $\{S[P] \mid S \text{ selector of } \Gamma\}$ of probabilities serves to define a partial order on the decisions with respect to the maximum expected utility criterion. If a single decision has to be found, or if a total order on the set of decisions is desired, some meta-expert has to specify those selectors he considers to be impossible and to quantify his preferences among the selectors of $\mathcal{S}$. Here once more the concept of second-order probabilities can be applied. If no such information is available we might choose a uniform distribution on $\mathcal{S}$. In this case the "expected probability" of ω is

$$\begin{aligned}
\frac{1}{|\mathcal{S}|} \sum_{S \in \mathcal{S}} S[P](\{\omega\}) &= \frac{1}{|\mathcal{S}|} \sum_{S \in \mathcal{S}} \sum_{\theta : S(\theta) = \omega} P(\{\theta\}) \\
&= \frac{1}{|\mathcal{S}|} \sum_{\theta : \omega \in \Gamma(\theta)} P(\{\theta\}) \cdot \prod_{\theta' : \theta' \neq \theta} |\Gamma(\theta')| \\
&= \sum_{\theta : \omega \in \Gamma(\theta)} \frac{P(\{\theta\})}{|\Gamma(\theta)|} \\
&= \sum_{A : \omega \in A} \frac{P(\{\theta \in \Theta \mid \Gamma(\theta) = A\})}{|A|}
\end{aligned} \tag{5.15}$$

So if we assume that the decisions $d_i \in D$ have the same utilities, the best decision is to choose d_i such that the quantity

$$\sum_{A:\omega\in A} \frac{P(\{\theta \in \Theta \mid \Gamma(\theta) = A\})}{|A|}$$

is maximal.

5.4 The Notion of an Information Source

The aim in this section is to develop an abstract description of expert systems with respect to static as well as dynamic aspects of their structure and functionality. So we have to tackle the problem of knowledge representation and of modeling the reasoning process. Since we focus, for the moment, on conceptual questions, the initial step is to identify and distinguish thoroughly the entities involved. Firstly there are sensors providing (possibly vague) data, secondly there are experts, who are able to rate the importance of sensors (in the form of probability distributions on sensor spaces) and thus to evaluate data obtained from the sensors. Thirdly we have meta-experts, whose knowledge concerns the mutual relationship of different experts, and who are relevant whenever common valuations have to be calculated.

Firstly we have to clarify the relation of the sensor and the evidence space. In our introductory example (Sect. 2.2) we considered for this purpose a mapping Γ that assigned to each sensor θ the corresponding relation $R_\theta \subseteq \Omega$. In the general case we obtain a mapping

$$\Gamma : \Theta \to \mathcal{F}_L(\Omega)$$

that attaches to each sensor a (possibly vague) datum of Ω. Let us firstly consider the simplest case of a sensor space Θ consisting of just one sensor θ which we encountered in our example and which often appears in practical applications. If there is only one sensor $\theta \in \Theta$ then we are forced to rely on its results, just as due to the lack of other elements of Θ, the absence of uncertainty is encoded by a probability distribution $P : \{\theta\} \to [0,1]$, where $P\big(\{\theta\}\big) = 1$. However, in the general case we have to take into account multiple elements of Θ, i.e. several human observers or technical devices, although note that throughout the whole book all data and sensor spaces are assumed to be *finite*. In contrast to our example, where we had no choice but to grant absolute reliability of our one and only sensor, we have to deal with various sensors each of which is reliable to some degree. This restricted reliability devolves to the information the sensors provide, thus also the uncertainty originally affecting the sensors devolves to the data space.

The quantification of importance makes use of probabilistic methods. If a probability P on Θ is specified by some expert, then we can imagine each sensor $\theta \in \Theta$ to be a representative speaking for a number of (virtual) subsensors providing the same output. P reflects the frequency of the different views among the subsensors, i.e. a random choice of some subsensor would yield with probability $P(\{\theta\})$ one of those represented by θ. The subsensors represented by some $\theta \in \Theta$ are considered to be indistinguishable with respect to their valuation of the facts in question. The probability P in some sense quantifies the importance of the elements of Θ, but it has to be stressed that if we follow a frequentistic approach the probabilistic interpretation of the numbers $P(\{\theta\})$ should rely on the relative frequency of subsensors in some "refined" sensor space Θ', or, if we use subjective probabilities, on the betting behavior of some human expert who is responsible for "his" set of sensors.

The sensor information, i.e. the results of technical devices or the observations of a human observer, is represented by a mapping Γ, that attaches to each element $\theta \in \Theta$ an element, a subset, or an L-set of Ω. As we know from this chapter, encoding information by a point-valued mapping $\Gamma : \Theta \to \Omega$ means dealing with precise data, while a set valued mapping $\Gamma : \Theta \to 2^{\Omega}$ allows us to cope with imprecise data, whereas in the most general case the knowledge representation has to rely on L-sets, i.e. an observation mapping of the form $\Gamma : \Theta \to \mathcal{F}_L(\Omega)$ which allows us to handle vague as well as imprecise or even precise data. So by choosing an appropriate observation mapping we take vagueness into account, whereas modeling uncertainty mainly consists in a reasonable assignment of numbers.

Obviously our approach to knowledge representation leads canonically to the concept of random sets considered in this chapter. We obtain a probability P on the sensor space Θ measuring the different sensors' importance and a mapping Γ that attaches to each sensor its observation $\Gamma(\theta)$. But as pointed out in Sect. 4.3, a single probability may not be sufficient to encode the experts' knowledge about the sensors. A statement like "I trust the sensor r2 to a degree of 0.45" should be represented by a class of probability distributions on Θ, each of which satisfies the condition $P(\{\text{r2}\}) = 0.45$. We obtain the set $\mathcal{P} = \left\{P : 2^{\Theta} \to 0,1 \;\middle|\; P(\{\text{r2}\}) = 0.45\right\}$. If, due to the absence of exhaustive information, a careful expert prefers to assign ranges instead of point values for probabilities, we have to use a more extensive set $\mathcal{P}$. This allows us to express ignorance or doubt. An example is the statement "My degree of confidence for sensor r2 is p, $p \in [0.4, 0.7]$". Here we would obtain the class

$$\mathcal{P} = \left\{P : 2^{\Theta} \to [0,1] \;\middle|\; P(\{\text{r2}\}) \in [0.4, 0.7]\right\}.$$

This justifies considering pairs of the form $(\mathcal{P}, \Gamma)$, where $\mathcal{P}$ is a set of probabilities and Γ is an observation mapping, instead of ordinary random sets. We define $(\mathcal{P}, \Gamma)$ to be an *information source*, since our interpretation of probabilities in the context of knowledge representation fits very well the notion of an information source, a notion well known in the field of information theory.

(5.22) Definition. *Let $\mathcal{P}$ be a set of probabilities on Θ and let Γ be an observation mapping, that takes its values within Ω, 2^Ω or $\mathcal{F}_L(\Omega)$, respectively. The pair $\mathcal{S} = (\mathcal{P}, \Gamma)$ is called an information source. (To be extended in Definition 5.24.)*

To enable us to model a meta-expert's uncertainty about the importance or reliability of the sensors at his disposal, he is allowed to specify a set $\mathcal{P}$ of probabilities. The structure of the set $\mathcal{P}$ depends on the meta-expert's actual aims; typical cases are listed below:

- If the meta-expert wants to rely on just one expert $\theta \in \Theta$ he should use a single probability P_θ, i.e.

$$\mathcal{P} = \{P_\theta\}, \quad \text{where } P_\theta(\{\theta\}) = 1.$$

- If the meta-expert is able to rate uniquely the importance of the elements of Θ, then this can also be modeled by a single probability distribution P, where $P(\{\theta\})$ measures the weight of expert θ.
- The (stochastical) independence of two sets $A, B \subseteq \Theta$ of experts can be expressed by the set

$$\mathcal{P} = \{P \mid P(A \cap B) = P(A) \cdot P(B)\}$$

of probabilities.

Note that the information *about* Ω we obtain *from* the experts' has to be distinguished thoroughly from the information concerning the importance of the experts. Throughout our book the experts' observations or opinions are encoded by the observation mapping Γ.

As pointed out in Sect. 5.1, each probability $P \in \mathcal{P}$ induces via Γ a mass distribution $\Gamma[P]$ on Ω, thus from $\mathcal{P}$ we obtain a set of mass distributions. We define

$$\Gamma[\mathcal{P}] \stackrel{\text{d}}{=} \{\Gamma[P] \mid P \in \mathcal{P}\}.$$

So if we consider only Ω we typically encounter sets of mass distributions, but the concept of an information source which we used up to now covers only the static part of the modeling task, thus in the next section we have to analyze the dynamic aspects.

5.4.1 Updating Information Sources

The next subject of interest is modeling the reasoning process, i.e. now we have to focus on the dynamic aspects of expert systems. In general the sensor outputs are not meant to be final valuations chiseled in stone but they do create restrictions which are subject to a continued tightening in the light of additional information, which we call *updating.* To clarify our intuitive idea of updating information sources let us once more return to our introductory example.

(5.23) Example. Recall that in our introductory example (Sect. 2.2) we obtained as the result of the knowledge acquisition process the observation mappings $\Gamma_1 : \{\theta_1\} \to \Omega$, $\Gamma_1(\theta_1) = R_{\theta_1}$, and $\Gamma_2 : \{\theta_2\} \to \Omega$, $\Gamma_2(\theta_2) = R_{\theta_2}$ (equation (2.6)) which summarize the structural knowledge of the experts regardless of any information about the location of the ship of actual interest. Since we assume that the experts' statements are not affected with uncertainty we obtain two mass distributions $m^{R_{\theta_1}}$ and $m^{R_{\theta_2}}$ on Ω. Of course, our final aim is the integration of these two pieces of information, but before we can tackle the integration problem, we have to specify explicitly which modifications of his original valuation of the situation each expert considers to be admissible. The underlying interpretation of the expert's statements is that to consider a specific ship means to restrict the set of possible situations covering the "true state of the world" with respect to the actual evidence. This meta-knowledge essentially determines the definition of two sets $\mathcal{G}_1$ and $\mathcal{G}_2$, where

$$\mathcal{G}_1 = \{\Gamma' : \{\theta_1\} \to 2^\Omega \mid \Gamma'(\theta_1) = \Gamma_1^E(\theta_1) = \Gamma_1(\theta_1) \cap E \text{ where } E \subseteq \Omega\}$$

$$\mathcal{G}_2 = \{\Gamma' : \{\theta_2\} \to 2^\Omega \mid \Gamma'(\theta_2) = \Gamma_2^E(\theta_2) = \Gamma_2(\theta_2) \cap E \text{ where } E \subseteq \Omega\}.$$

$\mathcal{G}_1$ and $\mathcal{G}_2$ contain besides Γ_1 and Γ_2 the "possible updates", reflecting what the experts would accept to be derived due to additional information. ◇

In more general terms, the sensor data are subject to a continous process of change. This process, which we call the *updating process*, is a consequence of the increasing amount of information about the actual situation. Of course it depends on the interpretation of the knowledge, in which way the original data are modified, i.e. which updating scheme is applied. So the task of specifying possible updates should be left to the expert, who provides the knowledge.

Typical updating schemes which will be analysed in the sequel are conditioning and data revision as introduced in Sect. 5.2. Data revision corresponds to a change of the observation mapping Γ, whereas conditioning reflects a change of the probabilities in P. Note that formally also conditioning can be characterized by changing Γ, so it is possible to describe the updating process in any case by a manipulation of the observation mapping(s).

There are (at least) two different ways to perform updates. Let $\mathcal{S} = (\mathcal{P}, \Gamma)$ be an information source.

- If some absolutely reliable but imprecise information in the form of a subset E of Ω has to be integrated, then the expert may "turn off" those sensors which provide observations contradicting E. He switches from Γ to the *conditional observation mapping* Γ_E. The same method can be applied if the meta-expert wants to neglect some particular sensor, therefore this type of updating corresponds to conditioning. If for all $\theta \in \Theta$ we have $\Gamma_E(\theta) = \emptyset$, then the new knowledge contradicts Γ. The updated source is $\mathcal{S}' = (\mathcal{P}, \Gamma_E)$. As an example consider the observation mapping $\Gamma : \Theta \to 2^{\Omega}$, which induces the set $\Gamma[\mathcal{P}]$ of mass distributions on Ω. If we require that all sensors θ with $\Gamma(\theta) \not\subseteq E$ have to be neglected, this leads to the set $\Gamma_E[\mathcal{P}] = \{m(\,.\,|E) \mid m \in \Gamma[\mathcal{P}]\}$.
- A completely different approach is used in our introductory example, where the information R_{θ_1} of the expert θ_1 was *revised* into $R_{\theta_1} \cap E$ and the information R_{θ_2} into $R_{\theta_2} \cap E$. This approach relies on the revision of Γ; semantically this means a modification of data and should for this reason be used very carefully. *Data revision* in any case requires the intervention of the expert, who evaluates the sensor information, therefore it is not possible to apply such an updating scheme purely on the sensor level. If we know for sure that all sensor data should lie within a given set E, and if we are authorized to change sensor output Γ to Γ^E, then we obtain the revised information source $\mathcal{S} = (\mathcal{P}, \Gamma^E)$ which induces the set $\Gamma^E[\mathcal{P}] = \{m_E \mid m \in \Gamma[\mathcal{P}]\}$ of mass distributions on Ω.

As we have seen, modeling the dynamic aspects of reasoning processes requires us to consider not only the information sources but also their behavior in the context of integrating additional knowledge. It turned out that for this purpose a set $\mathcal{G}$ of possible updates should be explicitly specified. Therefore we give the following extended definition of an information source.

(5.24) Definition. *A triple $\mathcal{S} = (\mathcal{P}, \Gamma, \mathcal{G})$ is called an information source, if and only if $\mathcal{P}$ is a class of probabilities on Θ, Γ is an observation mapping and $\mathcal{G}$ a set of observation mappings containing at least Γ. The information source $\mathcal{S}$ is called monotonic, if and only if $\mathcal{G} \subseteq \{\Gamma' \mid \Gamma' \sqsubseteq \Gamma\}$.*

Typical updates are conditional or revised observation mappings. Unfortunately the situation is complicated by the fact that the space which is subject to a sensor's perceptions may be a refinement (or a coarsening) of Ω, which is the space we use for the purpose of knowledge representation. In this case we have to "import" the information by projection (or extension). From a mathematical point of view it is of interest to determine the set of updates which are derivable from some observation mapping Γ. Given a set E representing some reliable, imprecise but unknown datum, this set may be a subset of Ω or it may be a subset of some refinement Ω' of Ω.

(5.25) Definition. *Let Γ be an observation mapping. We define the following notions for sets of possible updates:*

(a) $\mathcal{U}_{\text{cond}}(\Gamma) = \{\Gamma' \mid \Gamma' = \Gamma_E \text{ where } E \subseteq \Omega\}$

(b) $\mathcal{U}_{\text{cond}}^{\text{refin}}(\Gamma) = \{\Gamma' \mid \Gamma' \leq \Gamma\}$

(c) $\mathcal{U}_{\text{revis}}(\Gamma) = \{\Gamma' \mid \Gamma' = \Gamma^E \text{ where } E \subseteq \Omega\}$

(d) $\mathcal{U}_{\text{revis}}^{\text{refin}}(\Gamma) = \{\Gamma' \mid \Gamma' \sqsubseteq \Gamma\}$

(Note that in the case of point-valued observation mappings the distinction between conditioning and revision is not necessary, since the two concepts coincide.)

Updating according to Definition 5.25 is inherently monotonic. All possible updates Γ' are specializations of the original observation mapping, so because of the transitivity property of the relation $\sqsubseteq$ the subsequent use of updating schemes leads to a sequence of sets in a subset relationship. We have

$$\begin{array}{lllllll} \Gamma' & \in & \mathcal{U}_{\text{cond}}(\Gamma) & \Rightarrow & \mathcal{U}_{\text{cond}}(\Gamma') & \subseteq & \mathcal{U}_{\text{cond}}(\Gamma) \\ \Gamma' & \in & \mathcal{U}_{\text{cond}}^{\text{refin}}(\Gamma) & \Rightarrow & \mathcal{U}_{\text{cond}}^{\text{refin}}(\Gamma') & \subseteq & \mathcal{U}_{\text{cond}}^{\text{refin}}(\Gamma) \\ \Gamma' & \in & \mathcal{U}_{\text{revis}}(\Gamma) & \Rightarrow & \mathcal{U}_{\text{revis}}(\Gamma') & \subseteq & \mathcal{U}_{\text{revis}}(\Gamma) \\ \Gamma' & \in & \mathcal{U}_{\text{revis}}^{\text{refin}}(\Gamma) & \Rightarrow & \mathcal{U}_{\text{revis}}^{\text{refin}}(\Gamma') & \subseteq & \mathcal{U}_{\text{revis}}^{\text{refin}}(\Gamma) \end{array} \tag{5.16}$$

In order to evaluate an information source $\mathcal{S} = (\mathcal{P}, \Gamma, \mathcal{G})$ an expert has to specify explicitly which updates are admissible in the special context under consideration. Natural choices are, of course, the four sets $\mathcal{U}_{\text{cond}}(\Gamma)$, $\mathcal{U}_{\text{revis}}(\Gamma)$, $\mathcal{U}_{\text{revis}}^{\text{refin}}(\Gamma)$ or $\mathcal{U}_{\text{cond}}^{\text{refin}}(\Gamma)$, but often even subsets are reasonable. Recall that in Example 5.23 we considered two information sources $\mathcal{S} = (\mathcal{P}_1, \Gamma_1, \mathcal{G}_1)$, $\mathcal{S} = (\mathcal{P}_2, \Gamma_2, \mathcal{G}_2)$, where $\mathcal{G}_1 = \mathcal{U}_{\text{revis}}(\Gamma_1)$ and $\mathcal{G}_2 = \mathcal{U}_{\text{revis}}(\Gamma_2)$.

5.4.2 The Combination of Information Sources

Consider two information sources $\mathcal{S}_1 = (\mathcal{P}_1, \Gamma_1, \mathcal{G}_1)$, $\mathcal{S}_2 = (\mathcal{P}_2, \Gamma_2, \mathcal{G}_2)$, defined on different sensor spaces Θ_1, Θ_2 but both providing observations concerning Ω. In general we have to deal with n information sources but for simplicity we restrict ourselves to $\mathcal{S}_1$ and $\mathcal{S}_2$. From a conceptual as well as from a practical point of view it would be desirable if we could find some kind of joint source, comprising the different pieces of information encoded by the individual sources. Thus we have to solve two problems, firstly the combination of probabilities $P_1 \in \mathcal{P}_1$ and $P_2 \in \mathcal{P}_2$ and secondly the combination of observation mappings $\Gamma_1 \in \mathcal{G}_1$ and $\Gamma_2 \in \mathcal{G}_2$.

Let us consider first the case where $\mathcal{P}_1 = \{P_1\}$ and $\mathcal{P}_2 = \{P_2\}$. The basic idea of combining P_1 and P_2 is to encode the information borne by the probabilities as well as the information concerning their relation by a set $\mathcal{P}$ of probabilities on the product space $\Theta_1 \times \Theta_2$.

In order to ensure that no information is lost, each probability $P \in \mathcal{P}$ that corresponds to P_1 and P_2 should obey the *marginal conditions* $\Pi_{\{1,2\}}(P) = P_1$, $\Pi_{\{1,2\}}(P) = P_2$, i.e. the marginal probabilities of P should coincide with the original distributions P_1 and P_2.

- In the worst case no information on the relation of the sources is available, so $\mathcal{P}$ should contain all probabilities P which satisfy the marginal conditions.
- In the best case there is only one probability P in $\mathcal{P}$. If the elements of Θ_1 and Θ_2 are independently selected (in the stochastical sense), then the product probability is given by

$$(P_1 \otimes P_2)\Big(\{(\theta', \theta'')\}\Big) \stackrel{\text{d}}{=} P_1(\{\theta'\}) \cdot P_2(\{\theta''\}) \tag{5.17}$$

  for all $\theta' \in \Theta_1, \theta'' \in \Theta_2$. Note that also Pearl's approach ensures that only one probability distribution remains (for the details see Chap. 12).
- If the restrictions are too strong, then the set $\mathcal{P}$ may be empty. In this case the requirements are *contradictory.*

In practical applications the formulation of the conditions should be left to the expert who employs the respective sensors, since only he is able to provide a reasonable valuation of, for example, mutual perturbations.

(5.26) Example. Consider the sensor spaces $\Theta' = \{\theta'_1, \theta'_2, \theta'_3\}$, $\Theta'' = \{\theta''_1, \theta''_2, \theta''_3, \theta''_4\}$ and two probabilities P' and P'' on Θ' and Θ'', respectively. P' and P'' may be uniform distributions. Suppose the expert specifies the following requirements:

(*i*) the marginal conditions have to be satisfied,

(*ii*) $P(\{\theta'_3\} \mid \{\theta''_4\}) = 1$ and $P(\{\theta'_3\} \mid \{\theta''_3\}) = 0$, and

(*iii*) under the condition of $\{\theta''_1, \theta''_2\}$, θ'_1 and θ''_1 as well as θ'_2 and θ''_2 are statistically independent.

The three above requirements are simultaneously satisfied by the family

$$\mathcal{P} = \{P_\lambda \mid \frac{1}{24} \leq \lambda \leq \frac{4}{24}\}$$

of probabilities, the particular values are listed below:

	θ''_1	θ''_2	θ''_3	θ''_4
θ'_1	$\frac{5}{24} - \lambda$	$\frac{5}{24} - \lambda$	$2 \cdot \lambda - \frac{2}{24}$	0
θ'_2	λ	λ	$\frac{8}{24} - 2 \cdot \lambda$	0
θ'_3	$\frac{1}{24}$	$\frac{1}{24}$	0	$\frac{1}{4}$

◇

In principle the expert is free to specify the combined probability on the space $\Theta_1 \times \Theta_2$, just depending on his modeling aims, but nevertheless often it is reasonable to use $P_1 \otimes P_2$.

Up to now we considered the combination of single probabilities, but in general sets of probabilities have to be combined. Then we define

$$\text{comb}(\mathcal{P}_1, \mathcal{P}_2) \stackrel{\text{d}}{=} \bigcup_{P_1 \in \mathcal{P}_1, P_2 \in \mathcal{P}_2} \text{comb}(P_1, P_2)$$

where $\text{comb}(P_1, P_2)$ is a set of probabilities on $\Theta_1 \times \Theta_2$ satisfying the marginal conditions; an example is $\text{comb}(P_1, P_2) = \{P_1 \otimes P_2\}$.

Now we have to investigate the combination of two observation mappings Γ_1 and Γ_2. The result should be a joint observation mapping defined on $\Theta_1 \times \Theta_2$. If the observation mappings Γ_1 and Γ_2 map Θ_1 and Θ_2 to the sets $\mathcal{F}_{L_1}(\Omega)$ and $\mathcal{F}_{L_2}(\Omega)$, respectively, where $L_1 \neq L_2$, the first step should be to find a common basis for the description of vagueness. In the sequel we assume that L_1 and L_2 dovetail properly into each other, thus that it is possible to find a set L covering L_1 as well as L_2. So each L_1-set η' and each L_2-set η'' can be easily transformed into a corresponding L-set.

If the mappings Γ_1 and Γ_2 are compatible they can be combined *conjunctively* if the experts are forced to agree upon a common opinion or *disjunctively* if all experts' contributions have to be taken into account. Thus we define

$$\Gamma_1 \cup \Gamma_2 : \Theta_1 \times \Theta_2 \to 2^\Omega; \quad (\Gamma_1 \cup \Gamma_2)(\theta_1, \theta_2) \stackrel{\text{d}}{=} \Gamma_1(\theta_1) \cup \Gamma_2(\theta_2),$$

$$\Gamma_1 \cap \Gamma_2 : \Theta_1 \times \Theta_2 \to 2^\Omega; \quad (\Gamma_1 \cap \Gamma_2)(\theta_1, \theta_2) \stackrel{\text{d}}{=} \Gamma_1(\theta_1) \cap \Gamma_2(\theta_2).$$

If the information sources act on different data spaces Ω_1 and Ω_2, then in a first step the mappings have to be extended to the common data space $\Omega = \Omega_1 \times \Omega_2$ by defining

$$\widehat{\Gamma}_1 : \Theta_1 \to 2^{\Omega_1 \times \Omega_2}; \quad \widehat{\Gamma}_1(\theta_1) = \widehat{\Pi}_{\{1\}}^{\{1,2\}}(\Gamma_1(\theta_1)),$$

$$\widehat{\Gamma}_2 : \Theta_2 \to 2^{\Omega_1 \times \Omega_2}; \quad \widehat{\Gamma}_2(\theta_2) = \widehat{\Pi}_{\{2\}}^{\{1,2\}}(\Gamma_2(\theta_2)),$$

where $\widehat{\Pi}_1$ and $\widehat{\Pi}_2$ denote the cylindrical extensions as defined in Sect. 2.1. For the conjunctive combination we obtain

$$(\widehat{\Gamma}_1 \cap \widehat{\Gamma}_2)(\theta_1, \theta_2) = \widehat{\Pi}_{\{1\}}^{\{1,2\}}(\Gamma_1(\theta_1)) \cap \widehat{\Pi}_{\{2\}}^{\{1,2\}}(\Gamma_2(\theta_2)) = \Gamma_1(\Theta_1) \times \Gamma_2(\Theta_2).$$

(5.27) Example. Let $\Theta_1 = \{\theta'_1, \theta'_2, \theta'_3\}$ be a set consisting of three devices, each of which is able to determine the location of a ship with respect to the frame of discernment $\Omega_1 = \{\text{z3}, \ldots, \text{ld}\}$, whereas $\Theta_2 = \{\theta''_1, \theta''_2, \theta''_3\}$ consists of three human experts, providing information about the ship's type. We assume $\Omega_2 = \{\text{tv}, \text{cr}, \text{in}, \text{ya}\}$, where the possible types are "trading vessel" (tv), "cruiser" (cr), "inland ship" (in) and "yacht" (ya). Let $\Gamma_1 : \Theta_1 \to 2^{\Omega_1}$, $\Gamma_1(\theta') \neq \emptyset$ for all $\theta' \in \Theta_1$, be a mapping that assigns to each device its (possibly imprecise) observation of the ship's location, whereas $\Gamma_2 : \Theta_2 \to 2^{\Omega_2}$, $\Gamma_2(\theta'') \neq \emptyset$ for all $\theta'' \in \Theta_2$, describes the imprecise data concerning the ship's type, which we obtain from the human experts. The canonical "combined" mapping, that assigns to each pair (device, expert) the corresponding information, is $\Gamma : \Theta_1 \times \Theta_2 \to 2^{\Omega_1 \times \Omega_2}$, where

$$\Gamma(\theta', \theta'') = \Gamma(\theta') \times \Gamma(\theta'')$$

for all $(\theta', \theta'') \in \Theta_1 \times \Theta_2$. The mappings Γ_1 and Γ_2 may be given as listed below:

θ'	$\Gamma_1(\theta')$
θ'_1	{z2}
θ'_2	{z2, z1}
θ'_3	{z1, cr}

θ''	$\Gamma_2(\theta'')$
θ''_1	{tv, cr}
θ''_2	{cr, ya}
θ''_3	{ya}

As the corresponding joint mapping $\Gamma_1 \times \Gamma_2$ we obtain

	θ''_1	θ''_2	θ''_3
θ'_1	{z2} × {tv, cr}	{z2} × {cr, ya}	{z2} × {ya}
θ'_2	{z2, z1} × {tv, cr}	{z2, z1} × {cr, ya}	{z2, z1} × {ya}
θ'_3	{z1, cr} × {tv, cr}	{z1, cr} × {cr, ya}	{z1, cr} × {ya}

◇

Since in the general case we have to deal with sets $\mathcal{G}_1$ and $\mathcal{G}_2$ of observation mappings we define

$$\text{comb}(\mathcal{G}_1, \mathcal{G}_2) = \big\{\text{comb}(\Gamma_1, \Gamma_2) \;\big|\; \Gamma_1 \in \mathcal{G}_1, \Gamma_2 \in \mathcal{G}_2\big\},$$

where $\text{comb}(\Gamma_1, \Gamma_2)$ means either $\Gamma_1 \cap \Gamma_2$ or $\Gamma_1 \cup \Gamma_2$. For the combination of typical sets of updates we easily derive approximations, as the next theorem shows.

(5.28) Theorem. *Let $\Gamma_1 : \Theta_1 \to 2^{\Omega}$ and $\Gamma_2 : \Theta_2 \to 2^{\Omega}$ be two observation mappings.*

(a) For $\text{comb}(\Gamma_1, \Gamma_2) = \Gamma_1 \cap \Gamma_2$ *we derive*
- *(i)* $\text{comb}\big(\mathcal{U}_{\text{cond}}(\Gamma_1), \mathcal{U}_{\text{cond}}(\Gamma_2)\big) = \mathcal{U}_{\text{cond}}(\Gamma_1 \cap \Gamma_2)$
- *(ii)* $\text{comb}\big(\mathcal{U}^{\text{refin}}_{\text{cond}}(\Gamma_1), \mathcal{U}^{\text{refin}}_{\text{cond}}(\Gamma_2)\big) \subseteq \mathcal{U}^{\text{refin}}_{\text{cond}}(\Gamma_1 \cap \Gamma_2)$
- *(iii)* $\text{comb}\big(\mathcal{U}_{\text{revis}}(\Gamma_1), \mathcal{U}_{\text{revis}}(\Gamma_2)\big) = \mathcal{U}_{\text{revis}}(\Gamma_1 \cap \Gamma_2)$
- *(iv)* $\text{comb}\big(\mathcal{U}^{\text{refin}}_{\text{revis}}(\Gamma_1), \mathcal{U}^{\text{refin}}_{\text{revis}}(\Gamma_2)\big) \subseteq \mathcal{U}^{\text{refin}}_{\text{revis}}(\Gamma_1 \cap \Gamma_2)$

(b) For $\text{comb}(\Gamma_1, \Gamma_2) = \Gamma_1 \cup \Gamma_2$ *we derive*
- *(i)* $\text{comb}\big(\mathcal{U}_{\text{cond}}(\Gamma_1), \mathcal{U}_{\text{cond}}(\Gamma_2)\big) = \mathcal{U}_{\text{cond}}(\Gamma_1 \cup \Gamma_2)$
- *(ii)* $\text{comb}\big(\mathcal{U}^{\text{refin}}_{\text{cond}}(\Gamma_1), \mathcal{U}^{\text{refin}}_{\text{cond}}(\Gamma_2)\big) \subseteq \mathcal{U}^{\text{refin}}_{\text{cond}}(\Gamma_1 \cup \Gamma_2))$
- *(iii)* $\text{comb}\big(\mathcal{U}_{\text{revis}}(\Gamma_1), \mathcal{U}_{\text{revis}}(\Gamma_2)\big) = \mathcal{U}_{\text{revis}}(\Gamma_1 \cup \Gamma_2)$
- *(iv)* $\text{comb}\big(\mathcal{U}^{\text{refin}}_{\text{revis}}(\Gamma_1), \mathcal{U}^{\text{refin}}_{\text{revis}}(\Gamma_2)\big) \subseteq \mathcal{U}^{\text{refin}}_{\text{revis}}(\Gamma_1 \cup \Gamma_2))$

This theorem justifies using instead of the combined source

$$\mathcal{S} = \Big(\text{comb}(\mathcal{P}_1, \mathcal{P}_2), \text{comb}(\Gamma_1, \Gamma_2), \text{comb}(\mathcal{U}^{\text{refin}}_{\text{revis}}(\Gamma_1), \mathcal{U}^{\text{refin}}_{\text{revis}}(\Gamma_2))\Big)$$

the approximation

$$\mathcal{S} = \Big(\text{comb}(\mathcal{P}_1, \mathcal{P}_2), \text{comb}(\Gamma_1, \Gamma_2), \mathcal{U}^{\text{refin}}_{\text{revis}}(\text{comb}(\Gamma_1, \Gamma_2))\Big).$$

Now let us have a look at the structures, i.e. mass distributions, induced on Ω.

(5.29) Theorem. *Let $\Gamma_1 : \Theta_1 \to 2^{\Omega}$, $\Gamma_2 : \Theta_2 \to 2^{\Omega}$ be two observation mappings and let P_1, P_2 be two probabilities on Θ_1, Θ_2, respectively. If*

$$\text{comb}(P_1, P_2) = \{P_1 \otimes P_2\}$$

then we obtain

(a) $$(\Gamma_1 \cap \Gamma_2)[P_1 \otimes P_2](A) = \frac{\sum\limits_{A', A'' : A' \cap A'' = A} \Gamma_1 P_1(A') \cdot \Gamma_2 P_2(A'')}{\sum\limits_{A', A'' : A' \cap A'' \neq \emptyset} \Gamma_1 P_1(A') \cdot \Gamma_2 P_2(A'')},$$ *and*

(b) $(\Gamma_1 \cup \Gamma_2)[P_1 \otimes P_2](A) = \sum_{A',A'':A'\cup A''=A} \Gamma_1 P_1(A') \cdot \Gamma_2 P_2(A'')$

for all $A \subseteq \Omega$.

Proof.

(a)

$$(\Gamma_1 \cap \Gamma_2)[P_1 \otimes P_2](M) = \frac{\sum_{N',N'':(\Gamma_1\cap\Gamma_2)(N'\times N'')=M} (P_1 \otimes P_2)(N',N'')}{\sum_{N',N'':(\Gamma_1\cap\Gamma_2)(N'\times N'')\neq\emptyset} (P_1 \otimes P_2)(N',N'')}$$

$$= \frac{\sum_{N',N'':\Gamma_1(N')\cap\Gamma_2(N'')=M} P_1(N') \cdot P_2(N'')}{\sum_{N',N'':\Gamma_1(N')\cap\Gamma_2(N'')\neq\emptyset} P_1(N') \cdot P_2(N'')}$$

$$= \frac{\sum_{M',M'':M'\cap M''=M} (\sum_{N':\Gamma_1(N')=M'} P_1(N')) \cdot (\sum_{N'':\Gamma_2(N'')=M''} P_2(N''))}{\sum_{M',M'':M'\cap M''=M} (\sum_{N':\Gamma_1(N')=M'} P_1(N')) \cdot (\sum_{N'':\Gamma_2(N'')=M''} P_2(N''))}$$

$$= \frac{\sum_{M',M'':M'\cap M''\neq\emptyset} \Gamma_1[P_1](M') \cdot \Gamma_2[P_2](M'')}{\sum_{M',M'':M'\cap M''\neq\emptyset} \Gamma_1[P_1](M') \cdot \Gamma_2[P_2](M'')}.$$

$(\Gamma_1 \cap \Gamma_2)[P_1 \otimes P_2]$ is defined if and only if there are two sets M', M'' with $M' \cap M'' \neq \emptyset$ and $\Gamma_1[P_1](M') \cdot \Gamma_2[P_2](M'') \neq 0$.

(b) It follows directly from the definition that

$$(\Gamma_1 \cup \Gamma_2)[P_1 \otimes P_2](M) = \sum_{M',M'':M'\cup M''=M} \Gamma_1[P_1](M') \cdot \Gamma_2[P_2](M'')$$

for all $M \subseteq \Omega$.

□

If the observation mappings Γ_1 and Γ_2 act on different data spaces Ω_1 and Ω_2 we obtain a structure on $\Omega_1 \times \Omega_2$. In this case the focal elements of the induced mass distribution $\Gamma[P_1 \otimes P_2]$ are of the form $A' \times A''$, where A' and A'' are focal elements of $\Gamma_1[P_1]$ and $\Gamma_2[P_2]$, respectively. We obtain for $\Gamma \stackrel{\mathrm{d}}{=} \widehat{\Pi}_1(\Gamma_1) \cap \widehat{\Pi}_2(\Gamma_2)$

$$\begin{aligned}
&\Gamma[P_1 \otimes P_2](A' \times A'') \\
&= \frac{(P_1 \otimes P_2)(\{(\theta', \theta'') \in \Theta' \times \Theta'' | \Gamma_1(\theta') \times \Gamma_2(\theta'') = A' \times A''\})}{(P_1 \otimes P_2)(\{(\theta', \theta'') \in \Theta' \times \Theta'' | \Gamma_1(\theta') \times \Gamma_2(\theta'') \neq \emptyset\})} \\
&= \frac{(P_1(\{\theta' \in \Theta' | \Gamma_1(\theta') = A'\}) \cdot P_2(\{\theta'' \in \Theta'' | \Gamma_2(\theta'') = A''\})}{\sum_{A', A'': A' \neq \emptyset, A'' \neq \emptyset} (P_1(\{\theta' \in \Theta' | \Gamma_1(\theta') = A'\}) \cdot P_2(\{\theta'' \in \Theta'' | \Gamma_2(\theta'') = A''\})} \\
&= \frac{\Gamma_1[P_1](A') \cdot \Gamma_2[P_2](A'')}{\sum_{A', A'': A' \neq \emptyset, A'' \neq \emptyset} \Gamma_1[P_1](A') \cdot \Gamma_2[P_2](A'')}
\end{aligned}$$

for all A', A''. This means that the induced mass distribution $\Gamma[P_1 \otimes P_2]$ is determined by $\Gamma_1[P_1]$ and $\Gamma_2[P_2]$, i.e. by mass distributions on Ω_1 and Ω_2.

If we assume the information encoded by the sets $\eta' \in \mathcal{F}_L(\Theta_1)$ and $\eta'' \in \mathcal{F}_L(\Theta_2)$ to be independent, then we adequately represent the information borne by the pair (η', η'') by a vague datum $\eta' \times \eta'' \in \mathcal{F}_L(\Theta_1 \times \Theta_2)$, where

$$(\eta' \times \eta'')(\theta', \theta'') = \min\{\eta'(\theta'), \eta''(\theta'')\}$$

for all $\theta' \in \Theta_1$, $\theta'' \in \Theta_2$. In the special case of ordinary sets this notion coincides with that of cylinder sets. Whenever a set $A' \times A''$ or, in the case of L-sets, $\eta' \times \eta''$, carries the same information as its projections (A', A'') or (η', η''), respectively, then we talk of *non-interactivity.* This property was first considered in the context of fuzzy sets. Note that the use of $\mathcal{F}_L(\Theta_1 \times \Theta_2)$ instead of $\mathcal{F}_L(\Theta_1) \times \mathcal{F}_L(\Theta_2)$ dramatically reduces the complexity of handling the information given by (η', η'').

In the most typical case we have to deal with sources of the form

$$\mathcal{S}_1 = (\{P \mid P \leq P_1\}, \mathcal{U}^{\text{refin}}_{\text{revis}}(\Gamma_1)), \ \mathcal{S}_2 = (\{P \mid P \leq P_2\}, \mathcal{U}^{\text{refin}}_{\text{revis}}(\Gamma_2)).$$

Choosing $\text{comb}(P_1, P_2) = \{P_1 \otimes P_2\}$ and $\text{comb}(\Gamma_1, \Gamma_2) = \Gamma_1 \cap \Gamma_2$ we can use the source

$$\mathcal{S} = (\{P \mid P \leq P_1 \otimes P_2\}, \mathcal{U}^{\text{refin}}_{\text{revis}}(\Gamma_1 \cap \Gamma_2))$$

as an approximation for the combination of $\mathcal{S}_1$ and $\mathcal{S}_2$ since for each $(\Gamma' \cap \Gamma'')[P' \otimes P'']$, where $\Gamma' \sqsubseteq \Gamma_1$, $\Gamma'' \sqsubseteq \Gamma_2$, $P' \leq P_1$, $P'' \leq P_2$, there exist $P \leq P_1 \otimes P_2$ and $\Gamma \sqsubseteq \Gamma_1 \cap \Gamma_2$ such that $\Gamma[P] = (\Gamma' \cap \Gamma'')[P' \otimes P'']$.

5.5 Concluding Remarks

The concept of a random set, although known for a long time in connection with set-valued functions ([Aumann 1965], [Aubin and Frankowska 1990]) and geometric probabilities, was recently formalized in [Matheron 1975], [Kendall 1974], [Stoyan 1988]. In [Kruse and Meyer 1987] this approach is used as the foundation of statistical methods for the evaluation of data represented by sets. For conditioning we used the geometric rule of conditioning [Suppes and Zanotti 1977]. A.P. Dempster [Dempster 1967] analyzed random sets in terms of multivalued mappings, and originated the notion of upper and lower probabilities in this context. His concept of conditioning, well known as Dempster's rule of conditioning, relies on changing the multivalued mapping.

Random set theory in our approach deals with the handling of uncertain, imprecise data. This notion should not be confused with the treatment of classes of probabilities, which allow one to model only precise data, but with more flexibility with respect to the uncertainty aspects.

Our approach to decision making in the presence of imprecise data was inspired by random set theory, where the notion of a selector is of great importance. We assume additional information about the selectors to be provided by some meta-expert. If an equal distribution on the set of selectors is specified, we obtain a decision according to a "generalized insufficient reason principle".

In the field of probability theory the idea of "transporting" uncertainty about some space Θ to another space Ω is well known. So the notion of an information source (which is essentially a tuple (P, Γ)) is straightforward. We have taken into account here that in practice we have to deal with various updating schemes (besides conditioning) due to the complex requirements arising from knowledge representation and modeling the reasoning process.

Chapter 6

Mass Distributions

In this chapter we address once more the common treatment of imprecision and uncertainty. Our approach relies on the attachment of "masses" to vague data. Of course there are extensive similarities with the concept of belief functions [Shafer 1976] and even more with the transferable mass model [Smets 1978], but an important difference remains: we conceive mass distributions (in the case of imprecise data) just as a condensed representation of weighted (random) sets, where sets are used for the representation of imprecise data. Note that this does not necessarily require us to follow the purely frequentistic interpretation of uncertainty presented above, but allows also a "subjective" view. The practical value of the methods we develop in the sequel arises from the fact that they can be applied even if the underlying random set is unknown.

The necessity measures we consider in this chapter are belief functions with pleasant additional properties that make them especially easy to handle. These functions have to be thoroughly distinguished from those presented in Chap. 3 although they are identical from a formal point of view, since they have totally different semantics.

6.1 Basic Concepts

As we pointed out above, uncertainty structures on Ω are induced by uncertainty structures on the expert space Θ via an observation mapping Γ. Unfortunately the original space Θ in many cases is inaccessible, so we have to represent the experts' partial beliefs just in terms of the data space Ω.

6.1.1 Condensed Representations of Random Sets

In order to clarify what kind of structures on Ω we have to deal with and how to manipulate these structures, let us consider a random set (P, Γ). Each subset $A \subseteq \Omega$ corresponds to the set $\{\theta \in \Theta \mid \Gamma(\theta) = A\}$ of sensors which observe *exactly* A. Attaching their "probability mass" to A we obtain a *mass distribution*

$$\Gamma[P] : 2^{\Omega} \to [0,1]; \quad \Gamma[P](A) \stackrel{\mathrm{d}}{=} \begin{cases} \frac{P(\{\theta\in\Theta \mid \Gamma(\theta)=A\})}{P(\{\theta\in\Theta \mid \Gamma(\theta)\neq\emptyset\})} & \text{if } A \neq \emptyset, \\ 0 & \text{otherwise,} \end{cases} \tag{6.1}$$

where obviously

$$\text{(i) } \Gamma[P](\emptyset) = 0,$$
$$\text{(ii) } \sum_{A:A\subseteq\Omega} \Gamma[P](A) = 1$$

hold. Based on $\Gamma[P]$ the respective upper and lower probabilities can be determined only in terms of Ω. We have

$$(P_{\Gamma})_*(B) = \sum_{A:A\subseteq B} \Gamma[P](A),$$
$$(P_{\Gamma})^*(B) = \sum_{A:A\cap B\neq\emptyset} \Gamma[P](A). \tag{6.2}$$

Going the other way round, we can conceive a real-valued set function $m : 2^{\Omega} \to [0,1]$ satisfying (i) and (ii) to be induced by some unknown random set. This justifies applying equation (6.2) in order to calculate upper and lower probabilities from m. Analogously, for any random set operation a respective operation acting just on Ω can be derived. So the reader should keep in mind that the formalisms we develop in the sequel are nothing but condensed representations of the corresponding formalisms for random sets.

(6.1) Definition. *Each function $m : 2^{\Omega} \to [0,1]$ is called a mass distribution*[1]*, whenever*

(i) $m(\emptyset) = 0,$

(ii) $\sum_{A:A\subseteq\Omega} m(A) = 1$

hold.

To get an intuitive impression of what a mass distribution means, we may imagine that some human expert distributed a fixed amount of virtual "evidence mass" to certain results with respect to the reliability of the information sources. So on the formal level we have to deal with numbers attached to imprecise data, where the numbers reflect some human decision maker's valuation of evidence. The quantity $m(A)$ is understood to be the measure of support given to A but not to any strict subset of A. Those sets A with $m(A) > 0$ are called

[1] In [Shafer 1976] m is denoted as *basic probability assignment* (bpa). In order to prevent any confusion with probabilities we strictly avoid this notation.

focal elements. To obtain the *total* measure of support committed to some set A, we have to sum up the quantities $m(B)$ for all $B \subseteq A$.

(6.2) Definition. *If m is a mass distribution on Ω, then the function*

$$\mathrm{Bel}_m : 2^\Omega \to [0,1], \qquad \mathrm{Bel}_m(A) \stackrel{\mathrm{d}}{=} \sum_{B:B\subseteq A} m(B),$$

is called the belief function induced by m.

$\mathrm{Bel}_m(A)$ represents the degree to which the actual evidence supports A, i.e. it measures the *credibility* of A. We are also able to define the degree to which the evidence fails to refute A, i.e. the degree to which A remains *plausible*, by setting

$$\mathrm{Pl}_m(A) \stackrel{\mathrm{d}}{=} \sum_{B:A\cap B\neq\emptyset} m(B) = 1 - \mathrm{Bel}_m(\overline{A}). \tag{6.3}$$

We have

$$\mathrm{Bel}_m(A) \leq \mathrm{Pl}_m(A),$$
$$\mathrm{Bel}_m(A) + \mathrm{Bel}_m(\overline{A}) \leq 1,$$

for all $A \subseteq \Omega$. The next theorem shows how to construct a random set inducing a belief function Bel_m from the corresponding mass distribution m.

(6.3) Theorem. *Let Bel_m be a belief function on Ω where m is the corresponding mass distribution. Consider the sample space 2^Ω, the probability*

$$P : 2^{(2^\Omega)} \to [0,1], \qquad \begin{cases} P(T) = \sum_{T:T\in\mathcal{T}} m(T), & \textit{if } \mathcal{T} \neq \emptyset, \\ P(\emptyset) = 0 & \textit{otherwise,} \end{cases}$$

and the random set $\Gamma : 2^\Omega \to 2^\Omega$, $\Gamma(T) = T$. Then we have $m = \Gamma[P]$ and consequently $(P_\Gamma)_* = \mathrm{Bel}_m$, $(P_\Gamma)^* = \mathrm{Pl}_m$.

Proof. Q is a probability on 2^Ω. From Definition 5.11 we get

$$(P_\Gamma)_*(A) = \frac{Q(\Gamma_*(A))}{Q(\Gamma_*(\Omega))} = \frac{Q(\{T\subseteq\Omega \mid \emptyset\neq\Gamma(T)\subseteq A\})}{Q(\{T\subseteq\Omega \mid \emptyset\neq\Gamma(T)\})}$$
$$= \frac{\sum_{T:\emptyset\neq T\subseteq A} m(T)}{\sum_{T:\emptyset\neq T\subseteq\Omega} m(T)} = \frac{\mathrm{Bel}_m(A)}{\mathrm{Bel}_m(\Omega)} = \mathrm{Bel}_m(A)$$

for all $A \subseteq \Omega$.[2] □

[2] Now we are able to give a simple proof for Theorem 4.17. If $(P_\mathcal{L})_*$ is monotone of infinite order, then from $(P_\mathcal{L})_*(\emptyset) = 0$ and $(P_\mathcal{L})_*(\Omega) = 1$ it follows that $(P_\mathcal{L})_*$ is a belief function. Hence there is a random set Γ such that $(P_\Gamma)_* = (P_\mathcal{L})_*$ holds, and therefore $\mathcal{L}$ is of type 2.

(6.4) Example. Let us consider a situation where we want to evaluate our knowledge about the position of some ship with respect to the set $\Omega = \{z1, z2, z3\}$ containing its possible locations. Assume that the available evidence allows an allocation of amounts of belief to elements or subsets of Ω. Some portion of evidence might support zone1 and zone2 beyond zone3 without being divisible among the two zones. Suppose that there is further indivisible evidence concerning zone1 and zone3. Finally there may be little evidence that the ship's position is zone1 whereas it is possible to distribute the remaining evidence. The exact mass distribution and the corresponding belief and plausibility functions are listed below:

A	m	Bel_m	Pl_m
$\emptyset$	0	0	0
$\{z1\}$	0.2	0.2	1
$\{z2\}$	0	0	0.5
$\{z3\}$	0	0	0.5
$\{z1, z2\}$	0.3	0.5	1
$\{z1, z3\}$	0.3	0.5	1
$\{z2, z3\}$	0	0	0.8
Ω	0.2	1	1

◇

An additional possibility to characterize the information borne by a mass distribution is the concept of commonality functions. Let m be a mass distribution defined on Ω. The function

$$Q_m : 2^\Omega \to [0,1]; \quad Q_m(A) \stackrel{\text{d}}{=} \sum_{B: A \subseteq B} m(B) \tag{6.4}$$

measures the evidence mass that can freely move to every element or subset of A. This function is called the *commonality function.* Obviously $Q_m(A) = 0$ indicates that there is no evidence mass "above" A, i.e. attached to supersets of A.

By setting $m(\Omega) = 1$ and $m(A) = 0$ for all $A \neq \Omega$ we obtain $\text{Bel}_m(\Omega) = 1$ and $\text{Bel}_m(A) = 0$ for all $A \neq \Omega$. This belief function represents *total ignorance* about the domain under consideration, and is therefore called the *vacuous* belief function. On the other hand, setting $m(\{\omega_i\}) = p_i$, $i = 1, \ldots, n$ and $m(A) = 0$ for all non-elementary sets A leads to a *Bayesian belief function* or, in terms of probability theory, a discrete probability distribution.

6.1.2 Belief Functions

Up to now we have been concerned with belief functions induced by a given mass distribution. In the sequel we will have to ask whether a given set function is a belief function and, if so, by which mass distribution it is induced.

(6.5) Definition. *A set function* $\mathrm{Bel} : 2^{\Omega} \to [0,1]$ *is called a belief function if and only if there is a mass distribution* m *such that* $\mathrm{Bel} = \mathrm{Bel}_m$.

Obviously not each set function is a belief function. The following theorem gives the necessary and sufficient conditions a set function has to obey to be a belief function.

(6.6) Theorem. $\mathrm{Bel} : 2^{\Omega} \to [0,1]$ *is a belief function if and only if the following three conditions hold:*

(i) $\mathrm{Bel}(\emptyset) = 0$,

(ii) $\mathrm{Bel}(\Omega) = 1$,

(iii) $\mathrm{Bel}(\bigcup_{i=1}^{n} A_i) \geq \sum_{I:\emptyset \neq I \subseteq \{1,\ldots,n\}} (-1)^{|I|+1} \cdot \mathrm{Bel}(\bigcap_{i \in I} A_i)$
for all $n \in \mathbb{N}$ *and all* $A_1, \ldots, A_n \subseteq \Omega$.

Proof.
"$\Rightarrow$"Let $m : 2^{\Omega} \to [0,1]$ be a mass distribution such that $\mathrm{Bel}_m = \mathrm{Bel}$. From $m(\emptyset) = 0$ follows $\mathrm{Bel}(\emptyset) = 0$, and $\sum_{A:A \subseteq \Omega} m(A) = 1$ implies $\mathrm{Bel}(\Omega) = 1$.
So what remains is to check the third condition. Define $I(B) = \{i \in \{1,\ldots,n\} \mid B \subseteq A_i\}$ for all $B \subseteq \Omega$. Then for each $n \in \mathbb{N}$, and all $A_1, \ldots, A_n \subseteq \Omega$ it follows that

$$
\begin{aligned}
& \sum_{I:\emptyset \neq I \subseteq \{1,\ldots,n\}} (-1)^{|I|+1} \cdot \mathrm{Bel}(\bigcap_{i \in I} A_i) \\
= & \sum_{I:\emptyset \neq I \subseteq \{1,\ldots,n\}} ((-1)^{|I|+1} \cdot \sum_{B:B \subseteq \bigcap_{i \in I} A_i} m(B)) \\
= & \sum_{I:\emptyset \neq I \subseteq \{1,\ldots,n\}} ((-1)^{|I|+1} \cdot \sum_{B:I \subseteq I(B)} m(B)) \\
= & \sum_{B \subseteq \Omega : I(B) \neq \emptyset} (m(B) \cdot \sum_{I:\emptyset \neq I \subseteq I(B)} (-1)^{|I|+1}) \\
= & \sum_{B \subseteq \Omega : I(B) \neq \emptyset} \Big(m(B) \cdot (1 - \sum_{I:I \subseteq I(B)} (-1)^{|I|})\Big)
\end{aligned}
$$

$$\begin{aligned}
&\overset{(*)}{=} \sum_{B\subseteq\Omega:I(B)\neq\emptyset} m(B)\\
&= \sum_{B\subseteq\Omega:\exists i:B\subseteq A_i} m(B)\\
&\le \sum_{B:B\subseteq\bigcup_{i=1}^{n} A_i} m(B)\\
&= \mathrm{Bel}\Big(\bigcup_{i=1}^{n} A_i\Big)
\end{aligned}$$

For the verification of the equality $(*)$ we note that, if $G = \{g_1, \ldots, g_t\} \neq \emptyset$, we have

$$\sum_{H:H\subseteq G} (-1)^{|H|} = \binom{t}{0} - \binom{t}{1} + \binom{t}{2} + \ldots + (-1)^t\binom{t}{t} = (1-1)^t = 0.$$

"$\Leftarrow$"Let $\mathrm{Bel} : 2^{\Omega} \to [0,1]$ be a function that fulfils the conditions (i), (ii) and (iii). Define

$$m : 2^{\Omega} \to [0,1], \quad m(A) = \sum_{B:B\subseteq A} (-1)^{|A\cap\overline{B}|}\mathrm{Bel}(B).$$

For $A \subseteq \Omega$ we obtain

$$\begin{aligned}
\sum_{B:B\subseteq A} m(B) &= \sum_{B:B\subseteq A}\sum_{C:C\subseteq B} (-1)^{|B\cap\overline{C}|}\mathrm{Bel}(C)\\
&= \sum_{B:B\subseteq A} \Big((-1)^{|B|}\cdot \sum_{C:C\subseteq B} (-1)^{|C|}\mathrm{Bel}(C)\Big)\\
&= \sum_{C:C\subseteq A} \Big((-1)^{|C|}\mathrm{Bel}(C)\cdot \sum_{B:C\subseteq B\subseteq A} (-1)^{|B|}\Big)\\
&= (-1)^{|A|}\sum_{C:C\subseteq A} \Big((-1)^{|A\cap\overline{C}|}\mathrm{Bel}(C)\cdot \sum_{B:C\subseteq B\subseteq A} (-1)^{|B|}\Big).
\end{aligned}$$

By using the result in footnote 3 again we obtain

$$\begin{aligned}
\sum_{B:C\subseteq B\subseteq A} (-1)^{|B|} &= \sum_{B:B\subseteq C\cap\overline{A}} (-1)^{|A\cup B|}\\
&= (-1)^{|A|}\cdot \sum_{B:B\subseteq C\cap\overline{A}} (-1)^{|B|}\\
&= \begin{cases} 0, & \text{if } A\neq C\\ (-1)^{|A|}, & \text{if } A = C. \end{cases}
\end{aligned}$$

From $\mathrm{Bel}(\emptyset) = 0$ follows

$$\sum_{B:B\subseteq A} m(B) = (-1)^{|A|} \cdot \text{Bel}(A) \cdot (-1)^{|A|} = \text{Bel}(A)$$

for all $A \subseteq \Omega$.
From (i) and (ii) we get immediately $m(\emptyset) = 0$ and $\sum_{A:A\subseteq\Omega} m(A) = 1$. In order to prove that $m(A) \geq 0$ for all $A \subseteq \Omega$, we define $A_i = A \cap \overline{\{\omega_i\}}$ for $A = \{\omega_1, \ldots, \omega_n\}$ for all $i \in \{1, \ldots, n\}$. Then for each $B = \{\omega_{i_1}, \ldots, \omega_{i_k}\} \subseteq A$, $A \neq B$, the equality

$$\begin{aligned} B &= A \cap \bigcap_{j\in\{1,\ldots,n\}\cap\overline{J}} \overline{\{\omega_j\}} \\ &= \bigcap_{j\in\{1,\ldots,n\}\cap\overline{J}} A_j \end{aligned}$$

holds, where $J = \{i_1, \ldots i_k\}$. Therefore

$$\begin{aligned} m(A) &= \sum_{B:B\subseteq A} (-1)^{|A\cap\overline{B}|}\text{Bel}(B) \\ &= \text{Bel}(A) + \sum_{I:\emptyset\neq I\subseteq\{1,\ldots,n\}} (-1)^{|I|}\text{Bel}\Big(\bigcap_{i\in I} A_i\Big) \\ &= \text{Bel}(A) - \sum_{I:\emptyset\neq I\subseteq\{1,\ldots,n\}} (-1)^{|I+1|}\text{Bel}\Big(\bigcap_{i\in I} A_i\Big) \\ &\geq 0 \end{aligned}$$

which is a consequence of condition (iii).

□

(6.7) *Remark.* From this theorem it follows that if $\text{Bel} : 2^\Omega \to [0,1]$ is a belief function, then

$$m : 2^\Omega \to [0,1], \qquad m(A) = \sum_{B:B\subseteq A} (-1)^{|A\cap\overline{B}|}\text{Bel}(B) \tag{6.5}$$

is the *unique* mass distribution m such that $\text{Bel} = \text{Bel}_m$. Therefore we can say that a mass distribution m corresponds to a belief function Bel. ∘

The next topic we have to consider is the impact of some absolutely reliable information saying $\omega_0 \in E$, $E \subseteq \Omega$. One possibility is to focus on those sensors which provide observations that are compatible with E, i.e. to switch from (P, Γ) to $\big(P(\,.\mid \Gamma_*(E)), \Gamma\big)$. The corresponding mass distribution is given by

$$\begin{aligned}\Gamma[P](A|E) &\stackrel{\mathrm{d}}{=} \Gamma\Big[P(\,.\mid \Gamma_*(E))\Big](A)\\ &= \frac{P(\{\theta\in\Theta \mid \Gamma(\theta)\subseteq A\cap E, \Gamma(\theta)\neq\emptyset\})}{P(\{\theta\in\Theta \mid \Gamma(\theta)\subseteq E, \Gamma(\theta)\neq\emptyset\})}\\ &= \frac{\Gamma[P](A\cap E)}{\mathrm{Bel}_{\Gamma[P]}(E)} \text{ if } \emptyset\neq A\subseteq E\end{aligned}$$

and $\Gamma[P](A|E) = 0$ otherwise.

The above quantities exist, if and only if $\mathrm{Bel}_{\Gamma[P]}(E) \neq 0$. These results from the theory of random sets justify defining *conditional mass distributions* in the same way: If $\mathrm{Bel}_m(E) \neq 0$, then we define for all $A \subseteq E$

$$m(A|E) \stackrel{\mathrm{d}}{=} \frac{m(A)}{\mathrm{Bel}_m(E)}, \tag{6.6}$$

and $m(A|E) \stackrel{\mathrm{d}}{=} 0$, otherwise.

Thus we obtain

$$\mathrm{Bel}_m(A|E) = \frac{\mathrm{Bel}_m(A\cap E)}{\mathrm{Bel}_m(E)} \quad \text{and} \quad \mathrm{Pl}_m(A|E) = \frac{\mathrm{Pl}_m(A\cup \overline{E}) - \mathrm{Pl}_m(\overline{E})}{1-\mathrm{Pl}_m(\overline{E})}. \tag{6.7}$$

Note, that $\mathrm{Bel}_m(.|E)$ and $\mathrm{Pl}_m(.|E)$ are shorthand notations for $\mathrm{Bel}_{m(.|E)}$ and $\mathrm{Pl}_{m(.|E)}$, respectively.

Within the conditioning procedure the sensor outputs are considered to be fixed monolithic pieces of information; the decision maker is only free to accept or reject them as they are.

Belief revision[3] means, on the contrary, that the sensor information may be subject to some meta-expert's intervention. Revision corresponds to a change of the observation mapping, i.e. in view of E we switch from (P,Γ) to (P,Γ^E). We obtain

$$\begin{aligned}\Gamma^E[P](A) &= \begin{cases} \frac{P(\{\theta\in\Theta|\Gamma^E(\theta)=A\})}{P(\{\theta\in\Theta|\Gamma^E(\theta)\neq\emptyset\})} & \text{if } A\neq\emptyset\\ 0 & \text{otherwise}\end{cases}\\ &= \begin{cases} \frac{P(\{\theta\in\Theta|\Gamma(\theta)\cap E=A\})}{P(\{\theta\in\Theta|\Gamma(\theta)\cap E\neq\emptyset\})} & \text{if } A\neq\emptyset\\ 0 & \text{otherwise}\end{cases}\\ &= \begin{cases} \dfrac{\sum\limits_{B:B\cap E=A} \Gamma[P](B)}{\sum\limits_{B:B\cap E\neq\emptyset} \Gamma[P](B)} & \text{if } A\neq\emptyset\\ 0 & \text{otherwise}\end{cases}\end{aligned}$$

3 The concept of belief revision directly corresponds to "Dempster's rule of conditioning" [Shafer 1976].

This justifies defining the revised mass distribution $m_E : 2^{\Omega} \to [0,1]$ by setting

$$m_E(A) \stackrel{\mathrm{d}}{=} \begin{cases} \dfrac{\sum\limits_{C:E\cap C=A} m(C)}{1-\sum\limits_{C:C\cap E=\emptyset} m(C)}, & \text{if } A \neq \emptyset, \\ 0 & \text{otherwise.} \end{cases} \tag{6.8}$$

It is defined, if

$$\mathrm{Bel}_m(\overline{E}) = \sum_{C:C\subseteq\overline{E}} m(C) = \sum_{C:E\cap C=\emptyset} m(C) < 1.$$

The revision of m yields the belief function

$$\mathrm{Bel}_{m_E}(A) = \frac{\mathrm{Bel}_m(A \cup \overline{E}) - \mathrm{Bel}_m(\overline{E})}{1 - \mathrm{Bel}_m(\overline{E})}, \quad A \subseteq \Omega \tag{6.9}$$

and the plausibility function

$$\mathrm{Pl}_{m_E}(A) = \frac{\mathrm{Pl}_m(A \cap E)}{\mathrm{Pl}_m(E)}, \quad A \subseteq \Omega. \tag{6.10}$$

Revision should be done only under the responsibility of some real meta-expert. However, belief revision fits in a natural way with our intuitive idea of "belief" as partially movable evidence mass, where $m(A)$ is that amount of mass which can, in the light of new information, move to every subset of A but not to sets with elements outside of A. The main impact of supplementary knowledge saying "the truth is in the set $E \subseteq \Omega$ with certainty" is that all elements of $\overline{E}$ become impossible. All masses attached to subsets A of Ω by a mass distribution m after the revision with respect to the set $E \subseteq \Omega$ are allocated to the sets $A \cap E$. That means in the case of $A \subseteq E$ the masses remain where they are, and in the case of $A \cap E \neq \emptyset$ they move to $A \cap E$, whereas if $A \cap E = \emptyset$ they have to be neglected. In total, a set $C \neq \emptyset$ obtains the mass $\sum\limits_{A:A\cap E=C} m(A)$. The remaining mass $\sum\limits_{A:A\cap E=\emptyset} m(A)$ arises from partial contradictions of m and the knowledge of E. Since the masses should sum up to one, $1 - \sum\limits_{A:A\cap E\neq\emptyset} m(A)$ is distributed proportionally among those sets emerging as focal elements from the revision process.

(6.4) Example. (continued) Let us return to the previous example. For $E = \{\mathrm{z2}, \mathrm{z3}\}$ we derive

$$\sum_{C:C\cap E=\emptyset} m(C) = m(\emptyset) + m(\{\mathrm{z1}\}) = 0.2.$$

Since

$$\sum_{C:C\cap E=\{z2\}} m(C) = m(\{z2\}) + m(\{z1, z2\}) = 0.3,$$

we obtain

$$m_E(\{z2\}) = \frac{0.3}{1-0.2} = \frac{0.3}{0.8} = 0.375.$$

The complete revised mass distribution and the induced revised belief and plausibility functions are listed below:

A	$m_E(A)$	$\mathrm{Bel}_{m_E}(A)$	$\mathrm{Pl}_{m_E}(A)$
$\emptyset$	0	0	0
{z1}	0	0	0
{z2}	0.375	0.375	0.625
{z3}	0.375	0.375	0.625
{z1, z2}	0	0.375	0.625
{z1, z3}	0	0.375	0.625
{z2, z3}	0.25	1	1
Ω	0	1	1

◇

(6.8) *Remark.* From $Q_m(A) > 0$ it follows that there is a set $B \supseteq A$ such that $m(B) > 0$ holds. Supposing that $\mathrm{Bel}_m(\overline{E}) \neq 1$ holds, the revised mass distribution m_E exists, and we have $m_E(B \cap E) > 0$. Therefore we derive

$$Q_m(A) > 0 \Rightarrow Q_{m_E}(A \cap E) > 0.$$

○

(4.9) Example. (revisited) To compare the concepts of conditioning and revision, let us once more consider the simplified version of the Tversky-Shafer translator we introduced in Sect. 4.2. Modeling the situation by a mass distribution m on Ω we obtain

$$m\Big(\{(\theta_1, \mathrm{ps}, \mathrm{ps}), (\theta_1, \mathrm{tv}, \mathrm{tv})\}\Big) = \frac{1}{4}$$

$$m\Big(\{(\theta_2, \mathrm{ps}, \mathrm{tv}), (\theta_2, \mathrm{tv}, \mathrm{ps})\}\Big) = \frac{1}{4}$$

$$m\Big(\{(\theta_3, \mathrm{ps}, \mathrm{ps}), (\theta_3, \mathrm{tv}, \mathrm{ps})\}\Big) = \frac{1}{4}$$

$$m\Big(\{(\theta_4, \mathrm{ps}, \mathrm{tv}), (\theta_4, \mathrm{tv}, \mathrm{tv})\}\Big) = \frac{1}{4}.$$

Conditioning yields as in the case of random sets

$$\mathrm{Bel}_m(\text{ship type is ps} \mid \text{sensor message is ps}) = 0$$

$$\mathrm{Pl}_m(\text{ship type is ps} \mid \text{sensor message is ps}) = 1$$

whereas the revision leads to completely different results. We have

$$\mathrm{Bel}_{m_{\text{agent's message is ps}}}(\text{ship type is ps}) = \frac{1}{3}$$

$$\mathrm{Pl}_{m_{\text{agent's message is ps}}}(\text{ship type is ps}) = \frac{2}{3}.$$

This result is not surprising, because the semantics of data revision and conditioning are different. ◇

In the above considerations we always implicitly used some kind of *closed world assumption*. We supposed the frame of discernment Ω (or its refinements) to cover the actual problems completely, i.e. to be sufficient for the formulation of all relevant information. In the absence of such a closed world assumption we have an *open world* admitting the existence of additional unknown states of the world outside of Ω and the possibility that the true one is among them. The claim to allow a description of all phenomena therefore requires us to extend the frame of discernment Ω. A minimal extension is $\widetilde{\Omega} = \Omega \cup \{\alpha\}$, where α may represent the total set of unknown elements. By suitable refinements α can be split up to describe the situation more precisely, but if we decide to use $\widetilde{\Omega}$ any information is encoded in terms of a mass distribution $\widetilde{m} : 2^{\Omega \cup \{\alpha\}} \to [0,1]$. The analysis is facilitated if we assume that each focal set contains the unknown element α, i.e.

$$\forall \widetilde{A} \subseteq \widetilde{\Omega} : \left(\alpha \notin \widetilde{A} \Rightarrow \widetilde{m}(\widetilde{A}) = 0\right). \tag{6.11}$$

In so doing we can restrict ourselves to considering an *extended mass distribution* $n : 2^{\Omega} \to [0,1]$ defined by

$$n(A) = \widetilde{m}(A \cup \{\alpha\}) \text{ for all } A \subseteq \Omega. \tag{6.12}$$

As a consequence of this definition $n(\emptyset) = 0$ does not necessarily hold. Such a generalized mass distribution induces an *extended belief function*

$$\begin{aligned}\mathrm{Bel}_n(A) &\stackrel{\mathrm{d}}{=} \mathrm{Bel}_{\widetilde{m}}(A \cup \{\alpha\}) = \sum_{\widetilde{B}:\widetilde{B} \subseteq A \cup \{\alpha\}} \widetilde{m}(\widetilde{B}) \\ &= \sum_{B:B \cup \{\alpha\} \subseteq A \cup \{\alpha\}} \widetilde{m}(B \cup \{\alpha\}) = \sum_{B:B \subseteq A} n(A) \end{aligned} \tag{6.13}$$

as well as an *extended plausibility function*

$$\begin{aligned}\mathrm{Pl}_n(A) \stackrel{\mathrm{d}}{=} \mathrm{Pl}_{\widetilde{m}}(A) &= \sum_{\widetilde{B}: A\cap\widetilde{B}\neq\emptyset} \widetilde{m}(\widetilde{B}) \\ &= \sum_{B: A\cap B\neq\emptyset} \widetilde{m}(B\cup\{\alpha\}) = \sum_{B: A\cap B\neq\emptyset} n(B). \end{aligned} \tag{6.14}$$

We have $\mathrm{Pl}_n(A) = 1 - \mathrm{Bel}_n(\overline{A})$, and Bel_n also satisfies the properties (ii) and (iii) of Theorem 6.6, whereas $\mathrm{Bel}_n(\emptyset) = n(\emptyset)$ may be non-zero. If the assumption (6.11) is not satisfied, i.e., if we want to distinguish between the masses $\widetilde{m}(A)$ and $\widetilde{m}(A\cup\{\alpha\})$, then the extended mass distributions are not sufficient, and we have to consider the original space $\widetilde{\Omega}$.

6.2 Different Frames of Discernment

Let us firstly consider frames of discernments Ω and Ω' which are in a refinement relationship, where $\widehat{\Pi} : 2^{\Omega} \to 2^{\Omega'}$ is their refinement mapping. Such a situation raises the question, how and to what degree the knowledge coded by a mass distribution m' on Ω' can be expressed in terms of a mass distribution m on Ω. According to equation (5.9) we obtain

$$\Pi[m'] : 2^{\Omega} \to [0,1]; \quad \Pi[m'](A) = \sum_{\substack{A'\subseteq\Omega' :\\ \Pi(A')=A}} m'(A'), A \subseteq \Omega, \tag{6.15}$$

called the *projection* of m' to Ω. Whenever we have two mass distributions m and m' on Ω and Ω' respectively, where Ω' is a refinement of Ω with $\widehat{\Pi}$ as their refinement mapping, then in the case that $m = \Pi[m']$ holds we call m' a refinement of m.

Next we have to investigate the transfer of information in the reverse direction. Required is the formulation of a mass distribution m on Ω in terms of the refined frame of discernment Ω'. The definition is obvious (compare equation (5.8)); a suitable mass distribution is

$$\widehat{\Pi}[m] : 2^{\Omega'} \to [0,1]; \quad \widehat{\Pi}[m](A') = \begin{cases} m(A), & \text{if } A' = \widehat{\Pi}(A), \\ 0 & \text{otherwise} \end{cases} \tag{6.16}$$

denoted as the *vacuous extension* of m. From the definition it is clear that each vacuous extension is a refinement of the original mass distribution.

6.2.1 Specializations

The next aim of this chapter is to define the concept of specialization as we did in the case of the random sets.

(6.9) Definition. *Let s, t be mass distributions defined on Ω. s is a specialization of t (written $s \sqsubseteq t$) if and only if there is a probability space $(\Theta, 2^\Theta, P)$ and two observation mappings $S, T : \Theta \to 2^\Omega$ such that $s = S[P]$ and $t = T[P]$, where S is a specialization of T.*

According to Theorem 5.18 we obtain the following characterization.

(6.10) Theorem. *Let s, t be mass distributions defined on Ω. The following two statements are equivalent.*

(i) There are two mass distributions s' and t' defined on a refinement Ω' of Ω where $s = \Pi[s']$, $t = \Pi[t']$ and an event $E' \subseteq \Omega'$ such that $s'(B') = t'_{E'}(B')$ holds for each $B' \subseteq \Omega$.

(ii) $s \sqsubseteq t$.

Proof.
"$\Rightarrow$"Consider the space $\Theta = 2^\Omega \times 2^\Omega$ and the mappings S, T defined by

$$T(A,B) = B, \qquad S(A,B) = \begin{cases} A & \text{if } A \subseteq B \\ \emptyset & \text{otherwise.} \end{cases}$$

The probability P on Θ is determined as follows: for $h_B(\ . \ )$ given in Theorem 5.17,

$$P\Big(\{(A,B)\}\Big) = P(\{A\} \times 2^\Omega \mid 2^\Omega \times \{B\}) \cdot Q(2^\Omega \mid 2^\Omega \times \{B\})$$

where

$$Q(2^\Omega \times \{B\}) = t(B)$$

and

$$Q(\{A\} \times 2^\Omega \mid 2^\Omega \times \{B\}) = \begin{cases} h_B(A) & \text{if } t(B) \neq \emptyset \text{ and } A \subseteq B \\ 0 & \text{otherwise,} \end{cases}$$

for all $A, B \subseteq \Omega$. We have $S \sqsubseteq T$ by definition and

$$\begin{aligned} m_{(P_T)_*}(B) &= \frac{Q(\{(A,B) \in \Theta \mid T(A,B) = B\})}{1 - Q(\{(A,B) \in \Theta \mid T(A,B) = \emptyset\})} \\ &= \frac{Q(2^\Omega \times \{B\})}{1 - Q(2^\Omega \times \{\emptyset\})} = t(B) \end{aligned}$$

for all $B \subseteq \Omega$. Finally we obtain

$$m_{(P_S)_*}(B) = \frac{Q(\{(A,B) \in \Theta \mid S(A,B) = B\})}{1 - Q(\{(A,B) \in \Theta \mid S(A,B) = \emptyset\})}$$

$$= \frac{\sum_{B:B\subseteq\Omega} h_B(A)}{1 - \sum_{B:B\subseteq\Omega} h_B(\emptyset)} = s(A)$$

for all $A \neq \emptyset$.

"$\Leftarrow$"To prove the reverse direction we start with

$$m_{(P_S)_*}(B) = \frac{Q(\{\theta \in \Theta \mid S(\theta) = A\})}{1 - Q(\{\theta \in \Theta \mid S(\theta) = \emptyset\})},$$

$$m_{(P_T)_*}(B) = \frac{Q(\{\theta \in \Theta \mid T(\theta) = A\})}{1 - Q(\{\theta \in \Theta \mid T(\theta) = \emptyset\})},$$

for all $A \neq \emptyset$. Suppose the commonality function corresponding to $m_{(P_T)_*}$ has the value 0 for some set $B \subseteq \Omega$. Then $Q\big(\{\theta \in \Theta \mid T(\theta) \supseteq B\}\big) = 0$ by definition, and from $S(\theta) \subseteq T(\theta)$ for all $\theta \in \Theta$ it follows that $\{\theta \in \Theta \mid T(\theta) \supseteq B\} \supseteq \{\theta \in \Theta \mid S(\theta) \supseteq B\}$, i.e. $Q\big(\{\theta \in \Theta \mid S(\theta) \supseteq B\}\big) = 0$. Therefore the commonality function corresponding to $m_{(P_S)_*}$ has the value 0 for $B \subseteq \Omega$. From Theorem 5.12 we thus obtain $(P_S)_* \sqsubseteq (P_T)_*$, i.e. $s \sqsubseteq t$. □

The intuitive idea of a specialization is the projection of a revision. So it is obvious that the specialization relation is invariant with respect to further projections. The next theorem verifies this.

(6.11) Theorem. *Let Ω' be a refinement of the frame of discernment Ω, where $\widehat{\Pi} : 2^{\Omega} \to 2^{\Omega'}$ is the refinement mapping. For two mass distributions $s', t' : 2^{\Omega'} \to [0,1]$ we have*

$$s' \sqsubseteq t' \Rightarrow \Pi[s'] \sqsubseteq \Pi[t'].$$

Proof. From $s' \sqsubseteq t'$ it follows that there are two mass distributions s'' and t'' on Ω'', which is a refinement of Ω', and that there is a set $E'' \subseteq \Omega''$ such that $s''(B'') = t''_{E''}(B'')$ for all $B'' \subseteq \Omega''$, where s' and t' are the projections of s'' and t'' respectively. In this case Ω'' is also a refinement of Ω, so the projection of s'' and t'' onto Ω yields also $\Pi[s']$ and $\Pi[t']$ respectively. So we have $\Pi[s'] \sqsubseteq \Pi[t']$. □

In addition to the definition, the following theorem gives two equivalent characterizations for the specialization relationship. The first one provides a possibility to check easily whether $s \sqsubseteq t$ is valid or not. The second one reflects our intuitive idea of floating evidence masses, and describes the flow of the masses $t(A)$ onto the subsets of A.

(6.12) Theorem. *Let s and t be two mass distributions on Ω. The following three statements are equivalent*

(i) *s is a specialization of t,*

(ii) $\forall A \subseteq \Omega : \left(Q_t(A) = 0 \Rightarrow Q_s(A) = 0\right)$,

(iii) *For every $A \subseteq \Omega$ there are functions $h_A : 2^\Omega \to [0,1]$ such that*

a) $\sum_{B:B\subseteq\Omega} h_A(B) = t(A)$, *for all* $A \subseteq \Omega$,

b) $h_A(B) \neq 0 \Rightarrow B \subseteq A$, *for all* $A, B \subseteq \Omega$, *and*

c) $s(B) = \dfrac{\sum_{A:A\subseteq\Omega} h_A(B)}{1 - \sum_{A:A\subseteq\Omega} h_A(\emptyset)}$ *for all* $\emptyset \neq B \subseteq \Omega$.

($h_A(B)$ can be viewed as the amount of mass that moves from A to B)

Proof. (i) $\Rightarrow$ (ii): From $Q_t(A) = \sum_{C:C\supseteq A} t(C) = 0$ follows $t(C) = 0$ for all $C \supseteq A$. We have $\sum_{B'\subseteq\Omega':\Pi(B')=C} t'(B') = 0$ for all $C \supseteq A$, which yields

$$\Pi(B') \supseteq A \Rightarrow t'(B') = 0 \text{ for all } B' \subseteq \Omega'.$$

It follows that

$$\begin{aligned}
Q_s(A) &= \sum_{C:C\supseteq A} s(C) \\
&= \sum_{C:C\supseteq A} \sum_{\substack{B'\subseteq\Omega':\\ \Pi(B')=C}} s'(B') \\
&= \sum_{\substack{B'\subseteq\Omega':\\ \Pi(B')\supseteq A}} t'_{E'}(B') \\
&= \sum_{\substack{B'\subseteq\Omega':\\ \Pi(B')\supseteq A}} \frac{\sum_{D':D'\cap E'=B'} t'(D')}{1 - \sum_{D':D'\cap E'=\emptyset} t'(D')} \\
&= \frac{1}{c} \sum_{\substack{B'\subseteq\Omega':\\ \Pi(B')\supseteq A}} \sum_{D':D'\cap E'=B'} t'(D') \\
&= \frac{1}{c} \sum_{D':\Pi(D'\cap E')\supseteq A} t'(D') \\
&= 0
\end{aligned}$$

since if for D' we have $\Pi(D' \cap E') \supseteq A$ then this implies also $\Pi(D') \supseteq A$ and we derive $t'(D') = 0$.

(ii) $\Rightarrow$ (iii): Let m be a mass distribution on $\Omega = \{\omega_1, \ldots, \omega_n\}$. $\mathcal{A}_m \stackrel{\text{d}}{=} \{A \subseteq \Omega \mid Q_m(A) > 0 \text{ and } \forall B \neq A : B \supseteq A \Rightarrow Q_m(B) = 0\}$ contains the maximal focal sets with respect to m. As its most important properties, we have that $\mathcal{A}_m$ is not empty, that $\text{card}(A)_m \leq 2^n$ holds and that for each $B \subseteq \Omega$ with $Q_m(B) > 0$ there is at least one $A \in \mathcal{A}_m$ with $A \supseteq B$. Therefore we can define a mapping $K_m : 2^\Omega \to \mathcal{A}_m \cup \{\emptyset\}$ assigning to each $B \subseteq \Omega$ with $Q_m(B) > 0$ a set $K_m(B) \in \mathcal{A}$ where $K_m(B)$ is fixed and $K_m(B) \supseteq B$ holds. For sets B with $Q_m(B) = 0$ we define $K_m(B) = \emptyset$.

Let $C \in \mathcal{A}_s$. Then $Q_s(C) > 0$, and we also have $Q_t(C) > 0$. The set $K_t(C)$ is within $\mathcal{A}_t$ and we get $K_t(C) \supseteq C$. Define the mapping

$$h_A(B) = \begin{cases} s(B) \cdot \alpha_t, & \text{if } K_t\big(K_s(B)\big) = A,\ B \neq \emptyset, \\ t(A) - \sum\limits_{\substack{\emptyset \neq C \subseteq \Omega: \\ K_t(K_s(C)) = A}} s(C) \cdot \alpha_t, & \text{if } A \in K_t(\mathcal{A}_s),\ B = \emptyset, \\ t(A), & \text{if } A \notin K_t(\mathcal{A}_s),\ B = \emptyset, \\ 0 & \text{otherwise,} \end{cases}$$

where $\alpha_t = \min\{\frac{t(A)}{2^n} \mid t(A) > 0\}$. From the construction it is clear that $0 \leq h_A(B) \leq 1$ holds for all $A, B \subseteq \Omega$, and that (a), (b) are satisfied. Let $h_A(B) \neq 0$ and $B \neq \emptyset$. Then we have $A = K_t\big(K_s(B)\big) \supseteq K_s(B) \supseteq B$. Therefore for all A, B, it holds that $h_A(B) \neq 0 \Rightarrow B \supseteq A$, and (b) follows.

So it remains to show property (c): If $B \neq \emptyset$, then we have

$$\frac{\sum\limits_{A: A \subseteq \Omega} h_A(B)}{1 - \sum\limits_{A: A \subseteq \Omega} h_A(\emptyset)}$$

$$= \frac{\sum\limits_{A: A = K_t(B)} s(B) \cdot \alpha_t}{1 - \Big(\sum\limits_{A: A \notin K_t(\mathcal{A}_s)} t(A) + \sum\limits_{A: A \in K_t(\mathcal{A}_s)} t(A) - \sum\limits_{A: A \in K_t(\mathcal{A}_s)} \sum\limits_{\substack{\emptyset \neq C \subseteq \Omega: \\ K_t(K_s(C)) = A}} s(C) \cdot \alpha_t \Big)}$$

$$= \frac{s(B) \cdot \alpha_t}{\sum\limits_{C: C \neq \emptyset} s(C) \cdot \alpha_t}$$

$$= s(B).$$

(iii) $\Rightarrow$ (i): Let $\Omega = \{\omega_1, \ldots, \omega_n\}$, $2^\Omega = \{A_1, A_2, \ldots, A_i, \ldots, A_{2^n}\}$, and $\Omega' = \{\omega_1^1, \ldots, \omega_1^{2^n}, \omega_2^1, \ldots, \omega_2^{2^n}, \ldots, \omega_n^1, \ldots, \omega_n^{2^n}\}$, where the refinement mapping is defined by $\widehat{\Pi}\big(\{\omega_j\}\big) = \{\omega_j^1, \ldots \omega_j^{2^n}\}$ for $j = 1, \ldots, n$. Moreover let $A_i^k = \{\omega_j^k \mid \omega_j \in A_i\}$ for $i = 1, \ldots, 2^n$, $k = 1, \ldots, 2^n$. We obtain

$$\begin{aligned}\Pi(A_i^k) &= \{\omega_j \mid \widehat{\Pi}\{\omega_j\} \cap A_i^k \neq \emptyset\} \\ &= \{\omega_j \mid \omega_j \in A_i\} \\ &= A_i\end{aligned}$$

for $k = 1, \dots, 2^n$. With $E' = \bigcup\{A_j^j \mid \exists i : h_{A_i}(A_j) > 0\}$ from $h_{A_r}(A_s) > 0$ follows $E' \cap A_r^s = A_s^s$ for all $r, s \in \{1, \dots, 2^n\}$. We define $t'(A_i^k) = h_{A_i}(A_k)$ for $i, k \in \{1, \dots, 2^n\}$, and $t'(A') = 0$ otherwise. t' is a mass distribution; we have

$$\begin{aligned}\sum_{A':\Pi(A')=A_j} t'(A') &= \sum_{i,k:\Pi(A_i^k)=A_j} t'(A_i^k) \\ &= \sum_{k:\Pi(A_j^k)=A_j} t'(A_j^k) \\ &= \sum_k h_{A_j}(A_k) \\ &= t(A_j)\end{aligned}$$

for $j = 1, \dots, 2^n$. It follows that t' is a refinement of t.

Next we have to calculate the revised mass distribution $s'(B') \stackrel{\mathrm{d}}{=} t'_{E'}(B')$. Suppose $\sum_{A':A'\cap E'=\emptyset} t'(A') = 1$. Then we obtain

$$\sum_{A_s^r \cap E'=\emptyset} h_{A_r}(A_s) = 1$$

from the definition of t'. From $h_{A_r}(A_s) > 0$ follows $\emptyset = A_r^s \cap E' = A_s^s$, and therefore $A_s = \emptyset$. We conclude

$$1 = \sum_{A':A'\subseteq\Omega'} t(A') = \sum_{r,s\in\{1,\dots,2^n\}} h_{A_r}(A_s) = \sum_{r,s\in\{1,\dots,2^n\}} h_{A_r}(\emptyset)$$

which contradicts (c). So in the sequel we assume

$$\sum_{A':A'\cap E'=\emptyset} t(A') < 1.$$

We obtain

$$\begin{aligned}s'(B') = t'_{E'}(B') &= \frac{\sum_{A':A'\cap E'=B'} t'(A')}{1 - \sum_{A':A'\cap E'=\emptyset} t'(A')} \\ &= \frac{\sum_{i,k:A_i^k\cap E'=B'} t'(A_i^k)}{1 - \sum_{i,k:A_i^k\cap E'=\emptyset} t'(A_i^k)} = \frac{\sum_{i,k:A_i^k\cap E'=B'} h_{A_i}(A_k)}{1 - \sum_{i,k:A_i^k\cap E'=\emptyset} h_{A_i}(A_k)}\end{aligned}$$

for all $B' \subseteq \Omega$. It remains to show that s' is a refinement of s:

$$\begin{aligned}
&\sum_{B':\Pi(B')=A_j} s'(B') \\
&= \sum_{B':\Pi(B')=A_j} \frac{\sum\limits_{i,k:A_i^k \cap E'=B'} h_{A_i}(A_k)}{1-\sum\limits_{i,k:A_i^k\cap E'=\emptyset} h_{A_i}(A_k)} \\
&= \frac{\sum\limits_{i,k:\Pi(A_i^k\cap E')=A_j} h_{A_i}(A_k)}{1-\sum\limits_{i,k:A_k^k=\emptyset} h_{A_i}(A_k)} \\
&= \frac{\sum\limits_{i,k:\Pi(A_k^k)=A_j} h_{A_i}(A_k)}{1-\sum\limits_i h_{A_i}(\emptyset)} \\
&= \frac{\sum\limits_i h_{A_i}(A_k)}{1-\sum\limits_i h_{A_i}(\emptyset)} \\
&= s(A_j)
\end{aligned}$$

for $j = 1, \ldots, 2^n$. This completes the proof of Theorem 6.12. □

Property (iii) of Theorem 6.12 justifies talking about the existence of a *mass flow* that leads from t to s in the case $s \sqsubseteq t$. In this context $h_A(B)$ quantifies the mass flow from A to B, whereas mass $h_A(\emptyset)$ leaves the system. It is *not* assigned, say, to the empty set. The final renormalization is done just in order to keep "the world closed".

(6.13) *Remark.* We have

$$m_1 \sqsubseteq m_2 \Longrightarrow (m_1)_E \sqsubseteq (m_2)_E.$$

To prove this property starting with $Q_{m_1}(A) > 0 \Rightarrow Q_{m_2}(A) > 0$ we have to show $Q_{(m_1)_E}(B) > 0 \Rightarrow Q_{(m_2)_E}(B) > 0 \Rightarrow$. From $Q_{(m_1)_E}(B) > 0$ we know $\exists A' : (A' \cap E = B \wedge Q_{m_1}(A') > 0)$. This implies $Q_{m_2}(A') > 0$ and with Remark 6.8 we obtain

$$Q_{(m_2)_E}(A' \cap E) \Longrightarrow Q_{(m_1)_E}(B) > 0.$$

○

The concept of specialization allows us to compare mass distributions with respect to the information they bear; $s \sqsubseteq t$ means for example that s carries at least as much information as t. Therefore from a specialization point of view two mass distributions s and t are equivalent, if s is a specialization of t and t is a specialization of s; i.e.

$$s \equiv t \overset{\mathrm{d}}{\Longleftrightarrow} t \sqsubseteq s \quad \text{and } s \sqsubseteq t.$$

From property (ii) of Theorem 5.11 it is clear that $s \sqsubseteq s$ (reflexivity) and $(s \sqsubseteq t \wedge t \sqsubseteq u) \Rightarrow s \sqsubseteq u$ (transivity) hold for all mass distributions. The quotient set of the set of all mass distributions defined on Ω with respect to the equivalence relation $\equiv$ together with the relation $\sqsubseteq$ forms a partially ordered set. $s \sqsubseteq t \wedge \neg(s \equiv t)$ means that there is some refined space Ω', containing elements which are possibly true with regard to t' but certainly impossible with regard to s', where s' and t' are refinements of s and t respectively. This means that in a qualitative sense s carries more information than t.

(6.14) Example. Let $\Omega = \{\mathrm{z1}, \mathrm{z2}, \mathrm{z3}\}$ and s, t, u be mass distributions on Ω.

A	$s(A)$	$t(A)$	$u(A)$
$\emptyset$	0	0	0
$\{\mathrm{z1}\}$	0	0	0.1
$\{\mathrm{z2}\}$	0.4	0.4	0
$\{\mathrm{z3}\}$	0.1	0	0
$\{\mathrm{z1}, \mathrm{z2}\}$	0.2	0.5	0.1
$\{\mathrm{z1}, \mathrm{z3}\}$	0.3	0.1	0.4
$\{\mathrm{z2}, \mathrm{z3}\}$	0	0	0.4
Ω	0	0	0

We have the relations $s \equiv t$, $s \sqsubseteq u$, $t \sqsubseteq u$, but neither $s \equiv u$ nor $t \equiv u$. ◇

As in the case of probabilities the use of classes of belief functions which are compatible with some given information may be required. In terms of mass distributions this means that we have to consider for example the set of all specializations of a mass distribution t,

$$\{s \mid s \sqsubseteq t\}$$

or the set of all possible revisions of t

$$\{s \mid \exists B \, \forall A: \; t_B(A) = s(A)\}.$$

6.2.2 Strict Specializations

Now we turn to the notion of strict specialization.

(6.15) Definition. *Let s, t be mass distributions defined on Ω. s is a strict specialization of t (written $s \leq t$), if and only if there is a probability space $(\Theta, 2^\Theta, P)$ and two observation mappings $S, T : \Theta \to 2^\Omega$ such that $s = S[P]$ and $t = T[P]$, where S is a strict specialization of T.*

(6.16) Theorem. *Let s, t be mass distributions defined on Ω. The following two statements are equivalent.*

(i) There are two mass distributions s' and t' defined on a refinement Ω' of Ω where s' and t' are refinements of s and t, respectively and an event $E' \subseteq \Omega'$ such that $s'(B') = t'(B'|E')$ holds for each $B' \subseteq \Omega'$.

(ii) $s \leq t$.

The intuitive idea of a strict specialization is the projection of a conditioning. Note that like the specialization relation also the strict specialization relation is invariant against projection, i.e. we have

$$s' \leq t' \Rightarrow \Pi[s'] \leq \Pi[t']. \tag{6.17}$$

(The proof of this property runs by strict analogy to that of Theorem 6.11.)

We obtain two equivalent characterizations of this strict specialization relation:

(6.17) Theorem. *Let s, t be two mass distributions on Ω. The following three statements are equivalent.*

(i) s is a strict specialization of t

(ii) $\forall A \subseteq \Omega : \big(t(A) = 0 \Rightarrow s(A) = 0\big)$

(iii) There is a mass flow from t to s described by functions $h_A : 2^\Omega \to [0,1]$ such that $h_A(B) > 0$ implies $A = B$ or $B = \emptyset$.

Proof. (i) $\Rightarrow$ (ii): From $t(B) = 0$ follows $\sum\limits_{B':\Pi(B')=B} t'(B') = 0$. We have $t'(B')$ for all B' with $\Pi(B') = B$. So we obtain

$$t'(B'|E') = \begin{cases} 0, & \text{if } B' \not\subseteq E', \\ \dfrac{t'(B')}{\sum\limits_{B':B'\subseteq E'} t'(B)}, & \text{if } B' \subseteq E' \end{cases},$$

for all B' with $\Pi(B') = B$ (if the conditional mass distribution exists). It follows that $s'(B') = t'(B'|E') = 0$ for all B' with $\Pi(B') = B$, and therefore we have $s(B) = \sum_{B':\Pi(B')=B} s'(B') = 0$.

(ii) $\Rightarrow$ (iii): Define the real number α_t by setting $\alpha_t = \min\{t(A) \mid t(A) > 0, A \subseteq \Omega\}$. Let $h_A(A) = \alpha_t \cdot s(A)$ for $A \neq \emptyset$, $h_A(\emptyset) = t(A) - \alpha_t \cdot s(A)$ for $A \neq \emptyset$, and $h_A(B) = 0$ otherwise. From $t(A) \geq \alpha_t \cdot s(A)$ for *all* A we know that $h_A : 2^\Omega \to [0,1]$ such that $h_A(B) > 0$ implies $A = B$ or $B = \emptyset$. We have also

$$s(B) = \frac{\sum_{A:A\subseteq\Omega} h_A(B)}{1 - \sum_{A:A\subseteq\Omega} h_A(\emptyset)} = \frac{\alpha_t \cdot s(B)}{1 - 1 + \alpha_t \cdot \sum_{A:A\subseteq\Omega} s(A)}$$

(iii) $\Rightarrow$ (i): Consider the space $\Omega' = \Omega \times \{a,b\}$. Ω' is a refinement with the refinement mapping $\widehat{\Pi}(A) = A \times \{a,b\}$. Define t' by setting

$$\begin{aligned} t'\big(\widehat{\Pi}(A) \times \{a\}\big) &= h_A(A), \\ t'\big(\widehat{\Pi}(A) \times \{b\}\big) &= h_A(\emptyset), \\ t'(A') &= 0 \text{ otherwise.} \end{aligned}$$

This defines a mass distribution t' on Ω'. Conditioning with the set $E' = \Omega' \times \{a\}$ yields

$$\begin{aligned} t'\big(\widehat{\Pi}(A) \times \{a\} \mid \Omega \times \{a\}\big) &= \frac{t'\big(\widehat{\Pi}(A) \times \{a\}\big)}{\sum_{A':A'\subseteq\Omega\times\{a\}} t'(A')} = \frac{h_A(A)}{1 - \sum_{A':A'\subseteq\Omega} h_A(\emptyset)}, \\ t'\big(\widehat{\Pi}(A) \times \{b\} \mid \Omega \times \{a\}\big) &= 0, \\ t'\big(A' \mid \Omega \times \{a\}\big) &= 0 \text{ otherwise.} \end{aligned}$$

Consider the mass distribution s' defined by $s'(\ .\ ) = t'(\ .\ |E')$. We have

$$s'\big(\widehat{\Pi}(A) \times \{a\}\big) = \frac{h_A(A)}{1 - \sum_{A':A'\subseteq\Omega} h_{A'}(\emptyset)} = s(A)$$

for all $A \neq \emptyset$, and therefore s' is obviously a refinement of s. □

6.2.3 Orthogonal Extensions

At the beginning of this section we introduced the vacuous extension of some mass distribution $m : 2^{\Omega} \to [0,1]$ to the refined space Ω', and denoted the resulting mass distribution by $\widehat{\Pi}(m) : 2^{\Omega'} \to [0,1]$. This process preserves the information coded by m, but in general the refinement of the frame of discernment is done in order to obtain a more detailed representation of the domain under consideration. For this reason it should be possible to distribute the evidence mass $m(A)$ bound to $A \subseteq \Omega$ among those sets $B' \subseteq \Omega'$ with $\Pi(B') = A$, assuming that $\widehat{\Pi} : 2^{\Omega} \to 2^{\Omega'}$ is the appropriate refinement mapping. Moreover there should be the possibility to define a flow to the empty set in the course of extending m to Ω'.

(6.18) Definition. *Let Ω' be a refinement of Ω with $\widehat{\Pi}$ as the corresponding refinement mapping. A mass distributio m' on Ω' is called an orthogonal extension of m if m' is a specialization of the vacuous extension $\widehat{\Pi}(m)$ and if for all $A \subseteq \Omega$, $A' \subseteq \Omega'$ there is a mapping $h_{A'} : \Omega' \to [0,1]$ such that*

$$h_{\widehat{\Pi}(A)}(B') > 0 \Rightarrow \big(\Pi(B') = A \text{ or } B' = \emptyset\big).$$

(6.19) *Remark.* (i) In the context of orthogonal extensions we use instead of $h_{\widehat{\Pi}(A)}(B')$, $A \subseteq \Omega$, $B' \subseteq \Omega'$ the notation $h_A(B')$ reflecting our intuitive idea of a mass flow from A to the subsets B' of $\widehat{\Pi}(A)$.

(ii) Each mass distribution m' on Ω' with $\Pi(m') = m$ is obviously an orthogonal extension of m. ○

An orthogonal extension can bear information about the original mass distribution, for example there may be some set $A \subseteq \Omega$ with $h_A(\emptyset) > 0$, i.e. the evidence mass has to be neglected. In such cases the projection of an orthogonal extension yields not the original mass distribution but only a strict specialization.

(6.20) Theorem. *Let Ω' be a refinement of Ω with $\widehat{\Pi} : 2^{\Omega} \to 2^{\Omega'}$ as the corresponding refinement mapping. If m' is a orthogonal extension of some mass distribution $m : 2^{\Omega} \to [0,1]$, then we have*

$$\Pi[m'] \leq m.$$

Proof. The orthogonal extension m' may be determined by the functions $h_A(B')$, $A \subseteq \Omega$, $B' \subseteq \Omega'$. For the projection $\Pi(m')$ we obtain

$$\begin{aligned}
\Pi(m')(A) &= \sum_{B':\Pi(B')=A} m'(B') \\
&= \sum_{B':\Pi(B')=A} \frac{h_A(B')}{1 - \sum_{C:C\subseteq\Omega} h_C(\emptyset)} \\
&= \frac{1}{c} \cdot \sum_{B':\Pi(B')=A} h_A(B') \\
&= \frac{1}{c} \cdot m(A),
\end{aligned}$$

where $c = 1 - \sum_{C:C\subseteq\Omega} h_A(\emptyset) > 0$ for $A \neq \emptyset$. Obviously we have

$$m(A) = 0 \Rightarrow \Pi(m')(A) = 0.$$

□

As in the case of ordinary specializations we obtain that strict specializations are invariant with respect to further projections.

(6.21) Theorem. *Let Ω' be a refinement of Ω and let $\widehat{\Pi} : 2^{\Omega} \to 2^{\Omega'}$ be their refinement mapping. If $s', t' : 2^{\Omega'} \to [0,1]$ are two mass distributions on Ω' then*

$$s' \leq t' \Rightarrow \Pi[s'] \leq \Pi[t'].$$

Proof.

$$\begin{aligned}
\Pi(t')(A) = 0 &\Rightarrow \forall A' \subseteq \Omega' : \Pi(A') = A \Rightarrow t'(A) = 0 \\
&\Rightarrow \forall A' \subseteq \Omega' : \Pi(A') = A \Rightarrow s'(A) = 0 \\
&\Rightarrow \Pi(s')(A) = 0.
\end{aligned}$$

□

In the context of knowledge representation the most important mass distributions on product spaces are those for which each focal set is a cylinder, which means mass distributions $m : 2^{\Omega^N} \to [0,1]$, $N \subseteq M$, $N = \{i_1, \ldots, i_n\}$, where $m(A) > 0$ only if $A \in \mathcal{Q}(\Omega^N)$. In the sequel we use the notation

$$m : \mathcal{Q}(\Omega^N) \to [0,1]$$

to indicate that we restrict ourselves to mass distributions satisfying this condition. Note that if we restrict ourselves to sets $E^N \in \mathcal{Q}(\Omega^N)$ the class $\{m \mid m : \mathcal{Q}(\Omega^N) \to [0,1], m \text{ is a mass distribution}\}$ is closed with respect to conditioning and revision.

In the context of product spaces, orthogonal extensions are of special interest. The orthogonal extension of a mass distribution $s : \mathcal{Q}(\Omega^S) \to [0,1]$ onto Ω^T, $T \subseteq S$, is that mass distribution which distributes the evidence mass attached to $B \in \mathcal{Q}(\Omega^S)$ to those sets $C \in \mathcal{Q}(\Omega^T)$ with $\Pi_S^T(C) = B$, i.e. $C = B \times F$, $F \in \mathcal{Q}(\Omega^W)$, $W = T - S$, where $F \neq \emptyset$ is admissible. So we have a flow of masses from Ω^S to Ω^T, characterized by functions

$$h_B(C), \quad B \in \mathcal{Q}(\Omega^S), \quad C \in \mathcal{Q}(\Omega^T),$$

where $h_B > 0$ implies $\Pi_S^T(C) = B$ or equivalently $C = B \times F$, $F \in \mathcal{Q}(\Omega')$. The projection of an orthogonal extension $t : \mathcal{Q}(\Omega^T) \to [0,1]$ of a mass distribution $s : \mathcal{Q}(\Omega^S) \to [0,1]$ back to Ω^S yields a strict specialization of s, since there is possibly a flow to the empty set if there is $B \in \mathcal{Q}(\Omega^S)$ with $h_B(B \times \emptyset)$.

6.2.4 Conjunctive and Disjunctive Extensions

If we consider two different frames of discernment Ω_1 and Ω_2, related just by a mapping

$$I : 2^{\Omega_1} \to 2^{\Omega_2}$$

where $\omega_{01} \in A \subseteq \Omega_1$ implies $\omega_{01} \in I(A) \subseteq \Omega_2$ (note that $\omega_{01} \in \Omega_1$ and $\omega_{02} \in \Omega_2$ respectively denote the "true state of the world"), then a mass distribution $n : 2^{\Omega_1} \to [0,1]$ induces via I a mass distribution on Ω_2. We obtain

$$I[n] : 2^{\Omega_2} \to [0,1]; \quad I[n](B) = \begin{cases} \dfrac{\sum\limits_{A:I(A)=B} n(A)}{\sum\limits_{A:I(A)\neq\emptyset} n(A)} & \text{if } \exists A : I(A) = B \\ 0 & \text{otherwise,} \end{cases} \tag{6.18}$$

if there is $A \subseteq \Omega_1$ such that $n(A) > 0$ and $I(A) \neq \emptyset$.

Often such a mapping arises from the extension of a one-to-many correspondence between Ω_1 and Ω_2; i.e. from a mapping $I : \Omega_1 \to 2^{\Omega_2}$. We obtain reasonable extensions from a conjunctive and disjunctive combination of the images under I.

(6.22) Definition. *Let I be a mapping from Θ to 2^Ω. The conjunctive extension of I to 2^Θ is defined by*

$$I(A) = \bigcap_{\theta \in A} I(\theta)$$

for all $A \subseteq \Theta$ whereas the disjunctive extension of I to 2^Θ is defined by

$$I(A) = \bigcup_{\theta \in A} I(\theta).$$

Conjunctive extension should be applied if the different elements within a group $A \subseteq \Omega_1$ are considered to complete each other while disjunctive extension should be used if the different elements of a group represent alternative views, and a careful evaluation is required. (If, in the ideal case, there is a one-to-one correspondence, i.e. $I : \Omega_1 \to \Omega_2$, a disjunctive extension can be derived analogously.)

The transfer of mass distributions between different frames of discernment Ω_1 and Ω_2 related via a mapping I is, as far as disjunctive extensions are used, compatible with the specialization relation.

(6.23) Theorem. *Let $n, n' : 2^{\Omega} \to [0,1]$ be mass distributions on 2^{Ω_1} and let $I : \Omega_1 \to 2^{\Omega_2}$ be a mapping with $I : 2^{\Omega_1} \to 2^{\Omega_2}$ as its disjunctive extension. Then we have*

$$n \sqsubseteq n' \Rightarrow I[n] \sqsubseteq I[n'].$$

Proof. As a simple consequence of Theorem 6.12 for two mass distributions s, t on Ω we derive

$$\big(\forall C \subseteq \Omega : Q_s(C)>0 \Rightarrow Q_t(C)>0\big) \Leftrightarrow \big(\forall A \subseteq \Omega \exists B \supseteq A : s(A)>0 \Rightarrow t(B)>0\big).$$

Let $n' \sqsubseteq n$. Consider $M \subseteq \Omega$ where

$$0 < I[n'](M) = \frac{\sum\limits_{N:I(N)=M} n'(N)}{\sum\limits_{N:I(N)\neq\emptyset} n'(N)}.$$

Thus there is $N \subseteq \Omega$ with $I(N) = M$ and $n' > 0$. Since $n' \sqsubseteq n$ we obtain that there is $K \subseteq \Theta$ with $K \supseteq N$ and $n(K)>0$. This implies $I[n]\big(I(K)\big) > 0$, where $I(K) \supseteq I(N) = M$. So we have $I[n'] \sqsubseteq I[n]$.

Note that this part of the proof holds for mappings $I : \Theta \to \Omega$ as for mappings $I : \Theta \to 2^{\Omega}$. □

If we restrict ourselves to point-valued mappings we even obtain some kind of inversion.

(6.24) Theorem. *Let $n : 2^{\Omega_1} \to [0,1]$ and $n' : 2^{\Omega_2} \to [0,1]$ be two mass distributions. Given a point-valued mapping $I : \Omega_1 \to \Omega_2$ and its disjunctive extension $I : 2^{\Omega_1} \to 2^{\Omega_2}$ we obtain*

$$m' \sqsubseteq I[n] \Rightarrow \exists n' : \big(I[n'] = m' \wedge n' \sqsubseteq n\big).$$

Proof. Let $m' \sqsubseteq I[n]$. Consider $M \subseteq \Omega_2$ where $m'(M) > 0$. There is $\widetilde{M} \supseteq M$ with $I[n](\widetilde{M}) > 0$. Thus there is $\widetilde{N}$ such that $I(\widetilde{N}) = \widetilde{M}$ and $n(\widetilde{N}) > 0$. Choose $\widetilde{M}, \widetilde{N}$.

Define $N' \stackrel{\text{d}}{=} \widetilde{N} \cap I^{-1}(M)$. Then $I(\widetilde{N}) \cap I^{-1}(M) \subseteq I(\widetilde{N}) \cap I(I^{-1}(M)) \subseteq \widetilde{M} \cap M = M$. If $\omega \in M$, this implies $\omega \in \widetilde{M}$. Thus there is $\theta \in \widetilde{N}$ with $I(\theta) = \omega$. It follows that $\theta \in I^{-1}(M)$ and $\omega \in I(N')$. So we obtain $M = I(N')$.

Define $n' : 2^{\Omega_2} \to [0,1]$; $n'(A) \stackrel{\text{d}}{=} m'(A)$ if $\exists M \subseteq \Omega : N' = A$, $n' = A \stackrel{\text{d}}{=} 0$ otherwise. n' is well defined, since $N'_1 = N'_2 \Rightarrow M_1 = I(N'_1) = I(N'_2) = M_2$. n' is a mass distribution, since $\sum_{A:A\subseteq\Theta} n'(A) = \sum_{M:M\subseteq\Omega} m'(M) = 1$. Since $\widetilde{N} \supseteq N'$ and $n(\widetilde{N}) > 0$ we have $n' \sqsubseteq n$. We obtain

$$I[n'](M) = \frac{\sum_{A:I(A)=M} n'(A)}{\sum_{A:I(A)\neq\emptyset} n'(A)} = \frac{m'(M)}{\sum_{A:I(A)\neq\emptyset} n'(A)} = m'(M),$$

therefore there is $n' \sqsubseteq n$ such that $I[n'] = m'$. □

Unfortunately the result of Theorem (6.24) does not hold for set-valued mappings, as the next example shows.

(6.25) Example. Consider the sets $\Omega_1 = \{\omega^1\}$ and $\Omega_2 = \{\omega_1^2, \omega_2^2\}$, and the mapping $I : \Omega_1 \to 2^{\Omega_2}$ where $I(\omega^1) = \{\omega_1^2, \omega_2^2\}$. Let $n : 2^{\Omega_1} \to [0,1]$ a mass distribution with $n(\{\omega^1\}) = 1$. This yields $I[n](\{\omega_1^2, \omega_2^2\}) = 1$. The mass distribution $m : 2^{\Omega_2} \to [0,1]$, where $m(\{\omega_1^2\}) = 1$, is a specialization of $I[n]$ as n is the only mass distribution on Ω_1, and there is no $n' : 2^{\Omega_2} \to [0,1]$ such that $I[n'] = m'$. ◇

6.3 Measures for Possibility/Necessity

In Sect. 5.1 we considered consonant random sets, i.e. random sets where the observation mapping $\Gamma : \Theta \to 2^{\Omega}$ attaches subsets $A_1, \ldots, A_n \subseteq \Omega$ with $A_i \subseteq A_j$ whenever $i < j$ to the sensors with a non-zero importance, i.e. we obtain *nested sets* as observations. It is easy to check that we have for each consonant random set, in the case of existence,

$$(P_\Gamma)^*(A \cup B) = \max\{(P_\Gamma)^*(A), (P_\Gamma)^*(B)\}, \qquad A, B \subseteq \Omega \tag{6.19}$$

$$(P_\Gamma)_*(A \cap B) = \min\{(P_\Gamma)_*(A), (P_\Gamma)_*(B)\}, \qquad A, B \subseteq \Omega \tag{6.20}$$

This justifies talking about consonant mass distributions, if there is an ordering of the focal elements $A_1, \ldots, A_n$ such that $A_i \leq A_j$ holds, whenever $i \leq j$ holds. The evidence underlying a consonant belief function can be interpreted as pointing "in a single direction".

(6.26) Definition. *A set function* $\text{nec} : 2^\Omega \to [0,1]$ *is called a necessity measure if there is a consonant mass distribution* $m : 2^\Omega \to [0,1]$ *such that* $\text{nec} = \text{Bel}_m$. *The dual set function* $\text{poss} : 2^\Omega \to [0,1]$; $\text{poss}(A) \stackrel{\text{d}}{=} 1 - \text{nec}(\overline{A})$ *is called a possibility measure.*

The interpretation of the function poss is as follows: If for some set $A \subseteq \Omega$ $\text{poss}(A) = 0$ holds, then the elements in A are impossible. If on the other hand the value of $\text{poss}(\overline{A})$ is equal to 0 this implies that A covers the "true value" with certainty. $\text{poss}(\overline{A}) = 1$ and at the same time $\text{poss}(A) = 1$ means total ignorance about the truth or falsity of the elements in A.

(6.27) Theorem. *A necessity measure is completely characterized by the following three conditions:*

(i) $\text{nec}(\emptyset) = 0$,

(ii) $\text{nec}(\Omega) = 1$,

(iii) $\text{nec}(A \cap B) = \min\{\text{nec}(A), \text{nec}(B)\}$.

Proof.

"$\Rightarrow$" We have to demonstrate first that each necessity measure nec is a belief function. Let $A_1, \ldots, A_n$, $n \geq 1$ be subsets of Ω. Without loss of generality we assume that $\text{nec}(A_1) \leq \text{nec}(A_2) \leq \ldots \leq \text{nec}(A_n)$ hold. It follows that

$$
\begin{aligned}
&\sum_{I:\emptyset \neq I \subseteq \{1,\ldots,n\}} (-1)^{|I|+1} \text{nec}(A_i) \\
&= \sum_{1 \leq i \leq n} \text{nec}(A_i) - \sum_{1 \leq i < j \leq n} \text{nec}(A_i \cap A_j) \pm \ldots \\
&\qquad\qquad \ldots + (-1)^{n+1} \text{nec}(A_1 \cap \ldots \cap A_n) \\
&= \sum_{i=1}^{n} \text{nec}(A_i) - \sum_{i=1}^{n-1} \binom{n-i}{1} \text{nec}(A_i) \\
&\qquad + \sum_{i=1}^{n-2} \binom{n-i}{2} \text{nec}(A_i) \pm \ldots + (-1)^{n+1} \text{nec}(A_1 \cap \ldots \cap A_n)
\end{aligned}
$$

$$= \sum_{i=1}^{n} \text{nec}(A_i) \left\{ \binom{n-i}{0} - \binom{n-i}{1} \pm \ldots + (-1)^{n-i} \binom{n-i}{0} - \binom{n-i}{n-i} \right\}$$

$$= \text{nec}(A_n) \leq \text{nec}(A_1 \cup \ldots \cup A_n).$$

Suppose nec is not a consonant belief function. Then there exist focal elements A and B of nec neither of which is contained in the other. Let $a \in A$ and $b \in B$ such that $a \notin B$ and $b \notin A$. Then we have from $m_{\text{nec}}(A) > 0$ and $m_{\text{nec}}(B) > 0$

$$\text{poss}(\{a,b\}) = \sum_{C:\{a,b\} \cap C \neq \emptyset} m_{\text{nec}}(C) > \sum_{C=\{a\} \cap C \neq \emptyset} m_{\text{nec}}(C) = \text{poss}(\{a\}),$$

since $\{a\} \cap C \neq \emptyset$ implies $\{a,b\} \cap C \neq \emptyset$ and $B \cap \{a,b\} \neq \emptyset$ whereas $B \cap \{a\} = \emptyset$. Analogously we prove $\text{poss}(\{a,b\}) > \text{poss}(\{a\})$. It follows that $\text{poss}(\{a,b\}) > \max\left\{\text{poss}(\{a\}), \text{poss}(\{b\})\right\}$, which is a contradiction.

"$\Leftarrow$" Let A and B be arbitrary subsets of Ω. Let i_A be the largest integer i such that $A_i \subseteq A$ and let i_B be the largest integer i such that $A_i \subseteq B$. Then $A_i \subseteq A$ if and only if $i \leq i_A$, $A_i \subseteq B$ if and only if $i \leq i_B$, and $A_i \subseteq A \cap B$ if and only if $i \leq \min(i_A, i_B)$. Therefore we obtain

$$\begin{aligned} \text{nec}(A \cap B) &= \sum_{i=1}^{\min(i_A, i_B)} m_{\text{nec}}(A_i) \\ &= \min\left(\sum_{i=1}^{i_A} m_{\text{nec}}(A_i), \sum_{i=1}^{i_B} m_{\text{nec}}(A_i)\right) \\ &= \min\left(\text{nec}(A), \text{nec}(B)\right) \end{aligned}$$

□

A similar characterization for possibility measures can be easily derived from Definition 6.22. We obtain

(i) $\text{poss}(\emptyset) = 0$,

(ii) $\text{poss}(\Omega) = 1$,

(iii) $\text{poss}(A \cup B) = \max\{\text{poss}(A), \text{poss}(B)\}$. (6.21)

In the following we give a list of some important properties of a poss/nec pair.

- $A \subseteq B \Rightarrow \left(\text{nec}(A) \leq \text{nec}(B) \text{ and } \text{poss}(A) \leq \text{poss}(B)\right)$ for all $A, B \subseteq \Omega$,
- $\min\left(\text{nec}(A), \text{nec}(\overline{A})\right) = 0$ for all $A, B \subseteq \Omega$,
- $\max\left(\text{poss}(A), \text{poss}(\overline{A})\right) = 1$ for all $A, B \subseteq \Omega$,
- $\text{nec}(A) \leq \text{poss}(A)$ for all $A \subseteq \Omega$, (6.22)

- $\text{nec}(A \cap B) = \min(\text{nec}(A), \text{nec}(B))$ for all $A, B \subseteq \Omega$,
- From $\text{nec}(A) > 0$ follows $\text{poss}(A) = 1$ for all $A, B \subseteq \Omega$,
- From $\text{poss}(A) < 1$ follows $\text{nec}(A) = 0$ for all $A, B \subseteq \Omega$.

As in the case of finite probability measures, each possibility function poss : $2^\Omega \to [0,1]$ is completely determined by the values it awards to the singletons, i.e. to the subsets of Ω that contain only one element.

(6.28) Definition. *Let* poss : $2^\Omega \to [0,1]$ *be a possibility function. Then the mapping* $\mu_{\text{poss}} : \Omega \to [0,1]$ *defined by*

$$\mu_{\text{poss}}(\omega) \overset{\text{d}}{=} \text{poss}(\{\omega\}),$$

$\omega \in \Omega$, *is called the contour function of* poss.

We have

$$\text{poss}(A) = \max\{\mu_{\text{poss}}(\omega) \mid \omega \in A\} \tag{6.23}$$

$$\text{nec}(A) = \min\{1 - \mu_{\text{poss}}(\omega) \mid \omega \notin A\} \tag{6.24}$$

for all $A \subseteq \Omega$. Conversely if a mapping $\mu : \Omega \to [0,1]$ such that $\exists \omega : \mu(\omega) = 1$ is given, then there is exactly one possibility (necessity) measure on Ω such that $\mu = \mu_{\text{poss}}$ holds. It turns out that the calculation of the conditional possibility and the revised possibilities is very easy in terms of contour functions.

Conditioning in the framework of possibility/necessity functions is the same as in the more general context of belief functions. Let poss be a possibility measure with corresponding contour function μ and the necessity measure nec. Let $B \subseteq \Omega$ be a set such that $\text{nec}(B) \neq 0$. Then we have

$$\text{nec}(A|B) = \begin{cases} \frac{\text{nec}(A)}{\text{nec}(B)} = \frac{1-\max\{\mu(\omega)|\omega \in \overline{A}\}}{1-\max\{\mu(\omega)|\omega \in \overline{B}\}}, & \text{if } A \subseteq B, \\ 0 & \text{otherwise,} \end{cases} \tag{6.25}$$

$$\text{poss}(A|B) = \begin{cases} \frac{\text{poss}(A \cup \overline{B}) - \text{poss}(\overline{B})}{1-\text{poss}(\overline{B})} \\ = \frac{\max\{\mu(\omega)|\omega \in A \cup \overline{B}\} - \max\{\mu(\omega)|\omega \in \overline{B}\}}{1-\max\{\mu(\omega)|\omega \in \overline{B}\}}, & \text{if } A \subseteq B, \\ 0 & \text{otherwise.} \end{cases} \tag{6.26}$$

Similar formulas can be derived for belief revision in the sense of Dempster. Let B be a subset of Ω such that $\text{poss}(B) \neq \emptyset$. Then we have

$$\text{nec}_B(A) = \begin{cases} \frac{\text{nec}(A \cup \overline{B}) - \text{nec}(\overline{B})}{1-\text{nec}(\overline{B})} \\ = \frac{\max\{\mu(\omega)|\omega \in B\} - \max\{\mu(\omega)|\omega \in \overline{A} \cap B\}}{\max\{\mu(\omega)|\omega \in B\}}, & \text{if } A \cap B \neq \emptyset, \\ 0 & \text{otherwise,} \end{cases} \tag{6.27}$$

$$\text{poss}_B(A) = \begin{cases} \frac{\text{poss}(A)}{\text{poss}(B)} = \frac{\max\{\mu(\omega)|\omega\in A\}}{\max\{\mu(\omega)|\omega\in B\}}, & \text{if } A \cap B \neq \emptyset, \\ 0 & \text{otherwise.} \end{cases} \tag{6.28}$$

We see that the computational effort for the calculation with possibilities reduces to a minimum. This is also the case for the specialization concept. Whenever we have two possibilities poss and poss′ on Ω and Ω', respectively, where Ω' is a refinement of Ω with $\widehat{\Pi}$ as refinement mapping, we say that poss′ is a refinement of poss if and only if $\text{poss}(\{\omega\}) = \text{poss}'\Big(\widehat{\Pi}(\{\omega\})\Big)$ holds for all $\omega \in \Omega$. This coincides with the general case of measures, since we have for all $\omega \in \Omega$ the equation

$$\begin{aligned} \text{poss}(\{\omega\}) &= \sum_{A:\omega\in A} m_{\text{poss}}(A) \\ &= \sum_{A':\{y\in\Omega|\widehat{\Pi}(\{y\})\cap A'\neq\emptyset\}=A} m'(A') \\ &= \sum_{A':\widehat{\Pi}(\{\omega\})\cap A'\neq\emptyset} m'(A') = \text{poss}'\Big(\widehat{\Pi}(\{\omega\})\Big). \end{aligned}$$

We can also define the vacuous extension of $\text{poss}: 2^{\Omega} \to [0,1]$ to Ω' by setting $m_{\widehat{\Pi}(\text{poss})} \stackrel{\text{d}}{=} \widehat{\Pi}(m_{\text{poss}})$. It is clear that $\widehat{\Pi}(m_{\text{poss}})$ is a consonant mass distribution which corresponds to a possibility $\widehat{\Pi}(\text{poss})$ on Ω'. Conversely if m'_{poss} is the mass distribution of a possibility measure poss′ then $m_{\Pi(\text{poss}')} \stackrel{\text{d}}{=} \Pi(m'_{\text{poss}})$. We have the relation

$$A' \subseteq B' \Rightarrow \Pi(A') \subseteq \Pi(B'),$$

therefore $\Pi(m')$ is also a consonant mass distribution, i.e. the projection of a possibility is also a possibility.

As a particularly interesting example we state that the marginal possibilities in product spaces are indeed possibility measures. The concept of a specialization can also be considered in the framework of possibility measures.

(6.29) Definition. *Let* poss_1, poss_2 *be two possibilities defined on* Ω. *We call* poss_1 *a specialization of* poss_2 *(written* $\text{poss}_1 \sqsubseteq \text{poss}_2$*), if and only if there is a refinement mapping* $\widehat{\Pi}$, *and two possibility measures* poss'_1 *and* poss'_2 *on* Ω', *where* poss'_1 *(*poss'_2*) is a refinement of* poss_1 *(*poss_2*) and an event* $E' \subseteq \Omega'$ *such that* $\text{poss}'_1(A') = (\text{poss}'_2)_{E'}(A')$ *holds for all* $A' \subseteq \Omega'$.

We have the following simple characterization.

(6.30) Theorem. *Let* poss_1 *and* poss_2 *be two possibility measures on* Ω. *Then* poss_1 *is a specialization of* poss_2 *if and only if*

$$\forall \omega \in \Omega : [\mu_{\text{poss}_2}(\omega) = 0 \Rightarrow \mu_{\text{poss}_1}(\omega) = 0].$$

Proof.

"⇒"If poss_1 is a specialization of poss_2, then from Theorem 6.12(ii) it follows that

$$\forall A \subseteq \Omega : Q_{\text{poss}_2}(A) = 0 \Rightarrow Q_{\text{poss}_1}(A) = 0.$$

Q_{poss} denotes the commonality function corresponding to poss. Let $\omega \in \Omega$ be such that $\mu_{\text{poss}_2}(\omega) = 0$ holds. We have

$$\mu_{\text{poss}_2}(\omega) = \text{poss}_2(\{\omega\}) \sum_{A:\omega \in A} m_{\text{poss}}(A) = Q_{\text{poss}_2}\{\omega\} = 0.$$

Therefore $Q_{\text{poss}_1}(\{\omega\}) = 0$, and it follows that $\mu_{\text{poss}_1}(\omega) = 0$.

"⇐"Let $\Omega' = \bigcup_{\omega \in \Omega} \{\omega^1, \omega^2\}$. $\widehat{\Pi} : 2^{\Omega} \to 2^{\Omega'}$, $\{\omega\} \mapsto \{\omega^1, \omega^2\}$ is a refinement mapping. Define the possibility poss_2' on Ω' by setting $\mu_{\text{poss}_2'}(\omega^2) = \mu_{\text{poss}_2}(\omega)$, $\omega^2 \in \Omega'$, and $\mu_{\text{poss}_2'}(\omega^1) = \mu_{\text{poss}_1}(\omega) \cdot \min\{\mu_{\text{poss}_2}(\omega) \mid \mu_{\text{poss}_2}(\omega) > 0, \omega \in \Omega\}$, $\omega^1 \in \Omega'$. If $\mu_{\text{poss}_2}(\omega) = 0$ then $\mu_{\text{poss}_1}(\omega) = 0$ follows, and we have

$$\begin{aligned}
&\text{poss}_2'\left(\widehat{\Pi}(\{\omega\})\right) \\
&= \text{poss}_2'(\{\omega^1, \omega^2\}) \\
&= \max\{\mu_{\text{poss}_2'}(\omega^1), \mu_{\text{poss}_2'}(\omega^2)\} \\
&= 0 \\
&= \mu_{\text{poss}_2}(\omega).
\end{aligned}$$

If $\mu_{\text{poss}_2}(\omega) \neq 0$, then we have

$$\begin{aligned}
&\text{poss}_2'\left(\widehat{\Pi}(\{\omega\})\right) \\
&= \max\{\mu_{\text{poss}_2'}(\omega^1), \mu_{\text{poss}_2'}(\omega^2)\} \\
&= \max\left\{\mu_{\text{poss}_1}(\omega) \cdot \min\{\mu_{\text{poss}_2}(\omega) \mid \mu_{\text{poss}_2}(\omega) > 0\}, \mu_{\text{poss}_2}(\omega)\right\} \\
&= \mu_{\text{poss}_2}(\omega).
\end{aligned}$$

So poss_2' is a refinement of poss_2. Let $E' = \{\omega^1 \mid \omega \in \Omega\}$. Define $\text{poss}_1' = (\text{poss}_2')_{E'}$. The revised possibility exists, since $\text{poss}_2'(E') \neq 0$. We have $\mu_{\text{poss}_1'}(\omega^2) = 0$ for all $\omega^2 \in \Omega'$ and

$$\begin{aligned}
&\mu_{\mathrm{poss}'_1}(\omega^1) \\
&\quad = \frac{\mu_{\mathrm{poss}'_1}(\omega^1)}{\max\{\mu_{\mathrm{poss}'_2}(\omega^1) \mid \omega \in \Omega\}} \\
&\quad = \frac{\mu_{\mathrm{poss}'_1}(\omega^1) \cdot \min\{\mu_{\mathrm{poss}_2}(\omega) \mid \mu_{\mathrm{poss}_2}(\omega) > 0, \omega \in \Omega\}}{\max\{\mu_{\mathrm{poss}'_1}(\omega^1) \cdot \min\{\mu_{\mathrm{poss}_2}(\omega) \mid \mu_{\mathrm{poss}_2}(\omega) > 0, \omega \in \Omega\} \mid \omega \in \Omega\}} \\
&\quad = \frac{\mu_{\mathrm{poss}_1}(\omega)}{\max\{\mu_{\mathrm{poss}_1}(\omega) \mid \omega \in \Omega\}} \\
&\quad = \mu_{\mathrm{poss}_1}(\omega)
\end{aligned}$$

for all $\omega^1 \in \Omega'$. Therefore

$$\begin{aligned}
\mathrm{poss}_1(\{\omega\}) &= \mu_{\mathrm{poss}_1}(\omega) \\
&= \max\{\mu_{\mathrm{poss}'_1}(\omega^1), \mu_{\mathrm{poss}'_1}(\omega^2)\} \\
&= \mathrm{poss}'_1\left(\widehat{\Pi}(\{\omega\})\right),
\end{aligned}$$

and it is shown that poss'_1 is a refinement of poss_1.

□

6.4 Generalized Mass Distributions

Often the subject of an expert's valuation are vague, ill-defined entities which cannot be described adequately by crisp sets as pointed out in Chap. 2. In spite of those cases we introduce the notion of generalized mass distributions. They allow us to attach portions of evidence mass to the elements of $\mathcal{F}_L(\Omega)$, which is the set of L-sets of the frame of discernment Ω. Recall that we only deal with finite sets Ω and L. So we have to deal with observation mappings of the form

$$\Gamma : \Theta \to \mathcal{F}_L(\Omega).$$

Such a generalized observation mapping induces a generalized mass distribution of the form

$$\Gamma[P] : \mathcal{F}_L(\Omega) \to [0,1]; \quad \Gamma[P](\mu) = \begin{cases} \frac{P(\{\theta \in \Theta \mid \Gamma(\theta) = \mu\})}{P(\{\theta \in \Theta \mid \Gamma(\theta) \neq \mathbb{1}_\emptyset\})} & \text{if } \mu \neq \mathbb{1}_\emptyset \\ 0 & \text{otherwise.} \end{cases} \tag{6.29}$$

This justifies the following definition:

(6.31) Definition. *The mapping $m : \mathcal{F}_L(\Omega) \to [0,1]$ is called a generalized mass distribution, whenever*

(i) $m(\mathbb{1}_\emptyset) = 0$, *and*

(ii) $\sum_{\mu:\mu \in \mathcal{F}_L(\Omega)} m(\mu) = 1.$

The interpretation of a generalized mass distribution m is analogous to that of ordinary mass distributions. $m(\mu)$ is that amount of mass that supports μ but cannot be divided into portions supporting strict subsets of μ. By analogy with the case of ordinary mass distributions we define generalized belief, plausibility and commonality functions by defining

$$\mathrm{Bel}_m(\mu) \stackrel{\mathrm{d}}{=} \sum_{\nu:\nu \subseteq \mu} m(\nu),$$

$$\mathrm{Pl}_m(\mu) \stackrel{\mathrm{d}}{=} \sum_{\nu:\mu \cap \nu \neq \mathbb{1}_\emptyset} m(\nu),$$

$$Q_m(\mu) \stackrel{\mathrm{d}}{=} \sum_{\nu:\mu \subseteq \nu} m(\nu).$$

The concepts of conditioning and revision can also be applied to generalized mass distributions. For conditioning with respect to $\eta \in \mathcal{F}_L(\Omega)$ we obtain in the case of existence

$$m(\mu|\eta) = \begin{cases} \frac{m(\mu)}{\mathrm{Bel}(\eta)}, & \text{if } \mu \subseteq \eta \\ 0 & \text{otherwise,} \end{cases} \tag{6.30}$$

and thus

$$\mathrm{Bel}_m(\mu|\eta) = \frac{\mathrm{Bel}_m(\mu \cap \eta)}{\mathrm{Bel}_m(\eta)}, \tag{6.31}$$

$$\mathrm{Pl}_m(\mu|\eta) = \frac{\sum_{\nu \subseteq \mu:\nu \cap \mu \neq \mathbb{1}_\emptyset} m(\nu)}{\mathrm{Bel}_m(\eta)} \tag{6.32}$$

Revision yields in the case of existence

$$m_\eta(\mu) = \begin{cases} \frac{\sum_{\nu:\nu \cap \eta = \mu} m(\nu)}{\mathrm{Pl}_m(\eta)}, & \text{if } \mu \neq \mathbb{1}_\emptyset, \\ 0 & \text{otherwise,} \end{cases} \tag{6.33}$$

and therefore

$$\mathrm{Bel}_{m_\eta}(\mu) = \frac{\sum_{\nu:\nu \cap \eta \subseteq \mu} m(\nu)}{\mathrm{Pl}_m(\eta)} \tag{6.34}$$

$$\mathrm{Pl}_{m_\eta}(\mu) = \frac{\mathrm{Pl}_m(\eta \cap \mu)}{\mathrm{Pl}_m(\eta)}. \tag{6.35}$$

The next topic of interest is the concept of specialization. Note that the projection of some fuzzy set $\mu' \in \mathcal{F}_L(\Omega')$ to Ω where $\widehat{\Pi} : 2^{\Omega} \to 2^{\Omega'}$ is the corresponding refinement mapping in the case of finite sets Ω', Ω, and L is done by

$$\Pi(\mu')(\omega) = \max\{\mu'(\omega') \mid \omega' \in \widehat{\Pi}(\{\omega\})\}.$$

Since the underlying idea of specialization is the projection of a revision we define as in the case of ordinary mass distribution:

(6.32) Definition. *Let s, t be two generalized mass distributions defined on Ω. We call s a specialization of t ($s \sqsubseteq t$) if and only if there are two generalized mass distributions s', t' defined on a refinement Ω' of Ω with $\widehat{\Pi} : 2^{\Omega} \to 2^{\Omega'}$ as their refinement mapping where $s = \Pi[s']$, $t = \Pi[t']$, and an event $\eta' \in \mathcal{F}_L(\Omega')$ such that $s' = t'_{\eta'}$ holds.*

Once more we are interested in different equivalent characterizations of the specialization relation. We obtain the same results as in the case of ordinary mass distributions as the next theorem shows:

(6.33) Theorem. *Let s and t be two mass distributions on Ω. The following three statements are equivalent.*

(i) $s \sqsubseteq t$

(ii) $\forall \mu \in \mathcal{F}_L(\Omega) : \big(Q_t(\mu) = 0 \Rightarrow Q_s(\mu) = 0\big)$

(iii) For every $\mu \in \mathcal{F}_L(\Omega)$ there are functions $h_\mu : \mathcal{F}_L(\Omega) \to [0,1]$ such that

(a) $\sum_{\nu : \nu \in \mathcal{F}_L(\Omega)} h_\mu(\nu) = t(\mu)$ *for all* $\mu \in \mathcal{F}_L(\Omega)$

(b) $h_\mu(\nu) \neq 0 \Rightarrow \nu \subseteq \mu$ *for all* $\mu, \nu \in \mathcal{F}_L(\Omega)$

(c) $s(\nu) = \dfrac{\sum_{\mu : \mu \in \mathcal{F}_L(\Omega)} h_\mu(\nu)}{1 - \sum_{\mu : \mu \in \mathcal{F}_L(\Omega)} h_\mu(\mathbb{1}_\emptyset)}$ *for all* $\mathbb{1} \neq \nu \in \mathcal{F}_L(\Omega)$.

The proof of this theorem is analogous to that given for Theorem 6.12.

6.5 Decision Making with Mass Distributions

Next let us consider the case that just one mass distribution m on Ω is given, i.e. we have $m : 2^{\Omega} \to [0,1]$, $m(\emptyset) = 0$, $\sum_{A : A \subseteq \Omega} m(A) = 1$. Such a mass distribution we conceive to be induced by an unknown random set $\Gamma : \Theta \to 2^{\Omega}$ with $\Gamma(\theta) \neq \emptyset$ for all $\theta \in \Theta$ and $\Gamma[P] = m$, where the set Θ as well as the probability

P is hidden. Although the expert space lies in the dark, it turns out that the notion of selectors can be applied.

(6.34) Theorem. *If $m : 2^{\Omega} \to [0,1]$ is a mass distribution on Ω, then we have*

$$\{S[P] \mid \exists \Theta \neq \emptyset,\ \exists \Gamma : \Theta \to 2^{\Omega},\ \exists P : 2^{\Theta} \to [0,1],\ \exists S : \Theta \to \Omega \text{ such that } \Gamma(\theta) \neq \emptyset \text{ for all } \theta \in \Theta, m = \Gamma[P], \text{ and } S \text{ is selector of } \Gamma\}$$

$$= \{R \mid \mathrm{Bel}_m \leq R \leq \mathrm{Pl}_m, R \text{ probability on } \Omega\}.$$

Proof.

"$\subseteq$" We have

$$\begin{aligned}\mathrm{Bel}_m(A) &= \sum_{B:B\subseteq A} m(B) = \sum_{\theta:\Gamma(\theta)\subseteq A} R\{(\theta)\}\\ &\leq \sum_{\theta:S(\theta)\in A} P\{(\theta)\} = S[P](A) \leq \sum_{\theta:\Gamma(\theta)\cap A\neq\emptyset} P\{(\theta)\}\\ &= \sum_{B:B\cap A\neq\emptyset} m(B) = \mathrm{Pl}_m(A)\end{aligned}$$

for all selectors of S of Γ. Recall that assuming S to be a selector of Γ implies $S(\theta) \in \Gamma(\theta)$. Thus from $\Gamma(\theta) \subseteq A$ follows $S(\theta) \in A$ and from $S(\theta) \in A$ we obtain $\Gamma(\theta) \cap A \neq \emptyset$.

"$\supseteq$" Let R be a probability such that for all $A \subseteq \Omega$ we have $\mathrm{Bel}_m(A) < R(A) < \mathrm{Pl}_m(A)$. According to equation (4.20) we define $P^{(i)}_{d_1,\ldots d_n}$ such that

$$P(\{\omega_i\}) = \sum_{d_1,\ldots,d_n\in\{0,1\}} P^{(i)}_{d_1,\ldots,d_n}$$

$$m(S_{d_1,\ldots,d_n}) = \sum_{i=1}^{n} P^{(i)}_{d_1,\ldots,d_n}$$

hold, where $P^{(i)}_{d_1,\ldots,d_n} = 0$ unless $d_i = 1$. Consider the space

$$\Theta = \{0,1\}^n \times \{1,\ldots,n\},$$

the probability P on Θ defined by

$$P(\{(d_1,\ldots,d_n,i)\}) = P^{(i)}_{d_1,\ldots,d_n},$$

the mapping $S : \Theta \to \Omega$, where $S((d_1,\ldots,d_n,i)) = \omega_i$, and the random set $\Gamma : \Theta \to 2^{\Omega}$; $\Gamma((d_1,\ldots,d_n,i)) = S_{d_1,\ldots,d_n}$. Then S is a selector of Γ, $\Gamma[P] = m$ and $S[P] = R$.

□

The essential result of the above theorem is that *each* probability of the set $\{R \mid \mathrm{Bel}_m \leq R \leq \mathrm{Pl}_m\}$ could be the one induced by the unknown original. (Note that not only the mapping $S : \Theta \to \Omega$ is unknown but also the space Θ.) Of course it is rather difficult to handle such an infinite set, especially to formulate a priori knowledge concerning preferences among the various probabilities. Nevertheless, we may obtain a partial ordering by applying the maximum expected utility criterion in order to find a total ordering on the decisions.

Once more let us consider a set Θ of sensors and an observation mapping $\Gamma : \Theta \to 2^{\Omega}$. If a decision has to be made, then the expert θ (assume $\Gamma(\theta) \neq \emptyset$) specifies how "his" probability $P(\{\theta\})$ is to be distributed among the elements of $\Gamma(\theta)$. We obtain mappings $h_\theta : \Omega \to [0,1]$ where

(i) $h_\theta(\omega) = 0$, if $\omega \notin \Gamma(\theta)$, and

(ii) $$\sum_{\omega:\omega\in\Gamma(\theta)} h_\theta(\omega) = P(\{\theta\}). \tag{6.36}$$

The probability that we obtain for the purpose of decision making is given by

$$R_h(\{\omega\}) = \sum_{\theta:\theta\in\Theta} h_\theta(\omega). \tag{6.37}$$

This probability can be constructed just from the mass distribution $m : 2^{\Omega} \to [0,1]$ induced by P and Γ. The mappings h_θ characterize a *decision mass flow*, where in addition to the three conditions of Theorem 6.12 we have

$$h_A(B) > 0 \Rightarrow \mathrm{card}(B) = 1.$$

The resulting probability is denoted by $h(m)$.

If no preferences on elements of $\Gamma(\theta)$ are specified by θ, and if a decision has to be made, then a uniform distribution of the probability $P(\{\theta\})$ among the elements of $\Gamma(\theta)$ is reasonable. We obtain

$$R(\{\omega\}) = \sum_{A:\omega\in A} \frac{P(\{\theta \in \Theta \mid \Gamma(\theta) = A)}{\mathrm{card}(A)}, \tag{6.38}$$

a result that coincides with equation (5.12). If the underlying random set as well as the space Θ are unknown then the same result is obtained from the mass distribution m:

$$R(\{\omega\}) = \sum_{A:\omega\in A} \frac{m(A)}{\mathrm{card}(A)}. \tag{6.39}$$

This solution equals the one derived by the insufficient reason principle proposed by P. Smets and D. Dubois. Each expert $\theta \in \Theta$ distributes his "partial belief" uniformly within his datum $\Gamma(\theta)$. The decision mass flow corresponding to the insufficient reason principle is denoted by h_s; we have $R = h_s(m)$. This is, of course, just one possibility of many, since a meta-expert could, for example, advise the experts to choose just one element of $\Gamma(\theta)$ that has to be taken into account in the decision process, i.e. to choose one selector.

(6.35) *Remark.* Mass distributions on L-sets can be handled in a similiar way. Recall that we consider L-sets to be the vague projections of unknown crisp sets. Thus on each level l (of speculation) we have a non-empty crisp set $\mu_l \subseteq \Omega$. We obtain a mass distribution $m_l : 2^\Omega \to [0,1]$, $m_l(A) \stackrel{\mathrm{d}}{=} \sum_{\mu:\mu_l=A} m(\mu)$ for all $A \subseteq \Omega$, $l \in L$. This mass distribution refers to the level of l-necessity. We have $\{Q \mid \mathrm{Bel}_{m_l} < Q < \mathrm{Pl}_{m_l}\} \subset \{Q \mid \mathrm{Bel}_{m_{l'}} < Q < \mathrm{Pl}_{m'_l}\}$ if $l \geq l'$, i.e. it is possible to find several orders on the set of decisions. Thus in contrast to the case of imprecise data, where we find a unique decision according to the maximum expected utility criterion, here such a unique result is lacking.

It is clear that in the presence of uncertainty together with rather vague data it would be curious to obtain a unique "right" decision. In the literature these problems are not very extensively elaborated although some efforts have been made. ○

6.6 Knowledge Representation with Mass Distributions

This section explains how to represent and to integrate knowledge about some frame of discernment under consideration. In the most general case each piece of knowledge consists (1) of a (static) description in terms of sets of mass distributions and (2) of an explicit description of the possible updates of this piece of information (i.e. of its dynamic behavior). Based on such a compound structure, which is denoted as a knowledge representation structure, the integration of different pieces of information can be carried out. The reader should keep in mind that we assume each knowledge representation structure to be induced by a (possibly unknown) information source. If we have information about the relation of these pieces of knowledge to their source, or about their importance, then this allows us to obtain more substantiated results than in the absence of such "meta-knowledge". We consider especially those cases where the sources are stochastically independent, which lead (under additional assumptions) to a justification of Dempster's well known rule of combination and cases where there are quantitative or qualitative valuations concerning the importance of the underlying sources.

6.6.1 Encoding Knowledge by Mass Distributions

In practical applications the sensor space is often hidden, and only the space Ω is accessible. Thus we have to deal with sets of mass distributions which we conceive to be induced by unknown information sources.

(6.36) Example. (a) Once more, let $\Omega = \{\text{z3}, \text{z2}, \text{z1}, \text{ca}, \text{ld}, \text{rd}\}$. Then the information "The ship is in z1 with a certainty of 50%, and in any case the chance to be in z3 is bigger than the chance to be in z2, whereas the ship is definitely not in $\{\text{ca}, \text{ld}, \text{rd}\}$" contains imprecision as well as uncertainty. It can be represented by the set

$$\big\{m : 2^{\Omega} \to [0,1] \;\big|\, m(\{\text{z1}\}) = 0.5, m(\{\text{z3}\}) > m(\{\text{z2}\}) \\ \text{and } m(A) = 0 \text{ for } A \subseteq \{\text{ca}, \text{ld}, \text{rd}\}\big\}.$$

(b) The vague statement "The ship is far away from the coast" may be encoded by an L-set, where $L = \{l_0, l_1, l_2\}$. We obtain $\mu(\text{z3}) = l_2$, $\mu(\text{z2}) = l_1$, and $\mu(\omega) = l_0$ otherwise. If this is the only information, we have to represent the actual knowledge by a single generalized mass distribution $m : \mathcal{F}_L \to [0,1]$; where $m(\mu) = 1$ and $m(\mu') = 0$ if $\mu' \neq \mu$. ◇

But also if just Ω is considered we have to take into account the process of updating. So besides a set $\mathcal{M}$ of mass distributions encoding the a priori knowledge a second set $\mathcal{U}$, $\mathcal{M} \subseteq \mathcal{U}$, of derivable updates has to be specified.

(6.37) Definition. *A triple $\mathcal{K} = (\mathcal{F}_L(\Omega), \mathcal{M}, \mathcal{U})$, where $\mathcal{M}$ and $\mathcal{U}$ are sets of mass distributions on $\mathcal{F}_L(\Omega)$ such that $\mathcal{M} \subseteq \mathcal{U}$ and for all $m' \in \mathcal{U}$ there is $m \in \mathcal{M}$ with $m' \sqsubseteq m$, is called a knowledge representation structure on Ω.*

Any information source $\mathcal{S} = (\mathcal{P}, \Gamma, \mathcal{G})$ induces a knowledge representation structure $\mathcal{K}_{\mathcal{S}} = (\mathcal{F}_L(\Omega), \Gamma[\mathcal{P}], \mathcal{G}[\mathcal{P}])$. Thus if we consider the description of updating in terms of Ω we have to rely on the results obtained in Chap. 7 for information sources.

Typical sets of probabilities are conditionings or specializations of a single probability P_0, i.e. we have $\mathcal{P}_1 = \{P \mid \exists E \subseteq \Theta : P = P_0(\,.\,|E)\}$ or, in the most general case, $\mathcal{P}_2 = \{P \mid P \sqsubseteq P_0\}$. (Note that $\mathcal{P}_1 \subseteq \mathcal{P}_2$.)

(6.38) Theorem. *Let P_0 be a probability on Θ, and let $\Gamma : \Theta \to 2^{\Omega}$ be a set-valued mapping. In this case we have*

$$\Gamma[\{P \mid P \sqsubseteq P_0\}] = \{m \mid m \leq \Gamma[P_0]\}.$$

Proof.

"$\sqsubseteq$" Let $P \sqsubseteq P_0$. We have to show $\Gamma[P] \leq \Gamma[P_0]$. Assume $\Gamma[P](A) > 0$. This implies that there is $\theta \in \Theta$ with $\Gamma(\theta) = A$ and $P(\{\theta\}) > 0$. Since $P \sqsubseteq P_0$ we derive $P_0(\{\theta\}) > 0$ and therefore $\Gamma[P_0](A) > 0$. According to condition (ii) of Theorem 6.17 this yields $\Gamma[P] \leq \Gamma[P_0]$.

"$\sqsupseteq$" Let $m \leq \Gamma[P_0]$. Consider $A \subseteq \Omega$ with $m(A) > 0$. Since $m \leq \Gamma[P_0](A)$ we have $\Gamma[P_0](A) > 0$, thus there is $\theta \in \Theta$ such that $\Gamma(\theta) = A$ and $P_0(\{\theta\}) > 0$. Choose θ_A with $\Gamma(\theta_A) = A$ and $P(\{\theta_A\}) > 0$ for each A with $m(A) > 0$. Define $P : 2^\Theta \to 0,1$; $P(\{\theta_A\}) = m(A)$ if $\theta = \theta_A$ and $P(\{\theta\}) = 0$ otherwise P is well defined since $\theta_A = \theta_B \Rightarrow \Gamma(\theta_A) = \Gamma(\theta_B) = B$ and $\sum_{\theta \in \Theta} P(\{\theta\}) = \sum_{A:A \subseteq \Omega} P(\{\theta_A\}) = \sum_{A:A \subseteq \Omega} m(A) = 1$. Obviously $\Gamma[P] = m$. Since $P(\{\theta\}) > 0 \Rightarrow P_0(\{\theta\}) > 0$ we have $P \sqsubseteq P_0$.

□

(6.39) Example. Let $\Omega = \{\text{z3}, \text{z2}, \text{z1}, \text{ca}\}$, and let $m : 2^\Omega \to [0,1]$ be a mass distribution defined by $m(\{\text{z1}\}) = \frac{1}{2}$, $m(\{\text{z1}, \text{ca}\}) = \frac{1}{4}$, $m(\{\text{z2}, \text{z1}\}) = \frac{1}{4}$, and $m(A) = 0$ otherwise. We assume this information to be the imprecise result of some sensor's observation of a certain ship under consideration, i.e. $M = \{m\}$. If conditioning is used as an updating scheme we obtain

$$\begin{aligned} \mathcal{U}_c &= \{m_1, m_2, m_3, m_4\}, \text{ where} \\ m_1(\,.\,) &= m(\,.\,|\{\text{z1}\}), \\ m_2(\,.\,) &= m(\,.\,|\{\text{z1}, \text{z2}\}), \\ m_3(\,.\,) &= m(\,.\,|\{\text{z1}, \text{ca}\}), \\ m_4(\,.\,) &= m(\,.\,|\{\text{z2}, \text{z2}, \text{ca}\}) = m, \end{aligned}$$

while revision leads to

$$\begin{aligned} \mathcal{U}_r &= \{m'_1, \dots, m'_7\}, \text{ where} \\ m'_1 &= m_{\{\text{z1}\}}, m'_2 = m_{\{\text{z2}\}}, m'_3 = m_{\{\text{ca}\}}, \\ m'_4 &= m_{\{\text{z1},\text{z2}\}}, m'_5 = m_{\{\text{z1},\text{ca}\}}, m'_6 = m_{\{\text{z2},\text{ca}\}}, \\ m'_7 &= m\;. \end{aligned}$$

If conditioning and revision refer to a refinement of the space Ω we derive

$$\begin{aligned} \mathcal{U}'_c &= \left\{ m' \;\middle|\; m'(A) > 0 \Rightarrow A \in \{\{\text{z1}\}, \{\text{z2},\text{z1}\}, \{\text{z1},\text{ca}\}\} \right\}, \\ \mathcal{U}'_r &= \left\{ m' \;\middle|\; m'(\{\text{z2}, \text{ca}\}) = m'(\Omega) = 0 \right\}. \end{aligned}$$

◇

According to Definition 5.25 we introduce the following notations:

(6.40) Definition. *Let m be a mass distribution on Ω. We define*

$$\mathcal{U}_{\mathrm{cond}}(m) \stackrel{\mathrm{d}}{=} \{m' \mid m' = m(\,.\,|E) \text{ for } E \subseteq \Omega \text{ with } \mathrm{Bel}_m(E) > 0\}$$

$$\mathcal{U}_{\mathrm{revis}}(m) \stackrel{\mathrm{d}}{=} \{m' \mid m' = m(\,.\,|E) \text{ for } E \subseteq \Omega \text{ with } \mathrm{Pl}_m(E) > 0\}$$

$$\mathcal{U}^{\mathrm{refin}}_{\mathrm{cond}}(m) \stackrel{\mathrm{d}}{=} \{m' \mid m' \leq m\}$$

$$\mathcal{U}^{\mathrm{refin}}_{\mathrm{revis}}(m) \stackrel{\mathrm{d}}{=} \{m' \mid m' \sqsubseteq m\}$$

Since in the case of probabilities (on Ω) conditioning and revision coincide we define

$$\mathcal{U}_{\mathrm{prob}}(P) \stackrel{\mathrm{d}}{=} \{P' \mid P' = P(\,.\,|E) \text{ for some } E \subseteq \Omega \text{ with } P(E) > 0\}$$

$$\mathcal{U}^{\mathrm{refin}}_{\mathrm{prob}}(P) \stackrel{\mathrm{d}}{=} \{P' \mid P' \sqsubseteq P\}.$$

For sets $\mathcal{M}$ of mass distributions we define

$$\mathcal{U}_{\mathrm{cond}}(\mathcal{M}) \stackrel{\mathrm{d}}{=} \bigcup_{m\in\mathcal{M}} \mathcal{U}_{\mathrm{cond}}(m), \quad \mathcal{U}_{\mathrm{revis}}(\mathcal{M}) \stackrel{\mathrm{d}}{=} \bigcup_{m\in\mathcal{M}} \mathcal{U}_{\mathrm{revis}}(m),$$

$$\mathcal{U}^{\mathrm{refin}}_{\mathrm{cond}}(\mathcal{M}) \stackrel{\mathrm{d}}{=} \bigcup_{m\in\mathcal{M}} \mathcal{U}^{\mathrm{refin}}_{\mathrm{cond}}(m), \quad \mathcal{U}^{\mathrm{refin}}_{\mathrm{revis}}(\mathcal{M}) \stackrel{\mathrm{d}}{=} \bigcup_{m\in\mathcal{M}} \mathcal{U}^{\mathrm{refin}}_{\mathrm{revis}}(m),$$

and in case of a set $\mathcal{P}$ of probabilities (on Ω)

$$\mathcal{U}_{\mathrm{prob}}(\mathcal{P}) \stackrel{\mathrm{d}}{=} \bigcup_{P\in\mathcal{P}} \mathcal{U}_{\mathrm{cond}}(P), \quad \mathcal{U}^{\mathrm{refin}}_{\mathrm{prob}}(\mathcal{P}) \stackrel{\mathrm{d}}{=} \bigcup_{P\in\mathcal{P}} \mathcal{U}^{\mathrm{refin}}_{\mathrm{cond}}(P)$$

In analogy to equation (7.1) due to the transitivity of the relation $\sqsubseteq$ we obtain for updates of mass distributions

$$\begin{array}{lllllll}
m' & \in & \mathcal{U}_{\mathrm{cond}}(m) & \Rightarrow & \mathcal{U}_{\mathrm{cond}}(m') & \subseteq & \mathcal{U}_{\mathrm{cond}}(m) \\
m' & \in & \mathcal{U}^{\mathrm{refin}}_{\mathrm{cond}}(m) & \Rightarrow & \mathcal{U}^{\mathrm{refin}}_{\mathrm{cond}}(m') & \subseteq & \mathcal{U}^{\mathrm{refin}}_{\mathrm{cond}}(m) \\
m' & \in & \mathcal{U}_{\mathrm{revis}}(m) & \Rightarrow & \mathcal{U}_{\mathrm{revis}}(m') & \subseteq & \mathcal{U}_{\mathrm{revis}}(m) \\
m' & \in & \mathcal{U}^{\mathrm{refin}}_{\mathrm{revis}}(m) & \Rightarrow & \mathcal{U}^{\mathrm{refin}}_{\mathrm{revis}}(m') & \subseteq & \mathcal{U}^{\mathrm{refin}}_{\mathrm{revis}}(m) \\
P' & \in & \mathcal{U}_{\mathrm{prob}}(P) & \Rightarrow & \mathcal{U}_{\mathrm{prob}}(P') & \subseteq & \mathcal{U}_{\mathrm{prob}}(P) \\
P' & \in & \mathcal{U}^{\mathrm{refin}}_{\mathrm{prob}}(P) & \Rightarrow & \mathcal{U}^{\mathrm{refin}}_{\mathrm{prob}}(P') & \subseteq & \mathcal{U}^{\mathrm{refin}}_{\mathrm{prob}}(P)
\end{array} \qquad (6.40)$$

So far we have been concerned with the problem of updating pieces of information; next we have to focus on the *combination* of knowledge representation structures. Recall that we assume these knowledge representation structures to be induced by information sources but that due to the inaccessibility of the sources the combination has to be performed in terms of Ω. This is possible if we require some qualitative evaluation of the sources. The most interesting case leads to Dempster's well known rule of combination.

According to Theorem 5.29(a) we define the conjunctive combination of two mass distributions m_1 and m_2 by setting

$$(m_1 \oplus m_2)(A) \stackrel{\mathrm{d}}{=} \frac{\sum\limits_{A',A'':A'\cap A''=A} m_1(A') \cdot m_2(A'')}{\sum\limits_{A',A'':A'\cap A''\neq\emptyset} m_1(A') \cdot m_2(A'')} \tag{6.41}$$

if $A \neq \emptyset$, and $(m_1 \oplus m_2)(\emptyset) \stackrel{\mathrm{d}}{=} 0$. The conjunctive combination exists, if $\sum\limits_{A',A'':A'\cap A''=A} m_1(A') \cdot m_2(A'') > 0$. This rule is called *Dempster's rule of combination*. The operator $\oplus$ has very pleasant properties: in the case of existence we have

$$m' \oplus m'' = m'' \oplus m' \text{ (commutativity)} \tag{6.42}$$

and

$$(m' \oplus m'') \oplus m''' = m' \oplus (m'' \oplus m''') \text{ (associativity).} \tag{6.43}$$

For a mass distribution m^E on Ω we obtain

$$m^E \oplus m = m_E \tag{6.44}$$

where m_E denotes the result of the revision of m with respect to E.

In the case of imprecise data, i.e. if the focal sets of m' and m'' are elements of 2^Ω, the calculation of the Dempster combination can be facilitated by the use of commonality functions, since

$$Q_{m'\oplus m''}(M) = K \cdot Q_{m'}(M) \cdot Q_{m''}(M) \tag{6.45}$$

holds for all $M \neq \emptyset$. The normalizing constant K is given by

$$K^{-1} = \sum_{N:N\neq\emptyset} (-1)^{|N|+1} \cdot Q_{m'}(N) \cdot Q_{m''}(N). \tag{6.46}$$

Returning to our original problem concerning the combination of information sources, it turns out that in terms of Dempster's rule of combination, proposition (a) of Theorem 5.29 reads

$$Q_{(\Gamma_1\cap\Gamma_2)[P_1\otimes P_2]}(N) = Q_{\Gamma_1[P_1]\oplus\Gamma_2[P_2]}(N) = K \cdot Q_{\Gamma_1[P_1]}(N) \cdot Q_{\Gamma_2[P_2]}(N),$$

and for proposition (b) we obtain

$$\begin{aligned}\mathrm{Bel}_{(\Gamma_1\cup\Gamma_2)[P_1\otimes P_2]}(N) &= \sum_{M:M\subseteq N} (\Gamma_1\cup\Gamma_2)[P_1\otimes P_2](M)\\ &= \sum_{M',M'':M'\cup M''=M} \Gamma_1[P_1](M')\cdot\Gamma_2[P_2](M'')\\ &= \Big(\sum_{M':M'\subseteq M}\Gamma_1[P_1](M')\Big)\cdot\Big(\sum_{M'':M''\subseteq M}\Gamma_2[P_2](M'')\Big)\\ &= \mathrm{Bel}_{\Gamma_1[P_1]}(M)\cdot\mathrm{Bel}_{\Gamma_2[P_2]}(M)\end{aligned}$$

for all $N \subseteq \Omega$.

Obviously the operator $\oplus$ covers the conjunctive combination of independent sources.

In the case of different frames of discernment we obtain

$$\begin{aligned}&\Gamma[P_1\otimes P_2](A_1\times A_2)\\ &= \frac{(P_1\times P_2)(\{(\theta_1,\theta_2)\in\Theta_1\times\Theta_2 \mid \Gamma_1(\theta_1)\times\Gamma_2(\theta_2)=A_1\times A_2\})}{(P_1\times P_2)(\{(\theta_1,\theta_2)\in\Theta_1\times\Theta_2 \mid \Gamma_1(\theta_1)\times\Gamma_2(\theta_2)\neq\emptyset\})}\\ &= \frac{P_1(\{\theta_1\in\Theta_1 \mid \Gamma_1(\theta_1)=A_1\})\cdot P_2(\{\theta_2\in\Theta_2 \mid \Gamma_2(\theta_2)=A_2\})}{P_1(\{\theta_1\in\Theta_1 \mid \Gamma_1(\theta_1)\neq\emptyset\})\cdot P_2(\{\theta_2\in\Theta_2 \mid \Gamma_2(\theta_2)\neq\emptyset\})}\\ &= \Gamma_1[P_1](A_1)\cdot\Gamma_2[P_2](A_2) \qquad (6.47)\end{aligned}$$

for all $A_1 \subseteq \Omega_1$, $A_2 \subseteq \Omega_2$, and $\Gamma[P_1 \otimes P_2](B) = 0$ otherwise. This means that the induced mass distribution $\Gamma[P_1 \otimes P_2]$ is completely determined by $\Gamma[P_1]$ and $\Gamma[P_2]$, i.e. by mass distributions on Ω_1 and Ω_2. Note that the result means nothing but the Dempster combination of the vacuous extensions of $\Gamma[P_1]$ and $\Gamma[P_2]$, i.e. $\Gamma[P_1 \otimes P_2](B) = \widehat{\Pi}_1(\Gamma_1[P_1]) \oplus \widehat{\Pi}_2(\Gamma_2[P_2])$. Since the cylinders of focal sets always intersect, in this case the normalizing determinator always equals 1.

In order to define a reasonable combination of two *sets* of mass distributions $\mathcal{N}'$ and $\mathcal{N}''$ (written $\mathrm{comb}(\mathcal{N}',\mathcal{N}'')$), we define

$$\mathrm{comb}(\mathcal{N}',\mathcal{N}'') \stackrel{\mathrm{d}}{=} \bigcup_{\substack{n'\in\mathcal{N}'\\ n''\in\mathcal{N}''}} \mathrm{comb}(n',n''), \qquad (6.48)$$

where $\mathrm{comb}(n',n'')$ is a set of mass distributions on $\mathcal{F}_L(\Omega'\times\Omega'')$. It is reasonable to require that the elements $n \in \mathrm{comb}(n',n'')$ satisfy the following *weak marginal conditions*:

$$\Pi' n \sqsubseteq n' \qquad (6.49)$$

$$\Pi'' n \sqsubseteq n'' \qquad (6.50)$$

(Note that Π' and Π'' are projection mappings applied to mass distributions as pointed out in Sect. 5.2). The next theorem shows that the mass distribution $n' \otimes n''$ defined by

$$(n' \otimes n'')(\eta) \stackrel{\mathrm{d}}{=} \sum_{\substack{\eta',\eta'' \in \mathcal{F}_L(\Omega' \times \Omega''): \\ \eta' \cap \eta'' = \eta}} \widehat{\Pi}' n'(\eta') \cdot \widehat{\Pi}'' n''(\eta'')$$

$$= \sum_{\substack{\eta' \in \mathcal{F}_L(\Omega'), \eta'' \in \mathcal{F}_L(\Omega''): \\ \eta' \times \eta'' = \eta}} n'(\eta') \cdot n''(\eta'') \tag{6.51}$$

for all $\eta \in \mathcal{F}_L(\Omega' \times \Omega'')$ plays a special role for combining information sources.

(6.41) Theorem. *Let n' and n'' be two mass distributions defined on $\mathcal{F}_L(\Omega')$ and $\mathcal{F}_L(\Omega'')$, respectively, and let $n' \otimes n''$ be a mass distribution defined on $\mathcal{F}_L(\Omega' \times \Omega'')$ by*

$$(n' \otimes n'')(\eta) \stackrel{\mathrm{d}}{=} \sum_{\substack{\eta' \in \mathcal{F}_L(\Omega'), \eta'' \in \mathcal{F}_L(\Omega''): \\ \eta' \times \eta'' = \eta}} n'(\eta') \cdot n''(\eta'')$$

for all $\mathcal{F}_L(\Omega' \times \Omega'')$. In this case we have

(a) $n' \otimes n''$ satisfies the weak marginal conditions, and

(b) if m satisfies the weak marginal conditions, then $m \sqsubseteq (n' \otimes n'')$.

Proof. (a) Let $\eta' \in \mathcal{F}_L(\Omega')$ with $\Pi'(n' \otimes n'')(\eta') > 0$. We have to show that some set $\mu' \supseteq \eta'$, $n'(\mu') > 0$ exists. From the definition of $n' \otimes n''$ we know

$$0 < \Pi'(n' \otimes n'')(\eta') = \sum_{\eta' : \Pi'(\eta) = \eta'} \sum_{\mu', \mu'' : \mu = \mu' \times \mu''} n'(\eta') \cdot n''(\eta'').$$

Thus there are μ', μ'' such that $n'(\eta') \cdot n''(\eta'') > 0$ and $\Pi'(\mu' \times \mu'')(\eta') = \eta'$. Let $\theta' \in \Omega'$, we have

$$\eta'(\theta') = \Pi'(\mu' \times \mu'')(\theta') = \sup_{\theta'' \in \Omega''} \left\{ \inf\{\mu'(\theta'), \mu''(\theta'')\} \right\} \leq \mu'(\theta'),$$

which implies that means $\eta' \subseteq \theta'$.

(b) Let $\eta \in \mathcal{F}_L(\Omega' \times \Omega'')$ with $m(\eta) > 0$. We have to prove the existence of some set $\mu \supseteq \eta$ such that $(n' \otimes n'')(\eta') > 0$. Define $n' \stackrel{\mathrm{d}}{=} \Pi'(\eta)$, $n'' \stackrel{\mathrm{d}}{=} \Pi''(\eta)$. We derive

$$\Pi'(m)(\eta') \geq m(\eta) > 0 \text{ and } \Pi''(m)(\eta'') \geq m(\eta) > 0.$$

Since we assumed m to satisfy the weak marginal condition there exist sets $\mu' \supseteq \eta'$, $\mu'' \supseteq \eta''$ such that $n'(\eta') > 0$ and $n''(\eta'') > 0$. Define $\mu \stackrel{\mathrm{d}}{=} \mu' \times \mu''$, this leads to $(n' \otimes n'')(\mu) > 0$. What remains to be shown is that $\mu \supseteq \eta$: Letting $\eta' \in \Omega'$, $\eta'' \in \Omega''$,

$$\begin{aligned}\mu(\theta',\theta'') &= \inf\{\mu'(\theta'),\mu''(\theta'')\} \geq \inf\{\eta'(\theta'),\eta''(\theta'')\} \\ &= \inf\Big\{\sup_{w''}\{\eta'(\theta',w'')\},\sup_{w'}\{\eta'(w',\theta'')\}\Big\} \geq \eta(\theta',\theta'').\end{aligned}$$

□

(6.42) *Remark.* Note that in the case of ordinary mass distributions

$$n' : 2^{\Omega'} \to [0,1], \quad n'' : 2^{\Omega''} \to [0,1],$$

the combined mass distribution $n' \otimes n''$ satisfies even the marginal conditions, i.e. $\Pi'(n' \otimes n'') = n'$ and $\Pi''(n' \otimes n'') = n''$, while for generalized mass distributions this is in general *not* valid. ○

6.6.2 Integration of Different Pieces of Knowledge

In Chap. 5 we assumed the knowledge representation structures on Ω to be induced by explicitly specified information sources. Now we will consider the problem of integrating knowledge representation structures on Ω *without* the internal structure of the underlying information sources.

Although we do not know the original sources $\mathcal{S}_1, \ldots, \mathcal{S}_n$ we are aware of their existence and of the fact that they may be of different importance. In the sequel we assume that some *meta-expert* is able to assign a weight q_i to each source $\mathcal{S}_i$, $i = 1, \ldots, n$. We can imagine these numbers to be degrees of importance as well as to be the relative frequencies arising from a random choice among the sources. Of course these weights are strongly related to the reliability of the sources. In the general case, where we have n knowledge representation structures

$$\mathcal{K}_j = \big(\mathcal{F}_L(\Omega), \mathcal{P}_j, \mathcal{U}_{\text{prob}}(\mathcal{P}_j)\big), \quad j = 1, \ldots, n,$$

randomly selected with the probability q_i, as the integrated knowledge representation structure we obtain

$$\big(\mathcal{F}_L(\Omega), q_1 \cdot \mathcal{P}_1 + \ldots + q_n \cdot \mathcal{P}_n, \mathcal{U}_{\text{prob}}(q_1 \cdot \mathcal{P}_1 + \ldots + q_n \cdot \mathcal{P}_n)\big)$$

where

$$\begin{aligned}&q_1 \cdot \mathcal{P}_1 + \ldots + q_n \cdot \mathcal{P}_n = \\ &\qquad \{Q \mid Q(A) = \sum_{j=1}^{n} q_j \cdot Q_j(A), \text{ where } Q_j \in \mathcal{P}_j, j = 1, \ldots, n\}\end{aligned}$$

Note that we obtain

$$\mathcal{U}_{\mathrm{prob}}(q_1 \cdot \mathcal{P}_1 + \ldots + q_n \cdot \mathcal{P}_n)) \neq q_1 \cdot \mathcal{U}_{\mathrm{prob}}(\mathcal{P}_1) + \ldots + q_n \cdot \mathcal{U}_{\mathrm{prob}}(\mathcal{P}_n)).$$

From a semantical point of view we have various sensors, each providing vague data about the evidence space Ω. The concept of an information source comprises those sensors being under control of some *expert* who is able to valuate what he obtains from the sensors at his disposal. By the expert's valuation of the sensors under his control and these sensors' data a probability or, in the most general case considered here, a set of mass distributions on $\mathcal{F}_L(\Omega)$ is determined.

If the integration of different information sources is required, then we need some even higher instance, i.e. a *meta-expert*, who provides the probabilities q_i as well as the updating schemes. He has also to fix explicitly the sets of possible updates. If no meta-expert is able to give such detailed information, then an alternative approach may be used, which mainly relies on the sets of possible updates the experts directly specify.

Now assume that a meta-expert is unable to *quantify* the reliability of the different sources, i.e. knowledge representation structures, but he can provide statements like

> "The information we obtain from source 1 <u>and</u> the information of source 2 <u>or</u> source 3 should be taken into account, but don't care about any information from source 4"

Such items of meta-knowledge can be formalized by logical expressions. The following grammar defines the syntax of such expressions. Let $\{1, \ldots, n\}$ be the space of the knowledge representation structures $\mathcal{K}_1, \ldots, \mathcal{K}_n$, i.e. we have n sources:

(i) i is an expression for $i \in \{1, \ldots, n\}$
(ii) if A and B are expressions, then so are (A <u>and</u> B) as well as (A <u>or</u> B),
(iii) all expressions are defined by (1) and (2).

With this definition the above statement leads to the expression

$$(1 \ \underline{\text{and}}\ (2 \ \underline{\text{or}}\ 3)).$$

The integration of n structures, $i = 1, \ldots, n$, $(\mathcal{F}_L(\Omega), \mathcal{M}_i, \mathcal{U}_i)$ with respect to a logical expression relies mainly on set operations on the sets $\mathcal{U}_i$. We define

$$\begin{aligned} &(i)\ \mathcal{U}(i) = \mathcal{U}_i \\ &(ii)\ \mathcal{U}(A \ \underline{\text{and}}\ B) = \mathcal{U}(A) \cap \mathcal{U}(B) \\ &(iii)\ \mathcal{U}(A \ \underline{\text{or}}\ B) = \mathcal{U}(A) \cup \mathcal{U}(B). \end{aligned} \tag{6.52}$$

This clarifies the semantics of the logical expressions. In our above example we obtain $\mathcal{U}(1 \text{ \underline{and} } (2 \text{ \underline{or} } 3)) = \mathcal{U}_1 \cap (\mathcal{U}_2 \cup \mathcal{U}_3)$. The underlying basic idea is that an information source is essentially determined by the respective set of possible updates. Thus our aim is to obtain the set $\mathcal{U}(T)$, i.e. the set of possible updates of the integrated source, from the sets $\mathcal{U}_i$, $i \in \{1, \ldots, n\}$, where T denotes the logical expression which characterizes the mutual relations of the single sources. Doing so is reasonable, at least insofar as we always require $\mathcal{M}_i \subseteq \mathcal{U}_i$. Thus we define the integrated source by

$$\text{int}_T\Big((\mathcal{F}_L(\Omega), \mathcal{M}_1, \mathcal{U}_1), \ldots, (\mathcal{F}_L(\Omega), \mathcal{M}_n, \mathcal{U}_n)\Big) = \big(\mathcal{F}_L(\Omega), \mathcal{U}(T), \mathcal{U}(T)\big). \tag{6.53}$$

(6.43) Example. Suppose there are two knowledge representation structures $\mathcal{K}_1$, $\mathcal{K}_2$, encoding information about a ship under consideration with respect to the frame of discernment $\Omega = \{\text{z3}, \ldots, \text{z1}, \text{ca}\}$. From $\mathcal{K}_1$ we obtain the statement

"The ship is far away from the coast"

which is specified by the L-set μ, where $L = \{\text{white}, \text{grey}, \text{black}\}$:

Ω	z3	z2	z1	ca	rd	ld
$\mu(\omega)$	grey	grey	black	white	white	white

$\mathcal{K}_2$ relies on the data of four sensors, specifying (crisp) regions restricting the set of possible locations. The reliability of the sensor outputs is quantified by a mass distribution m'' on Ω, where

$$\begin{aligned} m''(\{\text{z1}\}) &= \frac{1}{2}, \\ m''(\{\text{z1}, \text{ca}\}) &= \frac{1}{4}, \\ m''(\{\text{z2}, \text{z1}\}) &= \frac{1}{4}, \\ m''(\mu) &= 0 \text{ otherwise.} \end{aligned}$$

In terms of our general approach we have $\mathcal{M}_1 = \{m'\}$, where $m' = m^{\mu}$ and $\mathcal{M}_2 = \{m''\}$. Assume the respective experts specify the sets $\mathcal{U}_1 = \mathcal{U}^{\text{refin}}_{\text{revis}}(\mathcal{M}_1)$, $\mathcal{U}_2 = \mathcal{U}^{\text{refin}}_{\text{revis}}(\mathcal{M}_2)$, containing the possible updates. Assume furthermore that the meta-expert provides the logical expression $T = (1 \text{ \underline{and} } 2)$ as his evaluation of the two sources. We obtain

$$\mathcal{U}(T) = \mathcal{U}_1 \cap \mathcal{U}_2 = \{m \mid m \sqsubseteq m'\} \cap \{m \mid m \sqsubseteq m''\},$$

containing those mass distributions which are admissible from expert 1's as well as expert 2's point of view. ◇

As the above example shows, the intersection of sets containing specializations of a mass distribution is a problem that naturally arises if we aspire to integrate knowledge representation structures. Therefore we provide the following theorem which solves this problem in its most general form.

(6.44) Theorem. *Let m' and m'' be two mass distributions on $\mathcal{F}_L(\Omega)$. Then*

$$\{m \mid m \sqsubseteq m'\} \cap \{m \mid m \sqsubseteq m''\} = \begin{cases} \{m \mid m \sqsubseteq m' \oplus m''\}, & \textit{if } m' \oplus m'' \textit{ exists,} \\ \emptyset & \textit{otherwise.} \end{cases}$$

Proof. (a) $m' \oplus m''$ exists, if and only if $\sum_{\mu',\mu'':\mu'\cap\mu''=\emptyset} m'(\mu') \cdot m''(\mu'') < 1.$

"$\subseteq$" Let $m \sqsubseteq m'$ and $m \sqsubseteq m''$, and let μ be an element of $\mathcal{F}_L(\Omega)$, such that $Q_{m'\oplus m''}(\mu) = 0$. We have to show that in this case the assumptions imply $Q_m(\mu) = 0$. We have

$$\sum_{\eta:\eta\supseteq\mu} (m' \oplus m'')(\eta) = 0,$$

and therefore

$$\sum_{\mu',\mu'':\mu'\cap\mu''=\emptyset} m'(\mu') \cdot m''(\mu'') = 0$$

for all $\eta \supseteq \mu$. Suppose $Q_{m'}(\mu) > 0$ and $Q_{m''}(\mu) > 0$. Then there are $\mu\prime$, μ'' with $\mu' \supseteq \mu$ such that $m'(\mu') \cdot m''(\mu'') = 0$ which is a contradiction. Therefore we have $Q_{m'}(\mu) = 0$ or $Q_{m''}(\mu) = 0$, i.e. $m \sqsubseteq m'$ or $m \sqsubseteq m''$. Thus we obtain $Q_m(\mu) = 0$.

"$\supseteq$" We have to prove that $m' \oplus m'' \sqsubseteq m'$ and $m' \oplus m'' \sqsubseteq m''$ hold. Then from the transitivity of the relation $\sqsubseteq$ it follows that $m \sqsubseteq m' \oplus m''$ implies $m \sqsubseteq m'$ and $m \sqsubseteq m''$. Consider $\mu \in \mathcal{F}_L(\Omega)$ with $Q_{m'}(\mu) = 0$. We obtain

$$Q_{m'\oplus m''}(\mu) = \frac{\sum_{\mu',\mu'':\mu'\cap\mu''\supseteq\mu} m'(\mu') \cdot m''(\mu'')}{1 - \sum_{\mu',\mu'':\mu'\cap\mu''=\emptyset} m'(\mu') \cdot m''(\mu'')} = 0,$$

since $m'(\mu') = 0$ for all $\mu' \supseteq \mu$, and it follows that $m' \oplus m'' \sqsubseteq m'$. $m' \oplus m'' \sqsubseteq m$ can be verified analogously.

(b) Suppose $\sum_{\mu',\mu'':\mu'\cap\mu''=\emptyset} m'(\mu') \cdot m''(\mu'') = 1$. Let m be an element of the set $\mathcal{M} = \{m \mid m \sqsubseteq m'\} \cap \{m \mid m \sqsubseteq m''\}$. From $m \sqsubseteq m'$, $m \sqsubseteq m''$, and

$\big(Q_{m'}(\mu) = 0$ or $Q_{m''}(\mu) = 0\big)$ follows $Q_m(\mu) = 0$. Consider $\mu \neq \emptyset$ with $m(\mu) > 0$. Then $Q_m(\mu) > 0$ implies $Q_{m'}(\mu) > 0$ and $Q_{m''}(\mu) > 0$. Therefore

$$\sum_{\mu', \mu'' : \mu' \cap \mu'' \supseteq \mu} m'(\mu') \cdot m''(\mu'') > 0$$

which contradicts our assumption. So for all $\mu \neq \emptyset$ we have $m(\mu) = 0$. Obviously this contradicts the definition of a mass distribution and shows that elements of $\mathcal{M}$ cannot exist, i.e. $\mathcal{M} = \emptyset$.

□

Applying this result to our original problem, i.e. the integration of two knowledge representation structures $\mathcal{K}_1$ and $\mathcal{K}_2$ with respect to the logical expression $T = (1 \ \underline{\text{and}} \ 2)$ where $U_1 = \mathcal{U}^{\text{refin}}_{\text{revis}}(\{m'\})$ and $U_2 = \mathcal{U}^{\text{refin}}_{\text{revis}}(\{m''\})$, we obtain

$$U(T) = \mathcal{U}^{\text{refin}}_{\text{revis}}(\{m'\}) \cap \mathcal{U}^{\text{refin}}_{\text{revis}}(\{m''\}) = \mathcal{U}^{\text{refin}}_{\text{revis}}(\{m' \oplus m''\}).$$

(6.45) Example. (continued) According to the above considerations the integration of the sources 1 and 2 with the respect to $T = 1 \ \underline{\text{and}} \ 2$ yields

$$\mathcal{U}(T) = \mathcal{U}_1 \cap \mathcal{U}_2 = \{m \mid m \sqsubseteq m' \oplus m''\},$$

We obtain

$$(m' \oplus m') : \mathcal{F}_L(\Omega) \longrightarrow 0,1; \quad (m' \oplus m'')(\nu) = \begin{cases} \frac{3}{4} & \text{if } \nu = \mu_1 \ , \\ \frac{1}{4} & \text{if } \nu = \mu_2 \ , \\ 0 & \text{otherwise,} \end{cases}$$

where

$$\mu_1 = 1\!\!\mathrm{I}_{\{\mathrm{z1}\}} \cap \mu = 1\!\!\mathrm{I}_{\{\mathrm{z1,ca}\}} \cap \mu, \quad \text{and} \quad \mu_2 = 1\!\!\mathrm{I}_{\{\mathrm{z1,z1}\}} \cap \mu,$$

which implies

	z3	z2	z1	ca
$\mu_1(\omega)$	white	white	black	white
$\mu_2(\omega)$	white	grey	black	white

In practical applications it is reasonable to use the mass distribution $m' \oplus m''$ as a representative of the set $\{m | m \sqsubseteq m' \oplus m''\}$, in order to facilitate the calculations.

◇

Besides $m' \oplus m''$ in general there are infinitely many mass distribution t, such that

$$\{m|m \sqsubseteq m'\} \cap \{m|m \sqsubseteq m''\} = \{m|m \sqsubseteq t\},$$

thus $m' \oplus m''$ is not a unique solution but it is a good choice since it can be easily derived. In order to prove this ambiguity we define, for all mass distributions m,

$$\mathcal{A}_m = \{\mu \mid \mu \neq \mathbb{I}_\emptyset \text{ and } Q_m(\mu) > 0\}. \tag{6.54}$$

The equality

$$\sum_{\mu',\mu'':\mu' \cap \mu'' \neq \emptyset} m'(\mu') \cdot m''(\mu'') = 0 \tag{6.55}$$

holds if and only if $\mathcal{A}_{m'} \cap \mathcal{A}_{m''} = \emptyset$. Otherwise, i.e. if $\mathcal{A}_{m'} \cap \mathcal{A}_{m''} \neq \emptyset$, the following properties are equivalent:

$$\begin{aligned} &(i)\ \{m \mid m \sqsubseteq m'\} \cap \{m \mid m \sqsubseteq m''\} = \{m \mid m \sqsubseteq t\} \\ &(ii)\ \mathcal{A}_{m'} \cap \mathcal{A}_{m''} = \mathcal{A}_t, \\ &(iii)\ \mathcal{A}_t = \mathcal{A}_{m' \oplus m''}, \\ &(iv)\ t \sqsubseteq m' \oplus m'' \text{ and } m' \oplus m'' \sqsubseteq t. \end{aligned} \tag{6.56}$$

If $\mathcal{A} \neq \emptyset$ is a subset of $\mathcal{F}_L(\Omega)$ such that $\mu \in \mathcal{A}$ and $\eta \subseteq \mu$ implies $\eta \in \mathcal{A}$, then a mass distribution t with the property $\mathcal{A}_t = \mathcal{A}$ can be easily constructed: the set

$$\mathcal{M}_{\mathcal{A}} = \{\mu \in \mathcal{A} \mid \neg(\exists \mu \in \mathcal{A} : \eta \subseteq \mu \text{ and } \eta \neq \mu)\}$$

contains the "maximal" elements of $\mathcal{A}$. We have to choose a mass distribution t such that $t(\mu) > 0$ for all $\mu \in \mathcal{M}_{\mathcal{A}}$. If $t(\eta) > 0$ implies that there is some $\mu \in \mathcal{M}$ with $\eta \subseteq \mu$, then we have $\mathcal{A} = \mathcal{A}_t$.

Roughly speaking, the sets $\mathcal{M}_{\mathcal{A}_t}$ carry the essential information with respect to the specialization relation carried by the mass distribution t. Thus we have the equivalence

$$(t' \sqsubseteq t'' \text{ and } t' \sqsupseteq t'') \Longleftrightarrow \mathcal{M}_{\mathcal{A}'_t} = \mathcal{M}_{\mathcal{A}''_t}. \tag{6.57}$$

(6.46) *Remark.* By analogy with Theorem 6.44 we obtain

$$\{m^A \mid A \subseteq R'\} \cap \{m^A \mid A \subseteq R''\} = \begin{cases} \{m^A \mid A \subseteq R' \cap R''\} & \text{if } R' \cap R'' \neq \emptyset, \\ \emptyset & \text{otherwise.} \end{cases}$$

Since $m^{R'} \oplus m^{R''} = m^{R' \cap R''}$ this implies that

$$\mathcal{U}_{\text{revis}}(\{m^{R'}\}) \cap \mathcal{U}_{\text{revis}}(\{m^{R''}\}) = \mathcal{U}_{\text{revis}}(\{m^{R'} \oplus m^{R''}\}) = \mathcal{U}_{\text{revis}}(\{m^{R' \cap R''}\}).$$

This relationship is of importance for handling either vagueness or imprecision alone. Thus in our introductory example the conjunctive integration of the two information sources

$$\left(2^{\Omega^{A\times L\times M}}, \{m^{R_{\theta_1}}\}, \mathcal{U}_{\text{revis}}\{m^{R_{\theta_1}}\}\right), \quad \left(2^{\Omega^{A\times L\times M}}, \{m^{R_{\theta_2}}\}, \mathcal{U}_{\text{revis}}\{m^{R_{\theta_2}}\}\right)$$

yields, according to equation (6.33),

$$(2^{\Omega^{A\times L\times M}}, \{m^{R_{\theta_1}} \cap m^{R_{\theta_2}}\}, \{m^A \mid A \subseteq R_{\theta_1} \cap R_{\theta_2}\}).$$

Without mentioning it we used $m^{R_{\theta_1} \cap R_{\theta_2}}$ as a representative of this set of mass distributions. ◦

Using the logical operator or, we have a logical expression of the form $T = A$ or B, and we have to calculate the union of sets $\mathcal{U}(A)$ and $\mathcal{U}(B)$. Once more let us consider the case of two sources 1 and 2, where the respective sets of possible updates are $\mathcal{U}(1) = \{m \mid m \sqsubseteq m'\}$ and $\mathcal{U}(2) = \{m \mid m \sqsubseteq m''\}$. The logical expression may be $T = 1$ or 2, thus we have to handle the set

$$\mathcal{U}(T) = \{m \mid m \sqsubseteq m'\} \cup \{m \mid m \sqsubseteq m''\}. \tag{6.58}$$

Again our interest is to find mass distributions t which can characterize this set. We obtain the following equality

$$\begin{aligned} \{m \mid m \sqsubseteq m'\} \cup \{m \mid m \sqsubseteq m''\} &\subseteq \{m \mid m \sqsubseteq t\} \\ \iff \mathcal{A}_{m'} \cup \mathcal{A}_{m''} &\subseteq \mathcal{A}_t. \end{aligned} \tag{6.59}$$

For the mass distribution u defined by

$$u(\mu) = \sum_{\mu',\mu'':\mu'\cup\mu''=\mu} m'(\mu') \cdot m''(\mu'') \quad \text{for all } \mu \in \mathcal{F}_L(\Omega) \tag{6.60}$$

we have $\mathcal{A}_{m'} \cup \mathcal{A}_{m''} \subseteq \mathcal{A}_u$, whereas the equality does not hold in general.

(6.47) Example. Consider the mass distributions m' and m'' on $\mathcal{F}_L(\Omega)$, where $\Omega = \{z3, z2, z1\}$, $m' = m^{\{z1\}}$ and $m'' = m^{\{z2,z3\}}$. Define furthermore the mass distribution r, where $r(\{z1\}) = r(\{z2, z3\}) = \frac{1}{2}$, $r(\mu) = 0$ otherwise. Obviously from ($m' \sqsubseteq t$ and $m'' \sqsubseteq t$) follows $r \sqsubseteq t$. For the mass distribution s defined by $s(\{z1\}) = s(\{z2, z3\}) = \frac{1}{2}$, $s(\mu) = 0$, otherwise we obtain $s \sqsubseteq r \sqsubseteq t$, but neither $s \sqsubseteq m'$ nor $s \sqsubseteq m''$ is valid. As this example shows, in general it is not possible to find a single mass distribution t where

$$\{m \mid m \sqsubseteq m'\} \cup \{m \mid m \sqsubseteq m''\} = \{m \mid m \sqsubseteq t\}.$$

So (in contrast to the conjunctive case) a *single* mass distribution is not sufficient to handle the disjunctive integration. ◇

As we pointed out in our introductory example, we assume that besides the expert knowledge there are the observations the user provides during the consultational dialogue. In order to model these pieces of knowledge we consider the consultation process as a special kind of information source. Since we assume that the consultational knowledge reflects just facts, we may have to deal with vagueness but not with uncertainty. Assume the user's vague observation is $\mu \in \mathcal{F}_L(\Omega)$. In this case we obtain

$$(\mathcal{F}_L(\Omega), \mathcal{M}_\mu, \mathcal{U}_\mu)$$

as the corresponding information source, where $\mathcal{M}_\mu = \{m^\mu\}$, and

$$\mathcal{U}_\mu = \{m \mid m \sqsubseteq m^\mu\}. \tag{6.61}$$

The integration of such a source and another source $(\mathcal{F}_L(\Omega), \mathcal{M}', \mathcal{U}')$ has to be done with respect to the logical expression $T = 1 \underline{\text{and}} 2$. We obtain

$$\mathcal{U} = \mathcal{U}' \cap \mathcal{U}_\mu \tag{6.62}$$

as the set of possible updates for the integrated source.

(6.48) Example. (a) Let $(\Omega, \{P\}, \mathcal{U})$ be an information source, where P is a probability, and let $\mathcal{U}_1 = \mathcal{U}_{\text{prob}}(P)$. If we know for sure that the result is within the set E, then we obtain a second source $(\mathcal{F}_L(\Omega), \{m^E\}, \mathcal{U}_E)$, where $\mathcal{U}_E = \{m \mid m \sqsubseteq m^E\}$. The integration leads to

$$\begin{aligned} \mathcal{U}(1 \underline{\text{ and }} 2) &= \mathcal{U}_1 \cap \mathcal{U}_E \\ &= \big\{Q \bigm| \exists E' \subseteq E, P(E') > 0 : Q = P(\,.\,|E'), \\ &\qquad \text{where } Q \text{ is a probability on } \Omega.\big\} \end{aligned} \tag{6.63}$$

In order to facilitate the calculations it is reasonable to use P and $P(\,.\,|E)$ as representatives of the sets $\mathcal{U}_1$ and $\mathcal{U}_1 \cap \mathcal{U}_E$, respectively.

(b) Recall that in our introductory example we considered two "experts" providing the relations R_{θ_1} and R_{θ_2}, respectively. As we pointed out in Remark 6.46 a conjunctive combination yields

$$\mathcal{U}(1 \underline{\text{ and }} 2) = \mathcal{U}_{\text{revis}}(\{m^{R_{\theta_1}}\}) \cap \mathcal{U}_{\text{revis}}\{m^{R_{\theta_2}}\}) = \mathcal{U}_{\text{revis}}(\{m^{R_{\theta_1} \cap R_{\theta_2}}\}).$$

The set E is a user observation, encoded by the knowledge representation structure $\mathcal{K}_3 = \Big(2^\Omega, \{m^E\}, \mathcal{U}_{\text{revis}}(\{m^E\})\Big)$. Integrating these three structures yields

$$\mathcal{U}((1 \underline{\text{ and }} 2) \underline{\text{ and }} 3) = \mathcal{U}_{\text{revis}}(\{m^{R_{\theta_1} \cap R_{\theta_2} \cap E}\}).$$

This set of mass distribution contains only simple support functions, therefore it corresponds to the power set of $R_{\theta_1} \cap R_{\theta_2} \cap E$. Once more we can use the mass distributions (or the corresponding sets) $m^{R_{\theta_1}\cap R_{\theta_2}}$ and $m^{R_{\theta_1}\cap R_{\theta_2}\cap E}$ instead of the sets $\mathcal{U}(1 \text{ \underline{and} } 2)$ and $\mathcal{U}((1 \text{ \underline{and} } 2) \text{ \underline{and} } 3))$ — as we did throughout Sect. 2.2 without mentioning it explicitly.

(c) Let $(2^{\Omega}, \{m_1\}, \mathcal{U}_1)$ be an information source on Ω, and let $\mathcal{U}_1 = \mathcal{U}_{\text{revis}}(\{m_1\})$. In order to integrate an observation $E \in 2^{\Omega}$ we have to calculate

$$\mathcal{U}(1 \text{ \underline{and} } 2) = \{m \mid \exists E' \subseteq E, \mathrm{Pl}(E') > 0 : m = (m_1)_{E'}\}$$

The representatives of $\mathcal{U}_1$ and $\mathcal{U}(1 \text{ \underline{and} } 2)$ are m_1 and $(m_1)_E$, respectively.

◇

The Examples 6.48 (a), (b) and (c) define typical special cases for which prototypical software tools have been implemented by the authors' working group. In the following, the theoretical basis of these implementations is considered thoroughly.

6.7 Simplifying Assumptions

As explained in the previous sections, we consider an expert system to consist of sensor spaces $\Theta_1, \ldots, \Theta_n$ and a data space or frame of discernment Ω which is subject to the sensors' observations. This knowledge is represented by knowledge representation structures defined on Ω; in the most general case we have to deal with a set $\mathcal{U}$ of mass distributions on $\mathcal{F}_L(\Omega)$ as pointed out in Sect. 6.6. In practice the appropriate type of knowledge representation structure depends on the type of uncertainty affecting the actual domain. In this section we will discuss assumptions concerning the structures of the expert spaces Θ_i and the types of uncertainty that appear. Additional assumptions concerning the evidence space Ω made in order to derive suitable algorithms will be considered later on.

The first assumption is that the domain is affected either with uncertainty (in connection with precise or imprecise data) or with vagueness but not with combinations of uncertainty and vagueness in its most general form. Therefore the knowledge representation structures on Ω, which we imagine to be induced by hidden information sources defined on $\Theta_1, \ldots, \Theta_n$, are of homogenous structure.

(a) Let us firstly consider the case of vagueness in the absence of uncertainty. We obtain knowledge representation structures $\mathcal{K}_i = (\mathcal{F}_L(\Omega), \mathcal{M}_i, \mathcal{U}_i)$, $i = 1, \ldots, n$, where $\mathcal{M}_i = \{m^{R_i}\}$. So in essence the respective sensors provide a vague relation R_i on $\mathcal{F}_L(\Omega)$ and the process of integrating and revising mass distribution is just set intersection, as we pointed out in Sect. 6.6.2.

The information is provided under the proviso, that additional knowledge arising from observations of the consulting user may require an updating process. According to Definition 6.5 we obtain

$$\mathcal{U}_i = \mathcal{U}_{\text{revis}}(\mathcal{M}_i) = \{m' \mid m' = (m^{R_i})_\eta \text{ for some } \eta \in \mathcal{F}_L\} \ .$$

Thus we have to deal with information sources of the form

$$\mathcal{K}_i = (\mathcal{F}_L(\Omega), \{m^{R_i}\}, \mathcal{U}_{\text{revis}}(\{m^{R_i}\})).$$

(b) In the case of uncertainty without any kind of vagueness we obtain knowledge representation structures of the form $\mathcal{K}_i = (\Omega, \mathcal{P}_i, \mathcal{U}_i)$, where $\mathcal{P}_i$ contains probabilities on Ω. This leads to the set

$$\mathcal{U}_i = \mathcal{U}_{\text{prob}}(\mathcal{P}_i) = \{P' \mid P' = P_i(\ . \mid E) \text{ for some } E \subseteq \Omega\} \ ,$$

i.e. we have to deal with knowledge representation structures of the form $\mathcal{K}_i = (\Omega, \{P_i\}, U_{\text{prob}}(\{P_i\}))$.

(c) If uncertainty and imprecision occur in parallel, we obtain knowledge representation structures of the form $\mathcal{K}_i = (2^\Omega, \mathcal{M}_i, \mathcal{U}_i)$, where $\mathcal{M}_i$ contains mass distrutions on Ω. Since in this case we require a general treatment we obtain

$$\mathcal{U}_i = \mathcal{U}^{\text{refin}}_{\text{revis}}(\mathcal{M}_i) = \{m' \mid \exists m \in \mathcal{M}_i : m' \sqsubseteq m\},$$

so we have $\mathcal{K}_i = (2^\Omega, \{m_i\}, \mathcal{U}^{\text{refin}}_{\text{revs}}(\{m_i\}))$.

As we pointed out initially we imagine the information sources to be induced by respective expert spaces $\Theta_1, \ldots, \Theta_n$. Each of these sources we assume to be equally reliable, so the appropriate method for the integration of the different sources is to perform a logical integration with respect to the logical expression

$$T = 1 \ \underline{\text{and}} \ \ldots \ \underline{\text{and}} \ n \ .$$

According to equation (6.52) we obtain

$$\mathcal{U} = \mathcal{U}(T) = \bigcap_{i=1}^{n} \mathcal{U}_i$$

The integration of the actual user observations has to be done according to Sect. 6.6.2 by the conjunctive integration of some special knowledge representation structures. In the case of vagueness such an additional information source is characterized by the set

$$\mathcal{U}_E = \{m' \mid m' = m^\eta \ , \eta \subseteq \mathbb{1}_E\},$$

whereas in the case of probabilities we obtain

$$\mathcal{U}_E = \{P' \mid P'(E) = 1\},$$

and if mass distributions are used this yields

$$\mathcal{U}_E = \{m' \mid \mathrm{Bel}_{m'}(E) = 1\} .$$

If the actual observation is $E = \{\omega_0\}$, which is an idealized but interesting case, then the L–set approach yields the simple support function $m^{\{\omega_0\}}$, the probabilistic approach leads to $P(\{\omega_0\}) = 1$ and the use of mass distribution yields also $m^{\{\omega_0\}}$.

With these considerations we have now passed the treatment of aggregating the experts' opinions and reached the problem of the combination of structural and consultational knowledge. Except in the idealized case of exact knowledge this requires additional assumptions on the evidence space, at least if we go beyond our previous efforts on a very high level of abstraction and aim at the development of practicable algorithms.

Thus the next step in the direction of applicable methods is to have a closer look at the evidence space Ω. Up to now our only assumption was that Ω is finite. In the sequel we will investigate the structure of Ω in more detail in order to use additional assumptions for a further development of suitable knowledge representation formalisms.

Our fundamental approach is to model the world, i.e. the actual domain, by a number of equal characteristics. These characteristics are the basic subject of observations, of questions, of knowledge, and of thinking about the domain. The characteristics of interest, which may be denoted as $X^{(1)}, \ldots, X^{(m)}$, are given canonically by the definition of the scope some system has to cover, for example the scales of measuring instruments, the output of sensors, and physical indices like distance, size, and duration. There may also be abstract characteristics like "type" or "activity". Each characteristic $X^{(i)}$, $i = 1, \ldots, m$, has its corresponding set of possible values $\Omega^{(i)}$, $i = 1, \ldots, m$; in the case of physical characteristics these sets contain, e.g., the range of a physical scale, in the case of abstract characteristics they contain abstract entities.

Our most important assumption concerns these sets $\Omega^{(i)}$, $i = 1, \ldots, m$. We assume them to be finite but fine enough to model all knowledge about $X^{(i)}$; the only simplifying coarsening lies in the renunciation of *additional characteristics.* Once more, we assume the sets $\Omega^{(i)}$, $i = 1, \ldots m$, to model the characteristics $X^{(i)}$ to *sufficient depth.*

The sets $\Omega^{(i)}$, $i \in M = \{1, \ldots, m\}$, form an m–dimensional product space $\Omega^M = \Omega^{(1)} \times \ldots \times \Omega^{(m)}$ as defined in Sect. 2.1, and each element of Ω^M is a vector ω^M, representing one state of that part of the world actually under consideration.

Additional knowledge which is neglected for simplicity concerns characteristics $X^{(m+1)}, X^{(m+2)}, \ldots$. The exhaustive representation of the domain would require us to consider a product space $\Omega^{M'}$, $M \subseteq M'$. Therefore we have to deal with projections of orthogonal extensions of the knowledge representation structures on Ω^M.

(6.49) Example. Let us return to our introductory example concerning the oil contamination of waters by trading vessels. In the terms of this section we have

$$\begin{aligned} X^{(1)} &= X_L, & \Omega^{(1)} &= \Omega^L = \{\text{z3, z2, z1, ca, rd, ld}\} \\ X^{(2)} &= X_A, & \Omega^{(2)} &= \Omega^A = \{\text{fd, hc, wp, np}\} \\ X^{(3)} &= X_M, & \Omega^{(3)} &= \Omega^M = \{\text{ad, ac, fi, im}\} \end{aligned}$$

So the appropriate evidence space is the product space $\Omega^{\{1,2,3\}}$.

In Sect. 2.2 we considered the case of purely imprecise knowledge. Two expert spaces $\Theta_1 = \{\theta_1\}$ and $\Theta_2 = \{\theta_2\}$ induce two information sources $(2^{\Omega^{\{1,2,3\}}}, \mathcal{R}_i, \mathcal{U}_i)$, $i = 1,2$, where $\mathcal{R}_1 = \{m^{R_1}\}, R_1 = \widehat{\Pi}^{\{1,2,3\}}_{\{1,2\}}(R_{\theta_1})$, $\mathcal{R}_2 = \{m^{R_2}\}, R_2 = \widehat{\Pi}^{\{1,2,3\}}_{\{2,3\}}(R_{\theta_2})$, $\mathcal{U}_1 = \mathcal{U}_{\text{revis}}(\mathcal{R}_1)$ and $\mathcal{U}_2 = \mathcal{U}_{\text{revis}}(\mathcal{R}_2)$. As we know from Example 8.8 by the logical combination of these two sources with respect to the logical expression 1 <u>and</u> 2 we obtain $\mathcal{U}(1 \text{ \underline{and} } 2) = \mathcal{U}_1 \cap \mathcal{U}_2 = \{m^R \mid R \subseteq R_1 \cap R_2\}$. In the case of vagueness in its most general form we obtain $\mathcal{U} = \{m^\rho \mid \rho \subseteq 1\!\!\mathrm{I}_{R_1} \cap 1\!\!\mathrm{I}_{R_2}\}$.

If our aim is to model uncertainty, then instead of the sets $\mathcal{R}_1, \mathcal{R}_2$ we would get sets $\mathcal{P}_1 = \{P \mid \Pi^{\{1,2,3\}}_{\{1,2\}}(P) = P_1\}$ and $\mathcal{P}_2 = \{P \mid \Pi^{\{1,2,3\}}_{\{2,3\}}(P) = P_2\}$, where P_1 and P_2 are the probability distributions on $\Omega^{\{1,2\}}$ and $\Omega^{\{2,3\}}$, respectively, which the expert provides. With $\mathcal{U}_i = \mathcal{U}_{\text{prob}}(\mathcal{P}_i)$, $i = 1,2$, we obtain

$$\mathcal{U} = \mathcal{U}_1 \cap \mathcal{U}_2 = \{P \mid \exists E \in \mathcal{Q}(\Omega^{\{1,2,3\}}) : P = Q(\,.\mid E), \quad Q = \frac{P_1 \cdot P_2}{\Pi^{\{1,2\}}_{\{2\}}(P_1)}\} \,.$$

The combination of uncertainty and imprecision leads to the consideration of mass distributions on Ω^M, but as we pointed out in Sect. 5.2 we restrict ourselves to using only cylinders of Ω^M as focal sets, i.e. we would have to consider the knowledge representation structures $(\mathcal{Q}(\Omega^M), \mathcal{M}, \mathcal{U}^{\text{refine}}_{\text{revis}}(\mathcal{M}))$, where $\mathcal{M} = \mathcal{M}_1 \cap \mathcal{M}_2$. ◇

Moving away from this example, in general we have knowledge representation structures on an m–dimensional product space Ω^M which we assume to be finite. The idealistic final state of knowledge after the consultation is a probability distribution P on Ω^M with $P(\{\omega_0^M\}) = 1$, or a mass distribution $m^{\{\omega_0^M\}}$ on Ω^M indicating that the tuple $\omega_0^M \in \Omega^M$ is the "true state of the world". In general a set of possible states will remain. How to determine the aposteriori knowledge representation structure, valuating the different possible tupels is the subject of the following chapters.

The next topics of interest are consequences of our assumption concerning the product structure of the evidence space for knowledge acquisition and consultation. Let us first consider the consultational knowledge. The observations of the user usually will concern single characteristics without any respect to their causal dependencies; in essence the user is able to give for every characterstic $X^{(i)}$, $i \in M$, a restriction $E^{(i)}$ of $\Omega^{(i)}$. If for some characteristic $X^{(k)}$ no information is available then this is encoded by $E^{(k)} \stackrel{d}{=} \Omega^{(k)}$; if, as the other extreme situation, the exact value of some characteristic $X^{(j)}$ is known to be $\omega_0^{(j)} \in \Omega^{(j)}$, this is represented by $E^{(j)} = \{\omega_0^{(j)}\}$. So the observations are always purely non-deterministic. This leads us to the construction of the total observation:

$$E = \bigcap_{i \in M} \widehat{\Pi}_{\{i\}}^{M}(E^{(i)}),$$

where the set E represents the aggregated observation of all characteristics.

The acquisition of expert knowledge gets a little more complicated. No expert will be able to specify directly m-dimensional dependencies; in general he will provide propositions concerning a small subset of characteristics, i.e. every piece of knowledge refers to some subspace Ω^T of Ω^M, $T \subseteq M$. The efficient aggregation of such information requires additional knowledge about the structure of dependencies among those subspaces. The next topic is the formal description and representation of such qualitative dependencies.

As we mentioned before, the formal basis of our exertions is a product space Ω^M, $M = \{1, \ldots, m\}$. Each piece of knowledge concerns some subspace Ω^T of Ω^M , $T \subseteq M$. If these pieces of knowledge completely specify the available information on Ω^M, then the union of the different subsets yields M. Furthermore we can assume $T' \not\subseteq T$ if $T' \neq T$, since every piece of knowledge concerning some subspace $\Omega^{T'}$, $T' \subseteq T$ can be expressed within Ω^T. The following definition formalizes these ideas.

(6.50) Definition. *Let $\Omega^{(1)}, \ldots, \Omega^{(m)}$ form a universe with index set M. A set $\mathfrak{T} \subseteq 2^M$ with $\bigcup_{T \in \mathfrak{T}} T = M$ and $T' \not\subseteq T$ for $T, T' \in \mathfrak{T}$ we call a modularization of M. $\mathfrak{T}$ defines a family of subspaces $\{\Omega^T \mid T \in \mathfrak{T}\}$.*

Every piece of knowledge concerning some subspace $T \in \mathfrak{T}$ we assume to be induced by an information source defined on some expert space Θ_T. These pieces of knowledge are encoded by information sources on Ω^M. In general the different pieces of knowledge are *not* independent, as this would imply $T \cap T' = \emptyset$ for $T, T' \in \mathfrak{T}$, i.e. the different subspaces would stand as isolated domains, independently, side by side. This is not the usual situation. For two characteristics $X^{(i)}$, $X^{(k)}$ there are two different kinds of dependency: firstly *direct* dependency, i.e. there is some $T \in \mathfrak{T}$ with $i, k \in T$; secondly an *indirect* dependence, i.e. there is a sequence $T_1, \ldots, T_q$ with $i \in T_1$, $T_\nu \cap T_{\nu+1} \neq \emptyset$, $\nu = 1, \ldots n-1$, $k \in T_q$. In the first case $X^{(i)}$ and $X^{(k)}$ are covered by the same piece of knowledge, and directly affect each other as every pair of characteristics $X^{(s)}$, $X^{(t)}$, $s, t \in T$ does. In the second case $X^{(i)}$ affects all the other characteristics covered by T_{j_1}, also the ones $X^{(s)}$ with $s \in T_1 \cap T_2$. These affect the other characteristics covered by T_2 including those $X^{(t)}$ with $t \in T_2 \cap T_3$ and so on. Finally also $X^{(k)}$ is affected. So for the dependence of characteristics the intersections of the sets $T \in \mathfrak{T}$ are of great importance. The characteristics can be classified with respect to the property of mutual dependency.

(6.51) Definition. *Let $\mathfrak{T} = \{T_1, \ldots, T_r\}$ be a modularization of M. The relation $\overset{\mathfrak{T}}{\equiv}$ on $M \times M$ is defined as follows*

$$i \overset{\mathfrak{T}}{\equiv} k \iff \forall T \in \mathfrak{T} : (\{i, k\} \subseteq T \text{ or } \{i, k\} \subseteq \overline{T}) \quad .$$

The relation $\overset{\mathfrak{T}}{\equiv}$ formalizes our idea of indistinguishability with respect to causal relationship. This relation induces a partition of M which is compatible with $\mathfrak{T}$ since it is an equivalence relation, as the next theorem shows.

(6.52) Theorem. *Let $\mathfrak{T} = \{T_1, \ldots, T_r\}$ be a modularization of M. The relation $\overset{\mathfrak{T}}{\equiv}$ on $M \times M$ induced by $\mathfrak{T}$ is an equivalence relation.*

Proof. We have to show that $\overset{\mathfrak{T}}{\equiv}$ is (a) reflexive (b) transitive and (c) symmetrical.

(a) For every $i \in M$ we have $\{i\} \subseteq T$ or $\{i\} \in \overline{T}$ for each $T \in \mathfrak{T}$, so $i \overset{\mathfrak{T}}{\equiv} i$ is valid.

(b) Since $\{i, k\} = \{k, i\}$ the relation $\overset{\mathfrak{T}}{\equiv}$ is trivially symmetrical.

(c) If $i \overset{\mathfrak{T}}{\equiv} s$ and $s \overset{\mathfrak{T}}{\equiv} k$ then we have for each T either $\{i, s\} \subseteq T$ which implies $\{s, k\} \subseteq T$ and therefore $\{i, k\} \subseteq T$ or $\{i, s\} \subseteq \overline{T}$ implying $\{s, k\} \subseteq \overline{T}$ and therefore $\{i, k\} \subseteq \overline{T}$.

□

(6.53) *Remark.* Let $\widehat{M} = \{S_1, \ldots, S_n\}$ be the quotient set $^M/_{\overset{\mathfrak{T}}{\equiv}}$, i.e. the set of all equivalence classes, which is a partition of M. This partition we call *induced* by the modularization $\mathfrak{T}$. Let $\kappa : M \to \widehat{M}$ be the mapping that assigns to every element of M its equivalence class. The set $\widehat{\mathfrak{T}} = \{\widehat{t} \mid \widehat{T} = \kappa(T),\ T_j \in \mathfrak{T}\}$ is a modularization of $\widehat{M}$. ∘

Having once constructed this induced partition $\widehat{M}$, then in order to decide in which way two characteristics $X^{(i)}$, $X^{(k)}$ depend on each other it is sufficient to consider $\kappa(i)$ and $\kappa(k)$ and to investigate whether there is some $\widehat{T} \in \widehat{\mathfrak{T}}$ with $\kappa(i),\ \kappa(k) \in \widehat{T}$ or some sequence $\widehat{T}_1, \ldots, \widehat{T}_q$ with $\kappa(i) \in \widehat{T}_1$, $\widehat{T}_p \cap \widehat{T}_q \neq \emptyset$, $\kappa(k) \in \widehat{T}_q$, $p = 1, \ldots, q-1$. For this reason the two sets $\widehat{\mathfrak{T}}$ and $\widehat{M}$ completely determine the qualitive dependency structure of the frame of discernment Ω^M, given partial information about subspaces Ω^{T_j}, $j = 1, \ldots, r$.

The subject of the next chapter is to develop suitable formalisms to talk about subspaces, their causal relations, their intersection and the sets of characteristics they cover. So the main purpose of the following discussion of graphs and hypergraphs is essentially to give the reader an intuitive understanding of the graph structures we will use later on and to show that those structures as well as the notations, which may seem to be a little complicated, arise canonically from the problem we are considering, i.e. the aggregation of pieces of partial knowledge.

6.8. Concluding Remarks

In this book we consider belief functions to be lower probabilities induced by a (possibly unknown) multivalued mapping. Semantically we assume the data to be imprecise and the uncertainty to be described by a single probability.

Unfortunately this approach is often confused with a different interpretation, where belief functions are viewed as the lower probabilities induced by a family of probability distributions of type 2. Here the data are precise, although a set of probabilities is involved. Both interpretations are contained in Dempster's original paper [Dempster 1967].

The interpretation of belief functions as a condensed method for representing sets of probabilities [Kyburg 1987], [Fagin and Halpern 1989], [Shafer 1981] is rather restrictive, since we can only handle convex sets of probabilities of type 2, an assumption which is often too strong for knowledge representation purposes (compare Sect. 4.7).

Shafer's interpretation [Shafer 1976] abandons the idea that belief functions can be viewed as lower probabilities. In his view "belief in a hypothesis does not measure the chance that it is true, but rather the strength of the arguments we have in favour of the hypothesis"[Pearl 1989]. This interpretation as the model of evidential reasoning is also shared by P. Smets [Smets 1989] where the impact of a piece of evidence consists in allocating parts of an initial unitary amount of belief among subsets of a frame of discernment. His model is therefore called the "Transferable Belief Model". Other interpretations have been presented by [Ruspini 1987], [Wilson 1990], and [Kohlas 1990].

The interpretation of belief functions as a general model for the quantification of subjective, personal beliefs and opposition to the Bayesian model lead to some problems and counterintuitive results ([Pearl 1990], [Weichselberger and Pöhlmann 1990]) especially when only Dempster's rules of conditioning and combination (compare Sect. 6.5) are used as updating schemes. One of the main problems with this approach is that Dempster's rule of conditioning is in essence a data revision that should be used very carefully. Another serious difficulty comes from the fact that for decision making purposes belief functions are not directly suitable and have to be transformed into probability functions, where well developed decision making procedures exist. In this interpretation, belief functions have a rather limited scope of application.

In our treatment we prefer to use possibly unknown belief functions as a "condensed representation" of a random set. Consequently there are different reasonable updating schemes, such as belief revision (equation (6.8)), which is often denoted (somewhat misleading) as Dempster's rule of conditioning as well as the "geometric" conditioning defined in equation (6.6). An overview of different updatings is given in [Planchet 1988].

Possibility measures are often used as an uncertainty management technique since their handling is extremely simple and homogeneous [Dubois and Prade 1988]. One should view possibility measures as plausibility measures with special properties – the consonant plausibility measures [Shafer 1976], where the essential information turns out to be treatable in the terms of so-called *contour functions*. A contour function is considered as the "trace" of a possibility measure on the elements of the frame of discernment [Dubois 1986]. Contour functions are often considered directly as vague data or fuzzy sets, which may lead to some confusion, since the definition of operators depends on the interpretation. For an introduction to the application of possibility theory we recommend [Dubois 1986] and [Dubois 1988].

Although his interpretation is often different from ours, we recommend [Shafer 1976] and [Shafer 1990] as an introduction to belief functions and the article [Smets 1988] as an overview of the transferable belief model. Here the closed world assumption is abandoned by allowing the attachment of positive masses to the empty set, a point of view we do not share. This model is analyzed in detail in Chap. 6.

The treatment of different frames of discernment was first addressed in [Shafer 1976] and our specialization concept has some similarities with approaches considered in [Dubois and Prade 1988], [Yager 1986].

Decision theoretic aspects of the "transferable belief model" are discussed in [Smets 1989].

In the literature, updating of probabilistic models is almost always embedded in conditioning concepts, although some other approaches have been presented [Lewis 1976]. We consider updating processes to be the sequential restriction of sets, containing "possible expansions". Thus these sets reflect the undeterminedness, i.e. the diversity of possibilities, we mentioned above.

A probability Q on Ω may serve as an example in order to illustrate our idea of conditioning with respect to some set $\emptyset \neq E \subseteq \Omega$. In the light of E the set

$$\mathcal{P} = \{P | \exists A \subseteq \Omega : P = Q(\ .\ |A), Q(A) > 0\}$$

of probabilities compatible with Q reduces to

$$\mathcal{P}_E = \{P | \exists A \subseteq E : P = Q(\ .\ |A), Q(A) > 0\}.$$

Knowing $\mathcal{P}$ means that starting with the state of ignorance about the actual probability an expansion may lead to each element of $\mathcal{P}$, whereas after conditioning with respect to E only elements of $\mathcal{P}_E$ remain admissible.

The combination and integration of information sources addresses the problem of achieving consensus in the presence of uncertainty. In the range of probabilistic models there are several publications on group decision making, the aggregation of probabilities and pooling methods [Genest, Zidek 1986]. In any case a careful analysis of dependency assumptions is indispensible [Morris 1986].

The classical rule of combination, which is considered to be a global updating scheme by other authors (see e.g. [Shafer 1976], [Smets 1980]), in the context of our approach loses much of its importance. It turns out that $m_1 \oplus m_2$ emerging from the application of this rule to m_1 and m_2 spreads an equivalence class of mass distributions, each of which is likewise capable of representing the intersection of the sets of updates, determined by the original mass distributions m_1 and m_2. But although $m_1 \oplus m_2$ is reduced to being one of many, the fact that it allows us to obtain easily a representative of the class means that it is still of practical value.

In our terminology *combination* refers to those aggregation problems where information on the dependencies is available. If such information is lacking we use the notion of *integration*. In the latter case we have to deal with uncertainty about uncertainty or second-order uncertainty. In the range of probability theory there are several proposals for second-order concepts [Koopman 1940], [Kyberg 1961], [Smith 1961]. Although in the probability community the need for higher-order probabilities is challenged [de Finetti 1977], [Spiegelhalter 1986], [Pearl 1988], [Cheesea 1986], we use these techniques for the purpose of *probabilistic* integration. Besides probabilistic approaches, non-probabilistic second-order structures have also been developed by [Shackle 1952] and others. In contrast, the *logical* integration of information sources as well as the integration of user observations relies on the concept of possible updates. The basic idea is to consider the sets of updates explicitly specified by the expert, and to perform set operations on these sets, according to the relations of the sources.

Chapter 7

On Graphical Representations

In order to provide reasonable notations for dependency relations we summarize firstly some results on graph theory. The study of hypergraphs leads to the consideration of dependency networks, which form a very powerful tool for analyzing qualitative aspects of knowledge representation.

7.1 Graphs and Trees

In this section we provide a brief survey of some wellknown results from graph theory. We restrict ourselves to those topics which are of interest for our further considerations.

7.1.1 Undirected Graphs

The basic subjects of graph theory are graphs. Roughly speaking, a graph consists of a set of *vertices* (nodes) connected by a set of *edges* (links). For the present we restrict ourselves to a special type of edge, which we assume to connect exactly two distinct vertices and to have no direction. In terms of the usual graph theoretic notation that means dealing with *undirected* graphs.

(7.1) Definition. *An (undirected) graph G is a pair $(V, \mathfrak{E})$, where $V = \{v_1, \ldots, v_n\}$ contains the vertices and $\mathfrak{E} = \{E_1, \ldots, E_r\}$ contains the edges. Each element $E \in \mathfrak{E}$ has the property*

$$(i) \quad E \subseteq V \quad \text{and}$$
$$(ii) \quad |E| = 2.$$

A subset $C \subseteq V$ is called a *clique*, if all vertices in C are mutually directly connected, that means $\{v, w\} \in \mathfrak{E}$ holds for all $v, w \in C$. Note that in order to simplify the nomenclature, whenever we talk of cliques we mean maximal cliques, i.e. cliques which are not covered by more comprehensive ones.

In a graph $G = (V, \mathfrak{E})$ a *chain* connecting the vertices v_1 and v_{q+1} is a sequence $\langle E_1, \ldots, E_q \rangle$ of edges such that $E_k = \{v_k, v_{k+1}\}$ for $k = 1, \ldots, q$. In this context v_1 is called the initial endpoint, whereas v_{q+1} is called the terminal endpoint. Since a chain is uniquely determined by the sequence of vertices v_k, $k = 1, \ldots, q+1$, sometimes the notation $\langle v_1, v_2, \ldots, v_{q+1} \rangle$ is used.

A *cycle* is a chain where (i) no edge appears twice in the sequence and (ii) the two endpoints are the same vertex. A graph $G = (V, \mathfrak{E})$ is called *connected* if for every pair v, w of nodes in V there is a chain with these vertices as its endpoints.

(7.2) Example. Let us consider the graph $G = (V, \mathfrak{E})$ where

$$V = \{a, b, c, d, f, g\},$$
$$E = \{\{a, b\}, \{b, c\}, \{b, d\}, \{c, f\}, \{d, f\}, \{f, g\}\} .$$

In Fig. 7.1 we give two equivalent graphical representations where the second one will be used in the following considerations.

As we see from Fig. 7.1 the graph G is connected. $\langle \{a, b\}, \{b, c\}, \{c, f\} \rangle$ is a chain of length 3 connecting a and f. $\langle \{b, c\}, \{c, f\}, \{d, f\}, \{d, b\} \rangle$ is a cycle of length 4. ◇

7.1.2 Trees

In the sequel we will focus our interest on a special type of graphs. We consider only those graphs which are connected and have no cycles. These requirements define what we know as *trees*.

(7.3) Definition. *A graph $G = (V, \mathfrak{E})$ is a tree, if G is connected and has no cycles.*

In a tree, a chain connecting two nodes is always uniquely determined, the removal of one edge will split the tree into two unconnected components. Thus we are able to provide an equivalent characterization of trees which relies on the process of their construction. We will show that subsequently adding leaves to an existing tree will preserve the tree property. But before we can turn to the respective theorem we have to define what we mean by a "leaf".

(7.4) Definition. *Let $G = (V, \mathfrak{E})$ be a graph. A vertice $v \in V$ is called a leaf, if and only if it is contained in only one edge $E \in \mathfrak{E}$.*

Now let us consider the stepwise construction of a tree. We start with some edge $E_1 \in \mathfrak{E}$. The graph consisting only of this one edge and the two vertices

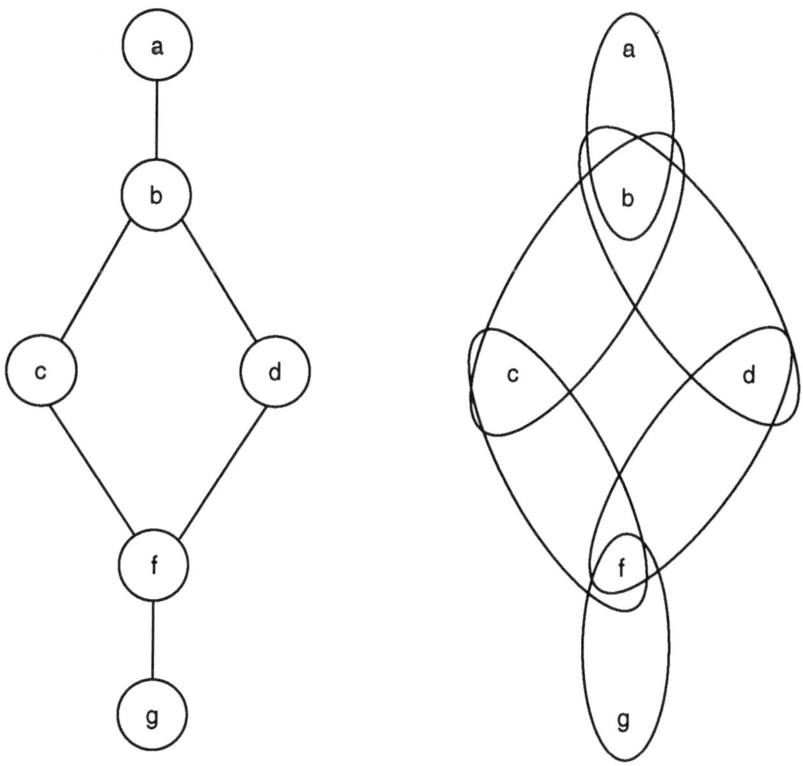

Fig. 7.1. Representations of undirected graphs

contained in E_1 obviously is a tree. Adding a leaf means adding some edge E_2 containing one vertice within the current graph (at the moment $(E_1, \{E_1\})$) and one outside of it. Doing so we obtain $(E_1 \cup E_2, \{E_1, E_2\})$. As the next theorem shows, the graph resulting from this procedure is assured to be a tree.

(7.5) Theorem. *A graph $G = (V, \mathfrak{E})$ is a tree, if and only if there is a numbering $\rho : \{1, \ldots, r\} \to \mathfrak{E}$, satisfying the following condition: If $\rho(i+1) = \{v, w\}$, then either v or w is contained in the set $V^i \stackrel{d}{=} \rho(1) \cup \ldots \cup \rho(i)$ for $i = 1, \ldots, r-1$. The sequence $\rho(1), \ldots, \rho(r)$ is called a construction sequence of G.*

In the sequel we use the notation $G^i = (V^i, \mathfrak{E}^i)$, where $\mathfrak{E}^i \stackrel{d}{=} \{\rho(1), \dots, \rho(i)\}$, for the graph emerging from the first i steps of the construction sequence.

Since our aim is to consider $(\widehat{M}, \widehat{\mathfrak{T}})$ (compare Remark 6.53) as some kind of graph, it is not sufficient to restrict our attention to edges containing only two vertices but it is necessary to permit edges consisting of three or more vertices. For this purpose the next section introduces "hypergraphs" which allow edges to be arbitrary sets of vertices.

7.2 Hypergraphs and Hypertrees

Also the notation of a hypergraph is a well known concept in graph theory. Again we consider a set $V = \{v_1, \dots, v_n\}$ of vertices and a set $\mathfrak{E} = \{E_1, \dots E_r\}$ of edges, but from now on we permit the edges to be *hyperedges*, i.e. to be arbitrary subsets of V.

(7.6) Definition. *A hypergraph H is a pair $(V, \mathfrak{E})$, where $V = \{v_1, \dots, v_n\}$ contains the vertices and $\mathfrak{E} = \{E_1, \dots, E_r\}$, $\mathfrak{E} \subseteq 2^V$ contains the (hyper–)edges, satisfying the following conditions:*

(i) $E \neq \emptyset$ *for all* $E \in \mathfrak{E}$,

(ii) $\bigcup_{E: E \in \mathfrak{E}} E = V.$

We furthermore assume the hypergraphs we consider to be *reduced* or *skeletal*, i.e. no hyperedge is a subset of another hyperedge.

(7.7) *Remark.* Obviously $(\widehat{M}, \widehat{\mathfrak{T}})$ as defined in Remark 6.53 is a hypergraph. Since in this interpretation the hyperedges $\hat{T} \in \widehat{\mathfrak{T}}$ define subspaces Ω^T of Ω^M which are subject to pieces of knowledge, $(\widehat{M}, \widehat{\mathfrak{T}})$ can be assumed to be reduced without loss of generality. Once having represented some subspace Ω^T there is no need to represent $\Omega^{T'}$, $T' \subseteq T$. ○

7.2.1 Hypertrees

In a hypergraph $H = (V, \mathfrak{E})$ a chain connecting the nodes $v, w \in V$ is a sequence of edges $\langle E_1, \dots, E_q \rangle$ such that

$$\begin{array}{ll} (i) & v \in E_1, w \in E_q \quad \text{and} \\ (ii) & E_k \cap E_{k+1} \neq \emptyset \text{ for } k = 1, \dots q-1. \end{array} \tag{7.1}$$

A *component* of a hypergraph is a maximal connected set of vertices. In the sequel we consider only connected hypergraphs, i.e. hypergraphs consisting of one component.

(7.8) *Remark.* Once more our interpretation of the hypergraph $(\widehat{M}, \widehat{\mathfrak{T}})$ we defined in Remark 6.53 justifies this assumption. Of course the domain of an expert system can be supposed to be connected, i.e. all pieces of knowledge are directly or indirectly in interaction with each other. ○

The next step would be to define *hypertrees* to be "acyclic" hypergraphs, but unfortunately the definition is not as straightforward as the definition of trees in the case of ordinary graphs. Thus we rely on the idea of a "leaf–by–leaf" construction method. For this purpose we firstly define the notion of a *twig*. Twigs are, roughly speaking, the leaves of hypergraphs.

(7.9) Definition. *Let $H = (V, \mathfrak{E})$ be a hypergraph. A hyperedge $T \in \mathfrak{E}$ is called a twig, if there is another hyperedge $B \in \mathfrak{E}$ with $B \neq T$ and $B \cap T \neq \emptyset$ such that if $E \in \mathfrak{E}$, $E \cap T \neq \emptyset$ and $E \neq B$, then $v \in E \cap T$ implies $v \in B$.*

The hyperedge B screens T against the rest of the hypergraph, so if the hypergraph is a graph, in the case of ordinary graphs the notion of a twig coincides with that of a leaf. B is called a *branch* for T.

(7.10) Example. Let us consider the hypergraph $H = (V, \mathfrak{E})$, where

$$V = \{\text{a,b,c,d,e,f}\},$$
$$\mathfrak{E} = \big\{\{\text{a,b,c,d}\}, \{\text{b,c,e}\}, \{\text{c,d,f}\}\big\}.$$

From Fig. 7.2 we conclude easily that {b,c,e} and {c,d,f} are twigs, whereas {a,b,c,d} is not.

◇

(7.11) Definition. *A hypergraph $H = (V, \mathfrak{E})$ is called a* hypertree *if there is a numbering $\rho : \{1, \ldots, r\} \longrightarrow \mathfrak{E}$ of the hyperedges in $\mathfrak{E}$ such that $\rho(i)$ is a twig with respect to the hypergraph $(\bigcup_{k=1}^{i} \rho(k), \{\rho(k) \mid k = 1, \ldots, i\})$ for $i = 1, \ldots, r$.*

The sequence $\rho(1)$, $\rho(2)$, ..., $\rho(r)$ is called a hypertree construction sequence. As in the case of graphs, $H^i = (V^i, \mathfrak{E}^i)$ denotes the hypertree emerging from the first i steps of the construction sequence.

Note that this definition of a hypertree is equivalent to the one used in data base theory [Maier 1983].

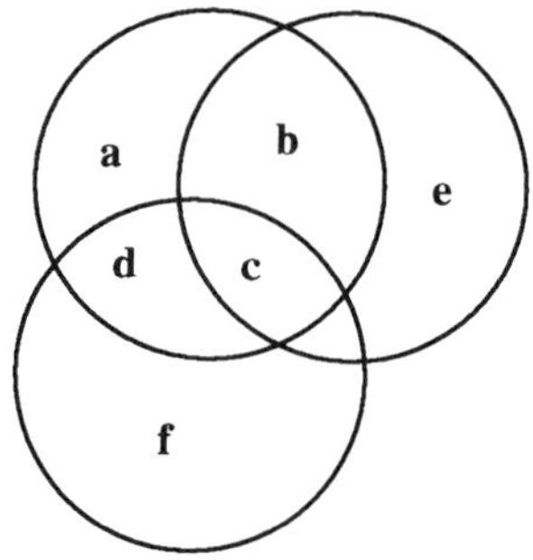

Fig. 7.2. The hypergraph of Example 7.10

If some hypergraph $H = (V, \mathfrak{E})$ under consideration is not a hypertree, then a *covering hypertree* $H' = (V, \mathfrak{E}')$ can be found. H' covers H if for every $E \in \mathfrak{E}$ there is some $E' \in \mathfrak{E}'$ such that $E' \supseteq E$. While it is easy to show that for every hypergraph H a covering hypertree H' exists (note that $H' = (V, \{V\})$ is such covering hypertree), the construction of "minimal" hypertrees leads to serious problems of complexity. In [Tarjan and Yannakakis 1984] an efficient algorithm for the construction of suboptimal hypertree covers is presented, it relies on a fill-in procedure acting on $G_H = (V, \{\{v, w\} \mid \exists E \in \mathfrak{E} : v, w \in E\})$, see Sect. 7.5.

(7.12) *Remark.* In the sequel we assume $(M, \mathfrak{T})$ to be a reduced, connected hypertree, which implies that also $(\widehat{M}, \widehat{\mathfrak{T}})$ is a reduced, connected hypertree. ○

(7.13) Example. Consider the hypergraph[1] $H = (V, \mathfrak{E})$, where

$$V = \{\text{a,t,l,f,x,s,b,d}\},$$
$$\mathfrak{E} = \big\{\{\text{a,t}\}, \{\text{t,l,f}\}, \{\text{f,x}\}, \{\text{l,s}\}, \{\text{s,b}\}, \{\text{f,b,d}\}\big\}.$$

H is depicted in Fig. 7.3. H is not a hypertree, since there is no hypertree construction sequence for this hypergraph. ◇

[1] This example originally comes from [Lauritzen and Spiegelhalter 1988] and is also considered in [Shafer and Shenoy 1988].

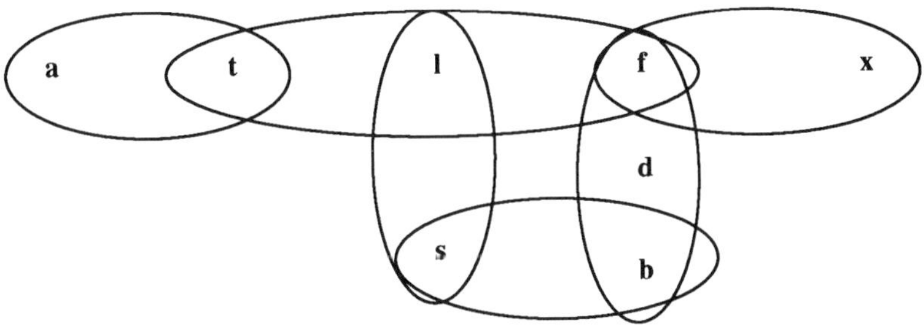

Fig. 7.3. A hypergraph, but not a hypertree

In general a direct application of Definition 7.6 in order to check whether a given hypergraph is a hypertree or not is impracticable. An easy way to test the hypertree property is the Graham Reduction, see e.g. [Maier 1983]; the formulation we present here is taken from [Tarjan and Yannakakis 1984].

H is a hypertree, if all vertices can be deleted by repeatedly applying the following two operations:

(*i*) delete a vertex that occurs in only one edge;
(*ii*) delete an edge that is contained in another edge. (7.2)

Note that a construction sequence of some hypertree defines a deletion sequence in the sense of the Graham Reduction.

(7.13) Example. (continued) The application of the Graham Reduction to the hypergraph H confirms what we already know, i.e. that H is not a hypertree. In contrast to H the hypergraph $H' = (V, \mathfrak{E}')$, where

$$\mathfrak{E}' = \{\{\mathrm{a,t}\}, \{\mathrm{t,l,f}\}, \{\mathrm{f,x}\}, \{\mathrm{l,f,b}\}, \{\mathrm{l,s,b}\}, \{\mathrm{f,b,d}\}\},$$

covers $\mathfrak{E}$ and is a hypertree, since the sequence in which the edges are listed above may serve as construction sequence. ◇

For our purposes it is of special interest to know which common vertices v couple edges E_1, E_2. Recall that in our interpretation the hyperedges represent subspaces, thus we are dealing with the question whether two of them are

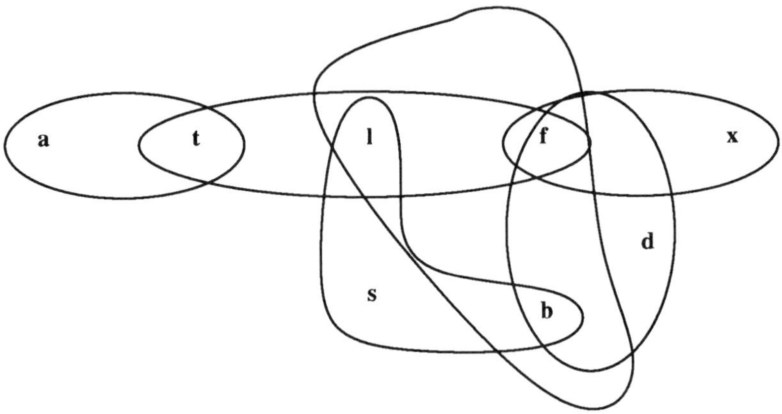

Fig. 7.4. A hypertree

overlapping, and, if so, which characteristics they share. Therefore we consider the graph $G(H)$ induced by a hypergraph H. $G(H)$ is defined as follows.

(7.14) Definition. *Let* $H = (V, \mathfrak{E})$ *be a hypergraph. The graph* $G(H) = (V', \mathfrak{E}')$ *where* $V' = V \cup \mathfrak{E}$, $\mathfrak{E}' = \{\{v, E\} \mid E \in \mathfrak{E}, v \in E\}$, *is called the graph of* H.

Note that the graph $G(H)$ of some hypertree H may have cycles, as the next example shows.

(7.15) Example. Let $H = (V, \mathfrak{E})$ be a hypergraph where

$$V = \{a, b, c, d, f, g, h\},$$
$$\mathfrak{E} = \{U, V, W, X\} ,$$

$U = \{a, b, c\}$, $V = \{b, c, d, f\}$, $W = \{c, d, f, g\}$, $X = \{f, g, h\}$. H is depicted in Fig. 7.5.

H is a hypertree: a valid hypertree construction sequence is U, V, W, X, as indicated in Fig. 7.6(a)–(d). The construction of $G(H)$ is depicted in Fig. 7.7. ◇

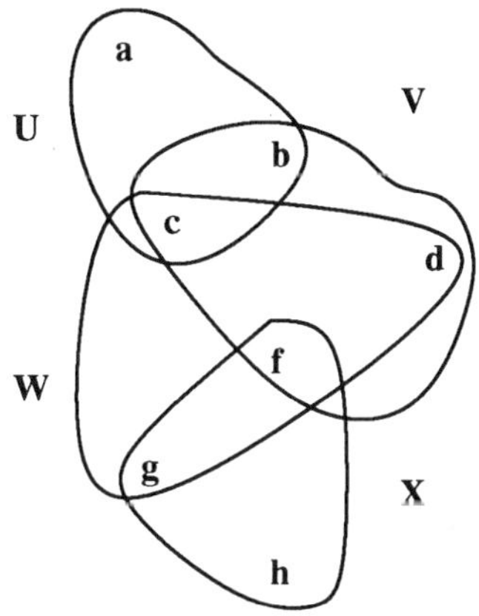

Fig. 7.5. The hypertree of Example 7.15

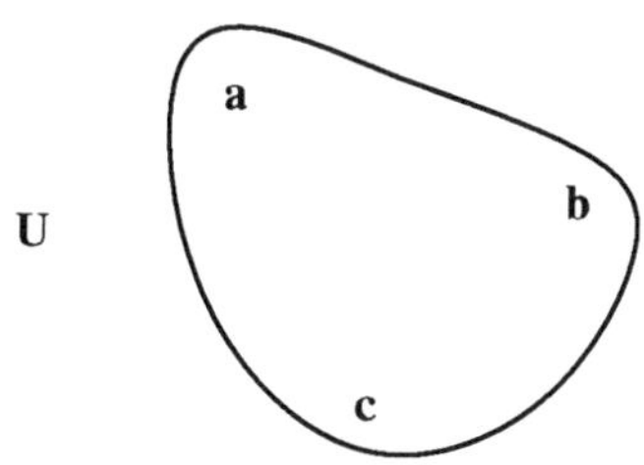

Fig. 7.6. (a) A single hyperedge always forms a hypertree

7.2.2 Simple Hypertrees

(7.16) Definition. *A hypertree $H = (V, \mathfrak{E})$ is called a simple hypertree if $G(H)$ contains no cycles.*

In the sequel we will show that from $H = (\widehat{M}, \widehat{\mathfrak{T}})$ as defined in Remark 6.53, if H is assumed to be a reduced, connected hypertree, we can construct a simple hypertree which in some sense "covers" H. Since we restrict our interest to the question whether two subspaces Ω^{T_1}, Ω^{T_2}, $\hat{T}_1, \hat{T}_2 \in \mathfrak{T}$ are in direct interaction with each other or not, vertices $\hat{v}, \hat{w} \in \widehat{M}$ where $\{\hat{v}, \hat{w}\} \subseteq \hat{T}_1 \cap \hat{T}_2$

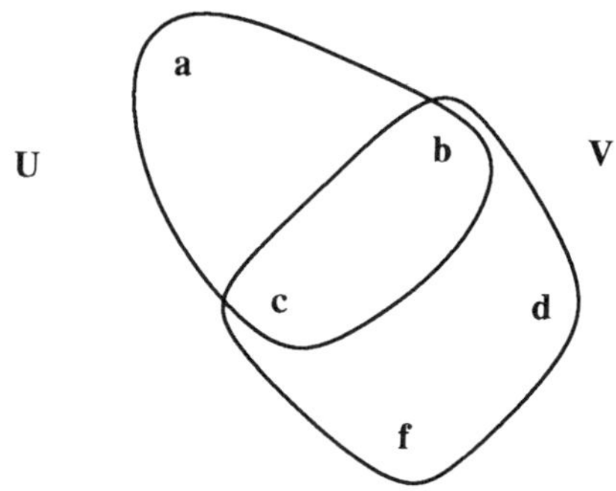

Fig. 7.6. (**b**) Two hyperedges also always form a hypertree

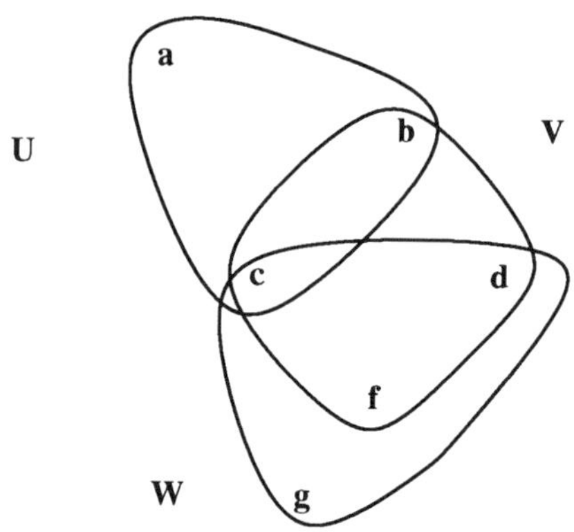

Fig. 7.6. (**c**) V screens W

need not be distinguished. In order to formalize this idea we define a relation on $\widehat{M} \times \widehat{M}$.

(7.17) Definition. *Let $H = (V, \mathfrak{E})$ be a reduced connected hypertree. The relation $\stackrel{\mathfrak{E}}{=}$ is defined as follows:*

$$v \stackrel{\mathfrak{E}}{=} w \quad \stackrel{d}{\Longleftrightarrow} \quad \begin{cases} (v = w) \text{ or} \\ (\exists E_1, E_2 \in \mathfrak{E} : \{v, w\} \subseteq E_1 \cap E_2 \text{ and } E_1 \neq E_2) \text{ or} \\ (\exists x \in V : v \stackrel{\mathfrak{E}}{=} x \text{ and } x \stackrel{\mathfrak{E}}{=} w) \,. \end{cases}$$

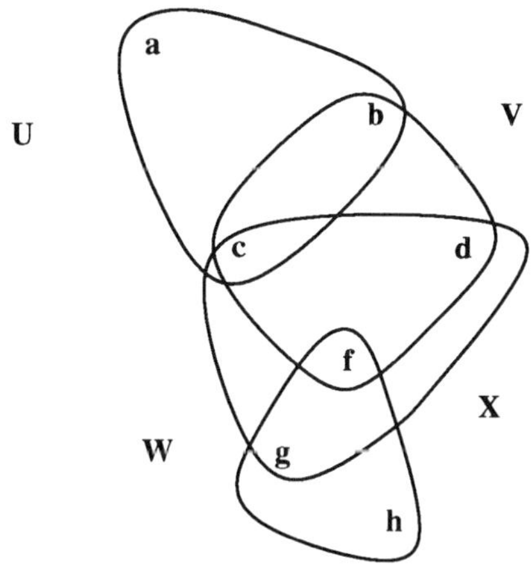

Fig. 7.6. (d) W screens X

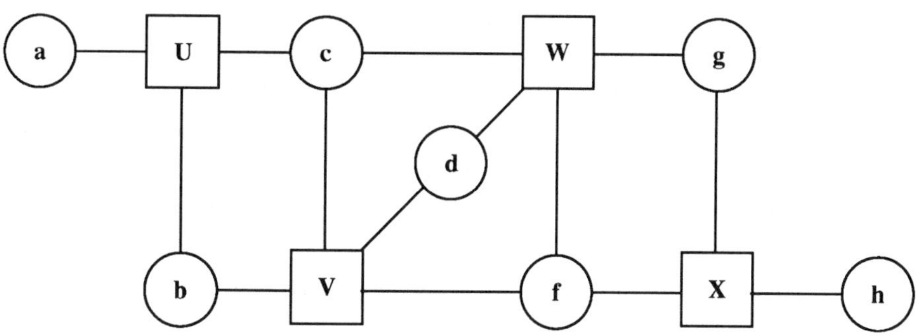

Fig. 7.7. $G(H)$ contains various cycles

Obviously the relation $\stackrel{\mathfrak{E}}{=}$ is reflexive, symmetrical and transitive, so it is an equivalence relation. Let $\widetilde{v}$ denote the equivalence class of $v \in V$. Define

$$\widetilde{E} = \{v \in V \mid \exists w \in \widetilde{v} : w \in E\} \text{ for all } E \in \mathfrak{E} \text{ and}$$

$$\widetilde{\mathfrak{E}} = \{\widetilde{E} \mid E \in \mathfrak{E}\} .$$

In this way we obtain a hypergraph $\widetilde{H} = (V, \widetilde{\mathfrak{E}})$ which is a hypertree covering $H = (V, \mathfrak{E})$ as the next theorem shows.

(7.18) Theorem. *Let ρ define the construction sequence of some hypertree $H = (V, \mathfrak{E})$. Define $H^i = (V^i, \mathfrak{E}^i)$ where $V^i = \rho(1) \cup \ldots \cup \rho(i)$, $\mathfrak{E}^i = \{\rho(1), \ldots, \rho(i)\}$. If there is a path connecting $v, w \in V^i$ in $G(H)$ then this path already exists in $G(H^i)$ or there is a hyperedge in $\mathfrak{E}^i$, covering v and w.*

Proof.

(1) All nodes of V^1 are covered by the hyperedge $\rho(1)$ which is the only element of $\mathfrak{E}^1$.

(2) Suppose the theorem to be proven for $i = 1, \ldots, k$. Adding $\rho(k+1)$ to H^k we obtain H^{k+1}.

Assume $\rho(k+1)$ "shortcuts" two nodes $v, w \in V^k$. Since $\rho(k+1)$ is a twig with respect to H^i there is an edge $\rho(q)$, $q \leq k$, that screens $\rho(k+1)$ and thus also contains v and w.

Then the theorem follows by induction.

□

(7.19) Theorem. *If $H = (V, \mathfrak{E})$ is a hypertree, then $\widetilde{H} = (V, \widetilde{\mathfrak{E}})$ is a hypertree covering H.*

Proof. Obviously $\widetilde{H}$ covers H, thus it remains to show that $\widetilde{H}$ is a hypertree. The ordering ρ may determine a hypertree construction sequence of H. From ρ we obtain a construction sequence of $\widetilde{H}$. There may be edges E_1, E_2, where $\widetilde{E}_1 = \widetilde{E}_2$. Assuming $\rho^{-1}(E_1) < \rho^{-1}(E_2)$, in this case, step $\rho^{-1}(E_2)$ of the orginal construction sequence is skipped.

To prove that in this way a valid construction sequence is determined, we have to show that $\widetilde{\rho(j)}$ is a twig with respect to the hypergraph $\widetilde{H}^j$. Since $\rho(j)$ is a twig with respect to the hypergraph H^j, we know that there is some hyperedge $\rho(k)$, $k < j$, that screens $\rho(j)$. We have to show that $\widetilde{\rho(k)}$ screens $\widetilde{\rho(j)}$.

If there is some edge $\widetilde{\rho(q)}$, $q < j$, such that $\widetilde{\rho(q)} \cap \widetilde{\rho(j)} \neq \emptyset$, then for each node $x \in \widetilde{\rho(q)} \cap \widetilde{\rho(j)}$ there is a path, consisting of a sequence of $\stackrel{\mathfrak{E}}{=}$ equivalent nodes including x, with say $v \in \rho(q)$ and $w \in \rho(j)$ as its endpoints. From Theorem 7.18 we know that either (a) this path already exists in $G(H^j)$ or (b) there is a hyperedge $\rho(q)$, $q \leq j$, with $\{v, w\} \subseteq \rho(q)$. In both cases we obtain $w \in \rho(k)$ since $\rho(j)$ is a twig in H^j, and since $x \stackrel{\mathfrak{E}}{=} w$ this yields $x \in \widetilde{\rho(k)}$. □

Next we have to provide a simple algorithm that allows us to determine the hypertree $\widetilde{H}$ by repeated modification of H.

(*i*) Define $H^{(1)} \stackrel{d}{=} (V, \mathfrak{E}^{(1)})$, where $\mathfrak{E}^{(1)} \stackrel{d}{=} \mathfrak{E}$.

(*ii*) Select $E, F, G \in \mathfrak{E}^{(k)}$ and $v, w \in V$ such that $\{v, w\} \subseteq E \cap F$ but $v \in G$ and $w \in \overline{G}$.

$$\text{Define } \mathfrak{E}^{(k+1)} \overset{d}{=} (\mathfrak{E}^{(k)} - \{G\}) \cup \{G \cup \{w\}\}. \tag{7.3}$$

This algorithm terminates since V as well as $\mathfrak{E}$ are finite sets. It stops after say l applications of step (ii) and we obtain $H^{(l)} = \widetilde{H}$. In order to verify this there are two things to be proved:

(a) Correctness: If step (ii) is applicable, then $v \overset{\mathfrak{E}}{=} w$.

(b) Completeness: If $v \overset{\mathfrak{E}}{=} w$ and there is $E \in \mathfrak{E}^{(k)}$ such that $v \in E$ but $w \in \overline{E}$ then step (ii) is applicable.

Proof. (a) Assume $\{v, w\} \subseteq E^{(k)} \cap F^{(k)}$. This implies that v as well as w are connected to vertices $E, F \in \mathfrak{E}$, where the nodes along the path are $\overset{\mathfrak{E}}{=}$ equivalent. If the paths have nodes in common, then the $\overset{\mathfrak{E}}{=}$ equivalence of v and w is obvious.

Otherwise the two paths together with the connections caused by E and F form a cycle in G_H. Define

$$m \overset{d}{=} \max\{\rho^{-1}(D) \mid D \in \mathfrak{E} \text{ is contained in the cycle}\}.$$

Thus the cycle is also contained in $G(H^m)$, where $\rho(m)$ is a twig. If (α) $\rho^{-1}(m) \in \{E, F\}$ then E or F is a twig and thus together with its branch "double"– connects the two paths. Otherwise (β) $\rho^{-1}(m)$ can be removed without breaking the cycle for the same reason, and (β) can applied to $G(H^{m-1})$ until, in the worst case, only E and F remain and we obtain $\{v, w\} \subseteq E \cap F$.
(b) Either there are two edges E, F with $\{v, w\} \subseteq E \cap F$, then step (ii) can be applied as soon as an edge $G^{(k)}$, $v \in G^{(k)}$, $w \in \overline{G^{(k)}}$, appears, or there are two paths in $G(H)$ connecting v and w via $\overset{\mathfrak{E}}{=}$ equivalent nodes to x and y respectively, such that $\{x, y\} \subseteq E \cap F$. If so, step (ii) can be applied along each of these paths until $E^{(k)}$, $F^{(k)}$ contain v and w. Then, if still necessary, step (ii) can be applied to $E^{(k)}$, $F^{(k)}$ and $G^{(k)}$. □

(7.20) *Remark.* The proof of correctness also shows that the intersection of any two edges $\widetilde{E}$, $\widetilde{F}$ contains exactly one equivalence class of $\overset{\mathfrak{E}}{=}$. ∘

The hypertree $H = (M, \mathfrak{T})$ induces a hypertree $\widetilde{H} = (M, \widetilde{\mathfrak{T}})$. Applying the equivalence relation $\overset{\widetilde{\mathfrak{T}}}{\equiv}$ (compare Remark 6.53) to $\widetilde{H}$ this yields a simple hypertree $H^* = (M^*, \mathfrak{T}^*)$. From the construction sequence of H^* we obtain a construction sequence of $G(H^*)$. If H^* is connected, then so is $G(H^*)$. Let $\widetilde{\rho}$ determine the construction sequence of $\widetilde{H}$. From $\widetilde{\rho}(1)$ we obtain the first vertices $(\widetilde{\rho}(1), \widetilde{v})$, $\widetilde{v} \in \widetilde{\rho}(1)$, of $G(H^*)$. Adding the next vertices $(\widehat{\rho}(2), \widehat{w})$, $\widehat{w} \in \widetilde{\rho}(1) \backslash \widetilde{\rho}(2)$, we know that the intersection $\widetilde{\rho}(1) \cap \widetilde{\rho}(2)$ contains exactly one equivalence class screening $\widetilde{\rho}(2)$ against the rest of the existing graph, thus no cycles can arise.

7.3 Analysis of Simple Hypertrees

Before we turn to those properties of simple hypertrees which are of importance for our further considerations we firstly clarify the relation between simple hypertrees and Markov trees, which are extensively used for local computation purposes.

7.3.1 Markov Trees

In case of a Markov tree, the vertices are nonempty sets:

(7.21) Definition. *We call a tree $(\mathfrak{E}, \xi)$ a Markov tree, if the following three conditions are satisfied:*

(i) $\mathfrak{E}$ is a set of hyperedges forming a hypergraph,
(ii) if $\{E, F\} \in \xi$, then $E \cap F \neq \emptyset$, and
(iii) if E, F are distinct elements of $\mathfrak{E}$, and if v is contained in E and F, then v is contained in every vertex (i.e element of ξ) along the path from E to F.

The notion of a Markov tree as defined above coincides with that of a join tree, well known in the field of data base theory (see e.g. [Maier 1983]).

(7.22) Theorem.

(a) If $H = (V, \mathfrak{E})$ is a hypertree then there exists ξ such that $(\mathfrak{E}, \xi)$ is a Markov tree.
(b) If $(\mathfrak{E}, \xi)$ is a Markov tree, then $H = (V, \mathfrak{E})$, where $V = \bigcup_{E:E \in \mathfrak{E}} E$ is a hypertree.

For a proof see e.g. [Maier 1983]. The keypoint is that a leaf in a Markov tree is a twig in the hypertree. If the dependency structure of the domain is represented by a simple hypertree, this ensures a *unique representation* of each vertex, i.e. each characteristic, within the graphical structure (in contrast to Markov tree representations).

(7.13) Example. (continued) In Fig. 7.8 a Markov tree representation of the hypertree $H' = (V, \mathfrak{E}')$ is depicted. The simple hypertree $\widetilde{H}$ emerging from the application of the algorithm (7.3) leads to the tree $G(\widetilde{H})$ shown in Fig. 7.9. Note that the vertex-nodes of $G(\widetilde{H})$ define a partition of V, thus each element of V is associated uniquely with one node in $G(\widetilde{H})$, whereas for example f appears in three nodes of the Markov tree. ◇

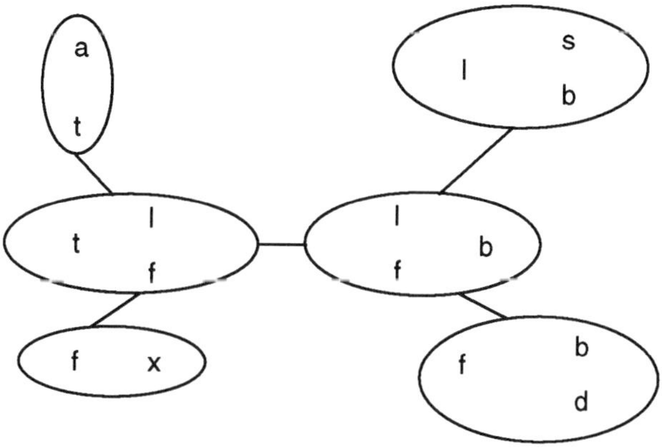

Fig. 7.8. A Markov tree representative for the hypertree of Fig. 7.4

7.3.2 Knowledge Representation with Hypergraphs

Now let us return to our original problem of knowledge representation. Once we have determined the dependency structure of the actual domain represented by a simple hypertree $H^* = (M^*, \mathfrak{T}^*)$, we can attach each piece of knowledge we obtain during the knowledge acquisition process to some subspace Ω^T, $T = \bigcup_{\hat{v} \in T^*} \hat{v}$, $T^* \in \mathfrak{T}^*$, which corresponds to some hyperedge T of H. Recall that the elements of M^* are *equivalence classes*, i.e subsets, of M. The hyperedges link the vertices $v^* \in T^*$, where each vertex v^* also represents some subspace Ω^{v^*}. These subspaces are subject to the direct factual knowledge provided by the user during the consultational dialogue.

The algorithms we present in the sequel will be based on a representation of the underlying (simple) hypertree by a system of link and node processors. Each hyperedge $T^* \in \mathfrak{T}^*$ is represented by a link processor K^{T^*} and each vertex $v^* \in M^*$ by a node processor X^{v^*}. The connection itself we imagine to be a bidirectional information exchange channel which ends at each connected processor in the form of a pair of input/output ports. At the node processors we assume there are additional ports for communication with the consulting user, i.e. an input port to receive the observations and an output port to provide

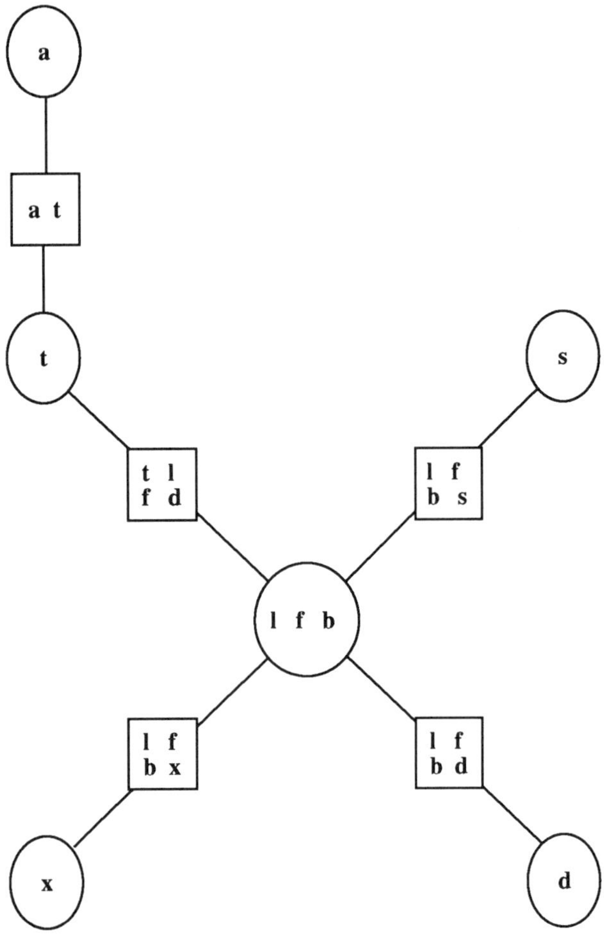

Fig. 7.9. The graph $G(\widetilde{H})$ which is a tree

the projection of the result that the inference process yields for the respective subspace.

Since we want to perform local computations we have to focus our interest on subhypertrees of the original simple hypertree. In the sequel we will provide a detailed notation for subhypertrees arising from the removal of one hyperedge.

Let $H = (V, \mathfrak{E})$ be a simple hypertree. Since $G(H)$ is a tree the removal of any edge $E \in \mathfrak{E}$ splits H into at least two unconnected subhypertrees (except when $E \in \mathfrak{E}$ is a twig).

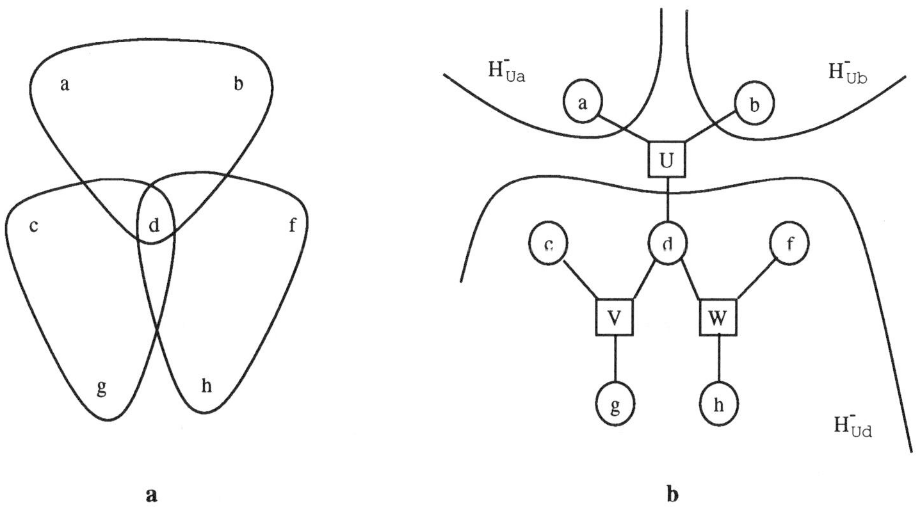

Fig. 7.10. Cutout of a simple hypertree: (a) H and (b) $G(H)$

(7.23) Example. Let us consider a cutout of a simple hypertree $H = (V, \mathfrak{E})$, where

$$V = \{\ldots, \text{a,b,c,d,f,g,h}, \ldots\},$$
$$\mathfrak{E} = \{\ldots, \text{U,V,W}, \ldots\},$$

$$U = \{\text{a,b,d}\}, \quad V = \{\text{c,d,g}\}, \quad W = \{\text{d,f,h}\}$$

The graphical representation of vertices (cycles) and hyperedges (squares) connected by lines relies on our intuive idea of node and link processors communicating along the lines (compare Fig. 7.10). The removal of the edge U would yield three remaining subhypertrees, one (H_{Ua}^-) above vertex a, one (H_{Ub}^-) above vertex b and one (H_{Ud}^-) below vertex d. ◇

(7.24) *Remark.* In general we denote the subhypertree arising from the removal of the hyperedge $E \in \mathfrak{E}$ and containing $v \in E$ by H_{vE}^-. Let

$$H_{vE}^- \stackrel{d}{=} (V_{vE}^-, \mathfrak{E}_{vE}^-) \tag{7.4}$$

where $V_{vE}^- \subseteq V$ and $\mathfrak{E}_{vE}^- \subseteq \mathfrak{E}$ are defined as follows:

$V_{vE}^{-} \stackrel{d}{=} \{w \in V \mid \text{in } G(H) \text{ there is a path from } v \text{ to } w \text{ not containing } E\}$

$\mathfrak{E}_{vE}^{-} \stackrel{d}{=} \{E \in \mathfrak{E} \mid E \subseteq V_{vE}^{-}\}.$

The complementary hypertree is denoted by

$$H_{vE}^{+} \stackrel{d}{=} (V_{vE}^{+}, \mathfrak{E}_{vE}^{+}) \tag{7.5}$$

where

$$V_{vE}^{+} \stackrel{d}{=} \overline{V_{vE}^{-}},$$
$$\mathfrak{E}_{vE}^{+} \stackrel{d}{=} \overline{\mathfrak{E}_{vE}^{-} \cup \{E\}}.$$

○

(7.25) *Remark.* Let $\widetilde{V}_{vE}^{+}$ and $\widetilde{\mathfrak{E}}_{vE}^{+}$ denote $V_{vE}^{+} \cup \{v\}$ and $\mathfrak{E}_{vE}^{+} \cup \{E\}$, respectively.

○

The next theorem proves some simple relationships of the subhypertrees defined above.

(7.26) Theorem. *Let $H = (V, \mathfrak{E})$ be a simple hypertree. We have*

(a.1) $V_{Ev}^{+} = \bigcup_{\substack{x: x \in E \\ x \neq v}} V_{Ex}^{-}$

(a.2) $\mathfrak{E}_{Ev}^{+} = \bigcup_{\substack{x: x \in E \\ x \neq v}} \mathfrak{E}_{Ex}^{-}$

(b.1) $V_{Ev}^{-} = \{v\} \cup \bigcup_{\substack{R: v \in R \\ R \neq E}} V_{vR}^{+} = \bigcup_{\substack{R: v \in R \\ R \neq E}} \widetilde{V}_{vR}^{+}$

(b.2) $\mathfrak{E}_{Ev}^{-} = \bigcup_{\substack{R: v \in R \\ R \neq E}} (\{R\} \cup \mathfrak{E}_{vR}^{+}) = \bigcup_{\substack{R: v \in R \\ R \neq E}} \widetilde{\mathfrak{E}}_{vR}^{+}$

(c.1) $V = \{v\} \cup \bigcup_{R: v \in R} V_{vR}^{+}$

(c.2) $\mathfrak{E} = \bigcup_{R: v \in R} (\{R\} \cup \mathfrak{E}_{vR}^{+})$

Proof.

(a.1) From the definition of V_{vE}^{-} we directly obtain $V = \bigcup_{x: x \in E} V_{xE}^{-}$, where the sets V_{xE}^{-} , $x \in E$, are mutually disjoint. So we obtain

$$V = V_{vE}^{-} \cup \bigcup_{\substack{x: x \in E \\ x \neq v}} V_{xE}^{-},$$

$$V_{vE}^{+} = \overline{V_{vE}^{-}} = \bigcup_{\substack{x: x \in E \\ x \neq v}} V_{xE}^{-}.$$

(a.2) In a similar way from the definition of $\mathfrak{E}_{vE}^{-}$ we have $\mathfrak{E} = \{E\} \cup \bigcup_{x:x\in E} \mathfrak{E}_{xE}^{-}$, again the sets $\{E\}$, $\mathfrak{E}_{xE}^{-}$, $x \in E$ are mutually disjoint and we get

$$\mathfrak{E} = \{E\} \cup \mathfrak{E}_{vE}^{-} \cup \bigcup_{\substack{x:x\in E\\ x\neq v}} \mathfrak{E}_{xE}^{-},$$

$$\mathfrak{E}_{vE}^{+} = \bigcup_{\substack{x:x\in E\\ x\neq v}} \mathfrak{E}_{xE}^{-}.$$

(b.1)
$$\begin{aligned}\{v\} \cup \bigcup_{\substack{R:v\in R\\ R\neq E}} V_{vR}^{+} &= \{v\} \cup \bigcup_{\substack{R:v\in R\\ R\neq E}} \bigcup_{\substack{x:x\in R\\ x\neq v}} V_{xR}^{-}\\ &= \{v\} \cup \bigcup_{\substack{R:v\in R\\ R\neq E}} \bigcup_{\substack{x:x\in R\\ x\neq v}} \{w \mid \text{ there is a path from } x \text{ to } w \text{ not via } R\}\\ &= \{v\} \cup \bigcup_{\substack{R:v\in R\\ R\neq E}} \{w \mid \text{ there is a path from } v \text{ to } w \text{ via } R\}\\ &= \bigcup_{\substack{R:v\in R\\ R\neq E}} \{w \mid \text{ there is a path from } v \text{ to } w \text{ via } R\}\\ &= \{w \mid \text{ there is a path from } v \text{ to } w \text{ not via } E\}\\ &= V_{vE}^{-}\end{aligned}$$

(b.2)
$$\begin{aligned}\bigcup_{\substack{R:v\in R\\ R\neq E}} (\{R\} \cup \mathfrak{E}_{vR}^{+}) &= \bigcup_{\substack{R:v\in R\\ R\neq E}} \bigcup_{\substack{x:x\in R\\ x\neq v}} (\{R\} \cup \{E \mid E \subseteq V_{RS_l}^{-}\})\\ &= \bigcup_{\substack{R:v\in R\\ R\neq E}} \{E \mid E \subseteq \widetilde{V}_{Rv}^{+}\} = \{E \mid E \subseteq V_{Ev}^{-}\} = \mathfrak{E}_{Ev}^{-}\end{aligned}$$

(c.1) and (c.2) are proven analogously. □

(7.27) *Remark.* In the sequel, whenever we refer to Remark 6.53, we use the notation

$$M_{vT}^{-} \stackrel{d}{=} \bigcup_{x^*:x^*\in(M^*)_{vT}^{-}} x^*$$

$$M_{vT}^{+} \stackrel{d}{=} \bigcup_{x^*:x^*\in(M^*)_{vT}^{+}} x^*$$

$$\widetilde{M}_{vT}^{+} \stackrel{d}{=} \bigcup_{x^*:x^*\in(\widetilde{M}^*)_{vT}^{+}} x^*$$

○

7.4 Dependency Networks

Knowledge turns out to be rather relative in the sense, that the expert determines the dependencies of *causes* and *effects* (compare [Pearl 1986]). In order to formalize this idea, we have to consider a further specification of simple hypertrees.

Let us consider the simple hypertree $H = (V, \mathfrak{E})$. For every edge $E \in \mathfrak{E}$ we define a *direction* by marking one vertex $h \in E$. We denote this causal target by h_E and call it the *head* of E. The remaining vertices of E are denoted by E_h, i.e. $E_h \stackrel{d}{=} E \cap \overline{\{h_E\}}$, and called the *body* of E. In this way we obtain a *directed simple hypergraph.*

(7.28) Definition. *The pair* $H_d = (V, \mathfrak{E}_d)$, *where* V *is a finite set of vertices and* $\mathfrak{E}_d \subseteq 2^V \times V$, *is called a directed simple hypertree if for all* $E_d = (E_h, h_E)$ *we have* $h_E \notin E_h$ *and if* $H = (V, \mathfrak{E})$, *where* $\mathfrak{E} = \{E \mid \exists E_d \in \mathfrak{E}_d$ *such that* $E = E_h \cup \{h_E\}\}$, *is a simple hypertree.* E_h *is called the body of* E_d *and* h_E *is called the* head *of* E_d.

(7.29) Example. Let $H_d = (V, \mathfrak{E}_d)$ where $V = \{a, b, c, d, f, g, h\}$ and

$$\mathfrak{E}_d = \{(\{a\}, c), (\{b\}, d), (\{c, d\}, f), (\{f\}, g), (\{f\}, h)\}.$$

As we easily see from Fig. 7.11, H_d is a directed simple hypertree. ◇

In order to be sure that the lines of argumentation dovetail properly into each other we furthermore assume that any vertex v which is contained in the body E_h of some directed hyperedge is the head of some other directed hyperedge, i.e. there is $F_d \in \mathfrak{E}_d$ such that $v = h_F$. So a directed hypertree has to be completed with "roots", i.e. directed hyperedges of the form $(\emptyset, v)$.

From the use of directed simple hypertrees arises one more problem. Assume there are two directed hyperedges $E_d, F_d \in \mathfrak{E}_d$ where $h_E = h_F$. If the directed simple hypertree is considered to represent "causal" dependencies, then this situation means that from at least two sources causal influences affect node $v = h_E = h_F$. As within a probabilistic calculus additional knowledge is required in order to integrate these two "streams" of information, in this case the two edges $E_d = (E_h, v)$ and $F_d = (F_h, v)$ should be replaced by a single edge $(E_h \cup F_h, v)$. (Note that such an operation does not affect the acyclicity of $G(H)$.)

(7.30) Definition. *Let* $H_d = (V, \mathfrak{E}_d)$ *be a directed simple hypertree.* $(V, \mathfrak{E}'_d)$, *where* $\mathfrak{E}'_d = \mathfrak{E}_d \cup \{(\emptyset, v) \mid \forall E_d \in \mathfrak{E}_d : v \neq h_E\}$, *is called a dependency network induced by* H_d *if* $\forall E_d, F_d \in \mathfrak{E}_d : [E_d \neq F_d \Longrightarrow h_E \neq h_F]$.

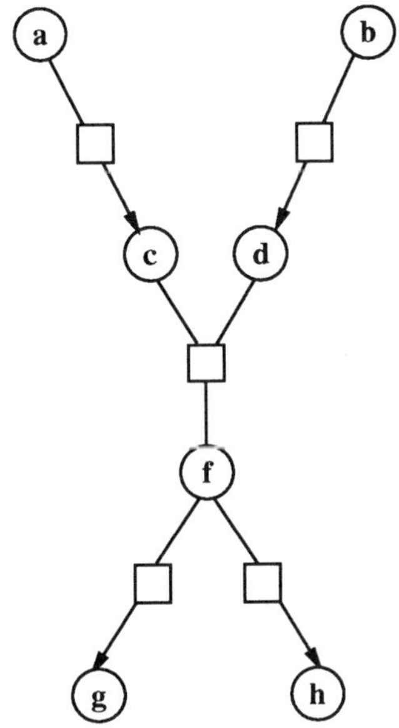

Fig. 7.11. A directed simple hypertree

(7.31) *Remark.* Because of the "roots" $(\emptyset, v)$ a dependency network corresponds to a hypertree but not to a reduced hypertree (compare Section 7.2). ∘

We can translate Theorem 7.26 into the notation of dependency networks. This yields

(a.1) $$V^+_{Eh_E} = \bigcup_{F:h_f \in E_h} V^-_{Eh_F}$$

(b.1) $$V^-_{Eh_E} = \{h_E\} \cup \bigcup_{F:h_E \in F_h} V^+_{Fh_E}.$$

In the case of a vertice $v \in E_h$ we have

(a.2) $$V^+_{Ev} = \bigcup_{\substack{x:x \in E \\ x \neq v}} V^-_{Ex}$$

(b.2) $$V^-_{Ev} = \{v\} \cup \bigcup_{\substack{F:v \in F \\ F \neq E}} V^+_{Fv}.$$

Now let us turn to those dependency networks that are based on a simple directed hypertree of the form $H^* = (M^*, \mathfrak{T}^*)$; compare Remarks 6.53 and

7.27. We know that the respective conditions hold for M^+_{Tv} and M^-_{Tv}. Thus we obtain

$$\begin{aligned} M &= M^-_{Th_T} \cup M^+_{Th_T} \\ &= \bigcup_{S:h_T\in S_h} M^+_{h_T S} \cup \{h_T\} \cup \bigcup_{v:v\in T_h} M^-_{T_v} \\ &= \bigcup_{S:h_T\in S_h} \left[\bigcup_{\substack{x:x\in S\\ x\neq h_T}} M^-_{S_x} \right] \cup \{h_T\} \cup \bigcup_{v:v\in T_h} \left[\{v\} \cup \bigcup_{\substack{S:v\in S\\ S\neq T}} M^+_{S_v} \right] . \end{aligned}$$

Using the shorthand notations

$$vT\uparrow \stackrel{d}{=} M^+_{T_v} \text{ and } vT\downarrow \stackrel{d}{=} M^-_{T_v},$$

for $v \in T_h$, $h_T \in S_h$, we obtain

$$\begin{aligned} M &= vT\downarrow \cup\ vT\uparrow \\ &= \bigcup_{S:h_T\in S_h} h_T S\uparrow \cup\ \{h_T\} \cup \bigcup_{v:v\in T_h} vT\downarrow . \end{aligned}$$

Defining

$$h_T\downarrow \stackrel{d}{=} \bigcup_{S:h_T\in S_h} h_T S\uparrow \text{ and } T\uparrow \stackrel{d}{=} \bigcup_{v:v\in T_h} vT\downarrow$$

yields

$$M = h_T\downarrow \cup \{h_T\} \cup T\uparrow .$$

Since in this way M is divided into *disjoint* sets we derive

$$\begin{aligned} h_T S\downarrow = M - h_T S\uparrow &= \bigcup_{\substack{S':h_T\in S'_h\\ S'\neq S}} h_T S'\uparrow \cup \{h_T\} \cup T\uparrow , \\ vT\uparrow = M - vT\downarrow &= h_T\downarrow \cup \{h_T\} \cup \bigcup_{\substack{x:x\in T_h\\ x\neq v}} xT\downarrow \end{aligned}$$

for $v \in T_h$, $h_T \in S_h$.

(7.32) Example. Consider the dependency network $H = (M, \mathfrak{T})$, where

$$M = \{\text{a,b,c,d,f,g,h}\}$$
$$\mathfrak{T} = \{\text{U,V,W}\} \ ,$$

U = ({a,b},f), V = ({c,d},g), W = ({f,g},h).

We have for example $\text{aU}\downarrow = \{a\}$, $\text{dV}\uparrow \ = \{a, b, c, f, g, h\}$, as we easily see from Fig. 7.12. ◇

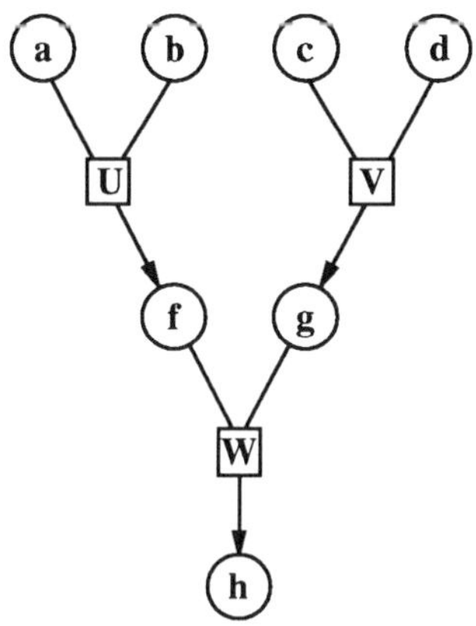

Fig. 7.12. Dependency network of Example 7.32

A dependency network induces a *consistent numbering* of the hyperedges $E \in \mathfrak{E}$, which reflects the flow of argumentation and thus the flow of knowledge acquisition. Consider the following sequence of subsets $\mathfrak{E}^k$ of $\mathfrak{E}$:

$$\mathfrak{E}^0 \stackrel{d}{=} \emptyset,$$
$$\mathfrak{E}^{k+1} \stackrel{d}{=} \mathfrak{E}^k \cup \{E \mid \forall v \in E_h : \exists F \in \mathfrak{E}^k : v = h_F\}$$

Obviously $\mathfrak{E}^1$ contains the roots of the dependency network, $\mathfrak{E}^2$ the direct successors of the roots, etc. After a finite number s of steps we have $\mathfrak{E}^s = \mathfrak{E}$. Each hyperedge is reached by this sequence, since from each node of the dependency network there is a path to at least one root. Each node except the root has at least one predecessor, and this is valid for every node, for its predecessor, for its prepredecessor and so on until the root is reached. No hyperedge is reached twice, since a dependency network has no cycles, either directed or undirected. For this reason the hyperedges can be enumerated starting with the elements in $\mathfrak{E}^1$, continuing with those in $\mathfrak{E}^2 - \mathfrak{E}^1$, in $\mathfrak{E}^3 - \mathfrak{E}^2$ and so on. Such ordering $\rho : \{1,\ldots,r\} \longrightarrow \{1,\ldots,r\}$ we call a consistent ordering (note that dependency consistent ordering is not to be confused with the hypertree construction sequence).

If we consider the paths in a dependency network, these paths indicate the flow of (mediate) influence. This flow of influence can be *blocked*, if the actual value of some (joint) characteristic along the path is known for sure.

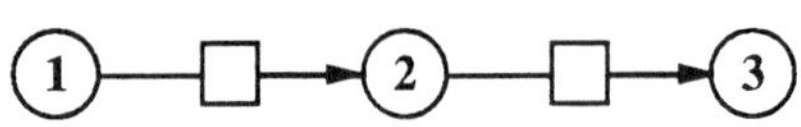

Fig. 7.13. Dependency network of Example 7.33

(7.33) Example. Let us once more consider our introductory example concerning the consequences of oil contamination. As the characteristics $X^{(1)}$ (location), $X^{(2)}$ (authority) and $X^{(3)}$ (measure) we have

$$M = \{1,2,3\} ,$$
$$\mathfrak{T} = \{(\emptyset,1),(\{1\},2),(\{2\},3)\} ,$$

which means the dependency network has the structure depicted in Fig. 7.13.

The location of the oil accident has an indirect influence (via the competent authority) on the measure. If the competent authority is known (say the harbor administration) then additional knowledge on the location has no further impact to the actual measure, since the flow of influence is indirect and would require an impact on $X^{(2)}$ which is impossible since the value of $X^{(2)}$ is known. ◇

(7.34) Definition. *Let $C = (V,\mathfrak{E})$ be a dependency network. The vertex $v \in V$ blocks a path $<\ldots,u,S,v,T,w,\ldots>$ if*

(*i*) $v = h_S$ *and* $v \in T_h$ *(head–to–tail blocking) or*
(*ii*) $v = h_T$ *and* $v \in S_h$ *(tail–to–head blocking) or*
(*iii*) $v \in T_h \cap S_h$ *(tail–to–tail blocking).*

A set $V_c \subseteq V$ of vertices is said to separate two nodes $v, w \in V$ if there is a node $x \in V_c$ such that x blocks the path connecting v and w.

The above definition is illustrated in Fig. 7.14.

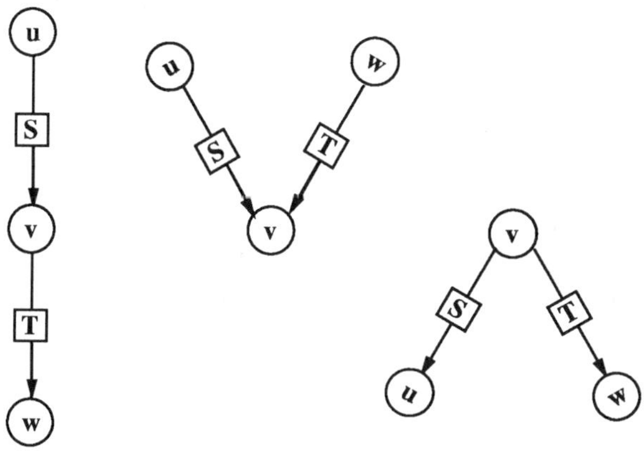

Fig. 7.14. Types of blocking

7.5 Triangulated Graphs

In the probabilistic approaches to uncertainty considered in Sect. 12.4 the notion of a triangulated graph will play a central role: If cliques are considered in order to characterize indivisible parts of some domain (see e.g. [Lauritzen and Spiegelhalter 1988]), then the property of *chordality* is of importance. A graph $G = (V, \mathfrak{E})$ is called *chordal* if every cycle with a length of at least four has a chord, i.e. an edge joining two nonconsecutive vertices in the cycle. Chordal graphs are sometimes said to be *triangulated.*

(7.35) Definition. *An undirected graph $G = (V, \mathfrak{E})$ is called triangulated if every cycle $\langle v_0, \ldots, v_k \rangle$ of length strictly greater than 3 possesses a chord, i.e., if $\{v_i, v_j\} \subseteq \mathfrak{E}$ for some i, j with $j \neq i \pm 1$ mod$(k+1)$.*

The undirected graph of Fig. 7.15(a) is triangulated whereas the graph (b) is not triangulated because the cycle $\langle a, b, d, c, a \rangle$ does not possess a chord.

Fig. 7.15. Two undirected graphs

An efficient method for checking whether a graph is triangulated or not is based on numbering the vertices in a special way. Let $G = (V, \mathfrak{E})$ be a finite graph with n vertices. An ordering $<_\alpha$ of the vertices is obtained by a one-to-one mapping

$$\alpha : \{1, \ldots, n\} \longrightarrow V,$$

where we define $\alpha(1) <_\alpha \alpha(2) <_\alpha \ldots <_\alpha \alpha(n)$.

The ordering $<_\alpha$ is called *perfect* if and only if for all $i = 1, \ldots, n$ the set

$$A_i = \big\{\alpha(1), \ldots, \alpha(i-1)\big\} \cap \big\{v \in V \mid \{v, \alpha(i)\} \subseteq \mathfrak{E}\big\} \tag{7.6}$$

of vertices is complete, i.e.,

$$\big\{\{v, v'\} \mid v, v' \in A_i, v \neq v'\big\} \subseteq \mathfrak{E}.$$

(7.36) Theorem. *An undirected graph is triangulated if and only if it admits a perfect numbering.*

The graph depicted in Fig. 7.16 is triangulated, and a perfect ordering is indicated. A fast algorithm for testing whether a graph is triangulated or not is based on the maximum cardinality search. It has the time complexity $O(n+e)$, where n is the number of vertices and e the number of edges. Informally the algorithm runs as follows:

Assign number 1 to an arbitrary vertex. For the next vertex to number, select the vertex adjacent to the largest number of previously numbered vertices, break ties arbitrarily. The complete algorithm is given in [Tarjan and Yannakakis 1984].

(7.37) Theorem. *If the graph is triangulated then the maximum cardinality search gives a perfect ordering.*

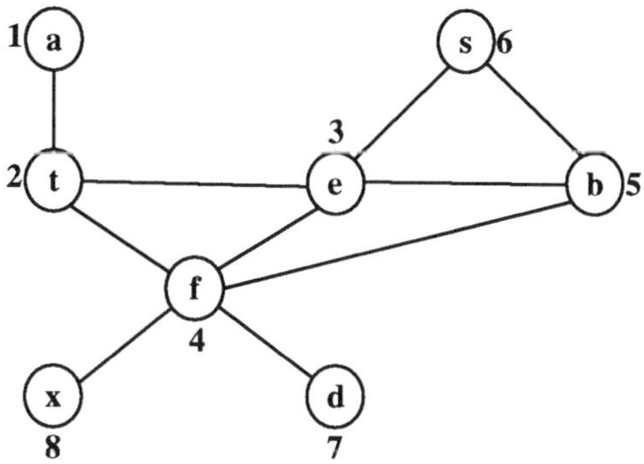

Fig. 7.16. A perfect ordering α of the graph

If a numbering is not perfect, i.e., if the graph is not triangulated, it can be made so by adding (filling-in) edges. A fast (but not optimal) method for computing this "fill-in" set is described in [Tarjan and Yannakakis 1984]. A fill-in set is called *minimal* if no renumbering of the nodes gives a fill-in set which is a subset of the given fill-in set; a fill-in set is minimal if it contains a minimal number of edges. The problem of computing a minimal fill-in set is NP-complete [Yannakakis 1981].

Testing for triangulatedness is done by computing the fill-in set for the numbering and checking whether or not this set is empty.

The minimum fill-in set of graph G in Fig. 7.14 a is $\{\{b, c\}\}$, i.e., in order to triangulate the graph G we have to add an edge between b and c. Fig. 7.17 b shows the corresponding triangulated graph with numbering according to the maximum cardinality search.

Triangulated graphs (also called chordal graphs) can be also considered from a hypergraph point of view.

If $H = (V, \mathfrak{E})$ is a hypergraph then the graph $G_H = (V, \mathfrak{E}^*)$ is an ordinary, undirected graph with vertices V and edges $\mathfrak{E}^* = \{\{v, w\} \mid \exists E \in \mathfrak{E} : \{v, W\} \subseteq E \text{ and } v, w \in V\}$. G_H should not be confused with the graph $G(H)$ considered in Def. 7.14. H is called conformal if every clique of G_H is contained in an edge of H. Furthermore H is said to be *acyclic* if H is conformal and G_H is triangulated (see [Tarjan and Yannakakis 1984], [Maier 1983], [Lauritzen and Spiegelhalter 1988]). An equivalent formulation is as follows: A hypergraph $H = (V, \mathfrak{E})$ is acyclic if and only if no elements in $\mathfrak{E}$ are subsets of other elements, and if the elements of $\mathfrak{E}$ can be ordered (i.e., $E_1 < \ldots < E_p$) to have the running intersection property

$$\forall j \geq 2 \quad \exists i < j : E_i \supseteq E_j \cap (E_1 \cup \ldots \cup E_{j-1}) \tag{7.7}$$

[Lauritzen, Speed, and Vijayan 1984]. In our nomenclature acyclic hypergraphs are called hypertrees [Shafer and Shenoy 1988]. Note that the running intersection property inspired the introduction of the intuitive notions of a twig and the construction sequence in Def. 7.9 and Def. 7.11.

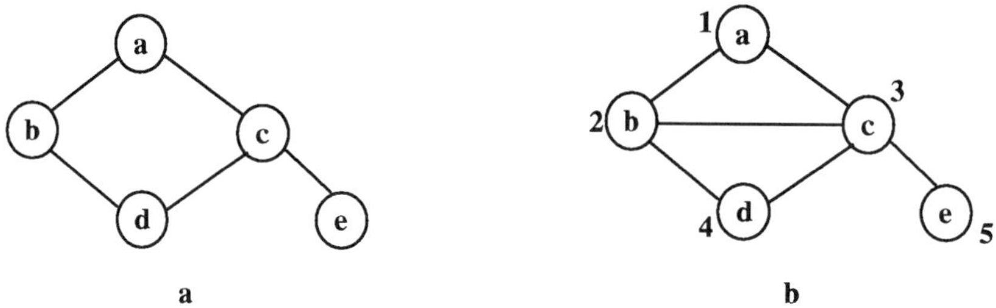

Fig. 7.17. A triangulated graph

(7.38) Theorem. *Let $G = (V, \mathfrak{E})$ be a triangulated graph, and let $C_1, \ldots, C_p$ be the maximal cliques in G. Then $H = (V, \{C_1, \ldots, C_p\})$ is a hypertree.*

The hypertree corresponding to the graph of Fig. 7.16 is depicted in Fig. 7.4.

7.6 Directed Acyclic Graphs

In the following directed acyclic graphs are introduced as a subclass of dependency networks.

(7.39) Definition. *A directed graph G is a pair $(V, \mathfrak{D})$, where $V = \{v_1, \ldots, v_n\}$ contains the vertices and $\mathfrak{D} = \{D_1, \ldots, D_r\}$ contains the edges. $\mathfrak{D}$ is a subset of the cartesion product $V \times V$. We assume that $(v, v) \notin \mathfrak{D}$ for all $v \in V$, i.e., we prohibit loops.*

As in the case of undirected graphs we say that a chain of length k from v to w in a directed graph $G = (V, \mathfrak{D})$ is a sequence of vertices $\langle v_0, v_1, \ldots, v_k \rangle$

with $v_0 = v$ and $v_k = w$ and $(v_i, v_{i+1}) \in \mathfrak{D}$ for $i = 0, \ldots, k-1$. A cycle is a chain $\langle v_0, \ldots, v_k \rangle$ with $v_0 = v_k$.

(7.40) Definition. *A graph is called a directed acyclic graph if G is directed and has no cycles.*

Directed acyclic graphs can be viewed as the simplest case of dependency networks as considered in Sect. 7.4.

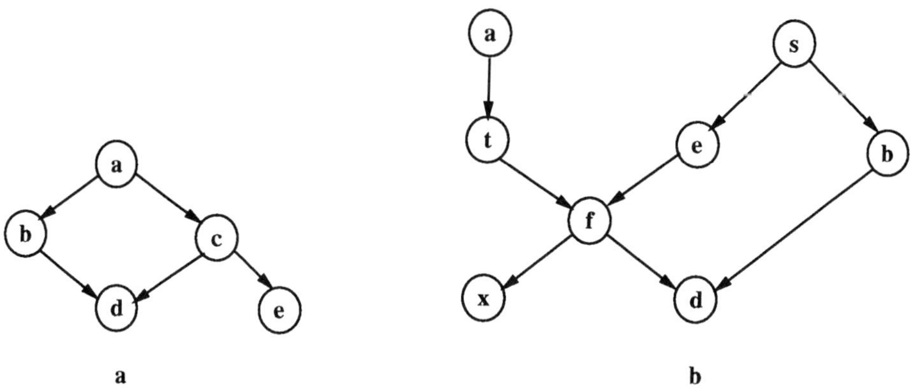

Fig. 7.18. Directed acyclic graphs

(7.41) Definition. *Let* $G = (V, \mathfrak{D})$ *be a directed graph. The graph* $G_m = (V, \mathfrak{E})$ *defined by*

$$\mathfrak{E} = \{\{u, w\} \mid (u, w) \in \mathfrak{D}\} \cup \{\{u, w\} \mid (u, v) \in \mathfrak{D}, (w, v) \in \mathfrak{D}, v \in V\}$$

is called the moral graph *relative to G.*

In order to find G_m one has to "marry" the parents and then to drop the directions. In general the moral graph is not triangulated.

The triangulated graph of Fig. 7.19 is depicted in Fig. 7.16. Moral graphs turned out to be useful in the decomposable models presented in Sect. 14.2.

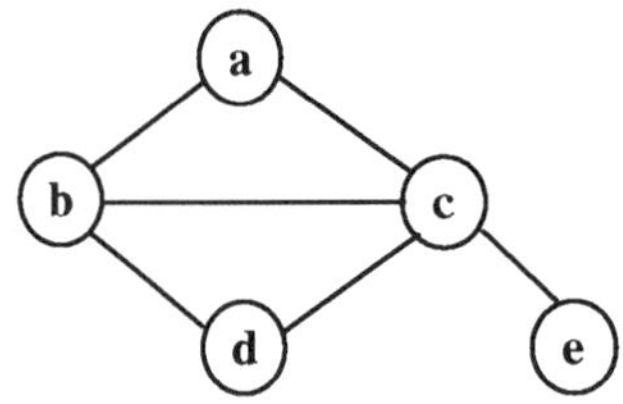

Fig. 7.19. (a) Moral graph corresponding to Fig. 7.18 (a)

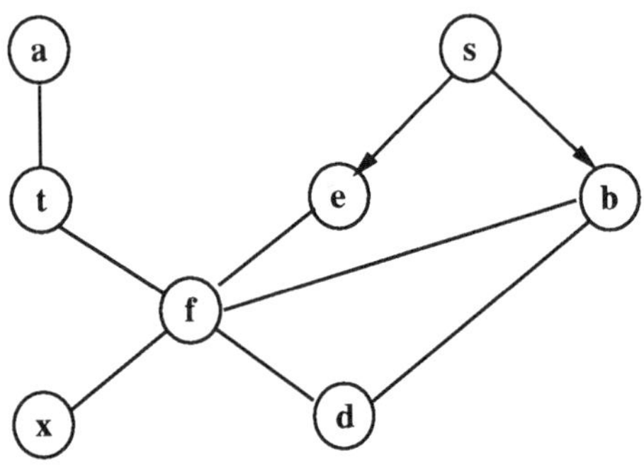

Fig. 7.19. (b) Moral graph corresponding to Fig. 7.18 (b), not triangulated

7.7 Concluding Remarks

Throughout this chapter we have restricted ourselves to those aspects of the theory of graphs and hypergraphs which are indispensible for our further considerations. For a more detailed presentation see for example [Berge 1973] or [Golumbic 1980]). Hypergraphs and hypertrees are well known concepts of great importance in the field of relational databases [Maier 1983]. A hypertree is what other authors call an acyclic hypergraph. Markov trees are also called join trees. We used the intuitive notions provided by [Shafer and Shenoy 1988].

Triangulated graphs have been discussed under various names, such as rigid circuit graphs [Dirac 1961], chordal graphs [Gavril 1972] and decomposable graphs [Lauritzen et al. 1984].

In [Jensen 1988] junction graphs, being strongly related to the intersection graphs considered in [Maier 1983] and the graphs $G(H)$ we applied for the representation of dependencies, are used in order to provide a characterization of decomposable hypergraphs. Note that the special property of simple hypertrees (a private notation) is that their junction graph is a tree.

Directed acyclic graphs (in short: DAG's) are also analyzed under various names, such as influence diagrams [Smith 1989], recursive models [Wermuth and Lauritzen 1983] or Bayesian Networks (causal networks) [Pearl 1986]. Dependency networks as considered in Definition 7.30 are introduced to give a compact description for propagation algorithms implemented by the authors.

Knowledge based systems that use a decision-theoretic model as a framework for knowledge representation and inference are referred to as normative expert systems. These models are based on the notion of influence diagrams. The evaluation of influence diagrams with respect to decision making is discussed in [Shachter 1986]. In our context we use influence diagrams that contain no decision nodes and no utility nodes. Those are called belief networks [Pearl 1988] or knowledge maps [Howard 1989].

The study of causation is central to the understanding of human reasoning. Temporal precedence is normally assured essential for defining causation [Suppes 1970]. Nontemporal semantics that determines the directionality of causal influences are considered in [Glymour et al. 1987]. In order to model the qualitative structure of causal theories, often directed acyclic graphs are used [Verma and Pearl 1990b]. These so called causal networks also allow to represent dependency knowledge of the form "x is independent of y once z is known". In our book we do not consider the notion of causality but analyze instead the term dependency.

Chapter 8

Modeling Aspects

The modeling and representation of expert knowledge as well as reasoning and decision making based on this knowledge have often to be done in real-world environments where one is ignorant about certain aspects of an actual problem.

In the preceding chapters we have learned that, in particular, the knowledge may be uncertain or vague: While uncertainty covers cases where an agent has a maybe subjective opinion on a proposition that is not definitely established for him, vagueness is due to the use of different contexts for describing facts and rules. In this sense, a proposition will be called uncertain if its truth or its falsity cannot be established definitely. Uncertainty can be present both in factual knowledge and – as a consequence of weak conditional statements or partial causal relationships between facts – in pieces of information such as "expert rules". The existence of uncertain and/or vague statements about the real world has as one consequence that decisions have to be made under risk. How to handle knowledge of this kind in an "intelligent" manner as well as on a formal basis was the topic of the previous chapters where different numerical models were introduced.

The aim of this chapter is to consider the *general aspects* underlying the most important numerical methods and applications developed during recent years for the representation and the processing of uncertainty in knowledge based systems. We further consider several aspects which are often discussed in relation to formal models for uncertainty handling and which, during the last two decades, have triggered the development of several more or less heuristic *certainty calculi.*

8.1 Rule Based Approaches

Early models for handling uncertain or vague knowledge were developed in the context of *rule-based systems*. In such systems explicit expert rules have to be formulated and are entered by them into the system. In the simplest case expert rules are of the form

$$\underline{\text{if}}\ X\ \underline{\text{then}}\ H\ \underline{\text{with}}\ cert(X,H),$$

where X and H are the rule's premise and conclusion, respectively. The semantics and the range of the numerical quantity $cert(X, H)$ heavily depends on the mathematical framework for representing the uncertain knowledge.

Concerning the semantics of the expert rule, we shall explain some of the difficulties in the context of precise data; in the case of vague data the situation is even worse. There are at least two different interpretations of the above rule. The first one reads

$$\underline{\text{if}}\ X\ \underline{\text{then}}\ belief(H) = cert(X, H),$$

which means that if we observe the fact X then the (partial) belief on H (which may be the probability, the possibility, or the credibility respectively, depending on the underlying mathematical model) should be updated to $cert(X, H)$. This interpretation is typically used in rule-based systems; here the *change in belief* is measured by $cert(X, H)$.

A totally different interpretation is obtained if we position this parentheses in the way:

$$belief(\underline{\text{if}}\ X\ \underline{\text{then}}\ H) = cert(X, H).$$

In our book we consider only multivariate approaches, where X is a subset of an underlying set Ω_X and H is a subset of Ω_H. Therefore we represent the rule $\underline{\text{if}}\ X\ \underline{\text{then}}\ H$ by using the subset $(X \times H) \cup (\overline{X} \times \Omega_H)$ of $(\Omega_X \times \Omega_H)$. So in this interpretation we obtain a measure for the *absolute belief* but not for the change of belief as in the above interpretation. For instance, using probabilistic models, in its first interpretation $cert(X, H)$ is always represented as a conditional probability $P(H|X)$, whereas in the second interpretation the absolute probability $P(X \times H \cup \overline{X} \times \Omega_H) = cert(X, H)$ is fixed. In the remainder of this section we consider the first interpretation.

In rule-based expert systems, rules of this kind are applied to infer (uncertain) knowledge about H on the basis of (uncertain) knowledge about X which itself is computed from some evidence E given, in the form of observations, for instance. In the following, the process of combining the uncertainty $cert(E, X)$ of X and the rule's uncertainty $cert(X, H)$ to obtain the conclusion's uncertainty $cert(E, H)$ will be called *sequential combination*.

Possibly, the conclusion of an expert rule does not represent the final hypothesis but arises as premise in one or more other expert rules. For instance, in the case

$$\begin{array}{l} \underline{\text{if}}\ X\ \underline{\text{then}}\ Y\ \underline{\text{with}}\ cert(X, Y) \\ \underline{\text{if}}\ Y\ \underline{\text{then}}\ Z\ \underline{\text{with}}\ cert(Y, Z) \\ \underline{\text{if}}\ Z\ \underline{\text{then}}\ H\ \underline{\text{with}}\ cert(Z, H) \end{array}$$

the final evaluation $cert(E, H)$ of hypothesis H is obtained from a sequence of sequential combinations starting with the information $cert(E, X)$.

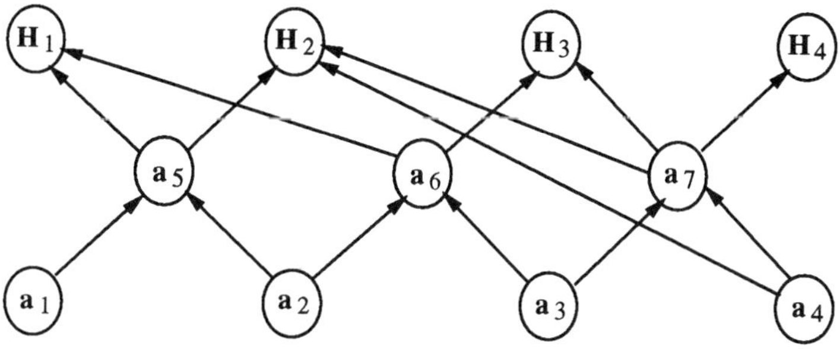

Fig. 8.1. A typical inference network

Generally, a hypothesis H may be supported by more than one expert rule as well as by several possibly independent information sources. Consequently, in the case of n expert rules

$$\underline{\text{if}}\ X_1\ \underline{\text{then}}\ H\ \underline{\text{with}}\ cert(X_1, H)$$
$$\underline{\text{if}}\ X_2\ \underline{\text{then}}\ H\ \underline{\text{with}}\ cert(X_2, H)$$
$$\vdots$$
$$\underline{\text{if}}\ X_n\ \underline{\text{then}}\ H\ \underline{\text{with}}\ cert(X_n, H),$$

first, n sequential combinations have to be performed to get the uncertainty values $cert(E_i)$, $i = 1, \dots, n$. Next, the quantities measuring the uncertainty of hypothesis H on the basis of different pieces of evidence must be combined to get the final uncertainty of H. This process will be called *parallel combination.* The consistency of the chosen parallel combination rule in the context of the mathematical uncertainty formalism and the assumptions made is an important criterion when evaluating a model for uncertain reasoning.

The set of expert rules as well as the process of sequential and parallel combinations may be represented in the form of an *inference network*, in which the arrows point from evidence to hypothesis (see Fig. 8.1). Note that in general the sequential and parallel combinations are based on *functions* that derive a "degree of belief" (e.g., the probability) in the conclusion for *any* score from the items in the premise. Other possibilities are to make use of *tabular combining functions* — tables that specify degrees of belief in conclusions for each combination of scores from items of evidence — or to make use of *modifiable combining functions* [Cohen, Shafer, and Shenoy 1987]. Functions of this kind are not reviewed in this book.

One basic assumption for the above considerations is a relatively simple rule structure consisting of atomic premises and propositions only. However, for practical applications more complex premises containing th logical connectives "and", "or", and "not" are of greater interest. In the simplest case these more complex expert rules may have the following structure

$$\underline{\text{if}}\ X\ \underline{\text{and}}\ Y\ \underline{\text{then}}\ H\ \underline{\text{with}}\ cert_1$$
$$\underline{\text{if}}\ X\ \underline{\text{or}}\ Y\ \underline{\text{then}}\ H\ \underline{\text{with}}\ cert_2$$
$$\underline{\text{if}}\ \underline{\text{not}}\ X\ \underline{\text{then}}\ H\ \underline{\text{with}}\ cert_3$$

One consequence of allowing more complex rule premises is that, before performing the sequential combination, the uncertainty of the whole premise must be established. The associated process will be called *premise evaluation.* For instance, assuming the known elementary uncertainty values $cert(E|X)$ and $cert(E|Y)$, the algorithm that can often be found when examining uncertainty models computes the uncertainty of "X and Y", "X or Y", and "not X" making use of the operators min, max, and difference from 1, respectively, applied to elementary uncertainty values.

Although rule-based systems turned out to be very successful in the representation of categorical knowledge they are often inadequate both to represent uncertain knowledge and to draw inferences on the basis of such knowledge. One problem is that mutually exclusive and exhaustive alternatives can be badly handled. But the most serious problem with these models is the incorrect handling of dependencies or causalities. This comes from the fact that in a rule which represents uncertainty the implication can go on in either direction (in contrast to logical approaches, where the implications are always directed). But this is only one typical aspect of uncertain rules. Early probabilistic systems failed because they did not correctly analyse the dependency structure. But even if one avoids imposing too many independence assumptions, each naive rule-based approach without the analysis of the dependency structure must fail [Heckerman and Horvitz 1988].

(8.1) Example. Let $\Omega^{(1)}, \Omega^{(2)}, \Omega^{(3)}, \Omega^{(4)}$ be four spaces each of them having two elements $\omega_1^{(i)}, \omega_2^{(i)} \in \Omega^{(i)}$. Suppose we want to specify rules between $\Omega^{(1)}$ and $\Omega^{(2)}$, $\Omega^{(1)}$ and $\Omega^{(3)}$, $\Omega^{(2)}$ and $\Omega^{(4)}$, $\Omega^{(3)}$ and $\Omega^{(4)}$, described by the following directed graph.

A straightforward probabilistic solution according to the first interpretation of rule certainties would be to specify a conditional probability for each connection, i.e., to characterize f.e. the links $\Omega^{(1)} \rightarrow \Omega^{(2)}$ by probabilities $P(\omega^{(1)}|\omega^{(2)})$. Then this system is in general contradictory, i.e., there is no probability distribution P on $\Omega^{(1)} \times \Omega^{(2)} \times \Omega^{(3)} \times \Omega^{(4)}$ with the corresponding conditional probabilities. A similar straightforward solution corresponding to the second interpretation is to specify a set of possible marginal probabilities for the subspaces $\Omega^{(1)} \times \Omega^{(2)}, \ldots, \Omega^{(3)} \times \Omega^{(4)}$. Unfortunately these quantities may

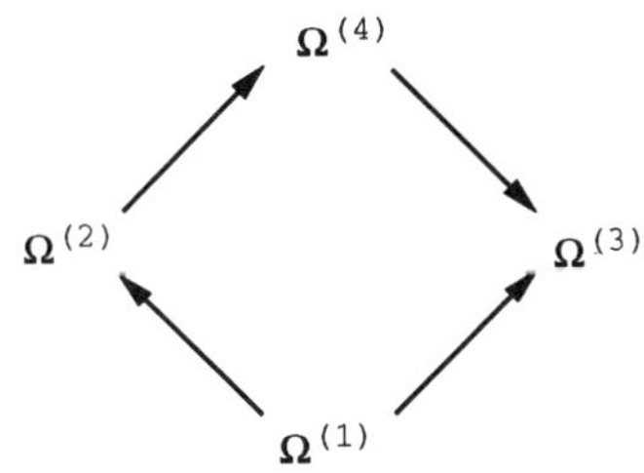

Fig. 8.2. Directed graph

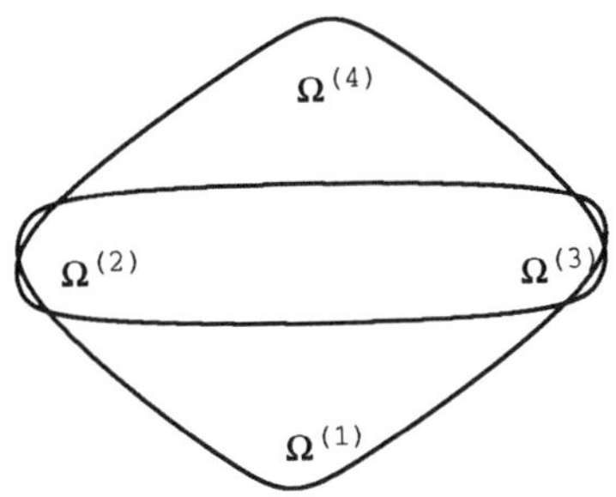

Fig. 8.3. Hypergraph

be also contradictory, as there may be no probability on the product space with corresponding marginal probabilities. The problem is again that the marginals on $\Omega^{(3)} \times \Omega^{(4)}$ obtained from $\Omega^{(1)} \times \Omega^{(2)} \times \Omega^{(3)}$ and $\Omega^{(2)} \times \Omega^{(3)} \times \Omega^{(4)}$ do not coincide. This shows that the modularity obtained by using two-dimensional structures is not sufficient to describe the dependencies. In our example we must handle the three-dimensional spaces $\Omega^{(3)} \times \Omega^{(2)} \times \Omega^{(4)}$ and $\Omega^{(1)} \times \Omega^{(2)} \times \Omega^{(3)}$ such that there is marginal coincidence on the space $\Omega^{(2)} \times \Omega^{(4)}$; i.e., we have to analyse the hypergraph.

A correct reformulation as a rule-based system would be use higher-dimensional rules such as

$$\underline{\text{if}}\ \omega^2 \text{ and } \omega^3\ \underline{\text{then}}\ \omega^4\ \underline{\text{with}}\ P(\omega^4|\omega^2, \omega^3).$$

For the other two rules we could impose conditional independence assumptions $P(\omega^2, \omega^3|\omega^1) = P(\omega^2|\omega^1) \cdot P(\omega^3|\omega^1)$ or use also in this case higher-dimensional rules. ◇

Each method for uncertainty handling is very sensitive with respect to the dependency problems, and it is not only the probabilistic approaches that must be used very carefully with respect to this topic.

It should be emphasized that the representation of categorical knowledge does not cause conceptual problems, whereas a naive extension of a rule-based approach to uncertainty has a good chance of failing. Unfortunately, in the earliest approaches to uncertainty handling, MYCIN and PROSPECTOR (see Chaps. 9 and 12), exactly this fact was not realized.

8.2 Model Based Representations

In order to solve the problems mentioned above it should be clear that we have to find adequate overall formal models for the representation of uncertainty and vagueness that allow

(i) knowledge representation, i.e. the capture and encoding of expert knowledge,
(ii) inference, i.e. the generation of recommendations or provision of relevant information based on the user input and expert knowledge, and
(iii) the communication of the results to the user.

8.2.1 Requirements on Models

An *overall formal model* allows us to represent in detail the exact theoretical background of all facets of the desired application. These models are often numerically intractable, and the use is often unable to understand them because they use higher-dimensional spaces (as in our book) or other abstract formalisms.

Therefore the formal model often must be transformed into a related but more vivid *model used for implementation* which is numerically tractable and efficient. The latter model is in general simpler as it only focuses on relevant information. In our book, for example, we do not directly use product probabilities, but a hypertree representation together with a concept of message exchange to build up the implementational model. Central to the success of an information system is its user-friendliness. As the user is often unable to work directly with complex models we have to define a special "user view". In the best case such a view gives an intuitive feeling of how to use the underlying model correctly. In our book we use graphical methods in this area.

It is essential to see that one should always start with the formal, mathematical model, and then specify the model used for implementation as well as the user view. In the "naive" rule-based approaches without uncertainty the underlying mathematical model was not considered, although it is easily obtained in the context of product spaces. From this point of view one can see why the canonical extensions to uncertainty performed in MYCIN or PROSPECTOR do not work in the general case. The reason is that for the generalized approaches a suitable formal model is not straightforward, as the respective authors unconsciously assumed.

8.2.2 On the Structure of Models

In the following we describe the kind of mathematical models analyzed in our book. Since the assumptions used here are rather weak we can use these models for analyzing all facets of numerical "measure theoretic" models in the area of knowledge based systems. Each model has a static part (which describes general domain knowledge) and a dynamic part (which serves as a method for integrating the casual knowledge, i.e., the user knowledge).

In order to model the *static part*, we always assume that there is a finite frame of discernment Ω, where the elements are considered to be mutually exclusive and exhaustive. The problem considered here is that exactly one element $\omega_0 \in \Omega$ is the "true" element which should be found by expert and user information. Typical examples are systems for diagnosis or finding a best configuration, etc. We assume that Ω is a product space Ω^M in order to simplify the notations.

Next the entities of interest, which may be subject to statements of the experts, have to be specified. We restrict our consideration to precise, imprecise or vague entities; the sets of possible entities are Ω, 2^{Ω}, $\mathcal{F}_L(\Omega)$, respectively. Note that in the last case we must be especially careful to provide semantics for vague data, i.e., to choose an adequate scale L. Finally a quantification of uncertainty has to be given. In our book we distinguish single-valued uncertainty measures from multiple-valued measures. A typical single-valued measure is a probability whereas multivalued measures are, for example, interval-valued probabilities. So the models considered in this book can be characterized by the four components

- the frame of discernment Ω,
- the set of entities of interest $\mathcal{D}$,
- the set of possible uncertainty values τ, and
- the mapping $cert : \mathcal{D} \rightarrow \tau$,

where *cert* is a mapping that assigns to each entity $d \in D$ its degree $cert(d) \in \tau$ of uncertainty.

The *dynamic part* of the model serves as a method for the integration of user knowledge: If the user provides a piece of information then the above quadruple has to be changed in general. We mainly consider the case that the user provides categorical knowledge, i.e., that the unknown true value ω_0 is definitely within a subset of E of Ω. But other information could be integrated as well.

Depending on the information the mapping *cert* has to be changed, but it may turn out that the set of entities of interest or even the measure of uncertainty and the frame of discernment should be changed.

The conditioning in the case of probability distributions or the different kinds of mass flow in the case of mass distributions are examples for the simplest kinds of change.

In the case of decision making, often a small part of the model, a *submodel*, is of relevance. If for example $(\Omega^M, \mathcal{D}, \tau, cert)$ is an uncertainty structure on the product space, we are often interested in an induced uncertainty structure on the subspace Ω^T. Therefore we have to consider marginalizing operators which allow us to calculate an induced uncertainty structure $(\Omega^T, \mathcal{D}', \tau', cert')$. Typically one uses the projection mappings Π_T^M for calculating Ω^T, $\mathcal{D}'$, and suitable marginalizing operations such as the ones for determing marginal probabilities.

From our point of view knowledge representation can be conceived as the description of the *statics* of epistemic states, whereas the updating process corresponds to their dynamics. [Gärdenfors 1988] distinguishes between two basic types of updating. *Expansion* means that a state where the truth of some proposition A is undetermined changes into one where A is accepted or A is rejected. *Contraction*, on the other hand, denotes the reverse process. We restrict ourselves to expansions since the treatment of contractions requires the development of non-monotonic techniques.

8.2.3 On the Choice of Mathematical Models

The choice of a suitable mathematical model is a very delicate task as there are (at least) the three divergent aspects efficiency, sufficiency, and acceptance by the user. Since the user should be in the center of interest, one has to choose a model such that the user is able to handle it correctly. Therefore he must understand it, which should be facilitated by special user views mentioned in Sect. 8.1. On the other hand the model must of course be sufficient, i.e., it must model the relevant aspects of the problem correctly. In order to solve this problem we have to think of the corresponding implementational model. This, however, has to be computationally tractable. The resolution of these different aims depend on the domain which has to be modeled, so there are no general methods to attack this problem. But there are some well elaborated strategies.

- *Eliminate Uncertainty*
  The most frequently proposed solution is to eliminate the uncertainty by re-engineering parts of the task.
- *Reduce Uncertainty*
  Often one can apply more powerful methods to extract more information from the world. A mistake often made in the development of uncertainty handling is to give first an implementation for the precise, certain cases and then to "extend" the methods to uncertainty. Such a method was also used in the naive extension of rule-based systems and often leads to wrong results, as uncertainty handling turns out to be a "new quality" of reasoning.
- *Use formal models with underlying correct semantics*
  Simple systems such as MYCIN were rather successful in a specific domain, but the use of shells, such as EMYCIN, with certainty factors often give counter intuitive or even false results. So the use of formal models with an underlying sound semantics is appropriate.

If we have no vagueness and no imprecision in the data, then the use of a sound probabilistic model is a good choice. The advantages are that there are well elaborated probabilistic theories, and that ther are links to decision theory as well as to statistical methods which allow us to make the system normative, i.e., to give the user standards of behaviour. Problems occur if the data are vague, as it is often not natural to use pure probabilistic systems in this case. Another problem is the representation of ignorance, which is done by using uniform distributions – although this is justifiable it may be unnatural that the result based on this representation is one decision and not a set of possible decisions.

In our book we therefore consider beside probabilistic models formal theories based on mass distributions and L-sets. Although the general principle "garbage-in – garbage-out" also holds for sound methods, the advantage is that results are correct with respect to the input. A result such as "the probability that the ship is at the loading dock is between 0 and 1" may be helpful as it states that the system knows that it knows nothing.

The main disadvantages of these formal models are that they are often difficult to transfer to the user, that large amounts of data must be available and that the complexity of the model does not allow proper algorithmic solution. Therefore it is often reasonable to look for "correct" approximations.

If a mathematical model is available but computationally untractable, then one has to find correct approximations. This strategy is often fairly reasonable, as the user is often content with rough results, and decisions such as "go approximately 500m" or "the ship is either of type A or B" are sufficient. In turn one gains the advantage for the implementation that the complexity can be reduced.

Although heuristic systems such as MYCIN, PROSPECTOR, etc. (see Chap. 9) were rather successful in their knowledge domain, one should, if possible avoid ad hoc methods such as certainty factors or probabilistic inference nets. These methods will turn out to handle bidirectional inferences often incorrectly, have difficulties in retracting conclusion and fail to take into account the correlation of sources of evidence. The main fault is often a blind use of a successful heuristic technique in a different knowledge domain.

8.2.4 Selected Problems with Mathematical Models

In this section we discuss some problems that occur in using mathematical models in the area of uncertainty and vagueness.

In particular, *underspecification* as the first topic is related to the fact that in most applications some of the uncertainty values needed for the computations may be unknown. In the following we discuss this problem from a probabilistic view, but similar remarks hold for other uncertainty handling methods as well.

For applications it is often realistic to assume that at most the simple prior probabilities $P(A_i)$ may be known. In general, the joint probabilities $P(\bigcap_i A_i)$ will be unknown. For instance, assume that in medical diagnosis hundreds of symptoms would have to be examined concerning their (joint) probabilities. While in the above example the case of unconditional probabilities was examined, the same arguments apply to the case of conditional probabilities. There are a number of methods that have been tried to find solutions in such a situation.

One method is to make the assumption of *(conditional) independence* when probabilistic knowledge is being combined. This assumption is made in most of the models and applications examined in the chapters below. A second method is to acknowledge the existence of *underspecified constraints* and to use a range of possible values for the probability in computations.

A third approach is to make use of the *maximum entropy* method (see [Cheeseman 1983] and [Jaynes 1979], for instance). In the following we will elaborate on this approach and point out the essential concepts. The problem is to find a unique value for the probability of a predicate in an underspecified system. The *maximum entropy principle* looks for the value of the probability that maximizes the entropy function H

$$H = -\sum_i p_i \log(p_i)$$

The probabilities p_i are related to uncertain propositions (predicates) made for the domain of reference. For ease of reading, the probabilities are parametrized by one index, but an arbitrary number of indices is possible. Because the numbers p_i are probabilities, an additional constraint is that

$$\sum_i p_i = 1.$$

In addition, constraints on the probabilities p_i may be asserted. Such constraints can be given by experts and can involve (conditional) dependencies between predicates. These constraints can also involve values that significantly differ from the best estimate that would maximize the entropy function.

The maximum entropy principle can be viewed as making the most unbiased assumption with regard to the available evidence. Any other assumption will assume dependencies that are not available. The difference between an assumed probability and the probability value derived by means of the maximum entropy method is a measure for the amount of information that is needed to make the assumption. It is important to note that this does not mean to say that events are conditionally independent, but – as already mentioned above – that this expresses the most unbiased assumption that can be made with the available evidence. Jaynes [Jaynes 1979] discusses the reasons for choosing the method of maximum entropy as the best estimate of the true probability.

In some applications we have the contrary situation: after choosing a model we recognize that there are more pieces of information than needed to specify the parameters of the model, i.e. we obtain *overspecification.* A typical example is already considered in Example 8.1 where we cannot arbitrarily fix the marginal probabilities of the two-dimensional subspaces as they don't fit in general. In this case we have contradictory knowledge. In general we have three ways to cope with such a situation: either we change the structure of the model, if we are sure that each piece of information is reliable, or if this is not the case the knowledge is contradictory, and we could either adjust the knowledge in cooperation with the expert in order to resolve the inconsistency or tolerate partial inconsistency, say, if the inconsistency appears in an area that isn't so relevant for the decision making process.

Another problem is related to the question how to take into account a possible uncertainty about the uncertainty (e.g. probability) value itself. This kind of ignorance is related to the incompleteness of knowledge; it is called *second order uncertainty.*

For statistical cases this may be described by means of an example: assume that one asks a set of employees for their income: 40% of them answer that they have an "income smaller than 4000 DM" (i.e. the statement A), 35% have an "income equal to or larger than 4000 DM" (i.e. $\neg A$), but the remaining 25% of the employees give no answer.

Assume now that one is interested in the probability $P(A(x))$, i.e., the probability that a person x randomly chosen from the set of employees has an income smaller than 4000 DM. 40% of the employees are *known* to have such an income; however, we are *ignorant* about the income of 25% of the employees. Therefore, at best, an interval can be specified as range for the probability $P(A(x))$ which itself is unknown: $0.40 \leq P(A(x)) \leq 0.65$.

Cheeseman [Cheeseman 1985b] proposes using a second-order probability formalism in which a probability is the subject of another probability. He interprets ignorance as the absence of knowledge in assigning a probability value. The uncertainty in assigning a probability is reflected in the second-order probability, represented by a probability or a probability density, for instance. His approach elicits clearly the difference between the uncertainty in the assignment and the uncertainty in the fact itself.

A mathematical foundation for higher-order probabilities is given in [Gaifman 1986]. He also proposes a *probabilistic logic* obtained by adjoining sentential operands of the higher-order probability formalism with the propositional calculus. Another approach, which allows representation of this kind of ignorance and its consideration in uncertain reasoning techniques, is given in Sect. 13.1. Note that in Sect. 4.6 we considered also second-order probabilities. It is also possible to introduce second-order calculi for other representation schemes (such as fuzzy sets of order 2 [Mizumoto 1979]).

8.3 Dependency Network Based Systems

In the area of uncertainty handling and vague data each piece of information has to be encoded. So a suitable *knowledge representation* formalism has to be chosen. Our first task is to find a coherent structure over a suitable frame of discernment. We don't consider in the following the problem of how to integrate several pieces of information of different experts but just have a look at the case where one expert organizes his knowledge. It turns out that experts are often able to provide information about uncertainty structures of subspaces of the overall space, e.g. in the form of rules, but in general not about the whole space, as it is often too complex. In this case we have the danger of providing contradictory knowledge, as the pieces of information need not fit (compare Example 8.1). The best way to handle this problem is to study the *dependency structure of knowledge* first on a purely qualitative level: the expert has to specify for each piece of information the subspace that is affected by this information. This qualitative knowledge can be best encoded in form of hypergraphs, also called dependency networks, inference networks, or causal networks. It is for example possible to study dependencies that reflect information of the kind "x is independent of y once z is known" in the context of probabilistic models (see Chap. 12) as well as other kinds of dependency. In essence we obtain a modularization of the knowledge domain. Note that graphical methods are available to visualize this process. In another step we then have to choose a suitable uncertainty and vagueness handling mechanism that is able to carry the pieces of information, where especially ignorance should be modeled very carefully. Also one must distinguish between models for absolute credibility and models for change in credibility. The assessment of the numbers used in the chosen representation, i.e., the knowledge acquisition or *knowledge elicitation* is again a crucial point.

The result is an uncertainty quadruple $(\Omega, \mathcal{D}, \tau, cert)$, and the knowledge of the expert should be encoded adequately within this mathematical structure. In order to make inferences based on additional (casual) user knowledge we consider special *updating mechanisms*. This means that we want to integrate a categorical piece of knowledge which is known to be true. From a mathematical point of view we have to model this by considering adequate updating schemes for the uncertainty structure. It should be stressed that there may be several reasonable schemes in the case of vagueness and uncertainty. In classical probabilistic models the only reasonable updating mechanism seems to be conditioning, whereas in the case of mass distributions we studied the two different mechanisms conditioning and data revision to some detail. The result is again an uncertainty structure over the space Ω. In order to work with this updated structure we generate recommendations based on the updated information. This procedure is based on marginalization concepts as well as decision making mechanisms. With a marginalization operator we will map the knowledge to the subspace of interest, whereas the uncertainty structure induced on that suabspace is evaluated by a suitable decision making scheme. A desired decision would be of course a result such as "Use decision 1" but also recommendations such as "Take decision 1 or 2, but definitely not 3" or "From the available knowledge I can't deduce any preferences" may be adequate and very valuable results.

In the following chapter we consider three different tools implemented by the authors which are designed according to the above *informal* description. These tools are based on probability distributions, L-sets, and mass distributions, respectively. They are compared with existing tools implemented by various authors.

Chapter 9

Heuristic Models

In this chapter we examine three different "heuristic uncertainty models". The characteristic feature of heuristic models is that their mathematical foundations are not or only incompletely led back to some sound theory – as given by probability theory, for instance. This is because heuristic approaches aim at avoiding certain "problems" arising from the use of, e.g., probability theory. The reasons that are often mentioned in this context are the amount of data needed (prior and conditional probabilities, joint probability distributions, etc.), the inability to distinguish between absence of belief and doubt, and that it is impossible to represent ignorance.

One of the most important heuristic uncertainty models is the *certainty factor approach* developed by Shortliffe and Buchanan. This uncertainty calculus – and the well known expert system MYCIN that is associated with this calculus – are introduced in Sect. 9.1. Furthermore, extensions to and reformulations of the certainty factor approach that were proposed in the literature are discussed.

The second heuristic model we consider in the sequel is based on the concept of *triangular norms* and *conorms*. It is important to notice that, because of their generality, T-norms and T-conorms give an infinite number of different "calculi of uncertainty". The selection of a special pair describes a particular calculus uniquely and completely. In addition to this general uncertainty model introduced in Sect. 9.2, we present and discuss the expert system RUM ("Reasoning with Uncertainty Module") developed by Bonissone.

Finally, as the third heuristic model, in Sect. 9.3 we present the system INFERNO developed by Quinlan. One characteristic feature of INFERNO's architecture is that the inference model is mainly based on *bounds propagation*: it can be used for both forward and backward inferences. Since the model does not make assumptions about (in)dependencies of data all the propagation constraints are provably correct. Because of this philosophy and the resulting fact that computed bounds sometimes may be weak, INFERNO is also called a "cautious" approach to uncertain inference.

Note that there is a variety of successful models (such as the *rough set* approach [Pawlak 1982]) which cannot be described in this monograph due to the limitation of space.

9.1 MYCIN – The Certainty Factor Approach

Shortliffe and Buchanan ([Shortliffe and Buchanan 1975], [Shortliffe 1976]) have proposed the *certainty factor* uncertainty measure and – based on this measure – a model to perform uncertain reasoning. Their approach aims at avoiding the "problem" of probability theory of not being able to distinguish between *absence of belief* and *doubt* or to represent *ignorance* related to the lack of knowledge, as pointed out above. Additionally, their approach requires a less exhaustive amount of data as compared to probability theory.

The certainty factor model has to be seen in relation to the development of the well known expert system MYCIN ([Buchanan and Shortliffe 1984], [Shortliffe 1976]) which was built during the years 1972–1976. MYCIN is an expert system for advising physicians how to treat patients suffering from bacteriogenous infectious diseases. As input MYCIN gets patient data, and the output consists of proposals for diagnosis and therapy. Later systems related to MYCIN are EMYCIN ([Van Melle 1980], [Van Melle, Shortliffe, and Buchanan 1984]), a domain-independent system based on MYCIN's control mechanisms and data structures, and RMYCIN ([Cendrowska and Bramer 1984]) which is a reconstruction of MYCIN. Concerning the definition and the use of certainty factors these systems include certain improvements and extensions. By taking domain-dependent knowledge into account, the core system EMYCIN can be extended to an expert system for any special frame of discernment.

However, the "heuristic simplicity" of the algorithms MYCIN and related systems used for dealing with uncertain knowledge leads to the fact that, in popular AI books, they are introduced as prototypical examples of how to manage uncertain reasoning. Another consequence is that also commercially available tools which allow the drawing of inferences on the basis of uncertain data are based on methods identical or at least similar to those of MYCIN. Since, on the other hand, the certainty factor approach makes use of measures and algorithms that are heuristic and at most "syntactically similar" to probabilistic ones, it has often been criticized (e.g. in [Adams 1976], [Horvitz, Heckerman, and Langlotz 1986], [Heckerman 1985]).

Below, we introduce the original inference model as it is implemented in MYCIN. The defined measures are traced back to and examined on a probabilistic basis. Further, we give a brief overview on how uncertain knowledge is represented in MYCIN, and elaborate on extensions of the original certainty factor approach and a consistent probabilistic reformulation as proposed by Heckerman. Finally, the most important theoretical and practical aspects of the certainty factor approach are discussed.

9.1.1 The Mathematical Model

The definition of the *certainty factor* as a measure for the truth of a proposition or hypothesis h is based on the definition of two further measures which have to be introduced first – the *measure of belief* MB and *measure of disbelief* MD.

Assuming $\Omega_H = \{h, \overline{h}\}$ as the hypothesis frame of discernment and $E \subseteq \Omega_E$ as the available evidence based on observations, for instance, these two measures are related to probability theory as follows [Shortliffe and Buchanan 1975]:

$$MB(h|E) \stackrel{\mathrm{d}}{=} \begin{cases} \dfrac{\max\{P(h|E), P(h)\} - P(h)}{1 - P(h)} & \text{if } P(h) \neq 1, \\ 1 & \text{otherwise,} \end{cases}$$
$$MD(h|E) \stackrel{\mathrm{d}}{=} \begin{cases} \dfrac{P(h) - \min\{P(h|E), P(h)\}}{P(h)} & \text{if } P(h) \neq 0, \\ 1 & \text{otherwise.} \end{cases} \tag{9.1}$$

In spite of the fact that the two measures are formally based on probability theory, their definitions do not play a role for the acquisition and computation of MB and MD: in the original certainty factor approach, the conditional and prior probabilities are assumed not to be available in general. So, instead of taking the probabilistic interpretations into account the measures of belief and disbelief are associated with the following intuitive meanings which have to be considered when numerical "belief values" are acquired:

$MB(h|E) = x$: "In the light of evidence E, one's belief that h is true increases by x"

$MD(h|E) = x$: "In the light of evidence E, one's disbelief that h is true increases by x"

The *certainty factor* CF is defined to measure the difference between MB and MD:

$$CF(h|E) \stackrel{\mathrm{d}}{=} MB(h|E) - MD(h|E) \tag{9.2}$$

From the definitions of the measures the following ranges can be derived:

$$MB, MD : \Omega_H \rightarrow [0,1]$$
$$CF : \Omega_H \rightarrow [-1,+1]$$

Taking into account the probabilistic definitions (9.1), also the certainty factor given by (9.2) can be equipped with a probabilistic interpretation:

$$CF(h|E) = \begin{cases} \dfrac{P(h|E) - P(h)}{1 - P(h)} & \text{if } P(h|E) \geq P(h)\,, \quad P(h) \neq 1, \\ \\ \dfrac{P(h|E) - P(h)}{P(h)} & \text{if } P(h) \geq P(h|E)\,, \quad P(h) \neq 0. \end{cases} \tag{9.3}$$

Note that the quantities "measure of (*increased*) belief" and "measure of (*increased*) disbelief" are *relative measures* and not absolute ones – like, say, a probability measure. Therefore, the certainty factor also describes a *change in belief*. The weighty consequences of a misinterpretation of the measures and their meaning are pointed out in [Adams 1976]. For further discussions see also [Heckerman 1985], [Horvitz and Heckerman 1985], and [Horvitz, Heckerman, and Langlotz 1986], for instance. Interpreting the above definition, a positive CF indicates that, regarding both belief and disbelief, there are more reasons to change one's belief in favor of the hypothesis than in disfavor of it, as consequence of the new evidence E.

In contrast to a probability measure, for the certainty factor CF generally

$$CF(h|E) + CF(\overline{h}|E) \leq 1 \tag{9.4}$$

holds. In particular, note the visible difference from the equation stating

$$P(h|E) + P(\overline{h}|E) = 1$$

which has to be true for a probability measure P, but keep in mind the different meanings of the (relative) certainty factor on the one hand and an (absolute) probability measure on the other. So, evidence supporting a hypothesis h to some degree $CF(h|E)$ does *not* support the negated hypothesis $\overline{h}$ to degree $1 - CF(h|E)$. For instance, it is possible that newly arising evidence does not have any influence either on h or on $\overline{h}$.

If uncertain knowledge is formulated in terms of CFs only (as in MYCIN), the two measures MB and MD can be calculated as follows:

$$MB(\ldots) = \begin{cases} 0 & \text{if } CF(\ldots) < 0, \\ CF(\ldots) & \text{if } CF(\ldots) \geq 0, \end{cases}$$

$$MD(\ldots) = \begin{cases} -CF(\ldots) & \text{if } CF(\ldots) < 0, \\ 0 & \text{if } CF(\ldots) \geq 0. \end{cases}$$

Notice that Shortliffe describes the sequential and parallel combinations in terms of MB and MD.

Assume the more general frame of discernment $\Omega_H = \{h_1, ..., h_n\}$ consisting of *exhaustive* and *mutually exclusive* values. Then the following limiting conditions must hold for certainty factors given by a user or calculated during the inference process:

$$\begin{aligned} \sum_{i=1}^{k} CF(h_i|E) &\geq -k \qquad \text{(for the } k \text{ values } h_i \text{ disconfirmed by } E) \\ \sum_{i=1}^{l} CF(h_i|E) &\leq 1 \qquad \text{(for the } l \text{ values } h_i \text{ confirmed by } E) \end{aligned} \tag{9.5}$$

Additional to the measures $MB(h|E)$, $MD(h|E)$, and $CF(h|E)$ which measure one's belief in *facts* based on certain observations E, a certainty factor is associated also with each *expert rule*. This certainty factor plays a role for performing the *sequential combination* in rule-based reasoning techniques.

In the context of this section, $CF(h|a)$ quantifies the relative degree of belief in the hypothesis h under the condition that the antecedent proposition a is known to be true. So, for the antecedent frame of discernment $\Omega_A = \{a, \overline{a}\}$ the simplest case of expert rule is given by

$$\underline{\text{if}}\ a\ \underline{\text{then}}\ h\ \underline{\text{with}}\ CF(h|a).$$

For any single expert rule, only one of the measures $MB(h|a)$ and $MD(h|a)$ is non-zero: for expert rules confirming the conclusion, it is $MB(h|a)$, for disconfirming rules it is $MD(h|a)$ (actually written as $CF(h|a) = MB(h|a)$ or $CF(h|a) = -MD(h|a)$, respectively).

Propagating uncertain information on the basis of above expert rule means that the measures $MB(h|E)$ and $MD(h|E)$ measuring the (change of) belief in hypothesis h are calculated by simply multiplying the rule's certainty $CF(h|a)$ and the certainty $CF(a|E)$ of the antecedent, which is given in terms of $MB(a|E)$ and $MD(a|E)$. The sequential combination is performed by

$$\begin{aligned} MB(h|E) &= MB(h|a) \cdot \max\{0, CF(a|E)\}, \\ MD(h|E) &= MD(h|a) \cdot \max\{0, CF(a|E)\}. \end{aligned} \tag{9.6}$$

So, if $CF(a|E)$ is not negative one obtains the hypothesis' change of belief $CF(h|E)$ by simply multiplying $CF(h|a)$ and $CF(a|E)$.

Of special interest is the case of a negative value $CF(a|E) < 0$: application of the sequential combination formula (9.6) leads to the result $CF(h|E) = 0$.

This, however, is only true if there is no explicit rule strength $CF(h|\overline{a})$ attached to a rule like

$$\underline{\text{if}}\ \overline{a}\ \underline{\text{then}}\ h\ \underline{\text{with}}\ CF(h|\overline{a}).$$

If such an expert rule is given, the negative value $CF(a|E) < 0$ can be applied: since $CF(\overline{a}|E) = -CF(a|E) > 0$ holds, the hypothesis' change of belief $CF(h|E)$ is obtained again by applying formula (9.6), i.e. by multiplying $CF(h|\overline{a})$ and $CF(\overline{a}|E)$.

To introduce the *parallel combination* based on certainty factors, assume the frames $\Omega_A = \{a, \overline{a}\}$, $\Omega_B = \{b, \overline{b}\}$, and $\Omega_H = \{h, \overline{h}\}$. If a hypothesis h is supported by a *set of expert rules*

$$\begin{array}{l}\underline{\text{if}}\ a\ \underline{\text{then}}\ h\ \underline{\text{with}}\ CF(h|a)\\ \underline{\text{if}}\ b\ \underline{\text{then}}\ h\ \underline{\text{with}}\ CF(h|b)\end{array}$$

the certainty factors $CF(h|E_1)$ and $CF(h|E_2)$ have to be combined to get the final evaluation of the hypothesis. In MYCIN, the combination is obtained through the following associative function valid for two evidence sources E_1 and E_2 (other "less important" functions for performing the parallel combination can be found in [Shortliffe and Buchanan 1975]):

$$MB(h|E_1,E_2) = \begin{cases} 0 & \text{if } MD(h|E_1,E_2) = 1, \\ \begin{array}{l} MB(h|E_1) + MB(h|E_2) \\ -MB(h|E_1)\cdot MB(h|E_2) \end{array} & \text{otherwise.} \end{cases}$$

$$MD(h|E_1,E_2) = \begin{cases} 0 & \text{if } MB(h|E_1,E_2) = 1, \\ \begin{array}{l} MD(h|E_1) + MD(h|E_2) \\ -MD(h|E_1)\cdot MD(h|E_2) \end{array} & \text{otherwise.} \end{cases} \tag{9.7}$$

For the general case of knowledge sources $E_1, \ldots, E_n$, $n \geq 2$, the quantities $MB(h|E_1, \ldots, E_j, E_{j+1})$ and $MD(h|E_1, \ldots, E_j, E_{j+1})$ have to be accumulated separately by repeatedly applying (9.7) for $j = 1, \ldots, n-1$. Only *in the end* (i.e., for $j = n-1$) are MB and MD combined to produce the final $CF(h|E_1, \ldots, E_n)$ using (9.2). A premature computation of the $CF(h|E_j)$ (by applying (9.2)) and *their* direct combination by using a formula similar to (9.7) applied to certainty factors is generally a non-associative operation. When applying the combination function (9.7), the case $MB(h|E_1) = MD(h|E_2) = 1$, for instance, must be considered and should result in an *undefined* statement. This is because the two sources of information lead to absolutely conflicting conclusions.

In addition to the function described above which applies to the measures of belief and disbelief directly, another function

$$CF(h|E_1, E_2) = \begin{cases} s + t - s \cdot t & \text{if } s \text{ and } t \text{ positive,} \\ s + t & \text{if } s \cdot t \in (-1, 0], \\ \textit{undefined} & \text{if } s \cdot t = -1, \\ s + t + s \cdot t & \text{if } s \text{ and } t \text{ negative,} \end{cases} \tag{9.8}$$

where $s \stackrel{\mathrm{d}}{=} CF(h|E_1)$ and $t \stackrel{\mathrm{d}}{=} CF(h|E_2)$, can be formulated applying to certainty factors directly. The certainty factors $CF(h|E_1)$ and $CF(h|E_2)$ needed for the above formula are obtained by

$$CF(h|E_i) = MB(h|E_i) - MD(h|E_i), \qquad i = 1, 2,$$

directly after performing the sequential combination on the basis of each single rule.

When applying (9.8) it must be considered that, first of all, the certainty factors with the same sign have to be combined. After that the two remaining factors with different signs are combined to get the final certainty factor. This restriction is equivalent to that related to the application of algorithm (9.8) which prescribes the separate combination of the MBs and MDs to get the final CF. The two algorithms for parallel combination are commutative and associative – always taking into account, however, the kind of restriction related to the combination. Another property of both algorithms is that the combination with a "vacuous" information source (i.e. $CF(h|E_i) = 0$) does not alter the other certainty factor.

Notice that neither sequential combination (9.6) nor the parallel combinations can be derived from the probabilistic definitions (9.1) or (9.3), respectively. Instead, Buchanan and Shortliffe justify the formulas for sequential and parallel combination on the basis of certain intuitive desiderata.

9.1.2 Uncertainty Representation in MYCIN

In this section we give an impression of how the certainty factor model considered in the previous section is embedded in the expert system MYCIN and of how uncertain knowledge is represented and handled. The knowledge base of MYCIN is partitioned in the following way:

- *static knowledge* consists of patient data such as age, sex, and name of the patient. Static data are feeded into the system by the physician and do not change during the whole session.

```
VAL [ORGANISM-1, IDENT] = (( STREPTOCOCCUS .6)
                           ( STAPHYLOCOCCUS .4)
                           ( DIPLOCOCCUS-PNEUMONIAE -1))
VAL [PATIENT-1, ALLERGY]= ((PENICILLIN 1) (AMPICILLIN 1)
                           (CARBENICILLIN 1)(METHICILLIN 1))
RULE 35:
  IF:   1) THE GRAMSTAIN OF THE ORGANISM IS GRAMNEG, AND
        2) THE MORPHOLOGY OF THE ORGANISM IS ROD, AND
        3) THE AEROBICITY OF THE ORGANISM IS ANAEROBIC,
  THEN: THERE IS SUGGESTIVE EVIDENCE (0.6) THAT THE IDENTITY
        OF THE ORGANISM IS BACTEROIDES
```

Fig. 9.1 Objects with certainty factors, a MYCIN-rule (from [Shortliffe 1976])

- *dynamic data* are provisional results which, in general, will change during the session. One example of a dynamic datum is the certainty factor of a sub-hypothesis which is already confirmed by some but not all expert rules.
- *production rules* of the form

  <u>if</u> premise <u>then</u> action

  are used as carrier for the medical (expert) knowledge. How to draw inferences on the basis of these rules was described in the previous section.

Figure 9.1 shows two dynamic objects and a rule of MYCIN's knowledge base. For instance, the identity of organism ORGANISM-1 is either STREPTOCOCCUS (with certainty 0.6) or STAPHYLOCOCCUS (with certainty 0.4). Looking at the certainty factors 0.6, 0.4, and -1 attached to ORGANISM-1 we see that the condition (9.5) is fulfilled since "identity of the organism" is a so-called *single-valued* attribute for which exactly one of the (three) mutually exclusive values has to be true. An exception concerning the exclusivity of values is given by the so-called *multi-valued* attributes. In this case more than one value can be labelled with the certainty factor 1, as for the object-attribute tuple [PATIENT-1, ALLERGY], for instance. Consequently, the conditions (9.5) are not applicable for multi-valued objects.

The static and dynamic data are stored as associative triples in the nodes of so-called context trees. For example, the assertion "*E. Coli* bacteria are gram-positive" is represented by means of the associative triple

```
(E.COLI GRAMSTAIN GRAM-POSITIVE)
```

with the object E.COLI, the attribute GRAMSTAIN, and GRAM-POSITIVE as the value of the object attribute. The context tree structure allows control over the sequence in which data are considered when evaluating the rules. Further, the access to neighboring nodes of the context tree becomes easy. Beyond the pure data records, other attribute-value entries can

- control the access to father nodes in the context tree,
- contain patterns of text to formulate questions or answers to the user,
- trigger further knowledge acquisition, or
- point to a set of rules needed to determine certain attribute values (backward chaining) or to trigger special actions (forward chaining).

Whether a rule premise is evaluated to be "true" so that the rule's "action-part" can be evaluated depends on the way in which the sub-premises are logically connected and on the truth values of the sub-premises: in MYCIN a rule's action is performed if the certainty factor of the whole premise exceeds the limit 0.2. The sub-premises are evaluated by making use of functions such as `KNOWN`, `DEFINITE`, `SAME`, etc. Applied to one or more certainty factors the functions compute a truth value and, contingent on the function, also a certainty factor.

For further information on MYCIN concerning details of knowledge representation and operation, and special system components, the reader is referred to [Shortliffe 1976] and [Buchanan and Shortliffe 1984].

9.1.3 Related Models and Proposals

The certainty factor model introduced and discussed above is slightly different from that used later in the systems EMYCIN ([Van Melle 1980], p.31) and RMYCIN ([Cendrowska and Bramer 1984], p.253). The modifications made for the development of EMYCIN and RMYCIN concern the manner in which the certainty factor is defined and in which the parallel combination is performed in the case of multiple rules and evidences (see also [Buchanan and Shortliffe 1984]).

The scheme used by these systems defines the certainty factor as

$$CF(\ldots) = \frac{MB(\ldots) - MD(\ldots)}{1 - \min\{MB(\ldots), MD(\ldots)\}} \tag{9.9}$$

which is identical to the original one used in MYCIN and given with (9.2) when at least one of MB and MD is zero. But, as argued in [Van Melle 1980], it is no longer necessary to accumulate *separate* MBs and MDs before producing the final CF (compare the above remarks concerning the way to apply formula (9.7)).

The CFs of the same or opposite sign can be combined directly (i.e. without looking at MB and MD as in (9.7)) by the following commutative and associative formula:

$$CF(h|E_1, E_2) = \begin{cases} s + t - s \cdot t & \text{if } s \text{ and } t \text{ positive,} \\ \dfrac{s+t}{1 - \min\{|s|, |t|\}} & \text{if } s \cdot t \in (-1, 0], \\ \textit{undefined} & \text{if } s \cdot t = -1, \\ s + t + s \cdot t & \text{if } s \text{ and } t \text{ negative,} \end{cases} \tag{9.10}$$

where $s \stackrel{\text{d}}{=} CF(h|E_1)$ and $t \stackrel{\text{d}}{=} CF(h|E_2)$. Use of these definitions of certainty factor and combination rule was made in [Van Melle 1980] for the development of EMYCIN and in [Cendrowska and Bramer 1984] for RMYCIN. The combination rule (9.10) has the same result as the combination (9.7) used in MYCIN if certainty factors with the same sign are combined.

From an intuitive standpoint the rule (9.10) may be advantageous when combining different signed certainty factors. The reason is that when some factors deviate widely from the rest, this no longer has such a fatal influence on the final certainty factor. For instance, consider the combination of 10 certainty factors with value 0.9 and one factor with value -0.8: application of rule (9.8) leads to ≈ 0.2 while combination (9.10) has ≈ 1 as result.

The original definition of certainty factors is inconsistent with a probabilistic interpretation of the functions used in MYCIN to combine the quantities (see formula (9.7)). This has been clearly demonstrated by Heckerman [Heckerman 1985]. Instead of the original definition of CF – given by the equations (9.2) and (9.3) – he proposes a slightly different definition of a certainty factor given by (9.11). It should be mentioned that Heckerman proposes a *class* of probabilistic interpretations for certainty factors, which is based on those intuitively appealing desiderata underlying the original certainty factor approach. So, the definition below represents only one of the possible probabilistic interpretations:

$$CF(h|E) \stackrel{\text{d}}{=} \begin{cases} \dfrac{P(h|E) - P(h)}{P(h|E) \cdot (1 - P(h))} & \text{if } P(h|E) > P(h), \\ \\ \dfrac{P(h|E) - P(h)}{P(h) \cdot (1 - P(h|E))} & \text{if } P(h) \geq P(h|E). \end{cases} \tag{9.11}$$

An extra term has been added in the denominator of the original definition of certainty factors (compare equation (9.3)).

In [Grosof 1985b] it is shown that the above modification is isomorphic to the model of "inference networks" proposed by Duda, Hart, and Nilsson and also to a special case in the Dempster-Shafer theory. Further, Grosof describes the manner in which also *a priori probabilities* can be taken into account on the basis of the certainty factor approach. This possibility is not given by the original model.

9.1.4 Conclusions

When introducing the certainty factor model, we already mentioned its importance and fame. This however has to be seen in relation to the model's simplicity and to the real existence of the (medical) expert system MYCIN. Indeed, MYCIN was one of the first "working" expert systems and has been under further development for over a decade.

On the one hand, this importance led to the fact that "performing uncertain reasoning on the basis of certainty factors" was often copied and embedded in expert systems and shells. Some of them are indeed commercially available. On the other hand, the simplicity of the certainty factor approach and its algorithms is mainly based on the fact that the formulas for sequential and parallel combinations are of a heuristic nature only. They are based on intuitively appealing desiderata instead of a consistent probabilistic framework.

Below, we summarize and discuss the most important remarks and criticisms which can be found in the literature.

- A well known problem is the discrepancy between the definition of the "certainty factor" as a *relative measure* on the one hand, and the fact that it is often described and interpreted (also in user's guides) to be an *absolute measure* of uncertainty on the other ([Horvitz, Heckerman, and Langlotz 1986]). Considering definitions (9.3), a certainty factor $CF(h|E)$ indeed quantifies a *change in belief* and is therefore different from absolute measures, such as the probability measure $P(h|E)$. How essential this difference – often not taken into account – can be, is visualized by the following example taken from [Adams 1976]).

(9.1) Example. Assume a hypothesis h_1 with both an a priori probability and a conditional probability larger than the probabilities of a second hypothesis h_2. Applying (9.1) and (9.2) the certainty factor of h_1 can be smaller than that of h_2. For example, from the (conditional) probabilities

$$\begin{array}{llcll} P(h_1) & = 0.8 & \gg & P(h_2) & = 0.2 \\ P(h_1|E) & = 0.9 & > & P(h_2|E) & = 0.8 \end{array}$$

one obtains

$$CF(h_1|E) = MB(h_1|E) = \frac{0.9-0.8}{1-0.8} = 0.50$$
$$CF(h_2|E) = MB(h_2|E) = \frac{0.8-0.2}{1-0.2} = 0.75$$

and, therefore,

$$CF(h_1|E) < CF(h_2|E).$$

◇

- Following Adams ([Adams 1976]) the certainty factor combination rule (9.8) is consistent with the combination of probability measures if one takes into account definition (9.1), the conditional independency, the absolute independency of the evidences involved, and if one combines identically signed certainty factors only.
The combination of factors with different signs has no probabilistic counterpart but is of a heuristic nature. The same holds for the algorithm for (sequentially) propagating certainty factors.
- In the case of premises containing boolean combinations, the certainty factor of conjunctive (disjunctive) connected subpremises is calculated by computing the minimum (maximum) of the certainty factors involved.
- The certainty factor approach does not support the overall combination of the certainties of *all values* of a given characteristic feature. For instance, assume that for a frame of discernment $\Omega_H = \{h_1, h_2, \ldots\}$ there exist certainty factors $CF(h_1|E_1)$ and $CF(\overline{h_2}|E_2)$. In spite of the fact that $h_1 \in \overline{h_2}$ holds, in the original certainty factor approach the two belief values are not combined.
This aspect is considered in [Gordon and Shortliffe 1984], [Gordon and Shortliffe 1985] where an extension of the certainty factor model is proposed by making use of the "belief functions" developed by Dempster and Shafer.
- When propagating certainty factors over a set of expert rules, there is no guarantee that the condition
$$\sum_{i=1}^{l} CF(h_i|E) \leq 1 \quad \text{(for the } l\ h_i \text{ confirmed by } E\text{)}$$
always holds. If necessary the truth of the formula has to be forced by normalization.
- Heckerman ([Heckerman 1985]) mentions that on the one hand certainty factors are elicited without a clear operational definition (as examined

above), but on the other, "MYCIN performs as well as experts in the field". But also an analysis of the applications of MYCIN has shown that small variations in the CFs attached to the rules made little difference in the final outcome: when replacing the CFs by four or five "certainty ranges", the system behaved approximately the same as before.
This leads to the conclusion that a misunderstanding of the meaning of the certainty factors, e.g. their use as absolute measures, need not have some "unqualified behavior" of the system as consequence. Especially in the case of small inference chains and simple hypotheses (which is a property of size and complexity of the medical domain modeled and not of the inference model) errors in the values of the certainties may have only a small spread.

However, the only way to test for sensitivity is to build a system and then to evaluate its performance – which indeed is an unacceptable way to proceed as far as the theoretical and consistent development of an inference model for handling uncertain knowledge is concerned.

In [Heckerman 1986] a probabilistic redefinition of certainty factors is analyzed in detail. From this point of view it can be shown that if uncertainty is propagated through an inference net, then the evidence must be conditionally independent given the hypothesis and its negation, and the inference network must have a tree structure. Since this assumptions are rarely satisfied in practical applications, one has to be very careful to use certainty factor approaches in new knowledge domains.

9.2 RUM – Triangular Norms and Conorms

The so-called *triangular norms and conorms* (T-norms and T-conorms) (see e.g. [Schweizer and Sklar 1983], [Smets and Magrez 1987], [Magrez and Smets 1989], [Bonissone 1987b]) represent the most general families of binary functions that can be applied to model the conjunction and disjunction operators "and" and "or". In the framework of reasoning under uncertainty these two operators play a role for premise evaluation and sequential and parallel combinations. It is important to notice that T-norms and T-conorms do not specify one special uncertainty calculus as given by a probabilistic model or the certainty factor approach of MYCIN, for instance. Instead, by T-norms and T-conorms an *infinite number* of different calculi of uncertainty is given. However, the selection of a *special* T-norm and T-conorm (and a suitable negation operator) uniquely and completely describes a *particular calculus of uncertainty.*

Practical use of these general families of binary functions was made by Bonissone (see e.g. [Bonissone 1987a], [Bonissone 1987c], [Bonissone, Gans, and Decker 1987]). He developed the "Reasoning with Uncertainty Module" RUM which is based on the concept of T-norms and T-conorms.

We start with the definitions of the functions T-norm and T-conorm and an examination of their properties. Further, the concept of negation functions is introduced. Concerning the selection of special "instances of T-norms" (i.e., special uncertainty calculi), we refer to experiments by Bonissone and Decker. On the basis of the general T-norms we derive a general algorithm for performing the sequential combination. This algorithm is based on the classical implication interpretation of an expert rule and applies *modus ponens* and *modus tollens* as inference mechanisms. Becoming more concrete, we consider the general propagation algorithm for special instances of T-norms. In Sect. 9.2.2 we point out the main features of the system RUM of Bonissone et al. The conclusions can be found in Sect. 9.2.3.

9.2.1 Families of Uncertainty Calculi – Triangular Norms and Conorms

The concept of so-called *triangular norms and conorms* (T-norms and T-conorms) was developed by Schweizer and Sklar (see e.g. [Schweizer and Sklar 1983]). In general, T-norms and T-conorms are two-place functions from $[0,1] \times [0,1]$ to $[0,1]$ that are monotonic, commutative, and associative.

In particular, the two functions are defined as follows [Magrez and Smets 1989]:

(9.2) Definition. *A T-norm (triangular norm) is a function* $T: [0,1] \times [0,1] \to [0,1]$ *such that for all* $a, b, c, d \in [0,1]$,

(i)	$T(a,1)$	$=$	a	*(boundary condition)*
(ii)	$T(a,b)$	$\leq$	$T(c,d)$, *if* $a \leq c$ *and* $b \leq d$	*(monotonicity)*
(iii)	$T(a,b)$	$=$	$T(b,a)$	*(commutativity)*
(iv)	$T(a,T(b,c))$	$=$	$T(T(a,b),c)$	*(associativity)*

(9.3) Definition. *A function* $S: [0,1] \times [0,1] \to [0,1]$ *is called a T-conorm (triangular conorm) iff, for all* $a, b, c, d \in [0,1]$, S *has the following properties:*

(i)	$S(0,a)$	$=$	a	*(boundary condition)*
(ii)	$S(a,b)$	$\leq$	$S(c,d)$, *if* $a \leq c$ *and* $b \leq d$	*(monotonicity)*
(iii)	$S(a,b)$	$=$	$S(b,a)$	*(commutativity)*
(iv)	$S(a,S(b,c))$	$=$	$S(S(a,b),c)$	*(associativity)*

As may be seen, the definitions of the two functions are different in their boundary condition only. Following Bonissone, for example, we will restrict ourselves to the special T-norms and T-conorms that fulfill the additional boundary conditions

$$T(0,0) = 0 \qquad \text{and} \qquad S(1,1) = 1 \tag{9.12}$$

For the special case of mappings from $\{0,1\} \times \{0,1\}$ to $\{0,1\}$ we get the truth tables of logical *and* and *or*. This justifies our considering general T-norms for representing conjunctive combinations and T-conorms to represent disjunctive combinations.

For *any* T-norm T and T-conorm S, we have

$$\begin{array}{ccccc} T_W(a,b) & \leq & T(a,b) & \leq & \min(a,b) \\ \max(a,b) & \leq & S(a,b) & \leq & S_W(a,b) \end{array} \tag{9.13}$$

where T_W is a T-norm and S_W is a T-conorm such that

$$T_W(a,b) = \begin{cases} a & \text{if } b = 1, \\ b & \text{if } a = 1, \\ 0 & \text{otherwise,} \end{cases} \qquad S_W(a,b) = \begin{cases} a & \text{if } b = 0, \\ b & \text{if } a = 0, \\ 1 & \text{otherwise.} \end{cases} \tag{9.14}$$

as can be easily verified. Despite being defined as two-place functions, T-norms and T-conorms can be defined recursively, so that they also apply to larger numbers of clauses. Because of their associativity, for $a_1, \ldots, a_n, a_{n+1} \in [0,1]$ the functions can be defined thus:

$$\begin{aligned} T(a_1, \ldots, a_n, a_{n+1}) &= T(T(a_1, \ldots, a_n), a_{n+1}), \\ S(a_1, \ldots, a_n, a_{n+1}) &= S(S(a_1, \ldots, a_n), a_{n+1}). \end{aligned} \tag{9.15}$$

In relation to uncertainty propagation in rule-based systems, the general T-norm can be applied to compute the "degree of certainty" of the *conjunction* of two clauses in the same premise. Because a T-norm describes a conjunction operation, it can also be applied to perform *sequential combination*, i.e. to combine the "certainty" of the premise of a rule with the certainty factor attached to the rule to obtain the certainty of the conclusion. Since the general T-conorm $S(a,b)$ represents a disjunction operation, it can be used for *parallel combination*, which combines the certainties of the same conclusions derived from different rules.

Moreover, T-norms and T-conorms represent the most general families of binary functions that can be applied to conjunction and disjunction.

(9.4) Definition. *A function* $N : [0,1] \to [0,1]$ *is called a negation (function) iff, for all* $a, b \in [0,1]$,

(i) $N(1) = 0$
(ii) $N(0) = 1$
(iii) $N(a) \leq N(b)$, *if* $a \geq b$ *(monotonicity)*

A special negation operator N that plays an important role in this framework is defined by

$$N(a) = 1 - a. \tag{9.16}$$

For example, in the case of the special mapping from $\{0,1\}$ to $\{0,1\}$ the negation of "logically true" (i.e. 1) is given by "logically false" (i.e. 0).

For this negation function, the Definitions 9.2 and 9.3 of the functions T and S are interdependent in the following way:

$$S(a,b) = N(T(N(a),N(b))) \qquad T(a,b) = N(S(N(a),N(b))) \tag{9.17}$$

In such cases, T (S) is called the *N-dual* of S (T). This dependency shows that – on the basis of a given negation function – instances of the very general functions T and S cannot be independently defined.

Concerning "instances of T-norms", notice that until now no specific *uncertainty calculus* has been defined. Because of the very generality of the above definitions, obviously still an *infinite* number of calculi that match the above definitions is possible. By means of experiments, Bonissone and Decker [Bonissone and Decker 1985] have shown that only a small (finite) subset of the (infinite) set of possible extensions produces notably different results if one assumes the "finest level of specificity" (i.e. the range) of an uncertainty measure to be limited by so-called *term sets*. The basic idea underlying the use of term sets is that it is reasonable to expect the user to provide *linguistic* estimates of the likelihood of given statements: to describe the degree of certainty in an expert rule or an evidence, experts and users are offered a verbal scale of certainty expressions.

This, however, is directly related to the notion of fuzzy sets and membership functions. In [Bonissone and Decker 1985] an example is given for the nine-element term set

{impossible, extremely_unlikely, very_low_chance, small_chance,
it_may, meaningful_chance, most_likely, extremely_likely, certain}

together with its associated semantics. For our purpose it is sufficient to avoid the details here and refer to the results of the experiments only.

Concerning the experiments made by Bonissone and Decker two remarks are necessary: it is obvious that on one hand the terms above have fuzzy-valued semantics and, on the other, the T-norms and T-conorms apply to *crisp ranges only*. To enable crisply defined functions to be evaluated with fuzzy-valued arguments Bonissone and Decker make use of the so-called *extension principle*. Further, the result of such an evaluation must be an element of the term set again. The process that maps the result of the aggregation of two elements of the term set back into the term set is called *linguistic approximation* and takes the "semantic similarity" of two arguments into account. For a detailled description of the extension principle and linguistic approximation and specific details concerning the experiments the reader is referred to [Bonissone and Decker 1985].

For the experiments Bonissone and Decker use three term sets consisting of 5, 9, and 13 elements, respectively. They provide three different levels of granularity for quantifying uncertainty. For each of the three term sets 11 different uncertainty calculi are evaluated: by evaluating each of the 11 special T-norms on the cross product of the term sets, the closure of a T-norm is generated. For determining T-norm "equivalence classes", the maximum percentage of differences allowed between members of the same class is established. On the basis of this threshold value, the closures are compared and the T-norms without "significant" differences are assumed to be members of the same equivalence class.

Bonissone and Decker have shown that the infinite number of uncertainty calculi can be reduced to at most five distinct equivalence classes. Moreover, for term sets with no more than nine elements, the above experiments lead to only three equivalence classes that can be represented by three uncertainty calculi given by the pairs (T_1, S_1), (T_2, S_2), and (T_3, S_3), with

$$\begin{array}{lcllcl} T_1(a,b) & = & \max(0, a+b-1) & S_1(a,b) & = & \min(1, a+b) \\ T_2(a,b) & = & a \cdot b & S_2(a,b) & = & a+b-a \cdot b \\ T_3(a,b) & = & \min(a,b) & S_3(a,b) & = & \max(a,b) \end{array} \quad (9.18)$$

For any practical purpose, these three T-norms (or T-conorms, respectively) generate sufficiently distinct results for term sets containing no more than nine elements. It is obvious that, as before, the special negation operator $N_i(a) = 1 - a,\ i = 1,2,3$ has been chosen. In the notation introduced above, S_i is then called the N_i-dual of T_i.

Since also the T-norms and T-conorms defined above can be ordered we can extend the inequalities (9.13) and obtain

$$\begin{array}{ccccccc} T_W(a,b) & \le & T_1(a,b) & \le & T_2(a,b) & \le & T_3(a,b) \\ S_3(a,b) & \le & S_2(a,b) & \le & S_1(a,b) & \le & S_W(a,b) \end{array} \quad (9.19)$$

It is apparent that the operator T_2 presents the *classical probability measure* applied to the intersection of independent arguments. Similarly, S_2 is the usual additive probability measure applied to the union of arguments. Further, T_3 is closely related to the intersection operation of a *necessity measure*, and S_3 is related to the union operation of a *possibility measure*. For further information on the concepts of necessity and possibility the reader is referred to the Sect. 3.3 and Chap. 10. Therefore, because of this correspondence of the general T-norms and T-conorms to existing special uncertainty calculi, these functions will be examined in other sections too – in the form of more special instances and without being explicitely mentioned.

An examination of the three T-norms leads to the conclusion that T_1 assumes that the arguments are considered to be as mutually exclusive as possible (i.e. the "worst" case). T_2 assumes independence of the arguments, and T_3 assumes that one argument attempts to subsume the other one (i.e. the "best" case).

A discussion of the concept of T-norms and T-conorms in general, its evaluation in relation to concrete uncertainty calculi, as well as an overview and examinations of different extensions of T-norms and T-conorms can be found also in other publications of Bonissone et al. (e.g. [Bonissone 1983], [Bonissone and Tong 1985], [Bonissone and Decker 1985], [Bonissone 1987a], [Bonissone 1987b]).

As already mentioned above, in relation to uncertainty propagation in rule-based systems the general T-norm can be applied to perform *sequential combination.* So, the actual "certainty" of the premise of a rule is combined with the certainty factor attached to the rule to obtain the actual certainty of the conclusion. In particular, assume two general expert rules of the kind

$$\begin{array}{l} \underline{\text{if}}\ A\ \underline{\text{then}}\ B\ \underline{\text{with}}\ [r\,,\cdot] \\ \underline{\text{if}}\ B\ \underline{\text{then}}\ A\ \underline{\text{with}}\ [r',\cdot] \end{array}$$

In the framework of this section the expert rules will be considered in relation to "classical implication" where *modus ponens* and *modus tollens* represent the main inference mechanisms. In the following the two expert rules may be represented also in the form

$$\text{“}A\xrightarrow{r}B\text{” and “}B\xrightarrow{r'}A\text{”, or “}A\xleftrightarrow{r,r'}B\text{”,}$$

respectively, where r and r' are the lower bounds of the degree of *sufficiency* and *necessity*, respectively. The "dot" included in the two ranges means that upper bounds are not taken into account here. In classical two-valued logics application of *modus ponens* is formally written

$$(A \wedge (A \rightarrow B)) \rightarrow B$$

and states that from the two facts "A" and "A implies B" the fact "B" can be concluded.

Modus tollens on the other hand is formally written

$$((A \rightarrow B) \wedge \neg B) \rightarrow \neg A$$

and states that if "A implies B" is known to be true and "B" is known to be false (i.e. "$\neg B$" is true) the falsity of "A" (i.e. truth of "$\neg A$") can be concluded.

The formal basis of *modus ponens* and *modus tollens*, and the proof of the above assertions, are based on the interpretation in which the categorical expert rule "if A then B" is set equal to the logical expression "$\neg A \vee B$" so that the following equivalence holds:

$$A \rightarrow B \quad \equiv \quad \neg A \vee B$$

In addition to the above two (uncertain) expert rules assume that *cert* is a mapping from the set of propositions to the range $[0,1]$. Assume further that the uncertainty $cert(A)$ concerning the truth of A is given by the range $s \leq cert(A) \leq t$. Then – on the basis of *modus ponens* and *modus tollens* – the known data and the result of the sequential combination are summarized as follows:

$$\begin{array}{rcccl} & & cert \ (A \rightarrow B) & \geq & r \\ & & cert \ (B \rightarrow A) & \geq & r' \\ s & \leq & cert \ (A) & \leq & t \\ \Longrightarrow & & & & \\ T(r,s) & \leq & cert \ (B) & \leq & S(1-r',t) \end{array} \tag{9.20}$$

Since S is the N-dual of T, in the resulting inequality S can be substituted so that also

$$T(r,s) \leq cert(B) \leq 1 - T(r', 1-t) \tag{9.21}$$

holds. The proof of the above result of the sequential combination is given in [Bonissone 1987a], for instance, and can be simply transferred into the framework of our book. To improve readability we make use of the abbreviations l and u for a range's lower and upper limit. So, for example, $l_{A \rightarrow B}$ is the lower limit for the certainty of the assertion "if A then B". Let

$$\begin{aligned} s &= l_A \\ t &= u_A = 1 - l_{\neg A} \\ r &= l_{A \rightarrow B} \\ r' &= l_{B \rightarrow A} = l_{\neg A \rightarrow \neg B} \end{aligned}$$

Since *modus ponens* states "$(A \wedge (A \rightarrow B)) \rightarrow B$", the T-norm T can be directly applied to the conjunction. So, one obtains the lower bound of the resulting certainty range for conclusion B:

$$l_B \geq T(l_A, l_{A \rightarrow B}) = T(s,r)$$

From *modus tollens* "$(\neg A \wedge (\neg A \rightarrow \neg B)) \rightarrow \neg B$" and the application again of the T-norm T we obtain

$$\begin{aligned} l_{\neg B} &\geq T(l_{\neg A}, l_{\neg A \to \neg B}) \\ &= 1 - S(1 - l_{\neg A}, 1 - l_{\neg A \to \neg B}) \\ &= 1 - S(u_A, 1 - l_{B \to A}) \end{aligned}$$

The upper limit u_B of the resulting certainty range is computed by

$$u_B = 1 - l_{\neg B} \leq S(u_A, 1 - l_{B \to A})$$

so that we finally obtain the inequality

$$T(l_A, l_{A \to B}) \leq l_B \leq cert(B) \leq u_B \leq S(u_A, 1 - l_{B \to A})$$

or

$$T(s, r) \leq cert(B) \leq S(t, 1 - r')$$

respectively.

The following special premises' truth values may be of interest since they characterize the behavior of the sequential combination applied to general T-norms.

- The truth of A (i.e. $s = t = 1$) leads to

$$r \leq cert(B) \leq 1$$

  If furthermore the expert rule "if A then B with $[r, \cdot]$" is a categorical one (i.e., $r = 1$) we obtain $cert(B) = 1$ as result.

- Falsity of A (i.e. $s = t = 0$) leads to

$$0 \leq cert(B) \leq 1 - r'$$

  If furthermore the expert rule "if B then A with $[r', \cdot]$" is a categorical one (i.e., $r' = 1$) we obtain $cert(B) = 0$ as result.

On the other hand, also in the framework of sequential combinations *special* T-norms and T-conorms are of interest since the respective results of the sequential combination may represent implication operators that are already known in multi-valued logics. So, for instance,

- the use of the functions $T_1(\cdot, \cdot)$ and $S_1(\cdot, \cdot)$ leads to the upper bound

$$cert(B) \leq \min(1, 1 - r' + t) \tag{9.22}$$

  which is known as *Lukasiewicz implication.*

- the use of the functions $T_2(\cdot, \cdot)$ and $S_2(\cdot, \cdot)$ leads to the upper bound

$$cert(B) \leq 1 - r' + t \cdot r' \tag{9.23}$$

which is known as *Kleene-Dienes-Łukasiewicz implication.*

- the use of the functions $T_3(\cdot,\cdot)$ and $S_3(\cdot,\cdot)$ leads to the upper bound

$$cert(B) \leq \max(1-r',t) \tag{9.24}$$

which is known as *Kleene-Dienes implication.*

Concerning the *incompleteness of knowledge* which is measured by the distance of an interval's two boundaries and represented by the ranges, the results of the three special sequential combinations can be ordered as follows:

$$\begin{aligned} &[T_3(s,r), S_3(t,1-r')] \\ \subseteq\ &[T_2(s,r), S_2(t,1-r')] \\ \subseteq\ &[T_1(s,r), S_1(t,1-r')] \end{aligned} \tag{9.25}$$

Let us return to the the special T-norm T_1 (and the corresponding special T-conorm S_1) where the result of the sequential combination is given by the range

$$[\, T_1(r,s)\ ,\ S_1(1,1-r')\,] \ = \ [\, \max(0,r+s-1)\ ,\ \min(1,1-r'+t)\,] \tag{9.26}$$

9.2.2 RUM

RUM (Reasoning with Uncertainty Module) is an architecture developed by Bonissone et al. based on the concept of general T-norms and T-conorms [Bonissone 1987a], [Bonissone 1987c], [Bonissone, Gans, and Decker 1987]. The system RUM has been tested in data fusion and both naval and aerial situation assessment. The experiments were carried out on the basis of an object-oriented symbolic battle management simulator. They consisted of correlating reports and tracks, locating and classifying platforms, and identifying intents and threats.

RUM is based on the three layers *representation*, *inference*, and *control.* The representation layer represents information used in the other two layers i.e., lower and upper bounds of a certainty measure given by linguistic probabilities, ignorance measure, consistency measure, and meta-information describing the evidence source. Details of the inference and the control layer are given below.

In RUM, the uncertain reasoning process is performed in the *inference layer.* In this layer five uncertainty calculi, based on their underlying triangular norms, are used. In addition to the three calculi defined in equations (9.18) above, RUM's inference layer provides users and experts with two other calculi given by

$$T'(a,b) = \max(0, (a^{0.5} + b^{0.5} - 1)^2)$$
$$S'(a,b) = 1 - \max(0, ((1-a)^{0.5} + (1-b)^{0.5} - 1)^2)$$

$$T''(a,b) = \max(0, (a^{-1} + b^{-1} - 1)^{-1})$$
$$S''(a,b) = 1 - \max(0, ((1-a)^{-1} + (1-b)^{-1} - 1)^{-1})$$

Assume that a special uncertainty calculus (T, S) has already been selected. For this calculus the uncertainty propagation then consists of four operations. In the notation of this book the operations are *premise evaluation*, *sequential combination*, *parallel combination*, and the so-called *source consensus operation.* Concerning the last two operations – parallel combination and source consensus – it should be mentioned that RUM distinguishes between rule instances generated from the *same expert rule* and rule instances derived from *different expert rules* [Bonissone, Gans, and Decker 1987]. In particular:

- In the case of *complex premises*, e.g.,

  $$\underline{\text{if}}\ A_1\ \underline{\text{and}}\ A_2\ \underline{\text{and}}\ \ldots\ \underline{\text{and}}\ A_m\ \underline{\text{then}}\ \ldots$$

  and known certainty ranges $cert(A_i) \in [s_i, t_i]$, $i = 1, \ldots, m$, the certainty range $[s, t]$ of the whole premise A is computed on the basis of the T-norm

  $$[s,t] \stackrel{\text{d}}{=} [T(s_1, s_2, \ldots, s_m),\ T(t_1, t_2, \ldots, t_m)].$$

- To perform *sequential combination* the premises' certainty $cert(A)$ has to be combined with the expert rule's certainty to obtain the certainty range $cert(B) \in [u, v]$ of the rule's conclusion B. For expert rules

  $$\begin{array}{l} \underline{\text{if}}\ A\ \underline{\text{then}}\ B\ \underline{\text{with}}\ [r\ ,\cdot] \\ \underline{\text{if}}\ B\ \underline{\text{then}}\ A\ \underline{\text{with}}\ [r', \cdot] \end{array}$$

  as described in the previous section, the sequential combination is defined as

  $$[u,v] \stackrel{\text{d}}{=} [T(r,s),\ N(T(r', N(t)))].$$

  Notice that for the negation operator $N(a) = 1 - a$ this result is identical to that given by inequality (9.21).

- In the case of more than one instance of an expert rule confirming the same conclusion B, a *parallel combination* has to be performed to get the final certainty $cert(B)$ of this conclusion. For example, if sequential combinations applied to n rule instances led to the certainty ranges $[u_j, v_j], j = 1 \ldots, n$, the conclusion aggregation is defined as

$$[u, v] \stackrel{\mathrm{d}}{=} [N(T(N(u_1), \ldots, N(u_n))),\ T(N(v_1), \ldots, N(v_n))].$$

- The *source consensus operation* is used to fuse the certainty measures of the *same evidence* A provided by *different sources*. If A is an observed fact, the fusion occurs before the evidence is used in the inference process. If, on the other hand, the evidence is the result of a parallel combination, the source consensus operation reduces the *incompleteness* of knowledge about $cert(A)$. If some of the sources are inconsistent, the resulting certainty ranges are disjoint. Assuming $[l_1, u_1], \ldots, [l_n, u_n]$ as the certainty intervals of the same conclusion provided by different sources of information, the source consensus operation coincides with the intersection of the intervals:

$$[l_{tot}, u_{tot}] \stackrel{\mathrm{d}}{=} [\max_i(l_i), \min_i(u_i)]$$

In RUM the "uncertainty calculus selection" is performed in the *control layer* which includes the functions required for the selection. Calculus selection means

- indicating for each expert rule the T-norm needed for premise evaluation and sequential combination, and
- indicating for each subset of rules asserting the same conclusion the T-conorm. The T-conorm (or its N-dual T-norm) is needed for parallel combination.

Which uncertainty calculus is selected depends on how well its characteristics (compare the remarks on "instances of T-norms") fit the local assumptions and the context-dependent information describing the situation (e.g. nature, reliability, and characteristics of the source of evidence; the conditions under which the evidence has been gathered).

The selection of a rule's T-norm for premise evaluation and sequential combination is a function of one's (pessimistic or optimistic) attitude toward risk. It defaults to the min-function T_3. T_1 generates the smallest premise evaluation and the weakest sequential combination, i.e., the widest uncertainty interval for the conclusion.

The selection of a rule set's T-conorm (given by its dual T-norm) for parallel combination on the other hand is a function of the correlation among this subset's rules. For instance, selection of T_1 means mutually exclusiveness and therefore extreme negative correlation. Since rule instances derived from the same rule are often subsumptive, the selection's value defaults to the max-function S_3.

Concerning the experiments the system RUM has been tested in, we will restrict ourselves to one simple example rule only. For instance, "Rule 550" [Bonissone 1987c] aims at identifying submarines:

Assuming that a sonar was used to generate a sensor report (that with other reports generated by the same sensor has been used to define a track associated with a platform), if the detected platform has a low noise emission, and is located at a depth of at least twenty meters, then it is extremely likely that it is a submarine. Otherwise, it may not be a submarine.

The value r as the lower bound of the degree of sufficiency of the implication is given by the linguistic probability "extremely_likely". The lower bound of the degree of necessity r' takes the value "it_may".

9.2.3 Final Remarks

For information concerning more complex families of instances of T-norms and T-conorms (i.e., larger sets of uncertainty calculi than considered above), the reader is referred to the original literature of Bonissone et al. mentioned above. Detailed information on the system RUM can also be found there (for instance, measures for inconsistency and conflict, an entropy function, detailed information on the representation layer, etc.).

In RUM, a natural interpretation of the representation is offered to users and experts: in contrast to the internal representation where real-valued intervals are used, for the external uncertainty representation *terms* are visible. In most of the other models for uncertain reasoning exactly *one* uncertainty calculus is selected a priori and is used uniformly for any inference. In RUM, in contrast to these systems, different uncertainty calculi can be determined.

The problem of inteprating defeasible reasoning is addressed in the tool PRIMO (Plausible ReasonIng MOdule), RUM's successor [Bonissone 1990].

When comparing different approaches for handling uncertainty, especially the global assumptions of independence of evidences are of interest: in the above model based on *families of uncertainty calculi*, the T-norm T_2 is selected when the evidences are independent. If the independence assumptions are known to be violated, another uncertainty calculus can be chosen. In RUM no global assumptions about the exhaustiveness and exclusiveness of hypotheses are used.

Nevertheless, when looking at RUM in the framework of the other systems and models examined in this book one should keep in mind that the system's design is mainly based on *heuristic* and *intuitive* decisions.

That T-norms and T-conorms are also basic concepts in the framework of fuzzy logic becomes visible in Chap. 10.

9.3 INFERNO – A Bounds Propagation Architecture

The system INFERNO was proposed by Quinlan ([Quinlan 1983a] and [Quinlan 1983b]) and can be viewed as a *non-directed architecture*, where information can be propagated in "all directions". As in most expert systems it is possible to reason (forward) from observations to conclusions, but additionally, the system reasons backward from hypothetical situations, for instance to predict just the necessary observations.

In its basics, INFERNO's architecture is probabilistic. However, it makes no assumptions about joint probability distributions of pieces of knowledge. Therefore, the correctness and validity of the inferences can be guaranteed. In the absence of assumptions the inferences made may be weak. If more information is available they can be strengthened. Because of this conservative model, INFERNO also is called "a cautious approach to uncertain inference". In the case of inconsistent information entered into the system, the user is informed. Additionally, INFERNO can make suggestions about changing the information to make it consistent.

So, the four ideas Quinlan's system INFERNO was designed around are given by the following list (taken from [Quinlan 1983b]):

- General systems for uncertain inference are better off *without assumptions* such as conditional independence whose universal validity is suspect.
- On the other hand, it should be possible to assert that particular groups of propositions exhibit relationships such as *independence.*
- There should be no restrictions on the *direction of information flow* in the network.
- The *consistency* of the data should be checked and the system should be able to advise on alternative methods of *rectifying inconsistencies.*

As a first short characterization, the inference model of Quinlan is based on *bounds propagation*: under the assumption that most correlations are unknown, the certainty measures are combined under two extreme assumptions

- that the components have a high positive correlation, and
- that they have a high negative correlation.

Drawing inferences then leads to upper and lower bounds, which are entered as inputs to subsequent computations. Consequently, this results in new bounds also on the certainty of the conclusions.

We start by defining the two uncertainty measures on which INFERNO is based. How uncertainty propagation is performed first will be shown on the basis of the simple if–then rules that are used in typical expert systems. Additionally, in INFERNO various relations between collections of propositions are allowed. The propagation constraints which are associated with the validities of these relations are derived too. The process of locating inconsistencies and that of rectification are only briefly mentioned. We conclude our overview of INFERNO by extracting its pros and cons, discussing the architecture, and referring to recent extensions and reformulations of this architecture.

Quinlan's mathematical model for uncertainty propagation is based on two measures: the uncertainty of a proposition A is characterized by the two values

$t(A)$ – a lower bound on the probability $P(A)$ of A, and
$f(A)$ – a lower bound on the probability $P(\overline{A})$ derived from evidence against A.

Consequently, INFERNO uses probability bounds instead of a single point probability, i.e.

$$t(A) \leq P(A) \leq 1 - f(A). \tag{9.27}$$

As in the Dempster-Shafer theory the explicit use of two measures has the advantage that *incomplete knowledge* concerning the exact uncertainty (i.e. probability) value of A is apparent.

Initially, each proposition has the trivial bounds $t(A) = 0$ and $f(A) = 0$. The actual values for the two measures are derived from evidence that supports or denies A, respectively. During the whole inference process they are propagated separately. The information concerning A is *consistent* as long as

$$t(A) + f(A) \leq 1. \tag{9.28}$$

Concerning the *propagation of uncertain knowledge*, the models examined in the previous sections were typically based on "expert rules" of the kind

if A then S with strength x

where incoming evidence is propagated in one direction only – from observations to conclusions.

In INFERNO, relations are viewed as *constraints* on the respective validities of collections of propositions. In this way, non-directed propagation of inferences is achieved: when changing a probability bound of a proposition, then to preserve the constraint some other bound has to be altered. So, in INFERNO the above expert rule is viewed as the *inferential relation*

A enables S with strength x

with the interpretation $P(S|A) \geq x$ and the two associated *propagation constraints*

$$t(S) \geq t(A) \cdot x \tag{9.29a}$$

$$f(A) \geq 1 - \frac{1 - f(S)}{x} \,. \tag{9.29b}$$

To preserve these two constraints, $t(S)$ has to be increased when $t(A)$ is increased, and $f(A)$ has to be increased when $f(S)$ is increased. Other relations and their associated propagation constraints are summarized below.

Propagation constraint (9.29*a*) is simply based on the inequalitiy

$$P(S) \geq P(S \cap A) = P(S|A) \cdot P(A) \geq x \cdot P(A),$$

which gives $P(S) \geq t(A) \cdot x$ and $t(A) \cdot x$ as lower bound for $t(S)$.

On the other hand, the relation $P(A) \leq \frac{P(S)}{x}$ leads to

$$P(\overline{A}) = 1 - P(A) \geq 1 - \frac{P(S)}{x} \geq 1 - \frac{1 - f(S)}{x}$$

(because of $P(S) \leq 1 - f(S)$) which determines $1 - \frac{1-f(S)}{x}$ as the lower bound of $f(A)$ and therefore gives (9.29*b*).

As already mentioned above, initially each proposition has the trivial bounds $t(A) = 0$ and $f(A) = 0$. These values may be changed explicitly by the user or implicitly on the basis of the relations connecting the propositions. But the range $[t(A), 1 - f(A)]$, in which the probability $P(A)$ is known to lie, can only become smaller when more information is inferred.

A propagation constraint is activated when a bound contained in the constraint's right-hand side is changed. If then the value of the bound on the constraint's left-hand side is less than the value on the right-hand side, the bound is increased to this new value.

So, for instance, for the relation "A enables S with strength x", in constraint (9.29*a*), $t(S)$ has to be checked whenever $t(A)$ is increased. Further, for constraint (9.29*b*), $f(A)$ must be checked whenever $f(S)$ is increased. If in this way the left-hand side of a propagation constraint is increased, also this newly increased value has to be propagated since it may play a role for the correctness of other constraints.

Above, with inequalities (9.29*a*) and (9.29*b*) two simple constraints related to the classical type of expert rules have been introduced. But in INFERNO, assertions are made about various other relations between propositions. So, there are further propagation constraints which are introduced below:

"A negates S" (with interpretation $A \equiv \overline{S}$):

$$t(A) = f(S) \tag{9.30a}$$

$$f(A) = t(S) \tag{9.30b}$$

"A conjoins $\{S_1, S_2, \ldots, S_n\}$" (with interpretation $A \equiv \bigcap_{i=1}^{n} S_i$):

$$t(A) \geq 1 - \sum_i (1 - t(S_i)) \tag{9.31a}$$

$$f(A) \geq f(S_i) \tag{9.31b}$$

$$t(S_i) \geq t(A) \tag{9.31c}$$

$$f(S_i) \geq f(A) - \sum_{j \neq i} (1 - t(S_j)) \tag{9.31d}$$

"A conjoins-independent $\{S_1, S_2, \ldots, S_n\}$"
(with interpretation $A \equiv \bigcap_{i=1}^{n} S_i$, $P(S_i \cap S_j) = P(S_i) \cdot P(S_j)$ for all $i \neq j$):

$$t(A) \geq \prod_i t(S_i) \tag{9.32a}$$

$$f(A) \geq 1 - \prod_i (1 - f(S_i)) \tag{9.32b}$$

$$t(S_i) \geq \frac{t(A)}{\prod_{j \neq i}(1 - f(S_j))} \tag{9.32c}$$

$$f(S_i) \geq 1 - \frac{1 - f(A)}{\prod_{j \neq i} t(S_j)} \tag{9.32d}$$

"A disjoins $\{S_1, S_2, \ldots, S_n\}$" (with interpretation $A \equiv \bigcup_{i=1}^{n} S_i$):

$$t(A) \geq t(S_i) \tag{9.33a}$$

$$f(A) \geq 1 - \sum_i (1 - f(S_i)) \tag{9.33b}$$

$$t(S_i) \geq t(A) - \sum_{j \neq i} (1 - f(S_j)) \tag{9.33c}$$

$$f(S_i) \geq f(A) \tag{9.33d}$$

"A disjoins-independent $\{S_1, S_2, \ldots, S_n\}$"
(with interpretation $A \equiv \bigcup_{i=1}^{n} S_i$, $P(S_i \cap S_j) = P(S_i) \cdot P(S_j)$ for all $i \neq j$):

$$t(A) \geq 1 - \prod_i (1 - t(S_i)) \tag{9.34a}$$

$$f(A) \geq \prod_i f(S_i) \tag{9.34b}$$

$$t(S_i) \geq 1 - \frac{1 - t(A)}{\prod_{j \neq i} f(S_j)} \tag{9.34c}$$

$$f(S_i) \geq \frac{f(A)}{\prod_{j \neq i}(1 - t(S_j))} \tag{9.34d}$$

"A <u>disjoins-exclusive</u> $\{S_1, S_2, \ldots, S_n\}$"
(with interpretation $A \equiv \bigcup_{i=1}^{n} S_i$ and $P(S_i \cap S_j) = 0$ for all $i \neq j$):

$$t(A) \geq \sum_i t(S_i) \tag{9.35a}$$

$$f(A) \geq 1 - \sum_i (1 - f(S_i)) \tag{9.35b}$$

$$t(S_i) \geq t(A) - \sum_{j \neq i} (1 - f(S_j)) \tag{9.35c}$$

$$f(S_i) \geq f(A) + \sum_{j \neq i} t(S_j) \tag{9.35d}$$

"$\{S_1, S_2, \ldots, S_n\}$ <u>mutually exclusive</u>"
(with interpretation $P(S_i \cap S_j) = 0$ for all $i \neq j$):

$$f(S_i) \geq \sum_{j \neq i} t(S_j) \tag{9.36}$$

It should be mentioned that other relations used in INFERNO are <u>inhibits</u>, <u>requires</u>, and <u>unless</u>. They can be defined in terms of <u>enables</u> and <u>negates</u>.

Equations (9.30a) and (9.30b) are obvious consequences of the definitions of the measures t and f. Proofs of the inequalities (9.31a) to (9.32d) can be led back to the proofs of inequalities (9.33a) to (9.34d): this is because the assertion, for example, that A is the conjunction of $\{S_1, S_2, \ldots, S_n\}$ is equivalent to the assertion that $\overline{A}$ is the disjunction of $\{\overline{S_1}, \overline{S_2}, \ldots, \overline{S_n}\}$. Consequently, the conjunction constraints are derived as the disjunction constraints – by simply interchanging the signs as well as t and f.

Inequalities (9.33a) and (9.33d) are given by the relation

$$\max_i P(S_i) \leq P(A)$$

which is basically for A that disjoins $\{S_1, S_2, \ldots, S_n\}$. From the other basic relation

$$P(A) \leq \sum_i P(S_i)$$

(9.33b) and (9.33c) can be derived—the first inequality by writing

$$P(\overline{A}) \geq 1 - \sum_i P(S_i) \geq 1 - \sum_i (1 - f(S_i))$$

and the second by

$$P(S_i) \geq P(A) - \sum_{j \neq i} P(S_j) \geq t(A) - \sum_{j \neq i} (1 - f(S_j))$$

If A is the disjunction of independent $\{S_1, S_2, \ldots, S_n\}$ then $\overline{A}$ is the conjunction of independent $\{\overline{S_1}, \overline{S_2}, \ldots, \overline{S_n}\}$. The combination rule for conjoining independent propositions is

$$P(\overline{A}) = \prod_i P(\overline{S_i})$$

which directly leads to inequalities (9.34a) and (9.34b) by writing

$$P(A) \geq 1 - \prod_i (1 - t(S_i))$$

and

$$P(\overline{A}) \geq \prod_i f(S_i)$$

respectively. (9.34c) and (9.34d) can be derived by rewriting the combination rule as

$$P(\overline{S_i}) = \frac{P(\overline{A})}{\prod_{j \neq i} P(\overline{S_j})}$$

The basic rule that applies to "A disjoins-exclusive $\{S_1, S_2, \ldots, S_n\}$" is

$$P(A) = \sum_i P(S_i)$$

which gives (9.35a) and (9.35b), and the rule's rearrangement

$$P(S_i) = P(A) - \sum_{j \neq i} P(S_j)$$

giving (9.35c) and (9.35d).

If the $S_i, i = 1, \ldots, n$ are known to be mutually exclusive then the equivalent inequalities

$$\sum_i P(S_i) \leq 1$$

and

$$\sum_{j \neq i} P(S_j) \leq 1 - P(S_i)$$

hold. The second inequality proves (9.36).

To demonstrate how Quinlan's bounds propagation approach works, we simply make use of one of his example networks.

(9.5) Example. The three user-defined relations

C <u>conjoins</u> $\{A, B\}$

E <u>disjoins-independent</u> $\{C, D\}$

B <u>enables</u> E <u>with strength</u> 0.8

are visualized in Fig. 9.2. For a first short interpretation, the relations state that

- C is the conjunction of A and B,
- nothing is known about (in)dependence and/or exclusiveness of A and B,
- E is the disjunction of C and D which are known to be independent,
- and the conditional probability $P(E|B)$ is at least 0.8.

If we assume $P(C) \in [0.55, 0.65]$ we know $t(C) = 0.55$ and $f(C) = 0.35$. All other $t(\ldots)$ and $f(\ldots)$ are initially zero. On the basis of this knowledge the following inferences are made:

$t(A) \geq 0.55$ (from constraint (9.31c), and $t(C) = 0.55$)
$t(B) \geq 0.55$ (from constraint (9.31c), and $t(C) = 0.55$)
$t(E) \geq 0.55$ (from constraint (9.34a), and $t(C) = 0.55$)

Other bounds do not change since for other inequalities the bound on the right-hand side is less than or equal to the value on the left-hand side.

If now for node B the assertions $t(B) = 0.9$ and $f(B) = 0.1$ (i.e. $P(B) = 0.9$) are made, again propagation constraints must be checked:

$f(A) \geq 0.25$ (from constraint (9.31d), $t(B) = 0.9$, and $f(C) = 0.35$)
$t(E) \geq 0.72$ (from constraint (9.29a), and $t(B) = 0.9$)
$t(D) \geq 0.20$ (from constraint (9.34c), $f(C) = 0.35$ and $t(E) \geq 0.72$) ◇

The information about some proposition A is *inconsistent*, if

$$t(A) + f(A) > 1.$$

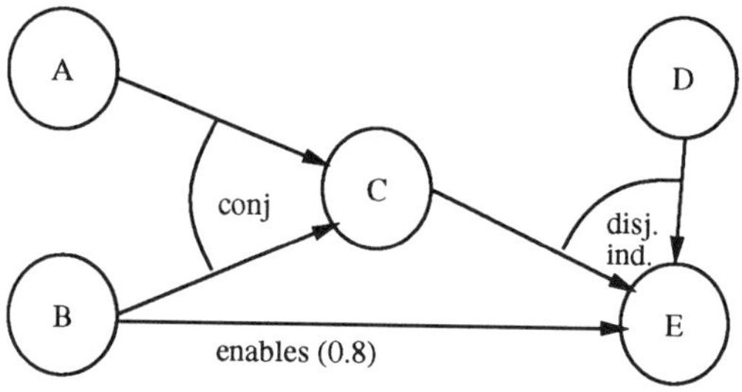

Fig. 9.2. Quinlan's inference network example (from [Quinlan 1983a])

In such cases, one or both values must be incorrect. Since the propagation constraints are provably correct, the information entered into the system must contain implicit contradictions. INFERNO can identify one or more assertions whose uncertainty value must be decreased and/or inferential relations that must be weakened. This process of locating a combination of changes that makes the bounds on all propositions consistent is called a *rectification.*

A detailed description of the process of finding and ranking all possible rectifications is included in [Quinlan 1983a]. The mathematical constraints for backing up inconsistencies are also listed there.

In [Quinlan 1983a] and [Quinlan 1983b], INFERNO's uncertainty handling model is also compared with other mathematical models, for instance, the PROSPECTOR approach and the Dempster-Shafer theory which are introduced later, and MYCIN's certainty factor model.

Coming now to the concluding remarks, a characteristic feature of INFERNO is that it can be used for both forward inferences and backward inferences, since it is based on *relations* between propositions and associated *propagation constraints.*

One main difference between the mathematical uncertainty model integrated in INFERNO and other models for handling uncertainty is the way, (independence) *assumptions* are handled or taken into account: in most expert systems the uncertainty models depend on certain independence assumptions concerning joint probability distributions. If these assumptions are unjustified in a certain application domain, the inferences drawn may be erroneous. In INFERNO, on the other hand, the validity of any inference can be guaranteed, since its uncertainty model is "cautious": it does not depend on assumptions about joint probability distributions of propositions.

However, in some cases the values of the bounds that are inferred on the basis of the propagation constraints may be weaker than those derivable by probabilistic interpretation. For illustration, Quinlan presented the example of the two relations "A enables B with strength x" (which has interpretation $P(B|A) \geq x$) and "C conjoins-independent $\{A, B\}$" (with interpretation $C \equiv A \cap B$, $P(A \cap B) = P(A) \cdot P(B)$). Because of the independence of A and B it can simply be deduced that $P(B) \geq x$ holds – an inference that cannot be drawn on the basis of the constraints.

As also mentioned by Pearl [Pearl 1988a], the bounds produced by the system INFERNO are too wide. For instance, in the case of given $P(A) = p$ and $P(B|A) = q$, the bounds obtained for $P(B)$ are

$$pq \leq P(B) \leq 1 - p(1 - q)$$

which for small p approach the unit interval $[0, 1]$.

Consequently, uncertainty propagation in INFERNO may lead to conclusions about the probability bounds of propositions that are weaker than the direct probabilistic interpretation of the INFERNO relations. But in each case the conclusions are provably correct consequences of the information entered by the user. Concerning the termination of propagation, like other inference network models INFERNO also prohibits changes from being propagated back to the source of original disturbance. INFERNO's concern for inconsistency detection and rectification was not described here.

Several reformulations and "improvements" have been proposed for the INFERNO architecture, just as they have been for other systems and models.

For instance, in [Liu and Gammerman 1988] an approach is introduced that is based on INFERNO but integrates *probabilistic logic* and *Dempster's rule of combination*. So, the original conditional probability concept $P(B|A)$ for rules is substituted by weights $P(A \rightarrow B)$ for rules which are represented in terms of logical implication through Boolean operations. The integration of a special form of Dempster's rule (for "singleton" hypotheses) is due to INFERNO's inability to combine the effects for a proposition from different sources. "Improvements to INFERNO" as well as a discussion of its basic concepts are also given by Saunders [Saunders 1989].

However, when discussing and sometimes criticizing the INFERNO approach of bounds propagation one should keep in mind that this early system was constructed as a response to systems that were based only on heuristic ideas of uncertainty propagation.

9.4 Other Heuristic Models

In the preceding sections three of the most important more or less heuristic models for uncertainty representation and propagation have been examined. The examination showed that the measures defined and the algorithms proposed are mainly influenced by practical criteria. This is the case especially for the systems MYCIN and RUM based on certainty factors and the general family of T-norms and T-conorms, respectively. Notice that T-norms and T-conorms are of interest also for fuzzy reasoning, considered in Chap. 10.

In its basics, the architecture of INFERNO makes use of probability measures as well as the axioms of probability theory. Further, it does not suffer under the artificial independence assumptions that are often criticized in relation to more formal approaches.

A characteristic feature of all these systems is that they try to overcome certain deficiencies, for instance of some probability-based models. So, they allow the representation of incomplete knowledge given by users and experts, and do not assume dependency on belief and doubt. A priori values for the certainties of "internal nodes" in drawing inferences are not needed. On the other hand, if available they cannot be taken into account in the original certainty factor approach or in the system RUM. Independence assumptions, either absolute or conditional, are not made.

Consequently, these systems and models are simple to use and widely accepted by users and experts who are not familiar with the deep theoretical, i.e. mathematical, foundations of probability theory, random sets, or belief functions, for instance as introduced in the earlier parts of this book. This simplicity may be the main reason for the fact that, for example, the certainty factor approach for handling uncertainty is still used in recent expert systems – some of which indeed are commercially available.

Consequently, when enjoying their simplicity one should always keep in mind that the systems' inferences and decision making are based on measures and algorithms that are more or less heuristic and intuitively developed – especially for real-world application where false decisions would have nontrivial consequences, as would be the case in military applications.

Finally, we want to refer to some other important heuristic approaches that were not considered in detail.

EXPERT [Weiss and Kulikowski 1979] is a system for designing and building models for consultation. When designing the EXPERT system, the authors integrated the experience they gained during the years of developing the CASNET system (see [Weiss et al. 1978]). Similarly to the MYCIN system, EXPERT makes use of *confidence measures* which are assigned on a scale of -1 to 1, with 1 as complete confirmation and -1 as complete denial. The system was used to develop consultation models in domains such as rheumatology, ophthalmology, and endocrinology. A detailed overview of both EXPERT and CASNET can be found in [Weiss and Kulikowski 1983].

Rollinger [Rollinger 1983] proposed an *evidence space* approach that represents the uncertainty of a statement as a point in a two-dimensional space. To verify the validity of any given premise in an expert rule Rollinger suggests the use of distances. The aspect of uncertainty propagation is only incompletely considered by him.

More heuristic numerical approaches can be found in the literature. For heuristic non-numerical models for handling uncertainty the reader is referred to Chapt. 15.

Chapter 10

Fuzzy Set Based Models

In Chap. 2 we used L-sets for modeling *vague observations* such as "The ship is *far away from the coast*". The vagueness in this statement arises from the use of the vaguely defined predicate "far away". A different approach consists in modeling directly the *vague concept* "far away" instead of modeling observed vague data.

The mathematical theory of fuzzy sets introduced by L. Zadeh [Zadeh 1965] was the first approach for handling vague concepts. In this chapter we consider the approaches for building knowledge based systems that rely on fuzzy sets and the related possibility theory. The latter theory describes principally the uncertainty that arises from the use of fuzzy predicates such as "tall", "far away" , etc. Note that there are several other interpretations of the notion of "possibility".

In the literature there are heuristic, certainty factor-like approaches using methods of fuzzy set theory such as in the RUM system described in Sect. 9.2, or in the shell FLOPS [Buckley 1986], which is an extension of OPS5 [Forgy 1981] and was successfully used to process echocardiogram images. These heuristic methods have to be designed corresponding to the domain of application.

Secondly there are models based on fuzzy quantifiers, but the respective algorithms have to be developed [Zadeh 1985b], [Dubois and Prade 1989].

Thirdly there are approaches that are founded directly on fuzzy set concepts or possibilistic approaches. We restrict ourselves to considering approaches in detail that rely on the later methods.

10.1 Fuzzy Sets

L. Zadeh suggested the use of fuzzy sets as a framework for designing knowledge based systems [Zadeh 1979b]. The main idea for introducing the idea of fuzzy sets was that in many applications the crisp membership degrees 1 (complete membership) and 0 (set exclusion) seem to be insufficient to describe vague concepts naturally. An example would be the "class of large numbers" where it is desirable to introduce several degrees of partial membership depending on the number considered. The partial membership degrees in fuzzy set theory are real numbers from the unit interval.

(10.1) Definition. *A fuzzy set f is defined by its membership function mapping a universe Ω to the unit interval, i.e.*

$$\mu_f : \Omega \to [0,1].$$

A fuzzy set f is called normalized, if there is an element $\omega \in \Omega$ with $\mu_f(\omega) = 1$.

The concept of a membership function generalizes the notion of the characteristic function of a conventional set. For an element $\omega \in \Omega$, the value $\mu_f(\omega)$ represents the degree of membership of ω in the fuzzy set f. Unlike conventional sets where elements either do belong or do not belong to a set, fuzzy sets admit partial membership ranging from 0 (non-membership) to 1 (full membership).

(10.2) Example. Let us consider the predicate "far away from the coast" in the context of location determination of ships. Suppose an evaluator wants to express that from his point of view a ship in area z1 is definitely far away from the coast; in ld, rd, ca a ship is considered definitely not far away from the coast, but in the areas z2 and z3 such a strict statement is not appropriate. Then a straightforward method is to consider membership values with which one is able to weight propositions such as "ships in area z2 are considered to be far away from the coast". Fuzzy set theory uses membership degrees in the unit interval, so the evaluator could attach the number 0.7 to the above statement. Note that bigger values correspond to greater confidence in the membership. 1 means definitely true, 0 means false. The "class of all locations that are considered far away from the coast" could then be described by a fuzzy set μ, that assigns to each location its corresponding membership value; in this example the evaluator might choose:

$$\mu(\mathrm{z1}) = 1, \ \mu(\mathrm{ld}) = \mu(\mathrm{rd}) = \mu(\mathrm{ca}) = 0, \ \mu(\mathrm{z2}) = 0.7, \ \mu(\mathrm{z3}) = 0.4.$$

◇

In particular fields of application, membership functions are often described by parameters, especially if the frame of discernment is infinite. So the trapezodial fuzzy set μ of $\mathbb{R}$, such as

$$\mu(x) = \begin{cases} 0 & \text{if } x < 1 \text{ or } x > 5 \\ x - 1 & \text{if } 1 \leq x \leq 2 \\ 5 - x & \text{if } 4 \leq x \leq 5 \\ 1 & \text{if } 2 < x < 4 \end{cases}$$

can be described by 4 parameters, here the vector $(1, 2, 4, 5)$.

The usual set-theoretic operations can be extended in a straightforward way by fuzzy logic to fuzzy sets. We obtain for all $\omega \in \Omega$

$$\begin{aligned} \mu_{f\cap g}(\omega) &\stackrel{\mathrm{d}}{=} (\mu_f \cap \mu_g)(\omega) \stackrel{\mathrm{d}}{=} \min\{\mu_f(\omega), \mu_g(\omega)\} && \text{(intersection)} \\ \mu_{f\cup g}(\omega) &\stackrel{\mathrm{d}}{=} (\mu_f \cup \mu_g)(\omega) \stackrel{\mathrm{d}}{=} \max\{\mu_f(\omega), \mu_g(\omega)\} && \text{(union)} \\ \mu_{\overline{f}}(\omega) &\stackrel{\mathrm{d}}{=} \quad (\overline{\mu_f})(\omega) \stackrel{\mathrm{d}}{=} 1 - \mu_f(\omega) && \text{(complement)} \end{aligned} \tag{10.2}$$

We say that μ is a subset of μ', written $\mu \subseteq \mu'$, if $\mu(\omega) \leq \mu'(\omega)$ holds for all $\omega \in \Omega$. Note that in defining, for example, the intersection operation Zadeh [Zadeh 1965] used the underlying assumption that there are truth degrees of the vague statements "ω is an element of f" and "ω is an element of g", and that the minimum of these values is equal to the truth degree of the statement "ω is an element of the intersection of f and g".

The above operations preserve several properties of a Boolean algebra such as distributivity, whereas the De Morgan laws are in general not satisfied. Of course we have for μ with $\mu(\omega) = 0.5$ for all $\omega \in \Omega$ the identity $\mu \cap \overline{\mu} \neq \mathbb{I}_{\emptyset}$, where $\mathbb{I}_{\emptyset} : \Omega \to [0,1]$, $\mathbb{I}_{\emptyset}(\omega) = 0$ for all $\omega \in \Omega$ is the null-element in this lattice. An axiomatic justification of the choice of these operations was given in [Bellman and Giertz 1973].

There are several possibilities for the representation of fuzzy sets. In contrast to the "vertical" view described above, a fuzzy set f can be represented in terms of its levelcuts $\{f_\alpha | \alpha \in [0,1]\}$, where

$$f_\alpha = \{\omega \in \Omega | \mu_f(\omega) \geq \alpha\} \tag{10.3}$$

(see Fig. 10.1). From the *representation theorem* [Zadeh 1971] we know that

$$\mu_f(\omega) = \sup\{\alpha \in [0,1] |\, \omega \in f_\alpha\} \tag{10.4}$$

holds for all $\omega \in \Omega$. Hence each family of sets $\{f_\alpha | \alpha \in [0,1]\}$ with the properties

$$\begin{aligned} &\text{(i)} \quad \alpha > \alpha' \implies f_\alpha \subseteq f'_\alpha, \\ &\text{(ii)} \quad f_0 = \Omega\,, \\ &\text{(iii)} \quad f_\alpha = \bigcap_{\beta:\beta<\alpha} f_\beta\,, \end{aligned} \tag{10.5}$$

can be considered as giving the level cuts of a fuzzy set f. This can be conceived to be a "horizontal" representation of a fuzzy set as a family of level cuts.

In Chap. 2 we used also some kind of "horizontal view" to introduce the concept of an L-set which is, on *the mathematical level*, of course strongly related to a fuzzy set.

Let A be a subset of Ω, and let $\phi : \Omega \to \Omega'$ be a function. Then the image of A with respect to ϕ is defined by

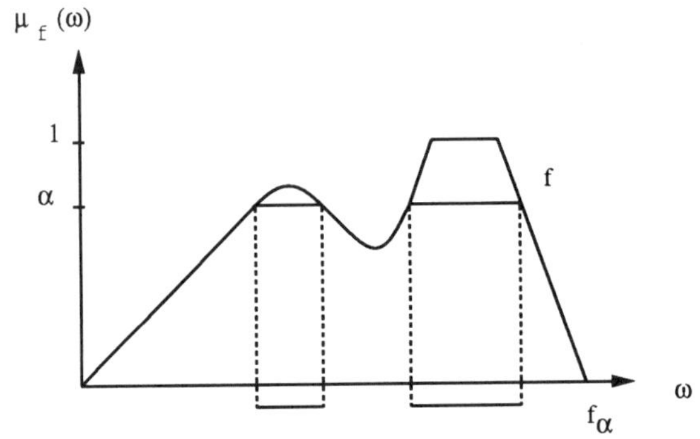

Fig. 10.1. Level cut

$$\phi(A) \stackrel{\mathrm{d}}{=} \{\omega' \in \Omega' |\, \exists \omega \in A : \phi(\omega) = \omega'\}.$$

The analogous definition of the image of a fuzzy set f of Ω with respect to ϕ is given by

$$\mu_{\phi(f)}(\omega') \stackrel{\mathrm{d}}{=} \sup\{\mu_f(\omega) |\, \omega \in \Omega \text{ and } \phi(\omega) = \omega'\} \tag{10.6}$$

for all $\omega' \in \Omega'$.

An important concept needed in fuzzy set theory is that of a *fuzzy relation*, which generalizes the conventional set-theoretic notion of relation. Let Ω_1 and Ω_2 be two universes. A fuzzy relation R has the membership function $\mu_R : \Omega_1 \times \Omega_2 \to [0,1]$. The projection of R on Ω_1 is the *marginal fuzzy set* defined by

$$\mu_{\Pi_{\{1\}}^{\{1,2\}}(R)}(\omega_1) \stackrel{\mathrm{d}}{=} \sup\{\mu_R(\omega_1, \omega_2) |\, \omega_2 \in \Omega_2\}, \tag{10.7}$$

for all $\omega_1 \in \Omega_1$. Other marginals are defined analogously. If f_1 is a fuzzy set on Ω_1, then μ_{f_1} can be extended to $\Omega_1 \times \Omega_2$ by

$$\mu_{\widehat{\Pi}_{\{1\}}^{\{1,2\}}(f_1)}(\omega_1, \omega_2) \stackrel{\mathrm{d}}{=} \mu_{f_1}(\omega_1) \tag{10.8}$$

for all $(\omega_1, \omega_2) \in \Omega_1 \times \Omega_2$.

Fuzzy sets are often used to model linguistic quantities such as "high" in the statement "the speed is high" by using the concept of a linguistic variable [Zadeh 1975b].

From a practical point of view, the question of membership function elicitation is crucial with fuzzy sets, and the lack of simple techniques has raised a lot of criticism [French 1984]. The problem comes from the fact that fuzzy set theory uses on an underlying simple multivalued logic, the *fuzzy logic*, for the definition of set theoretic operations. The modeling of vague observations with L-sets is of course simpler, especially if their meaning is that described in Chap. 2.

10.2 Possibility Distributions

The notion of possibility and necessity measures has already been studied twice so far in the framework of this book, in relation to vague data (Sect. 3.2) and mass distributions (Sect. 6.3). Here they are considered in connection with fuzzy sets. Although the mathematical properties are similar, the semantics of the approach considered here is totally different from that of the preceding definitions.

Suppose we want to use the vague concept "far away from the coast" – described by a fuzzy set f with membership function μ_f – to model the information "the ship is far away from the coast". This of course provides not a precise location of the ship but a partial belief to events such as "the ship is actually at {z2}". In the presence of fuzzy information we can intuitively give a kind of "conditional possibility distribution" for the values of Ω. Zadeh [Zadeh 1978] proposed the equality

$$\begin{aligned}&\text{Possibility (the ship is within }\{\text{z2}\}|\text{ the ship is far away from the coast)}\\ &\quad = \text{truth (the ship is far away from the coast }|\text{ the ship is within }\{\text{z2}\})\end{aligned}$$

which can be evaluated by using the membership degree $\mu_f(\text{z2})$ where μ_f is the membership function of the fuzzy set describing the "class of all locations that are far away from the coast". As a result we obtain a function $\text{Poss} : \Omega \to [0,1]$ defined by $\text{Poss}(\omega) = \mu_f(\omega)$. By setting $\text{Poss}(A) = \sup\{\mu_f(\omega)|\omega \in A\}$ for all $A \subseteq \Omega$ we obtain a possibility satisfying the properties of Definition 6.26.

Note that we have to distinguish carefully between the notion of partial truth used for the definition of fuzzy set context (vagueness) and the notion of partial belief (uncertainty) that arise from the fact that we have to handle incomplete information.

After this intuitive introduction of the notion of possibility we now give a formal description of the concepts.

Let ω_0 be an unknown value ranging over a set Ω, and let the piece of imprecise information be given as a set E, i.e. $\omega_0 \in E$ is known for sure, and $|E| \geq 2$ holds. If we ask whether another set A contains ω_0 or not there can be two possible answers

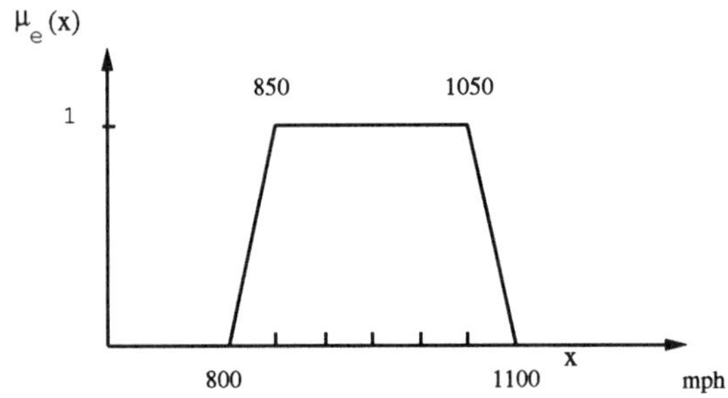

Fig. 10.2. Fuzzy set of Example 10.3

- if $A \cap E = \emptyset$ then it is impossible that $\omega_0 \in A$,
- if $A \cap E \neq \emptyset$ then it is possible.

On the formal level we obtain a mapping

$$\mathrm{Poss}_E : 2^\Omega \to [0,1], \quad \mathrm{Poss}_E(A) = \begin{cases} 1 & \text{if } A \cap E \neq \emptyset \\ 0 & \text{if } A \cap E = \emptyset \end{cases} \tag{10.9}$$

where 1 denotes "possible" and 0 denotes "impossible".

Poss_E is the simplest form of possibility measure. Possibility becomes a matter of degree when E becomes a fuzzy set. Let e be a fuzzy set of Ω where $\{e_\alpha | \alpha \in [0,1]\}$ gives the horizontal view of the membership function μ_e. We define canonically

$$\begin{aligned} &\mathrm{Poss}_e : 2^\Omega \to [0,1], \\ &\mathrm{Poss}_e(A) \stackrel{\mathrm{d}}{=} \sup\{\alpha | A \cap e_\alpha \neq \emptyset, \ \alpha \in [0,1]\} \\ &\qquad = \sup\{\mu_e(\omega) | \omega \in A\} \end{aligned} \tag{10.10}$$

$\mathrm{Poss}_e(\overline{A})$ tells us about the possibility of "not A", hence about the necessity of the occurrence of A. Thus it is natural to define the necessity of A as

$$\mathrm{Nec}_e(A) \stackrel{\mathrm{d}}{=} 1 - \mathrm{Poss}_e(\overline{A}) \tag{10.11}$$

in this context. Knowledge representation runs analogously to the exertions of Sect. 3.3, e.g. $\mathrm{Nec}_\Omega(A) = 0$ and $\mathrm{Poss}_\Omega(A) = 1$ imply total ignorance.

(10.3) Example. Suppose we have the information that the speed of an aircraft is approximately 1000 mph, but definitely not less than 800 mph and not more than 1100 mph. Figure 10.2 shows the corresponding fuzzy set e.

The possibility that the airplane's speed is less 850 mph is

$$\mathrm{Poss}_e\big(\{x|x \leq 850\}\big) = 0.5$$

whereas

$$\mathrm{Nec}_e\big(\{x|x \leq 850\}\big) = 1 - \mathrm{Poss}_e\big(\{x|x > 850\}\big) = 0$$

(in accordance with intuition).

◇

The problem of how to combine knowledge described by two possibility distributions can be reduced to the combination of the information contained in the underlying pieces of information. If this incomplete knowledge is described by two fuzzy sets, e.g.,

datum1: speed is high,
datum2: speed is more than 1000 mph,

then we *firstly* aggregate the information by using operators such as *min*, *max* (or other T-norms/T-conorms), and *secondly* establish the possibility distribution corresponding to "speed is high *and* more than 1000 mph".

So, in this context the combination causes no additional problems, whereas in the case when the underlying pieces of information are not described by fuzzy sets other combination mechanisms may be useful. This difference often leads to confusions.

In the above considerations we represented pieces of knowledge under the form of possibility distributions where the distributions are induced by fuzzy set specifications. It could be argued (but see [Cheeseman 1985]) that possibility measures may serve as a general uncertainty representation tool, where these measures just reflect an ordering between the relative possibilities of hypothetical values for a variable without being necessarily induced by a fuzzy set.

10.3 Approximate Reasoning

A new aspect arising in the case of reasoning under partial ignorance is associated with the question how to decide whether the premise of an "expert rule" can shown to be true with respect to a database. To characterize the situation, assume the following simple database consisting of only one datum and one rule

datum: speed is very low
rule: if speed is low then angle is small

The solution of course depends on the meaning of the information. A trivial solution is given in the case of classical two-valued logics: if the interpretation of the datum "speed is very low" is identical to the premise "speed is low" of the rule, the inference can be drawn and the implied fact is "angle is small" is added to the database. The process of comparing database entries and rule premises is often referred to as lexical matching. However, in order to be able to handle all possible situations, the knowledge base has to contain as many rules as values are possible for the speed assignment in the above simple rule structure.

Another situation is given if the terms "*very_low*" and "*low*" are characterized by fuzzy sets μ_{low} and μ_{very_low}. In this case, the expert rule presents a kind of fuzzy implication. While for two-valued logics a lexical matching process is sufficient, in the case of fuzzy reasoning involving vagueness, a semantic matching has to be performed where the "meaning" of terms or assertions are compared.

For instance, the two fuzzy sets μ_{very_low} and μ_{low} defined over $\Omega_1 = \{a, b, c\}$ and shown in Fig. 10.3 are neither identical nor completely exclusive.

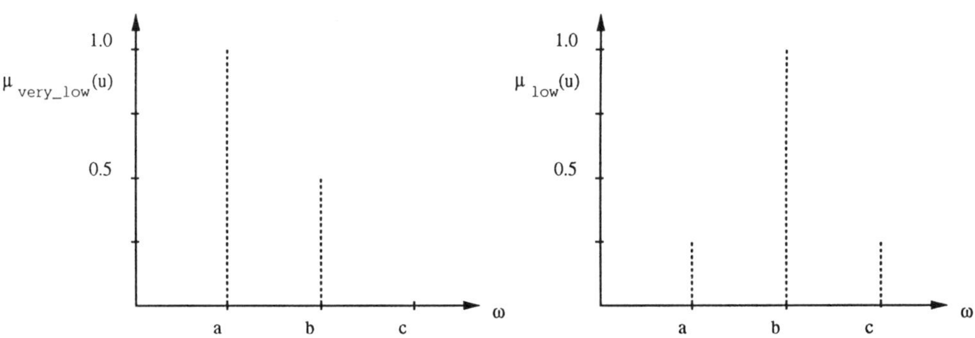

Fig. 10.3. Two simple fuzzy sets

Three different cases of semantical matching are shown in Fig. 10.4, where μ_P and μ_D are the membership functions of pattern and datum, respectively. In order to consider the simplest case again we consider the data base

datum: $X^{(1)}$ is A'
expert rule: if $X^{(1)}$ is A then $X^{(2)}$ is B

with fuzzy sets A, A' of Ω_1 and B is defined over Ω_2.

Drawing the inference on the basis of the above data base will result in

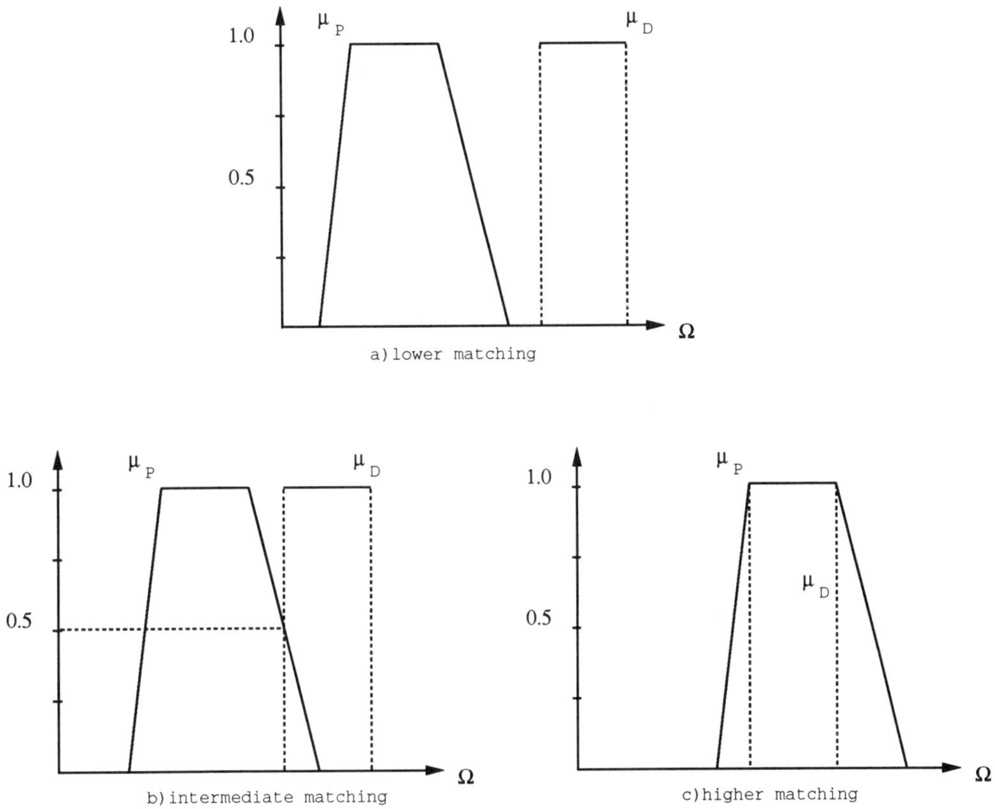

Fig. 10.4. Examples of matching cases

an entry "$X^{(2)}$ is B'", where B' is a fuzzy set of Ω_2. It is obvious that the structure as well as the information contents of B' will depend on the rule's conclusion as well as on the degree of matching of rule premise and datum.

The main idea to solve this problem is to encode each piece of information by possibility measures corresponding to suitable fuzzy sets and then to integrate this information. A suitable space is the product space $\Omega_1 \times \Omega_2$. The datum should be represented by the cylindrical extension of A' to $\Omega_1 \times \Omega_2$, i.e.

$$\widehat{\Pi}_{\{1\}}^{\{1,2\}}(\mu_{A'}) : \Omega_1 \times \Omega_2 \to [0,1], \quad \text{Poss}_{datum}(\omega_1, \omega_2) \stackrel{\text{d}}{=} \mu_{A'}(\omega_1) \tag{10.12}$$

A straightforward solution for the representation of the rule is given by the possibility distribution

$$\begin{aligned}\mathrm{Poss}^{Z}_{rule} &: \Omega_1 \times \Omega_2 \to [0,1],\\ \mathrm{Poss}^{Z}_{rule}(\omega_1,\omega_2) &= \left(\widehat{\Pi}^{\{1,2\}}_{\{1\}}(\mu_A) \cap \widehat{\Pi}^{\{1,2\}}_{\{2\}}(\mu_B)\right)\\ &\quad \cup \left(\widehat{\Pi}^{\{1,2\}}_{\{1\}}(\mu_{\overline{A}}) \cap \widehat{\Pi}^{\{1,2\}}_{\{2\}}(\mathbb{1}_{\{2\}})\right).\end{aligned} \tag{10.13}$$

This solution was proposed in [Zadeh 1979b], and it generalizes the correct solution for set-valued rules "if A then B" which have to be represented by $(A \times B) \cup (\overline{A} \times \Omega_2)$ as the subset of $\Omega_1 \times \Omega_2$ of tuples in accordance with the rule.

But there are other proposals for the representation [Mizumoto et al 1979], the most important of which in this context is the Gödel implication

$$\mathrm{Poss}^{G}_{rule}(\omega_1,\omega_2) = \begin{cases} 1 & \text{if } \mu_A(\omega_1) \le \mu_B(\omega_2),\\ \mu_B(\omega_2) & \text{otherwise.}\end{cases} \tag{10.14}$$

In the case of finite frames Ω_1 and Ω_2 it can simply be realized that the *implication matrix* $\mathrm{Poss}^{G}_{rule}(\omega_1,\omega_2)$ in any case contains the information on fuzzy set B: if A is normalized, the membership function μ_A takes value 1 for at least one $\omega_1 \in \Omega_1$, and for this ω_1 therefore is greater than or equal to μ_B for all values $\omega_2 \in \Omega_2$.

The two pieces of information – datum and rule, both given on the basis of $\Omega_1 \times \Omega_2$ – have to be integrated. We have to take dependencies of sources and so on into account, and there are several proposals for obtaining a combined possibility distribution.

Concerning the integration of datum and rule, the application of the principle of minimum specificity [Dubois and Prade 1990c] gives

$$\mathrm{Poss}_{product}(\omega_1,\omega_2) \stackrel{\mathrm{d}}{=} \min(\mathrm{Poss}^{G}_{rule}(\omega_1,\omega_2), \mathrm{Poss}_{datum}(\omega_1,\omega_2)).$$

Finally $\mathrm{Poss}_{product}$ has to be projected to the set Ω_2 by using (10.6). We obtain

$$\begin{aligned}&\mathrm{Poss}_{result} : \Omega_2 \to [0,1],\\ &\mathrm{Poss}_{result}(\omega_2) = \sup\{\mathrm{Poss}_{product}(\omega_1,\omega_2) | \omega_1 \in \Omega_1\}\,.\end{aligned} \tag{10.15}$$

Especially the use of the Poss^{Z}_{rule}-relation often gives results which contradict intuitive results. The Gödel implication in combination with the min-operator gives rather good results. We have the following properties

(i) if $\forall \omega_2 \in \Omega_2 : \mathrm{Poss}_B(\omega_2) \ge \sup\{\min(\mathrm{Poss}_{\mathrm{rule}}(\omega_1,\omega_2), \mathrm{Poss}_{A'}(\omega_1)) | \omega_1 \in \Omega_1\}$ then $\mathrm{Poss}_{rule} \subseteq \mathrm{Poss}^{G}_{rule}$, i.e. the Gödel relation is the least specific relation that satisfies the above desirable inequality,

(ii) if A' is normalized and $A' \subseteq A$ then $\mathrm{Poss}_{B'} = \mathrm{Poss}_B$,

(iii) we have $\mathrm{Poss}_{B'} \supseteq \mathrm{Poss}_B$ for all A' where A' is normalized.

So the inference has the following properties. In the case of identical fuzzy sets for pattern and datum (μ_A equals $\mu_{A'}$) the result is the membership function $\mu_{B'}$ identical to μ_B. An increasing deviation between A and A' has an increasing degree of vagueness of B' as consequence. In the extreme case the fuzzy sets for pattern A and datum A' may be disjoint: in this case one gets the membership function defined by $\mu_{B'}(\omega_2) = 1$ for all $\omega_2 \in \Omega_2$, and – concerning the associated possibility distribution – one is completely ignorant about the true value of $X^{(2)}$.

However, in spite of the fact that the implication matrix Poss^G_{rule} covers complete information on fuzzy set B it should also be noticed that the main information on rule-premise A gets lost. It follows that the matching process itselves which "compares" pattern A and datum A' is really vague.

A concrete example for the Gödel implication is given below.

(10.4) Example. Assume the database entries

datum:	X is very low
expert-rule:	If X is low then Y is small

with $\Omega_1 = \{a, b, c\}$ and $\Omega_2 = \{x, y, z\}$. The fuzzy sets for *very_low*, *low*, and *small* are given by the membership functions (for the graphical representation of *very_low* and *low* see Fig. 10.3)

$$\mu_{very\_low} : a \mapsto 1 \quad b \mapsto 0.5 \quad c \mapsto 0$$

$$\mu_{low} : a \mapsto 0.2 \quad b \mapsto 1 \quad c \mapsto 0.2$$

$$\mu_{small} : a \mapsto 0.1 \quad b \mapsto 0.9 \quad c \mapsto 1$$

Computation of the implication matrix $\text{Poss}^G_{rule}(\omega_1, \omega_2)$ consequently leads to

		Ω_2		
		x	y	z
Ω_1	a	0.1	1	1
	b	0.1	0.9	1
	c	0.1	1	1

It is obvious that the second row of the matrix presents the membership function associated with *small.* Concerning the information contained in the membership function μ_{low} one should notice that overwriting with $\mu_{low}(a) = \mu_{low}(c) = 0.8$, for instance, would change neither the implication matrix nor the result of the pattern matching process.

Finally, the computation of $\text{Poss}_{B'}(\omega_2)$ for all $\omega_2 \in \Omega_2$ leads to the membership function

$$\begin{aligned}\text{Poss}_{B'} : x &\mapsto \max(0.1,\ 0.1,\ 0) = 0.1\\ y &\mapsto \max(1,\ 0.5,\ 0) = 1\\ z &\mapsto \max(1,\ 0.5,\ 0) = 1\end{aligned}$$

◇

Note that the use of a T-norm for modeling a set theoretic "*and*" other than the main operator implies the use of a different fuzzy relation μ_{rule} [Dubois and Prade 1988c].

For example, in addition to the "Gödel-method" often the "Goguen-method" and the "Łukasiewicz-methods" are used [Dubois and Prade 1990c]:

$$\text{Poss}_{result}(\omega_2) = \sup_{\omega_1 \in \Omega_1} (\text{Poss}_{rule}(\omega_1, \omega_2) \cdot \text{Poss}_{datum}(\omega_1)), \quad \omega_2 \in \Omega_2$$

$$\text{Poss}_{rule}(\omega_1, \omega_2) = \begin{cases} 1 & \text{if } \mu_A(\omega_1) = 0 \\ \min(1, \frac{\mu_B(\omega_2)}{\mu_A(\omega_1)}) & \text{otherwise} \end{cases}, \quad \omega_1 \in \Omega_1, \omega_2 \in \Omega_2$$

and

$$\text{Poss}_{result}(\omega_2) = \sup_{\omega_1 \in \Omega_1} \max(0, (\text{Poss}_{rule}(\omega_1, \omega_2) + \text{Poss}_{datum}(\omega_1) - 1)), \quad \omega_2 \in \Omega_2$$

$$\text{Poss}_{rule}(\omega_1, \omega_2) = \min(1, 1 - \mu_A(\omega_1) + \mu_B(\omega_2)), \quad \omega_1 \in \Omega_1, \omega_2 \in \Omega_2.$$

These methods for sequential combinations based on fuzzy implications are closely related to the concept of T-norms and T-conorms considered in Sect. 9.2. For instance, the operators for computing the membership distribution Poss_{result} on the basis of Poss_{rule} and Poss_{datum} are given by the T-norms T_1, T_2, and T_3 defined in equation (9.18). The decision which of the alternatives one should choose depends on one's attitude towards risk.

10.4 Reasoning with Fuzzy Truth Value

Although we do not consider logical approaches in detail, we describe a method based on fuzzy truth values as it gives some insight into fuzzy set theory.

(10.5) Example. Consider the PNEUMON-I A application [Godo 1988] where the expert states the rule

if (i) community-acquired pneumonia is almost sure
(ii) bacterial disease is possible
(iii) no aspiration is very possible
then [possible]
Enterobacteria are quite possible,

and the observed facts are

(i)′ community-acquired pneumonia is very possible
(ii)′ bacterial disease is almost sure
(iii)′ aspiration is slightly possible

In order to perform inferences, nine expert-defined so-called "fuzzy truth values" are used:

impossible (0, 0)	quite-possible (0.35, 1)
almost-impossible (0, 0.05)	very possible (0.9, 1)
slightly-possible (0, 0.1)	almost sure (0.95, 1)
moderately-possible (0, 0.65)	sure (1, 1)
possible (0, 1)	

The fuzzy truth values ([Baldwin 1979] and [Zadeh 1976]) are described as fuzzy sets of the truth interval [0, 1]. Each fuzzy truth value is described by the two given parameters as shown in Fig. 10.5.

◇

Fuzzy truth values can be viewed as matching degrees expressed by fuzzy sets. Let A and B be two fuzzy sets of Ω. The compatibility of pattern A with respect to the datum B is usually described by the fuzzy truth value

$$\tau_{A,B} : [0,1] \rightarrow [0,1];$$

$$\tau_{A,B}(A) = \begin{cases} 0 & \text{if } \mu_A^{-1}(r) = \emptyset \\ \sup_x\{\mu_B(x) | r = \mu_A(x)\} & \text{otherwise.} \end{cases}$$

So, $\tau_{A,B}$ is the fuzzy set of the possible values of variable $\mu_P(x)$ when the possible values of x are restricted by μ_D.

The characteristics of a "compatibility" measure that presents a "fuzzy degree of truth" can at best clarified by examination of some special cases:

- if pattern P equals datum D (i.e. if $\mu_P = \mu_D$ holds) then the compatibilty of P with D is the special fuzzy set of [0, 1] given by

$$\forall r \in [0,1] : \tau(r) = r.$$

  Note that consequently the compatibility of fuzzy set A with A again is not 1!

- when μ_D is crisp, i.e., when $\forall x \in \Omega : \mu_D(x) \in \{0,1\}$ holds, then

$$\tau_{P,D} = \{\mu_P(x) \text{ such that } \mu_D(x) = 1\}$$

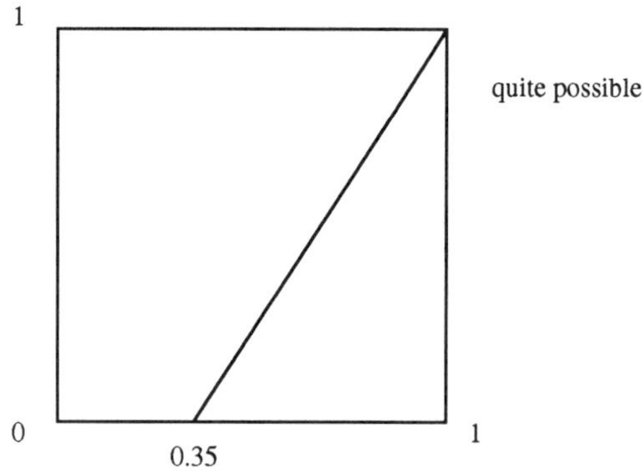

Fig. 10.5. Fuzzy truth value *quite possible*

- when μ_P is crisp, i.e., when $\forall x \in \Omega : \mu_P(x) \in \{0,1\}$ holds, then the definition of τ reduces to

$$\tau(r) = \begin{cases} \text{Poss}(P|D) & \text{if } r = 1 \\ 1 - \text{Nec}(P|D) & \text{if } r = 0 \\ 0 & \text{otherwise.} \end{cases}$$

Generally, τ subsumes the information on possibility measure $\text{Poss}(P|D)$ and necessity measure $\text{Nec}(P|D)$ since it can be shown that

$$\text{Poss}(P|D) = \sup_{r\in[0,1]} \min(r, \tau(r))$$
$$\text{Nec}(P|D) = \inf_{r\in[0,1]} \max(r, 1 - \tau(r))$$

holds.

On the other hand the application of a fuzzy truth value $\tau : [0,1] \to [0,1]$ to a fuzzy set A gives a transformed fuzzy set $A' \stackrel{\text{d}}{=} \tau \circ A$ defined by

$$\mu_{A'}(\omega) = \tau(\mu_A(\omega)) \text{ for all } \omega \in \Omega.$$

Therefore the evaluation of an inference pattern such as

datum	$(X \text{ is } A)$ is τ	
rule	if X is A then Y is B	
consequence		$(Y \text{ is } B)$ is η

can be reduced to the generalized *modus ponens*, where we have

datum	X is $\tau \circ A$
rule	if X is A then Y is B
consequence	Y is B'

Using the compatibility of patterns we could define a fuzzy truth value η by considering the compatibility of B and B'.

Of course, η depends on the fuzzy sets A and B, therefore in MILORD a simpler method is used in which the value η is considered to be independent of A and B and defined in analogy to equation (10.15) by setting

$$\eta : [0,1] \to [0,1]$$
$$\eta(r) = \sup\{\min(\tau(s), I_{\min}(s,r)) | s \in [0,1]\}, \ r \in [0,1]$$

where

$$I_{\min}(s,r) = \begin{cases} 1 & \text{if } s \leq r \\ 0 & \text{otherwise} \end{cases}$$

is the Gödel implication.

Various operations are needed in order to analyse complicated rule structures and so on, such as the *and*, *or*, *not* operations for truth values, the composition of fuzzy truth, and the compatibility calculations. This leads to problems, as the various results do not belong to the term set of truth values (which is a finite set of 9 elements). Therefore a "linguistic approximation" has to be performed in order to keep the operation over this set closed.

(10.5) Example. (Continuation) In our example the inference process is performed in the following steps (details in [Mantaras 1990]).

(a) The compatibility between (1) and (1′) gives: moderately possible
(b) The compatibility between (2) and (2′) gives: almost sure
(c) Negation of (3′) gives: no aspiration is very possible
(d) Compatibility between (3) and (c) gives: possible
(e) [a and b and d] gives: moderately possible
(f) forward reasoning M_I applied to (e) moderately possible and to the truth value of the rule (possible) gives: moderately possible
(g) composition between (f) (moderately possible) and the truth value of the conclusion (quite possible) gives: possible.

◇

10.5 Conclusions

Fuzzy reasoning means reasoning with fuzzy predicates on a precise database whereas *possibilistic reasoning* is a precise reasoning on a situation incompletely described in a database. The first kind of reasoning addresses the handling of vague concepts, and the results are consequently degrees of truth, whereas the second kind deals with uncertainty, therefore the results are degrees of uncertainty.

We classified fuzzy sets as a numerical approach, although one could argue that the whole theory should be embedded in a framework of a suitable multivalued logic, in which the acceptability degrees are viewed as truth values of vague propositions. For example, the degree of (partial) truth of the proposition "15 miles is a long distance" could be interpreted as the acceptability of 15 to be an element of the fuzzy set of "long distances". Zadeh [Zadeh 1979] proposed so-called fuzzy logic as a suitable multivalued logic. In this simple logic the logical operators and, or, not are represented by the following straightforward connectives: If S and T are sentences with $\text{truth}(S) = s$ and $\text{truth}(T) = t$, respectively then we obtain

$$\begin{aligned} \text{truth}(S \ \underline{\text{and}} \ T) &= \min\{s,t\}, \\ \text{truth}(S \ \underline{\text{or}} \ T) &= \max\{s,t\}, \\ \text{truth}(\underline{\text{not}} \ T) &= 1-t. \end{aligned}$$

There are several other criticisms concerning fuzzy sets and possibility theory. The concepts are mainly criticised for their lack of semantics. As a consequence there is a lack of simple techniques for elicitating membership functions and of methods for decision making based on fuzzy sets. These deficiencies have led several researchers, e.g. [Cheeseman 1986], [French 1984], [Haack 1979], and [Lindley 1987], to disqualify the fuzzy set approach.

A generalization of the fuzzy set calculus is given in [Goguen 1967] where lattices are considered as the range of membership functions.

A special property of fuzzy logic is its truth functionality, i.e. the degree of truth of a conjunction, disjunction or negation is a function of the degrees of truth of the elementary propositions involved. For the representation of the operators and and or, other T-norms and T-conorms are often used [Weber 1983]. The choice of the unit interval as the set of truth values is from this point of view somewhat arbitrary, and one should sometimes use other scales which are neither necessarily metric nor totally ordered. For possibility measures which are based on fuzzy sets the same objection holds.

Although the semantics of "classical" fuzzy set theory based on fuzzy logic is quite unsatisfactory there are some hints that the underlying intuition allows us to build efficient systems. The method for approximate reasoning has already been used for the development of so-called fuzzy controllers [Sugeno 1985] which turn out to be very successful in real world applications. Beside the suitability of the domain of application the success of these methods lies in the fact that fuzzy set theory or possibility theory can serve as a correctness preserving approximation of more sophisticated methods. Possibility theory is a good candidate for efficient realizations as it is easy to implement and has connections to vague data and logics and can be also viewed as a method for uncertainty handling. Nevertheless, clear semantics have to be provided. Beside our "measure theoretic" view described in Chap. 2 there are approaches which embed vagueness in suitable logics (such as intutionistic logics or Łukasiewicz logics) or even in probability theory (by considering special likelihood functions).

The main ideas in the field of fuzzy set theory are due to L. Zadeh. The book [Yager 1987] contains a collection of his most important papers including the first papers on fuzzy sets. Possibility theory in AI is described in [Dubois and Prade 1988] and [Smets et al. 1988], and the mathematical basis of fuzzy sets is discussed in the paper [Dubois and Prade 1989]. Recent results on approximate reasoning are contained in the survey paper [Dubois and Prade 1990c]. As a textbook for the classical fuzzy set theory we recommend [Klir and Folger 1989].

Chapter 11

Reasoning with L-Sets

In this chapter we develop a propagation algorithm for the case of vague information. In essence the ordinary set calculus described in Sect. 2.2 is extended to the case of L-sets. The final aim is to find an unknown true value $\omega_0 \in \Omega$ on the basis of a collection of vague data.

One can see the method considered in this chapter as a simple constraint satisfaction problem.

11.1 Knowledge Representation with L-Sets

As pointed out in the previous chapters, we assume that our frame of discernment is the product space Ω^M, i.e. we consider characteristics $X^{(1)}, \ldots, X^{(m)}$ taking their values within finite domains $\Omega^{(i)}$, $i \in M$. Each piece of information is encoded as an L-set of a suitable subspace of Ω^M. Vague observations are represented typically by L-sets of one-dimensional subspaces $\Omega^{(i)}$, $i \in M$, whereas vague rules are considered to be vague observations of non-vague causal relations. This means that a rule can also be represented as an L-set of an at least two-dimensional subspace of Ω^M. We assume furthermore that the qualitative structure of the domain is determined by a simple hypertree $H = (M, \mathfrak{T})$, where $\mathfrak{T}$ is the set of hyperedges. Recall that, according to Remark 7.20, in the general case the elements of M are equivalence classes of characteristics. Each hyperedge $T \in \mathfrak{T}$ corresponds to a subspace Ω^T, $T \subseteq M$ which is subject to one piece of information. If we assume that we have r experts, where expert j is competent for Ω^{T_j}, then we obtain from j a vague relation $R^{T_j} \in \mathcal{F}_L(\Omega^{T_j})$. Note that all experts rely here on the same degrees of acceptability. In order to integrate the information provided by all the experts *within* the L-set calculus, we have to define an L-set which summarizes these pieces of information. The "conjunctive" combination consists in just intersecting the corresponding L-sets. This proposal is justified at least in the case where the origin of the L-sets is as described in Chap. 3. Then the intersection of L-sets is correctness preserving. We obtain as the overall relation

$$R = \bigcap_{j=1}^{r} \widehat{\Pi}_{T_j}^{M}(R^{T_j}). \tag{11.1}$$

Taking into account the impact of observational knowledge, we have to integrate

$$E = \bigcap_{j:j\in M} E^j. \tag{11.2}$$

Finally we obtain the L-set $R \cap E$. Although the reasoning problem is solved in principle with these considerations, serious problems remain. A consulting user is in general not interested in some set $R \cap E \subseteq \Omega^M$ but in restrictions for the values of the different characteristics $X^{(1)}, \ldots, X^{(m)}$, so finally we have to calculate the projections of $R \cap E$ onto the axes $\Omega^{(i)}$, $i \in M$. We finally obtain for every $X^{(i)}$, $i \in M$, some set

$$A^{(i)} \stackrel{\mathrm{d}}{=} \Pi_{\{i\}}^{M}(R \cap E) \tag{11.3}$$

which restricts the set of possible values. (Note that on the formal level we obtain marginal mass distributions $m^{A^{(i)}} \stackrel{\mathrm{d}}{=} \Pi_{\{i\}}^{M}[m^{R\cap E}]$.)

Recall that we are dealing with L-sets, i.e. $E^{(i)} \in \mathcal{F}_L(\Omega^{(i)})$, $i \in M$, $R^T \in \mathcal{F}_L(\Omega^T)$, $T \in \mathfrak{T}$ and $E, R \in \mathcal{F}_L(\Omega^M)$, so the correct interpretation of the result is of importance. It should be based on the level sets. Each level set represents a crisp proposition on the respective level of acceptance. On the higher levels the experts as well as the observer provide more optimistic valuations whereas at the lower levels they provide conservative approximations. So if on the level $l_{\max}$ no consistent element remains, this means *not* that the whole result is inferior, but that its "value" is given by the "width" of the L-set.

The interpretation of L-sets in terms of level sets allows us to obtain not only a sound interpretation of the results of reasoning processes but also an interesting approach to the implementation. From a practical point of view it is sufficient to restrict ourselves to a small number of level sets; most people are not able to distinguish between more than 7 different levels of acceptance, so it is reasonable to consider only a finite number of levels (typically 10) and to perform parallel "crisp" computations on each of them, as we will discuss in detail later on.

11.2 On the Interpretation of Vague Rules

Since an expert or knowledge engineer is usually unable to specify vague relations in the form of multi-dimensional L-sets of the subspaces Ω^T, $T \in \mathfrak{T}$, some kind of "formal" language is needed. Since the *modus ponens* rule of classical logic fits very well with the way humans think about causal dependencies it is useful to provide a linguistic tool related to the well known concept of implication. This should enable an expert to express the assertion that from the fact that the value of some characteristic $X^{(i)}$ is within the set A we can conclude that the value of some other chracteristic $X^{(j)}$ is within some set $B \subseteq \Omega^{(j)}$. In the range of knowledge based systems, for this purpose mostly *rules* of the form

$$\underline{\text{if}}\ X^{(i)} \in A\ \underline{\text{then}}\ X^{(j)} \in B$$

are used. Note, that the premise $(X^{(i)} \in A)$ as well as the conclusion $(X^{(j)} \in B)$ can refer to multiple characteristics, but in order to avoid unnecessary complexity in the notation we restrict ourselves for the moment to the most simple form of rules.

If the value of the characteristic $X^{(i)}$ is not contained in A, then (at least from this rule) nothing can be concluded about the actual value of $X^{(j)}$. Such a rule typically represents structural knowledge, as we saw in Sect. 2.2. If we know, for example, from some observation of the actual problem that the value of characteristic $X^{(i)}$ is within some set E, then the question is, what can we conclude from this information about characteristic $X^{(j)}$. In the case of $E \subseteq A$ the rule should certainly yield $X^{(j)} \in B$, but if $E \cap \overline{A} \neq \emptyset$ then nothing can be concluded.

Up to now we have assumed the sets A, B, E to be *crisp*. This is useful to clarify the underlying intuitive idea. To formalize the process of reasoning we described above, we have to consider a relation $R^{(i,j)}$ on $\Omega^{(i,j)}$. Obviously all $\omega^{\{i,j\}} \in A \times B$ are elements of $R^{(i,j)}$. In addition we have also that all

$$\omega^{\{i,j\}} \in \widehat{\Pi}_{\{i\}}^{\{i,j\}}(\overline{A}).$$

We derive

$$R^{(i,j)} = \left(\widehat{\Pi}_{\{i\}}^{\{i,j\}}(A) \cap \widehat{\Pi}_{\{j\}}^{\{i,j\}}(B)\right) \cup \widehat{\Pi}_{\{i\}}^{\{i,j\}}(\overline{A}). \tag{11.4}$$

The pairs of values contained in $R^{(i,j)}$ are those which are compatible from the expert's point of view.

If the application of such a rule to an observation E is required, then the relation resulting from the rule and the subset of the frame of discernment representing the observation have to be intersected.

(11.1) Example. Once more let us consider our introductory example. Recall that in Sect. 2.2 we translated the original (intuitive) notation into our standard notation for product spaces, i.e.

$$\Omega^{(1)} = \Omega_L = \{\text{z3}, \ldots, \text{rd}\},$$
$$\Omega^{(2)} = \Omega_A = \{\text{wp, np, fb, hc}\},$$
$$\Omega^{(3)} = \Omega_M = \{\text{ad, ac, fi, im}\},$$

so the actual frame of discernment is the product space $\Omega^{\{1,2,3\}}$. The proposition

(P6): The fire department has no equipment for actions at sea

is then encoded by

$$\text{(R6): } \underline{\text{if}}\ X^{(1)} \in \{\text{z3, z2, z1}\}\ \underline{\text{then}}\ X^{(2)} \in \overline{\{\text{fb}\}}\ .$$

That is, we have the relation $R^{(1,2)} \subseteq \Omega^{\{1,2\}}$ given in the table below:

$R^{(1,2)}$	wp	np	fb	hc
z3	1	1	0	1
z2	1	1	0	1
z1	1	1	0	1
ca	1	1	1	1
ld	1	1	1	1
rd	1	1	1	1

where in the table we have the values of the characteristic function of the relation. If we know from some observation that the ship is outside the 3–mile zone (z1), then we have $X^{(1)} \in E^{(1)} = \{\text{z2, z3}\}$. Extended to $\Omega^{\{1,2\}}$, this yields

$\widehat{\Pi}_{\{1\}}^{\{1,2\}}(E^{(1)})$	wp	np	fb	hc
z3	1	1	1	1
z2	1	1	1	1
z3	0	0	0	0
ca	0	0	0	0
ld	0	0	0	0
rd	0	0	0	0

These two results may be the evaluation of an information source and some observation. The conjunctive combination yields the set $R^{(1,2)} \cap \widehat{\Pi}_{\{1\}}^{\{1,2\}}(E^{(1)})$ given in the table below:

$R^{(1,2)} \cap \widehat{\Pi}_{\{1\}}^{\{1,2\}}(E^{(1)})$	wp	np	fb	hc
z3	1	1	0	1
z2	1	1	0	1
z1	0	0	0	0
ca	0	0	0	0
ld	0	0	0	0
rd	0	0	0	0

By its projection to $\Omega^{(2)}$ we obtain $\overline{\{\text{fb}\}}$ as the result. The observation {z1,ca} would yield $\Omega^{(2)}$, i.e. no conclusion is possible.

◇

The application of if ... then rules correspond to *modus ponens*, which is well known from classic logic. Consider, for example, the proposition "All humans are mortal". What we called "observation" corresponds to a statement like "Socrates is human", and the projection of the resulting set $R \cap E$ onto the axis of interest corresponds to the conclusion "Socrates is mortal". Given a relation $R^{(i,j)} \subseteq \Omega^{\{i,j\}}$ and an observation $E^{(i)} \subseteq \Omega^{(i)}$, for the indicator function of the projected intersection we obtain

$$I_{\Pi_{\{j\}}^{\{i,j\}}(R^{(i,j)} \cap \widehat{\Pi}_{\{j\}}^{\{i,j\}}(E^{(i)}))}(\omega^{(j)}) = \max_{\omega^{(i)} \in \Omega^{(i)}} \{\min(I_{E^{(i)}}(\omega^{(i)}), I_{R^{(i,j)}}(\omega^{(i)}, \omega^{(j)}))\}.$$

The situation gets a little more complicated if the antecedent or the conclusion of some rule is a compound expression, i.e. if one rule concerns more than two characteristics. For example

$$\underline{\text{if}}\ X^{(i)} \in A_1\ \underline{\text{and}}\ X^{(j)} \in A_2\ \underline{\text{then}}\ X^{(k)} \in B$$

But in this case we have no difficulties, since we can consider the condition $X^{(i)} \in A_1$ and $X^{(j)} \in A_2$ as

$$(X^{(i)}, X^{(j)}) \in \widehat{\Pi}(A_1) \cap \widehat{\Pi}(A_2)$$

and similarly $X^{(i)} \in A_1$ or $X^{(j)} \in A_2$ as

$$(X^{(i)}, X^{(j)}) \in \widehat{\Pi}(A_1) \cup \widehat{\Pi}(A_2)$$

and handle the rule as described above.

To describe the dependencies in some subspace Ω^{T_j}, $T_j \in \mathfrak{T}$ in general, one rule will not be sufficient, but several rules, each of them concerning a subspace of Ω^{T_j}, will be required. We can imagine that we derive each rule from some subexpert, where the subexperts have to be conjunctively connected, so we have to intersect the single relations after extending them to Ω^{T_j}. Finally we have for every subspace Ω^{T_j} a relation R^{T_j}.

Up to now we have considered only crisp sets as conclusions, and as observations. The construction of a vague relation representing the knowledge encoded by a vague rule

$$\underline{\text{if}}\ X^{(i)}\text{“} \in \text{”}\alpha\ \underline{\text{then}}\ X^{(j)}\text{“} \in \text{”}\beta\ ,$$

where $\alpha \in \mathcal{F}_L(\Omega^{(i)})$, $\beta \in \mathcal{F}_L(\Omega^{(j)})$ in strict analogy to the case of crisp sets would yield

$$R = \widehat{\Pi}_{\{i\}}^{\{i,j\}}(\alpha) \cap \widehat{\Pi}_{\{i\}}^{\{i,j\}}(\beta) \cup \widehat{\Pi}_{\{i\}}^{\{i,j\}}(\text{“complement”}(\alpha))$$

but this definition would lead to serious problems as no complement exists for L-sets.

The vague rule has to be interpreted level by level, where on the more optimistic levels rules from lower levels are also accepted. Thus the level sets R_l, $l \in L$ of the vague relation R are determined by the sets $\rho(l) \subseteq \Omega^{\{1,2\}}$, where

$$\rho(l) \stackrel{\mathrm{d}}{=} \widehat{\Pi}_{\{1\}}^{\{1,2\}}(A_l) \cap \widehat{\Pi}_{\{1\}}^{\{1,2\}}(B_l) \cup \widehat{\Pi}_{\{1\}}^{\{1,2\}}(\overline{A_l}),\ l \in L,$$

by

$$R_l = \bigcap_{l':l' \leq l} \rho(l').$$

(11.3) Example. We may consider the causal relation of locations and competent administration boards. Consider the vague rule

$\underline{\text{if}}$ $X^{(1)} \in$ far_away_from_the_coast $\underline{\text{then}}$ $X^{(2)} \in$ weak

where “far_away_from_the_coast” is represented by the L-set α,

$$\alpha : \Omega \longrightarrow L; \quad \alpha(\omega) = \begin{cases} \textit{black} & \text{if } \omega \in \{\textit{z3},\textit{z2}\} \\ \textit{grey} & \text{if } \omega \in \{\textit{z1}\} \\ \textit{white} & \text{otherwise} \end{cases}$$

and “weak” is represented by the L-set β,

$$\beta : \Omega \longrightarrow L; \quad \beta(\omega) = \begin{cases} \textit{black} & \text{if } w \in \{np\} \\ \textit{grey} & \text{if } \omega \in \{hc\} \\ \textit{white} & \text{otherwise.} \end{cases}$$

We derive the following indicator functions:

ρ(black)	np	wp	fb	hc
z3	1	0	0	0
z2	1	0	0	0
z1	1	1	1	1
ca	1	1	1	1
ld	1	1	1	1
rd	1	1	1	1

ρ(grey)	np	wp	fb	hc
z3	1	0	0	1
z2	1	1	0	1
z1	1	0	0	1
ca	1	1	1	1
ld	1	1	1	1
rd	1	1	1	1

R	np	wp	fb	hc
z3	b	w	w	g
z2	b	w	w	g
z1	b	w	w	b
ca	b	b	b	b
ld	b	b	b	b
rd	b	b	b	b

From an observation "the ship is near_the_coast" represented by

$$\alpha' : \Omega \longrightarrow L; \alpha'(\omega) = \begin{cases} \textit{black} & \text{if } \omega \in \{\textit{z1}\} \\ \textit{grey} & \text{if } \omega \in \{\textit{z2},\textit{ca}\} \\ \textit{white} & \text{otherwise,} \end{cases}$$

we obtain

$$\Pi^{\{1,2\}}_{\{2\}}(R \cap \widehat{\Pi}^{\{1,2\}}_{\{1\}}(\alpha'))(\omega) = \begin{cases} \textit{black} & \text{if } \omega \in \{\textit{np},\textit{hc}\} \\ \textit{white} & \text{otherwise.} \end{cases}$$

◇

Note that with this combination of the relation R we obtain $\Pi^{\{i,j\}}_{\{i\}}(R_l) = \Omega^{(i)}$ and $\Pi^{\{i,j\}}_{\{j\}}(R_l) = \Omega^{(j)}$ for all $l \in L$ and thus $\Pi^{\{i,j\}}_{\{i\}}(R_l) = \mathbb{1}_{\Omega^{(i)}}$ and $\Pi^{\{i,j\}}_{\{j\}}(R_l) = \mathbb{1}_{\Omega^{(j)}}$.

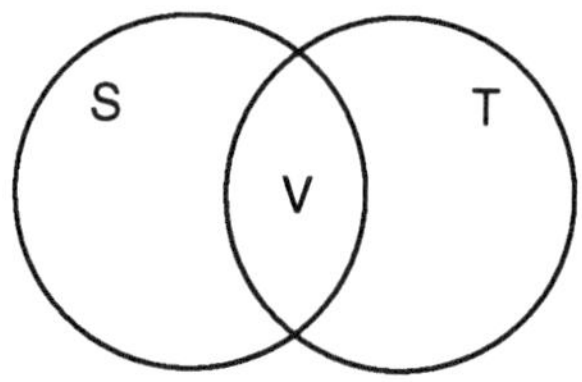

Fig. 11.1. Illustration of Theorem 11.5

11.3 L-Sets on Product Spaces

In this section we reconsider L-sets defined on product spaces from the special point of view of knowledge representation. One important question in this area concerns the relevance of knowledge about some subspace Ω^S for the subspace Ω^T and the combination of knowledge concerning different subspaces. We provide two theorems solving these problems in their most general form and illustrate them by an example.

First let us consider the case of some L-set μ of Ω^S, where Ω^S is a subspace of the frame of discernment Ω^M, $M = \{1, \ldots, m\}$, $S \subseteq M$. We now ask for an L-set η of Ω^T, $T \subseteq M$, representing the information about Ω^T that results from our knowledge of Ω^S. Let $V = S \cap T$ as in Fig. 11.1. We can imagine that the portion of the knowledge about Ω^S relevant for Ω^T is the portion concerning Ω^V.

There are two ways to construct the set $\eta \in \mathcal{F}_L(\Omega^T)$, as the first theorem shows, firstly to project μ to Ω^V and then to extend the result of Ω^T, or secondly to extend μ to $\Omega^{S\cup T}$ and then to project the result to Ω^T. They lead to the same set.

(11.5) Theorem. *Let Ω^M, $M = \{1, \ldots, m\}$ be the frame of discernment and let S, T and V subsets of M with $S \cap T = V$. Then*

$$\widehat{\Pi}_V^T(\Pi_V^S(\mu)) = \Pi_T^{S\cup T}(\widehat{\Pi}_S^{S\cup T}(\mu))$$

holds for all $\mu \in \mathcal{F}_L(\Omega^M)$.

(11.6) Example. In order to illustrate the above theorem let us once more consider our introductory example, i.e. the product space is given by the cartesian

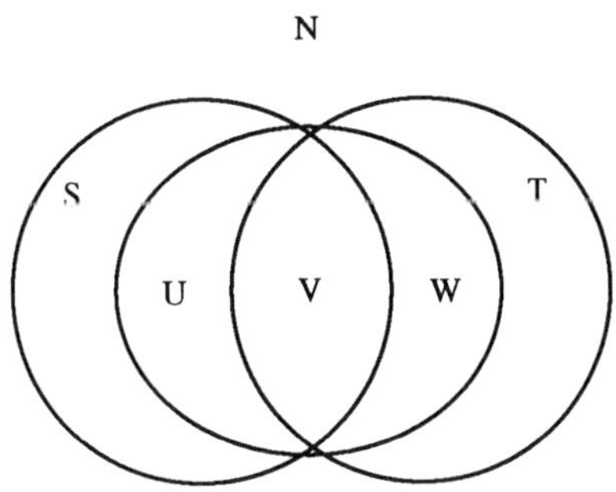

Fig. 11.2. Illustration of Theorem 11.7

product of the three sets $\Omega^{(1)}$, $\Omega^{(2)}$, $\Omega^{(3)}$. Let

$$\mu = \{(z3,fb),(z3,wp),(z3,hc),(z2,fb),(z1,wp),(z1,fb)\}.$$

We obtain

$$\Pi^{\{1,2\}}_{\{2\}}(\mu) = \{fb,wp,hc\},$$

$$\widehat{\Pi}^{\{2,3\}}_{\{2\}}(\{fb,wp,hc\}) = \{(fb,ad),(fb,ac),(fb,fi),(fb,im),\ldots,(hc,im)\},$$

and

$$\widehat{\Pi}^{\{1,2,3\}}_{\{1,2\}}(\mu) = \{(z3,fb,ad),(z3,fb,fi),(z3,fb,im),\ldots,(z1,fb,im)\},$$

$$\Pi^{\{1,2,3\}}_{\{2,3\}}(\{\{(z3,fb,ad),(z3,fb,fi),(z3,fb,im),\ldots,(z1,fb,im)\}\}) = \{\ (fb,ad),(fb,ac),(fb,fi),(fb,im),\ldots,(hc,im)\}$$

◇

The situation gets more complicated if we consider two sets $\mu \in \mathcal{F}_L(\Omega^S)$ and $\eta \in \mathcal{F}_L(\Omega^T)$ and ask for their information within some subspace Ω^N, $N \subseteq M$ (see Fig. 11.2). The next theorem gives an answer to this question.

(11.7) Theorem. *Let Ω^M, $M = \{1,\ldots m\}$ be the frame of discernment and let $S,T,U,V,W,N \subseteq M$, with $S \cap T = V$, $U \subseteq S \cap \overline{T}$, $W \subseteq T \cap \overline{S}$ and $N = U \cup V \cup W$. For $\mu \in \mathcal{F}_L(\Omega^S)$ and $\eta \in \mathcal{F}_L(\Omega^T)$ we have*

$$\widehat{\Pi}^N_{U\cup V}(\Pi^S_{U\cup V}(\mu)) \cap \widehat{\Pi}^N_{V\cup W}(\Pi^T_{V\cup W}(\eta\ )) = \Pi^{S\cup T}_N(\widehat{\Pi}^{S\cup T}_S(\mu) \cap \widehat{\Pi}^{S\cup T}_T(\eta\ ))$$

(11.8) *Remark.*

(a) In case of $N \cap T = \emptyset$ Theorem 11.7 reduces to

$$\Pi_N^S(\mu) = \Pi_N^{S\cup T}(\widehat{\Pi}_S^{S\cup T}(\mu) \cap \widehat{\Pi}_T^{S\cup T}(\eta\ ))$$

if there is some ω_T with $\eta\ (\omega_T) = l_{max}$.

(b) In case of $N = S$ we obtain

$$\mu \cap \widehat{\Pi}_V^S(\Pi_V^T(\eta\ )) = \Pi_S^{S\cup T}(\widehat{\Pi}_S^{S\cap T}(\mu) \cap \widehat{\Pi}_T^{S\cup T}(\eta\ ))$$

and if $N \cap T = \emptyset$ we derive

$$\mu = \Pi_N^{S\cup T}(\widehat{\Pi}_S^{S\cup T}(\mu) \cap \widehat{\Pi}_T^{S\cup T}(\eta\ ))$$

○

(11.9) Example. Once more let us consider our introductory example. Since $R_{\theta_1} \subseteq \Omega^{\{1,2\}}$, $R_{\theta_2} \subseteq \Omega^{\{2,3\}}$ from Theorem 11.7 we obtain

$$\Pi_{\{1,2\}}^{\{1,2,3\}}(\widehat{\Pi}_{\{1,2\}}^{\{1,2,3\}}(R_{\theta_1}) \cap \widehat{\Pi}_{\{2,3\}}^{\{1,2,3\}}(R_{\theta_2})) = R_{\theta_1} \cap \widehat{\Pi}_{\{2\}}^{\{1,2\}}(\Pi_{\{2\}}^{\{2,3\}}(R_{\theta_2}))$$

Note that in this case in terms of Theorem 11.7 we have

$$S = \{1,2\}, \quad T = \{2,3\}, \quad U = \{1\}, \quad V = \{2\}, \quad W = \emptyset,$$

which means $N = U \cap V \cap W = S = \{1,2\}$, so Remark 11.8 takes effect. ◇

11.4 Local Computation of Marginal L-Sets

As we know from Sect. 11.1, we have to calculate the intersection of the evaluation of the information source with respect to the logical expression 1 <u>and</u>... <u>and</u> r and the evaluation of the (compound) observation E. Let

$$R \stackrel{\mathrm{d}}{=} \bigcap_{j=1}^{r} \widehat{\Pi}_{T_j}^M(R^{T_j}),$$

$$E \stackrel{\mathrm{d}}{=} \bigcap_{x:x\in M} \widehat{\Pi}_{\{x\}}^M(E^x).$$

Let $H = (M, \mathfrak{T})$ be the underlying simple hypertree. We first consider some small hypertree defined by $M = \{a,b,c,d,f,g,h\}$ and $\mathfrak{T} = \{U,V,W\}$, where $U = \{a,b,d\}$, $V = \{c,d,g\}$, $W = \{d,f,h\}$. This simple hypertree has the structure depicted in Fig. 11.3. The square boxes denote the hyperedges.

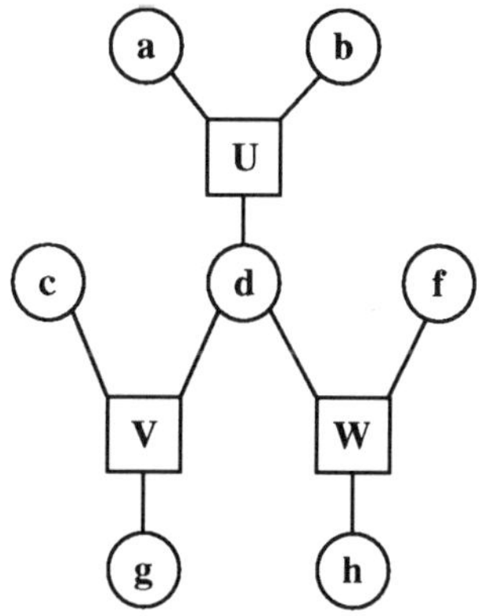

Fig. 11.3. A simple hypertree

The final goal is to represent the hypertree by a network of link and node processors, communicating along the connecting lines. In the sequel we assume the observations E^x, $x \in M$, to be available at the node processors and the relations R^T, $T \in \mathfrak{T}$, at the link processors. Recall again that according to Remark 7.20 in the general case the elements of M are equivalence classes of characteristics. We derive

$$
\begin{aligned}
&\Pi_d^M(R \cap E) \\
&\quad = \Pi_d^M\left(\bigcap_{x\in M} \widehat{\Pi}_x^M(E^x) \cap \bigcap_{T\in\mathfrak{T}} \widehat{\Pi}_T^M(R^T)\right) \\
&\quad = \Pi_d^M\left(\widehat{\Pi}_d^M(E^d) \cap \bigcap_{T\in\mathfrak{T}} \widehat{\Pi}_T^M(R^T) \cap \bigcap_{x:d\neq x\in T} \widehat{\Pi}_x^M(E^x)\right) \\
&\quad = \Pi_d^M\left(\widehat{\Pi}_d^M(E^d) \cap \bigcap_{T\in\mathfrak{T}} \widehat{\Pi}_T^M\left\{R^T \cap \bigcap_{x:d\neq x\in T} \widehat{\Pi}_x^T(E^x)\right\}\right) \\
&\quad = \Pi_d^M\left(\widehat{\Pi}_d^M(E^d) \cap \bigcap_{T\in\mathfrak{T}} \widehat{\Pi}_T^M\left\{R^T \cap \widehat{\Pi}_{T-d}^T\Big(\bigcap_{x:d\neq x\in T} \widehat{\Pi}_x^{T-d}(E^x)\Big)\right\}\right)
\end{aligned}
$$

$$\stackrel{(*)}{=} E^d \cap \Pi_d^M \left(\bigcap_{T \in \mathfrak{T}} \widehat{\Pi}_T^M \left\{ R^T \cap \widehat{\Pi}_{T-d}^T (\bigcap_{x: d \neq x \in T} \widehat{\Pi}_x^{T-d}(E^x)) \right\} \right)$$

$$\stackrel{(**)}{=} E^d \cap \bigcap_{T \in \mathfrak{T}} \Pi_d^T \left(R^T \cap \widehat{\Pi}_{T-d}^T (\bigcap_{x: d \neq x \in T} \widehat{\Pi}_x^{T-d}(E^x)) \right) \tag{11.5}$$

The transformation $(*)$ makes use of Theorem 11.7. In terms of this theorem we have

$$\mu = E^d \subseteq \Omega^d, \eta = \bigcap_{T \in \mathfrak{T}} \widehat{\Pi}_T^M \left(R^T \cap \widehat{\Pi}_{T-d}^T (\bigcap_{x: d \neq x \in T} \widehat{\Pi}_x^{T-d}(E^x)) \right) \subseteq \Omega^M$$

and obtain

$$\Pi_d^M(\widehat{\Pi}_d^M(\mu) \cap \widehat{\Pi}_M^M(\eta)) = \mu \cap \widehat{\Pi}_d^d(\widehat{\Pi}_d^M(\eta)) = \mu \cap \Pi_d^M(\eta).$$

The transformation $(**)$ is similar. We have

$$\mu_T = (R^T \cap \widehat{\Pi}_{T-d}^T (\bigcap_{x: d \neq x \in T} \widehat{\Pi}_x^{T-d}(E^x)))$$

and derive

$$\begin{aligned} &\Pi_d^M(\widehat{\Pi}_U^M(\mu_U) \cap \widehat{\Pi}_V^M(\mu_V) \cap \widehat{\Pi}_W^M(\mu_W)) \\ &\quad = \widehat{\Pi}_d^d(\Pi_d^U(\mu_U) \cap \Pi_d^V(\mu_V) \cap \Pi_d^W(\mu_W)) \\ &\quad = \Pi_d^U(\mu_U) \cap \Pi_d^V(\mu_V) \cap \Pi_d^W(\mu_W). \end{aligned}$$

Defining for $d \in T$

$$B^{T-d} \stackrel{\text{d}}{=} \bigcap_{x: d \neq x \in T} \widehat{\Pi}_x^{T-d}(E^x)\ , \quad T \in \mathfrak{T}\ , \tag{11.6}$$

$$A_{dT}^+ \stackrel{\text{d}}{=} \Pi_d^T(R^T \cap \widehat{\Pi}_{T-d}^T(B^{T-d}))\ , \quad T \in \mathfrak{T}\ , \tag{11.7}$$

from equation (11.5) we obtain

$$B^d = \Pi_d^M(R \cap E) = E^d \cap \bigcap_{T \in \mathfrak{T}} A_{dT}^+\ . \tag{11.8}$$

In our introductory hypergraph A_{dT}^+ summarizes the impact of all knowledge from the subgraph H_{dT}^+ on X^d. The node processor X^d receives messages from all neighboring link processors. To provide such messages the link processor R^T needs the sets E^x, $d \neq x \in T$.

Now imagine that the hypergraph considered is only a cutout of some more comprehensive hypergraph. Then the link processor R^T needs instead of E^x, $d \neq x \in T$, some message A^-_{dT} summarizing the impact of the subgraph H^-_{xT}.

Assuming

$$A^-_{xT} = \Pi_x^{M^-_{xT}}\Big(\bigcap_{y:y\in M^-_{xT}} \widehat{\Pi}_y^{M^-_{xT}}(E^y) \cap \bigcap_{T':T'\in \mathfrak{T}^-_{xT}} \widehat{\Pi}_{T'}^{M^-_{xT}}(R^{T'})\Big)$$

we obtain

$$\begin{aligned}
B^{T-d} &= \bigcap_{x:d\neq x\in T} \widehat{\Pi}_x^{T-d}(A^-_{xT}) \\
&= \bigcap_{x:d\neq x\in T} \widehat{\Pi}_x^{T-d}\left\{ \Pi_x^{M^-_{xT}} \left(\bigcap_{y\in M^-_{xT}} \widehat{\Pi}_y^{M^-_{xT}}(E^y) \cap \bigcap_{T'\in \mathfrak{T}^-_{xT}} \widehat{\Pi}_{T'}^{M^-_{xT}}(R^{T'}) \right)\right\} \\
&\overset{(*)}{=} \Pi_{T-d}^{M^+_{dT}} \left\{ \bigcap_{x:d\neq x\in T} \widehat{\Pi}_{M^+_{xT}}^{M^+_{dT}} \left(\bigcap_{x'\in M^-_{xT}} \widehat{\Pi}_{x'}^{M^+_{dT}}(E^{x'}) \cap \bigcap_{T'\in \mathfrak{T}^-_{xT}} \widehat{\Pi}_{T'}^{M^+_{dT}}(R^{T'}) \right)\right\} \\
&\overset{(**)}{=} \Pi_{T-d}^{M^+_{dT}} \left(\bigcap_{x'\in M^+_{dT}} \widehat{\Pi}_{x'}^{M^+_{dT}}(E^{x'}) \cap \bigcap_{T'\in \mathfrak{T}^+_{dT}} \widehat{\Pi}_{T'}^{M^+_{dT}}(R^{T'}) \right)
\end{aligned}$$

(*) From Theorem 7.26, we know $M^+_{dT} = \bigcup_{x:d\neq x\in T} M^-_{xT}$ and $\mathfrak{T}^+_{dT} = \bigcup_{x:d\neq x\in T} \mathfrak{T}^-_{xT}$.

(**) Compare Theorem 11.7.

So we derive

$$\begin{aligned}
&A^+_{dT} \\
&= \Pi_d^T(R^T \cap \widehat{\Pi}_{T-d}^T(B^{T-d})) \\
&= \Pi_d^T(R^T \cap \widehat{\Pi}_{T-d}^T(\Pi_{T-d}^{M^+_{dT}}(\bigcap_{y\in M^+_{dT}} \widehat{\Pi}_y^{M^+_{dT}}(E^y) \cap \bigcap_{T'\in \mathfrak{T}^+_{dT}} \widehat{\Pi}_{T'}^{M^+_{dT}}(R^{T'})))) \\
&\overset{(*)}{=} \Pi_d^T(\widehat{\Pi}_T^{\widetilde{M}^+_{dT}}(R^T) \cap \\
&\qquad \widehat{\Pi}_{M^+_{dT}}^{\widetilde{M}^+_{dT}}(\bigcap_{y:y\in M^+_{dT}} \widehat{\Pi}_y^{M^+_{dT}}(E^y) \cap \bigcap_{T':T'\in M^+_{dT}} \widehat{\Pi}_{T'}^{M^+_{dT}}(R^{T'})))) \\
&\overset{(**)}{=} \Pi_d^{\widetilde{M}^+_{dT}}(\bigcap_{y:y\in M^+_{dT}} \widehat{\Pi}_y^{M^+_{dT}}(E^y) \cap \bigcap_{T':T'\in M^+_{dT}} \widehat{\Pi}_{T'}^{M^+_{dT}}(R^{T'}))
\end{aligned}$$

(*) Again, from Theorem 7.26, in connection with Remark 7.25, we have $M^+_{dT} \cup T = \widetilde{M}^+_{dT}$.

(∗∗) We have also $\mathfrak{T}_{dT}^{+} \cup \{T\} = \widetilde{\mathfrak{T}}_{dT}^{+}$.

After these preliminary exertions we can now calculate the final result

$$\begin{aligned}
B^d &= E^d \cap \bigcap_{T:d\in T} A_{dT}^{+} \\
&= E^d \cap \bigcap_{T:d\in T} \Pi_d^{\widetilde{M}_{dT}^{+}} \left\{ \bigcap_{x:x\in M_{dT}^{+}} \widehat{\Pi}_x^{\widetilde{M}_{dT}^{+}}(E^x) \cap \bigcap_{T':T'\in\widetilde{\mathfrak{T}}_{dT}^{+}} \widehat{\Pi}_{T'}^{\widetilde{M}_{dT}^{+}}(R^{T'}) \right\} \\
&\overset{(*)}{=} E^d \cap \Pi_d^M \left(\bigcap_{T:d\in T} \widehat{\Pi}_{\widetilde{M}_{dT}^{+}}^{M} \bigcap_{x:x\in M_{dT}^{+}} \widehat{\Pi}_x^{\widetilde{M}_{dT}^{+}}(E^x) \cap \bigcap_{T':T'\in\widetilde{\mathfrak{T}}_{dT}^{+}} \widehat{\Pi}_{T'}^{\widetilde{M}_{dT}^{+}}(R^{T'}) \right) \\
&= E^d \cap \Pi_d^M \left\{ \bigcap_{x:x\in M-d} \widehat{\Pi}_x^{M}(E^x) \cap \bigcap_{T':T'\in\mathfrak{T}} \widehat{\Pi}_{T'}^{M}(R^{T'}) \right\} \\
&= \Pi_d^M \left\{ \bigcap_{x:x\in M} \widehat{\Pi}_x^{M}(E^x) \cap \bigcap_{T':T'\in\mathfrak{T}} \widehat{\Pi}_{T'}^{M}(R^{T'}) \right\}
\end{aligned} \tag{11.12}$$

(∗) From Theorem 7.26 we know $\bigcup_{T:d\in T} \widetilde{M}_{dT}^{+} = M$ and $\bigcup_{T:d\in T} \widetilde{\mathfrak{T}}_{dT}^{+} = \mathfrak{T}$.

As this result shows, we can calculate $\Pi_d^M(R \cap E)$ in large hypertrees by the same scheme as in our small example, if the messages the link processors obtain obey the introductory assumption. It remains to show that a node processor is able to provide the messages its neighboring link processor needs, but only by the messages it receives.

If X^x is a leaf node, then we have $M_{xT}^{-} = \{x\}$, $\mathfrak{T}_{xT}^{-} = \emptyset$, and $M_{xT}^{-} = x$, so we obtain

$$\begin{aligned}
A_{xT}^{-} &= \Pi_x^x \left(\bigcap_{x'\in\{x\}} \widehat{\Pi}_{x'}^{x}(E^{x'}) \cap \bigcap_{T'\in\emptyset} \widehat{\Pi}_{T'}^{x}(R^{T'}) \right) \\
&= E^x.
\end{aligned}$$

We have

$$\begin{aligned}
A_{dT'}^{-} &= E^d \cap \bigcap_{T:d\in T\neq T'} A_{dT}^{+} \\
&= \Pi_d^{M_{dT'}^{-}} \left(\bigcap_{x'\in M_{dT'}^{-}} \widehat{\Pi}_{x'}^{M_{dT'}^{-}}(E^{x'}) \cap \bigcap_{T''\in\mathfrak{T}_{dT'}^{-}} \widehat{\Pi}_{T''}^{M_{dT}^{-}}(R^{T''}) \right)
\end{aligned} \tag{11.13}$$

which indeed obeys our assumptions. So it is valid at the leaf nodes and remains valid throughout the propagation process.

Once more it has to be emphazised that the propagation process is *no* relaxation process, since the number of propagation steps is finite and well determined. There are no feedbacks, since the message

$$A^{-}_{dV} = E^{d} \cap \bigcap_{T:d \in T \neq V} A^{+}_{dT} \tag{11.14}$$

from the node processor X^d to the link processor R^V does *not* depend on the message A^{+}_{dT} being sent in the reverse direction. The same holds for the reverse direction

$$\begin{aligned} A^{+}_{dV} &= \Pi^{V}_{d}(R^{V} \cap \widehat{\Pi}^{V}_{V-d}(B^{V-d})) \\ B^{V-d} &= \bigcap_{x:d \neq x \in V} \widehat{\Pi}^{V-d}_{x}(A^{-}_{xV}) \ , \end{aligned} \tag{11.15}$$

A^{+}_{dV} does *not* depend on A^{-}_{dV}.

11.5 The Propagation Algorithm

The algorithm is formulated on the basis of independent node and link processors, performing local computations, using the information they store and the messages they receive, to obtain the information needed by the user and the messages they have to send to their neighbors. The node and link processors are of identical structure, and the only thing that differs is the number of neighbors and the set of possible values for the characteristics they deal with. In the case of compound characteristics these sets are product spaces. The "dialogue" with the user we can also imagine as the exchange of messages: the user sends his observation and the node processor sends its result.

First let us have a look at the structure and local algorithm of some node processor X^d, $d \in M$. Assume there are three neighboring link processors R^U, R^V, R^W, $U, V, W \in \mathfrak{T}$. Then the node processor has for each of these link processors one pair of "ports", i.e. an input and an output port. The input ports receives the messages A^{+}_{dT}, $T = U, V, W$, whereas the output port sends the message A^{-}_{dT}, $T = U, V, W$. For the communication with the user we have an input port to receive the observation E^d and to send the result $\Pi^{M}_{d}(R \cap E)$. Figure 11.4 shows the structure of such a node in principle.

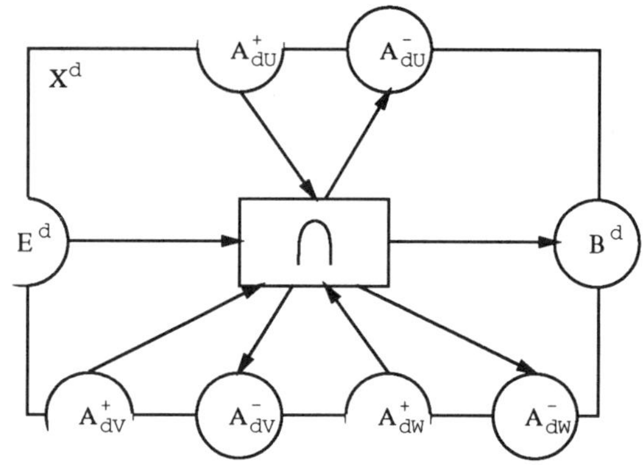

Fig. 11.4. The structure of a node processor

The local algorithm is very simple:

- If the node processor X^d receives a message A^+_{dT} from the link processor R^T
  (i) it updates all messages A^-_{dT} and sends them to its neighboring link processors (see equation 11.14),
  (ii) it updates the result B^d and sends it to the user (i.e. shows it numerically or graphically on the screen).

The structure as well as the algorithm are analogous, if the node X^d has more than 3 neighbors. If the node is a leaf of the hypertree, i.e. it has only one neighboring link processor, then the structure is as pointed out, and the only special feature (covered by the algorithm above) is the fact that receiving a message from this one neighbor will, except for a modified output, cause no further flow of messages.

The structure of a link processor is slightly more complicated. For the sake of simplicity we consider a link processor R^U with three neighboring node processors X^a, X^b, X^d. Like a node processor a link processor has for each of its neighbors a pair of input and output ports. They receive the messages A^-_{xU} from the neighboring node processor X^x, $x = a, b, d$, and send the message A^+_{xU}. We assume the process of knowledge acquisition to be completed, i.e. the relation R^U is determined and stored at the link processor so no further communication with the user is required. The structure of the link processor R^U is shown in Fig. 11.5. The local algorithm is as follows:

- If the link processor R^T receives the message A^-_{dT} from the link processor X^d, then it updates the message $A^+_{dT'}$, $T' \neq T$ and sends them to its neighboring node processors (see equation 11.15).

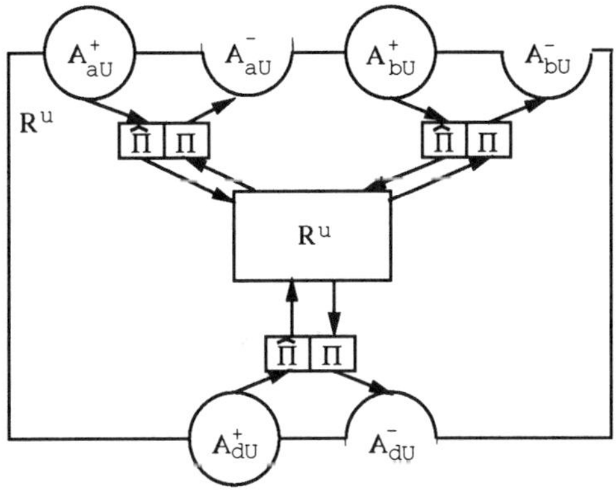

Fig. 11.5. Structure of a link processor

In this way, each processor works locally, controlled only by its local algorithm, responding only to messages it receives. But one thing has to be emphasized: although several messages may flow through the hypertree, each processor processes exactly *one* message at a time.

In general we start with $E^x = \Omega^x$ for all nodes X^x, which represents the state of total absence of observations. Actually we should have $\Pi^T_x(R^T) = \Omega^x$, $x \in T$, $T \in \mathfrak{T}$, for all x, T, otherwise Ω^x should be restricted, since in this case there are some elements of Ω^x which are a priori impossible, but in general it is neither realistic nor desirable to have such restrictions, so after the knowledge acquisition the network of processors should be "initialized" to get the "a priori restriction" of the characteristics. For this purpose all node processors X^x initially having the observation $E^x = \Omega^x$ should send this message to *all* their neighbors. (For leaf nodes this is correct, as we see directly from our effort to prove the algorithms for observation of non-leaf nodes X^x. Imagine that there is a leaf node $X^{x'}$ concerning a virtual subspace $\Omega^{x'} = \Omega^x$ connected with X^x by a virtual relation $R^{T'}$, $T' = \{x, x'\}$ defined by

$$R^{T'}(\omega^{x'}, \omega^x) = \begin{cases} 1 & \text{if } \omega^{x'} = \omega^x \\ 0 & \text{otherwise} \end{cases} ,$$

representing identity. Imagine the user message to X^x to be a message of $X^{x'}$ which obtains, as a leaf node, the user message.).

After this initialization procedure, each observation causes a flow of messages through the network, but it has to be emphasized that the number of messages is determined. As we pointed out before, there are no feedbacks, and the longest path a message has to pass is given by the longest path in the network, which is its diameter.

11.6 Aspects of Implementation

The implementation of the L-set reasoning algorithm presented in the previous sections should be based on an object oriented environment. The realization of the node and link processors by objects of some appropriate object class is obviously very efficient, especially with respect to the ability of "objects" to send and receive messages and to compute these messages by appropriate "methods". Moreover the implementation on independently working "processors" offers a possibility to parallelize the computations.

The representation of fuzzy sets is done efficiently by the use of some kind of layered architeture, i.e. we assume a finite and relatively small number of levels of acceptance to be sufficient and perform the reasoning process independently on each layer with crisp sets. This kind of representation of fuzzy sets already turned out to be very useful during the implementation of a software tool for statistics with vague data [Kruse and Meyer 1987]. Human experts as well as human users are in general unable to distinguish between more than 7 different levels of acceptance. Such layered architecture offers an additional possibility for parallelization, since each layer works independently of the others. At this point there should be a remark on the propagation algorithm. If a processor receives some messages and updating the messages to be sent yields no modification, then these messages should be omitted, since they would only yield a useless flow of messages without any new information. In contrast to the approach based on probability theory and belief functions, in the case of L-sets this can happen often. If the propagation algorithm is used to control technical processes directly, then the different results at the nodes will immediately influence the steering facilities. In such a case the surface of the system is subordinated, but if human decision makers are supported by the system it is of fundamantal importance. In any case there should be a graphical surface both for the expert during knowledge acquisition and for the user during consultation.

Let us first consider the process of knowledge acquisition. This process should be performed in two steps. Firstly the qualitative structure of the characteristics of interest should be determined. That means the expert generates "nodes" of interest and inserts "links" which, from this point of view, are direct influences. One possibility is here to prohibit cyclic dependencies by not allowing the insertion of such "short-cut" links. The other possibility is to allow cyclic dependencies, but to require that, before the quantitive knowledge

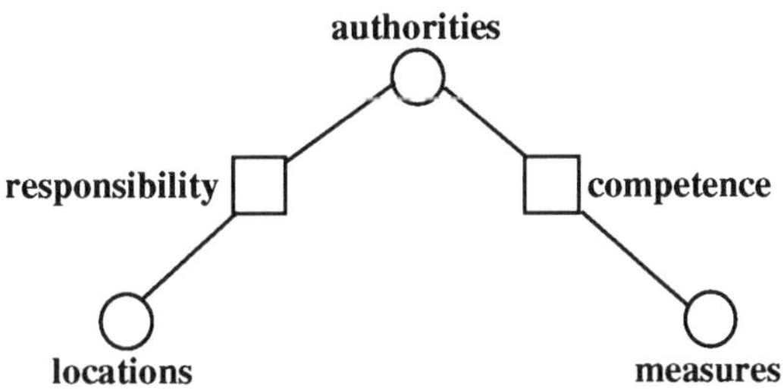

Fig. 11.6. The processor network for the introductory example

can be acquired, the cycles are removed, either automatically or by direct user intervention. It has to be emphasized that the joining of nodes has to be paid for with a growing complexity, since the representation of two spaces Ω^{T_1} and Ω^{T_2} requires a capacity of

$$\prod_{i \in T_1} |\Omega^{(i)}| + \prod_{i \in T_2} |\Omega^{(i)}|$$

but the representation of $\Omega^{T_1 \cup T_2}$ requires

$$\prod_{i \in T_1} |\Omega^{(1)}| \cdot \prod_{i \in T_2} |\Omega^{(1)}|.$$

In our example the result of this first step of the knowledge acquisition is the simple hypertree depicted in Fig. 11.6.

The second step concerns the quantification of the dependencies. That means the expert has to provide the relations for the links. In general he will use rules to determine these relations. Recall that the interpretation of rules was discussed in detail in Sect. 11.2. So each link processor has a relation rule store, accessible for knowledge acquisition. There should be a rule editor supporting the formulation of rules.

Having defined such a language, rules like

$$\underline{\text{if}}\ X^{(1)} \in \text{far\_away\_from\_the\_coast}\ \underline{\text{then}}\ X^{(2)} \in \text{weak}$$

can be used to determined causal relations on the various levels. This concept allows a modularization of the knowledge, since every rule is attached to that link processor it concerns. So the addition, removal or modification of rules can occur without problems.

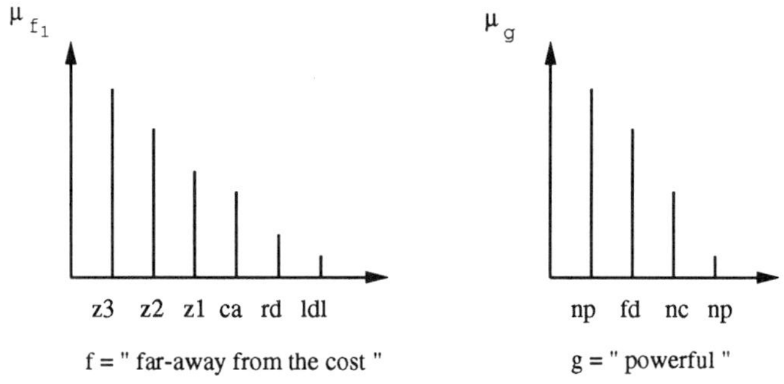

Fig. 11.7. Typical vague observations

The same aspects are of importance for the consulting user. He or she needs a graphical representation to understand the flow of influence. The interpretation of his observation he can attach to the node processors. For this purpose he should be able to use the language determined for the formulation of rules. For example his observation may be

$$X^{(1)} \in \underline{\text{not}}\ \text{near\_the\_coast}.$$

As the result he should get a graphical presentation of the projection of the resulting set onto each axis of interest. The graphical presentation should support the correct interpretation of the results by emphazising the level set interpretation of the L-sets.

(11.10) *Remark.* To provide a term "not" in connection with L-sets is very dangerous, since the complementation of L-sets is a quite crucial point of the L-set theory. ○

Chapter 12

Probability Based Models

As argued in the preceding parts of this book, *probability theory* offers a theoretically sound model for representing uncertainty and for considering it in reasoning techniques. Just in the last few years a revival of using probability theory in representing uncertainty has taken place. It was given considerable insight into the application of probability theory and pointed out some misconceptions about its applicability. Also, new theoretical results from statistics and probability theory present arguments for the utility of probabilities for reasoning.

Considering this important role that probability theory today plays for uncertain reasoning, the aim of this chapter is the examination and discussion of proposals for probabilistic reasoning published in the literature. Further, implementations and applications of such proposals are introduced.

In particular, in the first section we start with a discussion of general knowledge representation aspects in the framework of probabilities. Beyond the different possibilities for interpreting expert rules we consider the fact that, in general, conditional probability measures may be used as diagnostic or as predictive measures.

In Sect. 12.2 we continue with really simple models that may be viewed as *straightforward approaches* for making use of probabilities in rule-based uncertain reasoning. In particular, we introduce and discuss the inference model of Ishizuka et al. A characteristic feature of their approach is that diagnostic probabilistic reasoning can be performed on the basis of *explicit expert rules*. Further, the ideas of Adams are discussed. Adams proposed a model that in its basics can be mapped to the certainty factor approach of Buchanan and Shortliffe. So the (restrictive) assumptions made in the certainty factor approach and some inconsistencies are made visible.

The approach of "inference networks" proposed by Duda et al. is examined in Sect. 12.3. In this model expert rules are interpreted as directed links labelled with so-called *likelihood ratios*. As a concrete expert system based on inference networks we present the system PROSPECTOR.

The approach looked at in Sect. 12.4 is that of "decomposable graphic models" developed by Pearl, Lauritzen, Spiegelhalter et al. A characteristic feature

feature of this model is that uncertainty and belief is propagated through a network by local operations only: each node of the network is viewed as a single processor which exchanges messages with its neighbor nodes. A prototype expert system based on this model is the system MUNIN developed by Andreassen et al. and introduced in Sect. 12.4.2. In Sect. 12.4.3 the expert system shell HUGIN is introduced. Section 12.5 describes a model also based on dependency networks which relies, in contrast to the HUGIN-approach, on a constraint satisfaction point of view. So, although the algorithm turns out to be equivalent to that described in [Pearl 1988], its implementation is presented in detail. A general discussion of the probabilistic models can be found in the conclusions in Sect. 12.6.

12.1 The Interpretation of Rules

For the interpretation of the quantity p attached to an expert rule like

$$\underline{\text{if}}\ A\ \underline{\text{then}}\ B\ \underline{\text{with}}\ p$$

four different possibilities may be of interest.

First of all, p can be interpreted as a *conditional probability*, i.e. $p = P(B|A)$. This interpretation is used in the network approaches of Pearl and Spiegelhalter (see Sect. 12.4), but also the simple straightforward rule-based approaches of Ishizuka and Adams are based on conditional probabilities.

A second view is to interpret the degree p in the framework of *measure free conditioning*. For details on this view the reader is referred to the work of Nguyen.

The interpretation of the expert rule in the context of *probabilistic logics*, however, is directly related to so-called *material implication*. In this interpretation p is set equal to the probability $P(\neg A \vee B)$. Such approaches that are closely related to classical logics will be examined later. Notice that this general rule interpretation already played a role in Sect. 9.2, where classical implication was considered in relation to the heuristic system RUM.

For the fourth interpretation, which takes *sets of conditional probabilities* into account, we assume that the uncertainty p of an expert rule "$\underline{\text{if}}$ A $\underline{\text{then}}$ B" is measured by a range $[l, u]$ for the (conditional) probabilities $P(B|A)$. For the sake of generality, we additionally assume the knowledge on a range for the probabilities $P(B|\overline{A})$, such that

$$\{P \mid P(B|A) \in [l\ ,\ u\ ] \text{ and } P(B|\overline{A}) \in [l',\ u']\}. \tag{12.1}$$

Further, assume the uncertainty on (the truth of) A to be given by the ranges

$$\begin{aligned} &P(A) \in [s,\ t], \\ &P(\overline{A}) \in [1-t,\ 1-s]. \end{aligned} \tag{12.2}$$

Then the sequential combination can be performed by applying the "rule of total probability"

$$P(B) = P(B|A)P(A) + P(B|\overline{A})P(\overline{A})$$

to the bounds of the probability ranges for $P(A)$, $P(\overline{A})$, $P(B|A)$, and $P(B|\overline{A})$. This leads to a probability of B that has to meet the following boundary conditions:

$$\begin{aligned} &P(B) \geq \min(ls + l'(1-s),\ lt + l'(1-t)) \\ &P(B) \leq \max(us + u'(1-s),\ ut + u'(1-t)) \end{aligned}$$

So, applying the expert rules

$$\begin{array}{l} \underline{\text{if}}\ A\ \underline{\text{then}}\ B\ \underline{\text{with}}\ [l\ ,\ u\ ] \\ \underline{\text{if}}\ \overline{A}\ \underline{\text{then}}\ B\ \underline{\text{with}}\ [l',\ u'] \end{array}$$

to the knowledge $s \leq P(A) \leq t$ leads to the range

$$\begin{aligned} P(B) \in [\ &\min(ls + l'(1-s),\ lt + l'(1-t))\ , \\ &\max(us + u'(1-s),\ ut + u'(1-t))\ ] \end{aligned} \tag{12.3}$$

In the special case of a completely unknown probability range for probability $P(B|\overline{A})$, i.e. $P(B|\overline{A}) \in [l', u'] = [0, 1]$, the range (12.3) reduces to

$$P(B) \in [\ ls\ ,\ us + 1 - s\ ] \tag{12.4}$$

A property which is characteristic for a mathematical model for drawing inferences is the *direction* of the conditional probabilities (i.e. the "rule strength") which have to be entered by the expert for the domain of interest. For instance, in the case of conditional probabilities $P(H|X)$, an uncertainty in the occurrence of the manifestations X will lend a certain credence to the hypothesis H. The other way around, in the case of conditional probabilities $P(X|H)$, a strong belief in a given hypothesis H creates a certain expectation for the occurrence of its various manifestations. How the quantities will vary in dependence on the extension of this basic concept should be visualized by the following simple example. Notice that for this example the "classical interpretation" is applied to probability.

(12.1) Example. Assume two boxes box_1 and box_2, each containing 8 balls. While box_1 is filled with 4 white and 4 black balls, the other box contains 2 black and 6 white balls. Assume now a person who draws one ball from a box without looking.

To model this application we first define the two frames of discernment $\Omega_X = \{box_1, box_2\}$ and $\Omega_H = \{color_1, color_2\}$. Assume $color_1$ to be the color "black". The probabilistic dependency of "choosing box_i" and "the drawn ball has $color_j$" can be described by conditional probabilities, where we have to distinguish two different possibilities: in the case of rule

$$\underline{\text{if}}\ X\ \underline{\text{then}}\ H\ \underline{\text{with}}\ P(H|X)$$

the probability measure $P(H|X)$ gives the (conditional) probability that a ball with $color_j$ is drawn, under the condition that box_i is chosen. For instance, in the case of box_2 the probability of drawing a black ball is $\frac{2}{8}$ since 2 of the 8 balls of box_2 are black.

On the other hand, the probability measure $P(X|H)$ may be attached to the rule giving the conditional probability that box_i has been chosen, under the condition that a $color_j$ ball has been drawn:

$$\underline{\text{if}}\ X\ \underline{\text{then}}\ H\ \underline{\text{with}}\ P(X|H)$$

For example, the probability of box_2 under the condition that a black ball has been drawn is $\frac{2}{6}$ since 2 of the 6 existing black balls are contained in box_2. For the two ways to represent the uncertain knowledge the probability matrices are shown below.

$P(H\|X)$	black	white
box_1	$\frac{4}{8}$	$\frac{4}{8}$
box_2	$\frac{2}{8}$	$\frac{6}{8}$

$P(X\|H)$	box_1	box_2
black	$\frac{4}{6}$	$\frac{2}{6}$
white	$\frac{4}{10}$	$\frac{6}{10}$

◇

How the first type of expert rule in the above example can be used for probabilistic reasoning will be shown in Sect. 12.2: for instance, the algorithm proposed by Ishizuka et al. is based on conditional probabilities pointing from evidence to hypothesis. This rule convention is also used in many other expert systems.

The procedure with the second type of expert rule directly leads to Bayes' formula: assuming $\Omega_H = \{h_i | i = 1, \ldots, n\}$ as an exhaustive set of mutually exclusive events (e.g. diseases), *Bayes' formula* (compare Theorem 4.9) applies to all elements h_i obeying the condition $P(h_i) > 0$:

$$P(h_i|E) = \frac{P(E|h_i) \cdot P(h_i)}{P(E)} = \frac{P(E|h_i) \cdot P(h_i)}{\sum_{k=1}^{n} P(E|h_k) \cdot P(h_k)}$$

With Bayes' formula the *prior probabilities* $P(h_i)$ and the *conditional probabilities* $P(E|h_i)$ are related to the *posterior probabilities* $P(h_i|E)$. How to interpret Bayes' formula can simply be shown in the case of medical uncertain reasoning, for example:

Assuming h_i to be one of n different diseases or diagnoses, $P(h_i)$ represents the a priori probability that a "typical" patient has disease h_i – without taking the set E of symptoms into account. Then the quantity $P(E|h_i)$ is the conditional probability that a patient shows the set of symptoms E assuming that he has disease h_i. E is called the set of data relevant or evident for disease h_i.

However, it is important to notice that the "direction" in which the expert knowledge on conditional probabilities is coded has no direct influence on how the reasoning process itself is performed: for example, in the model proposed by Pearl, Lauritzen and Spiegelhalter (see Sect. 12.4), each directed link $X \rightarrow H$ is quantified by a conditional probability matrix with entries $P(H_j|X_i)$. However, in Bayesian networks evidential reasoning involves *both diagnostic (bottom-up)* and *predictive (top-down)* inferences.

As another example we cite the "inference networks" proposed by Duda, Hart, and Nilsson (see Sect. 12.3.1). In their approach the links (comparable to classical expert rules) are labelled by likelihood ratios (compare Sect. 4.3) such as

$$\lambda = \frac{P(X|H)}{P(X|\overline{H})}$$

In spite of the fact that the respective conditional probabilities "point" from H or its complement to X, incoming evidence is propagated in the other direction.

So, the expert's knowledge on dependencies of hypotheses and subhypotheses may be represented in different ways. Independent of the kind of modeling, evidential reasoning may involve diagnostic or predictive inferences. However, "classical" approaches and systems are applied to and built for diagnostic tasks so that uncertain information is propagated from evidence to hypotheses only.

12.2 The Straightforward Use of Probabilities

The models for handling uncertain knowledge examined in this chapter can be viewed as more or less direct and straightforward applications or realizations of probability theory applied to a *set of expert rules*. A main characteristic feature of these rule-based models is the fact that flow of (uncertain) knowledge is controlled by a global process. Information entered into the system is propagated in *one direction* to determine the probabilities of the final hypothesis.

So, such models aim at offering a diagnosis (e.g. a disease) on the basis of certain evidence (e.g. symptoms). Flow of information additionally in the other direction for predicting certain manifestations of a given hypothesis, for instance, is not considered. This is a special feature of "Bayesian networks" examined in Sect. 12.4.

So, the aim of this section is to show how to make use of probability theory (e.g. prior and conditional probabilities) in the simplest case. In particular, the inference procedure of Ishizuka and Fu [Ishizuka, Fu, and Yao 1982a] for uncertain reasoning on the basis of probability theory is considered. This procedure has been applied by Ishizuka et al. in SPERIL, which is a rule-based damage assessment system for existing structures (see [Ishizuka, Fu, and Yao 1981], [Ishizuka, Fu, and Yao 1982b], [Ishizuka 1983], for instance).

Further, the simple probability model of (medical) reasoning considered by Adams is introduced and differences to Ishizuka's approach are made explicit. In [Adams 1976] Adams also shows that his probability model is equivalent to a portion of the model used in the MYCIN system.

12.2.1 The Model of Ishizuka et al.

Ishizuka et al. propose a rule-based model which allows inferences to be drawn and evidence to be combined on the basis of probability theory. They take into account that knowledge base design experts as well as the users of the system assert not only simple elementary hypotheses but also more complex ones.

To introduce the formal model, assume the characteristics X, Y, and H and the *universes of discourse*

$$\begin{aligned} X_0 &\stackrel{\mathrm{d}}{=} \Omega_X = \{x_1, \dots, x_I\} , \\ Y_0 &\stackrel{\mathrm{d}}{=} \Omega_Y = \{y_1, \dots, y_J\} , \\ H_0 &\stackrel{\mathrm{d}}{=} \Omega_H = \{h_1, \dots, h_K\} \end{aligned} \tag{12.5}$$

of mutually exclusive elements. For every universe assume that its elements are jointly exhaustive. The conditional probability measures given by

$$\begin{aligned} &P(x_i|E_X)\,, \qquad i=1,\ldots,I\,, \qquad \sum_{i=1}^{I} P(x_i|E_X)=1 \\ &P(y_j|E_Y)\,, \qquad j=1,\ldots,J\,, \qquad \sum_{j=1}^{J} P(y_j|E_Y)=1 \end{aligned} \tag{12.6}$$

describe the probability distributions related to the characteristics X and Y on the basis of the information entered into the system. The information may be either directly entered by a user or processed by the system if X or Y are contained in the conclusions of other rules.

Before introducing the original approach of Ishizuka et al. we start with a very simple, straightforward formalism that explains how a procedure for probabilistic reasoning can work and that clarifies the characteristic features of rule-based probabilistic reasoning and the assumptions that have to be made. This special case involves a rule based approach to representing knowledge. It is essential that data as well as rules may have tags indicating their uncertainty. In this particular example all tags indicating uncertainty will be represented by means of (conditional) probabilities.

Suppose the spaces $\Omega_X = X_1 \cup \overline{X_1}$, $\Omega_Y = Y_1 \cup \overline{Y_1}$, and $\Omega_H = H_1 \cup \overline{H_1}$, each contain two jointly exhaustive and mutually exclusive elements. Further, assume the characteristics X, Y, and H, as before, and two expert rules with the following structure:

<u>if</u> X *is* X_1 <u>then</u> H *is* H_1 <u>with</u> $P(H_1|X_1)$
<u>else</u> [<u>if</u> X *is* $\overline{X_1}$ <u>then</u>] H *is* H_1 <u>with</u> $P(H_1|\overline{X_1})$

<u>if</u> Y *is* Y_1 <u>then</u> H *is* H_1 <u>with</u> $P(H_1|Y_1)$
<u>else</u> [<u>if</u> Y *is* $\overline{Y_1}$ <u>then</u>] H *is* H_1 <u>with</u> $P(H_1|\overline{Y_1})$

For example, in the case of the first rule the dependency of *conclusion* H_1 and *premises* X_1 and $\overline{X_1}$ is quantified by the conditional probabilities $P(H_1|X_1)$ and $P(H_1|\overline{X_1})$. The aim of an inference procedure applied to the two expert rules is to derive the uncertainty on H_1 if the conditional probabilities $P(X_1|E_X)$ and $P(Y_1|E_Y)$ are known. As in previous chapters, E is called the "evidence" covering the whole knowledge (may-be observations) which is evident to determine the probabilities $P(X_1|E_X)$ and $P(Y_1|E_Y)$.

Applying the *law of total probability*, the probabilities $P(H_1|E_X)$ and $P(H_1|E_Y)$ can be easily derived:

$$\begin{aligned} P(H_1|E_X) &= P(H_1|X_1,E_X)\cdot P(X_1|E_X) + P(H_1|\overline{X_1},E_X)\cdot P(\overline{X_1}|E_X) \\ &= P(H_1|X_1)\cdot P(X_1|E_X) + P(H_1|\overline{X_1})\cdot P(\overline{X_1}|E_X) \end{aligned} \tag{12.7}$$

$$
\begin{aligned}
P(H_1|E_Y) &= P(H_1|Y_1, E_Y) \cdot P(Y_1|E_Y) + P(H_1|\overline{Y_1}, E_Y) \cdot P(\overline{Y_1}|E_Y) \\
&= P(H_1|Y_1) \cdot P(Y_1|E_Y) + P(H_1|\overline{Y_1}) \cdot P(\overline{Y_1}|E_Y) \qquad (12.8)
\end{aligned}
$$

These two *sequential combinations* combine the evidential knowledge on premises and the knowledge about the uncertainty attached to rules. Note the assumed equalities

$$P(H_1|X_1, E_X) = P(H_1|X_1) \text{ and } P(H_1|\overline{X_1}, E_X) = P(H_1|\overline{X_1})$$

when deriving (12.7), and analogously

$$P(H_1|Y_1, E_Y) = P(H_1|Y_1) \text{ and } P(H_1|\overline{Y_1}, E_Y) = P(H_1|\overline{Y_1})$$

when deriving (12.8). They describe the assumptions that the evidences E_X and E_Y have no direct influence on hypothesis H_1.

After introducing a simple special case for performing the sequential combination based on conditional probabilities, we next examine the model of Ishizuka et al. Assume the two rules "R1" and "R2", each confirming a *subset* of the elementary hypotheses of H_0:

rule R1:

if	X *is* X_1	then	H *is* H_{X_1}	with	$P(H_{X_1}|X_1)$
else if	X *is* X_2	then	H *is* H_{X_2}	with	$P(H_{X_2}|X_2)$
			⋮		
else			H *is* H_0	with	1

rule R2:

if	Y *is* Y_1	then	H *is* H_{Y_1}	with	$P(H_{Y_1}|Y_1)$
else if	Y *is* Y_2	then	H *is* H_{Y_2}	with	$P(H_{Y_2}|Y_2)$
			⋮		
else			H *is* H_0	with	1

The elements $X_1, X_2, \ldots$ represent mutually exclusive *subsets* of X_0. Its union does not necessarily cover the complete universe of discourse X_0. In the same way the elements $Y_1, Y_2, \ldots$ are mutually exclusive subsets of Y_0. For example, in the case of rule R1 the set

$$\overline{X_1 \cup X_2 \cup \ldots} = X_0 - (X_1 \cup X_2 \cup \ldots)$$

covers those elements of X_0 not explicitly mentioned in the premise of the rule R1. They are considered by the else-part of the rule R1. Identical statements can be made for "rule R2" and the sets $Y_1, Y_2, \ldots$

Because of the additivity property of a probability measure, the conclusion of a rule can be viewed to be implicitly extended. For instance, in the case of rule R1 each conclusion

$$\ldots \underline{\text{then}}\ H\ is\ H_{X_l}\ \underline{\text{with}}\ P(H_{X_l}|X_l)$$

is associated with the equivalent conclusion

$$\ldots \underline{\text{then}}\ H\ is\ \overline{H_{X_l}}\ \underline{\text{with}}\ 1 - P(H_{X_l}|X_l) \tag{12.9}$$

The final else-clause

$$\ldots \underline{\text{else}}\ H\ is\ H_0\ \underline{\text{with}}\ 1$$

of each rule takes the fact into account that the expert may be *ignorant* about the effect of some premises.

To draw inferences on the basis, say, of rule R1, the evidential knowledge E_X concerning characteristic X has to be combined with the conditional probabilities attached to the rule. The aim of the sequential combination is to derive the a posteriori probabilities $P(h_k|E_X)$, $k = 1, \ldots, K$.

The actual probabilities of the subsets X_l are derived by simply adding the probabilities of its elements:

$$P(X_l|E_X) = \sum_{x_i \in X_l} P(x_i|E_X) \tag{12.10}$$

For drawing the inference, Ishizuka et al. propose the following formula:

$$\begin{aligned} P(h_k|E_X) = & \sum_{l:\ h_k \in H_{X_l}} \frac{P(H_{X_l}|X_l)}{|H_{X_l}|} \cdot P(X_l|E_X) \\ & + \sum_{l:\ h_k \in \overline{H_{X_l}}} \frac{1 - P(H_{X_l}|X_l)}{|\overline{H_{X_l}}|} \cdot P(X_l|E_X) \\ & + \frac{1}{K} \cdot (1 - \sum_l P(X_l|E_X)) \end{aligned} \tag{12.11}$$

In general, the a posteriori probability of the hypothesis h_k is influenced

- directly by the first of the three terms (12.11), if h_k is an element of a conclusion H_{X_l},
- indirectly by the implicit rule-extension (12.9) and the second term of (12.11), if h_k is an element of a complementary hypothesis $H_0 - H_{X_l}$, or

- by the third term of (12.11) which measures the proportional probability mass of the elements of X_0 not explicitly considered by the rule's premise.

The use of the denominators $|H_{X_l}|$, $|\overline{H_{X_l}}|$, and K in formula (12.11) is due to the possible use of *sets* of elements and to the possible *incompleteness* of the union of the premises' sets X_l. The consequence is that a portion of the probability mass has to be "distributed" to get a probability distribution for H_0, too. For the above formula Ishizuka et al. assume a uniform distribution because no more information is given and the uniform distribution has the maximum entropy.

Formula (12.11) is based on the *law of total probability* which can be simply derived

- if one assumes a very simple rule premise and rule conclusion consisting of elementary values only, instead of *sets* of elements, and
- if the rule premise is complete in the sense that all elements of the universe of discourse X_0 are explicitly contained in the premise.

If the first condition is met, the cardinality $|H_{X_l}|$ of the first term of (12.11) is 1. If the second condition is fulfilled, the third term of the formula will be zero. As already mentioned above when deriving equation (12.7), for instance, also equation (12.11) assumes that evidence E_X has no *direct* influence on the hypotheses h_k.

In a similar way to drawing the inference on the basis of rule R1 and the evidential knowledge on X, the second rule R2 can be applied. Next a procedure is needed to compute the probabilities $P(h_k|E_X, E_Y)$ from the probabilities $P(h_k|E_X)$ and $P(h_k|E_Y)$ due to the different information sources, i.e. to perform the *parallel combination.* A parallel combination procedure makes it possible to compute the uncertainty of a statement in the presence of *multiple* parallel inferences, as is the case with the above two example rules.

Returning to our simple introductory case based on the dichotomized frame of discernment $\Omega_X = H_1 \cup \overline{H_1}$, the application of Bayes' formula to the conditional probability $P(H_1|E_X, E_Y)$ leads to

$$P(H_1|E_X, E_Y) = \frac{P(E_X, E_Y|H_1) \cdot P(H_1)}{P(E_X, E_Y)}$$

The assumption of *conditional independence*

$$P(E_X, E_Y|H_1) = P(E_X|H_1) \cdot P(E_Y|H_1)$$

then leads to the equation

$$P(H_1|E_X, E_Y) = \frac{P(E_X|H_1) \cdot P(E_Y|H_1) \cdot P(H_1)}{P(E_X, E_Y)}$$

Again applying Bayes' formula to the conditional probabilities $P(E_X|H_1)$ and $P(E_Y|H_1)$ gives

$$P(H_1|E_X, E_Y) = \frac{P(H_1|E_X) \cdot P(E_X)}{P(H_1)} \cdot \frac{P(H_1|E_Y) \cdot P(E_Y)}{P(H_1)} \cdot \frac{P(H_1)}{P(E_X, E_Y)}$$

A rearrangement of terms gives the following desired result

$$P(H_1|E_X, E_Y) = \frac{P(H_1|E_X) \cdot P(H_1|E_Y)}{P(H_1)} \cdot \frac{P(E_X) \cdot P(E_Y)}{P(E_X, E_Y)} \tag{12.12}$$

Thus with the use of Bayes' formula and the assumption of conditional independence the *parallel combination formula* can be derived. The last term on the right in equation (12.12) is constant and serves as a renormalization factor. An explicit representation of the renormalization factor and the consideration that $P(H_1|E_X, E_Y) + P(\overline{H_1}|E_X, E_Y) = 1$ holds leads to the equation

$$P(H_1|E_X, E_Y) = \frac{\dfrac{P(H_1|E_X) \cdot P(H_1|E_Y)}{P(H_1)}}{\dfrac{P(H_1|E_X) \cdot P(H_1|E_Y)}{P(H_1)} + \dfrac{P(\overline{H_1}|E_X) \cdot P(\overline{H_1}|E_Y)}{P(\overline{H_1})}} \tag{12.13}$$

which applies to compute the probability $P(H_1|E_X, E_Y)$ from the probabilities $P(H_1|E_X)$ and $P(H_1|E_Y)$.

The parallel combination formula proposed by Ishizuka et al. for the more complex frames generalizes formula (12.13). Identical to the assumptions made for deriving formula (12.13), also here the conditional independence of the evidence involved under the condition of known h_k has to be assumed. For the parallel combination, additionally the a priori probabilities $P(h_k)$, $k = 1, \ldots, K$ must be known:

$$P(h_k|E_X, E_Y) = \frac{\dfrac{P(h_k|E_X) \cdot P(h_k|E_Y)}{P(h_k)}}{\displaystyle\sum_{l=1}^{K} \frac{P(h_l|E_X) \cdot P(h_l|E_Y)}{P(h_l)}} \tag{12.14}$$

12.2.2 The Model of Adams

The approach of Adams [Adams 1976] is—like that of Ishizuka et al.—directly based on probability theory. In [Adams 1976] Adams also shows that his probability model is equivalent to a portion of the model used in the MYCIN system: in its basics his model can be mapped to the certainty factor approach of Buchanan and Shortliffe. So the (restrictive) assumptions made in the certainty factor approach and some inconsistencies can be made visible.

The main difference between the two models is that Adams additionally assumes the *absolute* independence of the sources E_X and E_Y, or, in the general case of m pieces of evidence, the absolute independence of the "evidences" $E_1, \ldots, E_m$. Consequently, the equation

$$P(E_1, \ldots, E_m) = P(E_1) \cdot \ldots \cdot P(E_m)$$

is assumed to be true. If we additionally assume that there are only two hypotheses H and $\overline{H}$, equation (12.12) can be further simplified since the renormalization factor can be removed:

$$P(H|E_1, \ldots, E_m) = [\prod_{j=1}^{m} \frac{P(H|E_j)}{P(H)}] \cdot P(H) \tag{12.15}$$

This equation is closely related to the sequential combination used in the framework of inference networks (compare equation (12.26) in Sect. 12.3.1). Because of the equality

$$\frac{P(H|E_j)}{P(H)} = \frac{P(E_j|H)}{P(E_j)}$$

also the parallel combination

$$P(H|E_1, \ldots, E_m) = [\prod_{j=1}^{m} \frac{P(E_j|H)}{P(E_j)}] \cdot P(H) \tag{12.16}$$

can be derived from (12.15) and is used by Adams. In this way one gets a computationally simple method of serially adjusting new incoming information and evidence concerning the hypothesis H. For a new piece of evidence E_m one obtains

$$P(H|E_1, \ldots, E_{m-1}, E_m) = \frac{P(E_m|H)}{P(E_m)} \cdot P(H|E_1, \ldots, E_{m-1}) \tag{12.17}$$

with $P(H|E_0) \stackrel{\mathrm{d}}{=} P(H)$ before any evidence is introduced.

However, for his model Adams assumes the conditional independence of the evidences not only on H but also on $\overline{H}$. So additionally the equation

$$P(E_1, \dots, E_m|\overline{H}) = P(E_1|\overline{H}) \cdot \dots \cdot P(E_m|\overline{H})$$

is assumed to be true, and – similarly to the derivation of formulas (12.16) and (12.17) – the formulas (12.18) and (12.19) are obtained. With

$$P(\overline{H}|E_1, \dots, E_m) = [\prod_{j=1}^{m} \frac{P(E_j|\overline{H})}{P(E_j)}] \cdot P(\overline{H}) \qquad (12.18)$$

one gets

$$P(H|E_1, \dots, E_{m-1}, E_m) = 1 - \frac{P(E_m|\overline{H})}{P(E_m)} \cdot [1 - P(H|E_1, \dots, E_{m-1})] \quad (12.19)$$

with E_m as new evidence and the definition $P(H|E_0) \stackrel{\text{d}}{=} P(H)$, as before.

From the above considerations it is obvious that Adams mainly focuses on the parallel combination rule. In [Adams 1976], he also examines the relation of his probabilistic model to the certainty factor approach of Shortliffe et al. (compare Sect. 9.1). So the (restrictive) assumptions made in the certainty factor approach and some inconsistencies can be made visible.

He introduces the two parameters *measure of belief* MB and *measure of disbelief* MD which are defined as follows:

$$MD(H, E_j) \stackrel{\text{d}}{=} 1 - \frac{P(E_j|H)}{P(E_j)} \qquad (12.20)$$

$$MB(H, E_j) \stackrel{\text{d}}{=} 1 - \frac{P(E_j|\overline{H})}{P(E_j)} \qquad (12.21)$$

Both MB and MD are in the range $[0, 1]$ and are zero when the evidence has no bearing on the hypothesis. $MB(H, E_j) = 1$ holds when E_j implies H with certainty, $MD(H, E_j) = 1$ holds when E_j implies $\overline{H}$ with certainty. It is obvious that the definitions of the two measures are identical to those of the original certainty factor approach. It is also obvious that, for parallel combinations, the second terms of the left hand sides of equations (12.20) and (12.21) are parts of the combination rules (12.17) and (12.19), respectively. So, the measures MB and MD can be *directly* used for parallel combinations.

For a deeper analysis of the relations between Adams' probabilistic model and the certainty factor approach, the reader is referred to [Adams 1976]. Some of Adams' results have already mentioned in Sect. 9.1 when discussing the certainty factor model.

12.2.3 Discussion

The study of the mathematical model for probabilistic reasoning developed by Ishizuka et al. and the approach of Adams shows that under certain restrictions a simple control mechanism for drawing inferences on the basis of probability theory can be derived.

The most important properties and assumptions underlying the inference procedures may be summarized as follows: prior probabilities are taken into account by the parallel combination formulas (12.13) and (12.14). Possibly incomplete knowledge concerning the exact probability values cannot be considered by the above models.

Further, the control procedure works in one "direction" only: (uncertain) knowledge on *symptoms* is combined with uncertainty attached to a rule for determining the uncertainty of *hypotheses*. Application of the above procedures to predict possible symptoms is not considered.

While both models are directly based on probability theory they differ in the (independence) assumptions made. These are an important criterion when evaluating published models and applications which involve handling uncertain knowledge. In particular, the following independence assumptions were made visible:

- The *conditional independence of the evidences (given H_1)*, e.g.

$$P(E_1, \ldots, E_m | H_1) = P(E_1|H_1) \cdot \ldots \cdot P(E_m|H_1) \tag{12.22}$$

  is assumed when combining different sources of uncertainty.

- The assumption of *conditional independence of the evidences (given the complements $\overline{H_1}$)*, e.g.

$$P(E_1, \ldots, E_m | \overline{H_1}) = P(E_1|\overline{H_1}) \cdot \ldots \cdot P(E_m|\overline{H_1}) \tag{12.23}$$

  is made by Adams.

- The assumption of *conditional independence (in regard to evidence E_X)*

$$P(H_1|X_1, E_X) = P(H_1|X_1) \tag{12.24}$$

  requires that – for a given rule of the kind "<u>if</u> X_1 <u>then</u> H_1 <u>with</u> $P(H_1|X_1)$" – an existing evidence E_X influences the uncertainty of X_1 only, but not *directly* that of H_1.

- The assumption of *absolute independence of the evidences involved* requires that

$$P(E_1, \ldots, E_m) = P(E_1) \cdot \ldots \cdot P(E_m). \tag{12.25}$$

Coming back to the mathematical models of Ishizuka and Adams, in both models the independence (12.22) of the evidences involved under the condition of the given hypotheses is assumed. This assumption is made for the parallel combination rules. Further, they make use of assumptions (12.24) for the sequential combination algorithms. These two assumptions are typically made also in many other architectures of uncertainty. Moreover, Adams' model is also based on the absolute independence of the evidences (12.25). Also, the conditional independence (12.23) applied to the *complementary hypotheses* is assumed. So, the computational simplicity of his approach is vitiated by the restrictive nature of the whole set of independence assumptions made.

However, when discussing Adams' ideas one should keep in mind that his mathematical framework is aimed at being compared to the certainty factor approach, so that the remarks made by him may be directly related to the applicability and consistency of the *certainty factor model.*

12.3 PROSPECTOR – Inference Networks

In [Duda, Hart, and Nilsson 1976], Duda, Hart, and Nilsson propose a "subjective Bayesian method for rule-based inference systems" which aims at combining some of the advantages of both formal and informal (i.e. heuristic) methods for uncertain reasoning. This theoretical model of *inference networks* has the characteristic feature that an explicit expert rule as considered in previous chapters is now interpreted as a link in a directed graph. The uncertainty attached to the rule is attached to the link.

The theoretical model has been used as basis of the expert system PROSPECTOR described in, e.g., [Duda et al. 1978], [Duda, Gaschnig, and Hart 1981], [Gaschnig 1982]. PROSPECTOR is a computer consultant system intended to aid geologists in evaluating the favorability of an exploration site or region for occurrences of ore deposits of particular types [Duda, Gaschnig, and Hart 1981]. Often, PROSPECTOR is shown to be well functioning by referring to statistics, as done in [Gaschnig 1982], for instance. Further, it is mentioned that making use of PROSPECTOR enabled researchers to find ore deposits [Weiss and Kulikowski 1983].

12.3.1 The Inference Network Model

The starting point for introducing the formal basis of "inference networks" as proposed by Duda et al. [Duda, Hart, and Nilsson 1976] is Bayes' formula for the simple case of two dichotomous (i.e. two-valued) frames of discernment. Assuming

$$\Omega_X = X \cup \overline{X} \qquad \text{and} \qquad \Omega_H = H \cup \overline{H}$$

and given the simple expert rule "if X then H", according to equation (4.8) we derive

$$P(H|X) = \frac{P(X|H) \cdot P(H)}{P(X)} \tag{12.26}$$

for $P(X) > 0$. If X is known to be true then the premise of the rule will match the knowledge base and formula (12.26) applies to compute the a posteriori probability $P(H|X)$ on the basis of the probabilities $P(H)$, $P(X) = 1$, and the conditional probability $P(X|H) = 1$, i.e. $P(H|X) = P(H)$.

Dividing formula (12.26) by the corresponding formula for $P(\overline{H}|X)$ leads to the following form of Bayes' formula

$$\frac{P(H|X)}{P(\overline{H}|X)} = \frac{P(X|H)}{P(X|\overline{H})} \cdot \frac{P(H)}{P(\overline{H})} \tag{12.27}$$

which applies to compute the so-called *posterior odds* on hypothesis H under the condition that the premise matches, i.e. that X is *known to be true*. Notice that a possible uncertainty concerning the premises' truth is not considered at this stage of derivation. So, taking up the terminology introduced in Sect. 4.3, each of the three terms in the above equation is associated with an unique interpretation:

$$\begin{aligned} O(H) &\stackrel{\mathrm{d}}{=} \frac{P(H)}{P(\overline{H})} = \frac{P(H)}{1-P(H)} && \text{prior odds,} \\ O(H|X) &\stackrel{\mathrm{d}}{=} \frac{P(H|X)}{P(\overline{H}|X)} = \frac{P(H|X)}{1-P(H|X)} && \text{posterior odds,} \\ \lambda &\stackrel{\mathrm{d}}{=} \frac{P(X|H)}{P(X|\overline{H})} && \text{likelihood ratio.} \end{aligned} \tag{12.28}$$

Taking the measure O into account we get the *odds-likelihood form*

$$O(H|X) = \lambda \cdot O(H) \tag{12.29}$$

of Bayes' theorem (compare Theorem 4.9). In the context of inference networks this equation describes an *update process* of the odds on hypothesis H, assuming that premise X matches the knowledge base and that the likelihood ratio λ as well as the prior odds $O(H)$ are known. For instance, a "high" value $\lambda \gg 1$ reflects the fact that X is *sufficient* for H since the truth of X would lead to an increase of the (posterior) odds attached to hypothesis H. If, on the other hand, λ is much less than 1 (e.g. $\lambda \approx 0$) then the truth of X has a decrease of the certainty of H as consequence.

The probability P can simply be reconstructed from the odds O. That means that the two measures carry the same information. For example, equation (12.30) shows the relationship between posterior probabilities and posterior odds:

$$P(H|X) = \frac{O(H|X)}{1 + O(H|X)} \tag{12.30}$$

By analogy we can characterize the update process for hypothesis H if premise X is *known to be false*:

$$O(H|\overline{X}) = \xi \cdot O(H) \quad \text{with} \quad \xi = \frac{P(\overline{X}|H)}{P(\overline{X}|\overline{H})} = \frac{1 - P(X|H)}{1 - P(X|\overline{H})} \tag{12.31}$$

In the case of a very low value $0 \leq \xi \ll 1$ and the knowledge that premise X is false, indifferent prior odds on H are transformed into low posterior odds on H. In this sense, premise X may be viewed to be *necessary* for H. Under the same conditions, a high likelihood ratio ξ leads to a confirmation and an increase of the odds on H. The separate treatment of the *two* cases "X is true" and "X is false" has the consequence that an expert has to know both values, λ and ξ, when formulating the domain rules. Consequently, the expert rules will have the following structure:

<u>if</u> premise X <u>then</u> hypothesis H <u>with</u> degree (λ, ξ)

For the following assume E to be the conjunction of observations containing the information relevant to determine either the truth or the falsity of X. Then the algorithm for computing the posterior odds on hypothesis H may be formulated as follows: if, on the basis of the observations E, it is known that X is true, the posterior odds $O(H|X)$ can be processed on the basis of the prior odds $O(H)$ (or the prior probabilities $P(H)$, respectively) and the ratio λ. If, on the other hand, observations E lead to the fact that X is false, the posterior odds $O(H|\overline{X})$ are obtained by formula (12.31). Consequently, the posterior probabilities on H are given by

$$P(H|E) = \begin{cases} P(H|X) & = \dfrac{O(H|X)}{1 + O(H|X)} \quad \text{if } P(X|E) = 1, \\ \\ P(H|\overline{X}) & = \dfrac{O(H|\overline{X})}{1 + O(H|\overline{X})} \quad \text{if } P(X|E) = 0. \end{cases} \tag{12.32}$$

To summarize, the update process can be performed if the rules' ratios λ and ξ can be obtained from the knowledge base, and if in the real environment it is clear that premise X is either *true* or *false*. However, the above procedure would not fit most real world applications since the following requirements are not met:

- the update process should cover the situation where the "user" of the expert system (as well as the system itself when evaluating sub-hypotheses) is *uncertain* about the truth or falsity of X such that

$$0 < P(X|E) < 1.$$

- From the view of the experts who formulate the rules, it should be possible to create instances of the measures λ and ξ as well as prior odds in a *subjective way*, i.e. without certain underlying formal restrictions. Indeed, although both likelihood ratios λ and ξ have to be given separately by the expert, they are not completely independent of each other: from the definitions (12.28) and (12.31) of λ and ξ the following equation can be derived:

$$\xi = \frac{1 - \lambda \cdot P(X|\overline{H})}{1 - P(X|\overline{H})} \qquad \text{with} \qquad 0 < P(X|\overline{H}) < 1. \tag{12.33}$$

  Therefore, most of the quantities for λ and ξ will be inconsistent since they are based on the expert's "experience" more than on the strict formula (12.33). As a *necessary* condition for the truth of (12.33), the following equivalences have to be fulfilled:

$$\begin{aligned} \xi > 1 &\Leftrightarrow \lambda < 1 \\ \xi < 1 &\Leftrightarrow \lambda > 1 \\ \xi = 1 &\Leftrightarrow \lambda = 1 \end{aligned} \tag{12.34}$$

  Not only intuition but also facts in the real world may lead to the consequence that, concerning the relation of premise X and hypothesis H, an expert comes to the following opinion (compare also the quantities attached to the links in the PROSPECTOR model shown in Fig. 12.2): *"The presence of X enhances the odds on H, but the absence of X has no significance."* [Duda, Hart, and Nilsson 1976]
  The assertion implies $\lambda > 1$ and $\xi = 1$ which obviously does not meet the equivalences (12.34).

The first of the above requirements, i.e. taking uncertain evidence into account, is fulfilled by the following linear interpolation which takes the two cases $P(X|E) = 0$ and $P(X|E) = 1$ into account, too (compare equation (12.32)):

$$P(H|E) = P(H|X) \cdot P(X|E) \; + \; P(H|\overline{X}) \cdot P(\overline{X}|E) \tag{12.35}$$

The measures $P(H|X)$ and $P(H|\overline{X})$ are obtained by applying (12.30) to (12.29) and (12.31). To guarantee the correctness of the interpolation (12.35), it has to be assumed that observations E relevant for X do not carry any *direct* information on H (i.e., $P(H|X,E) = P(H|X)$ and $P(H|\overline{X},E) = P(H|\overline{X})$).

A (necessary) consequence of the linear interpolation can be asserted as follows: if the observations E carry no better information on the certainty of X than already given by the a priori probability $P(X)$, then the application of (12.35) leaves the probability of H unchanged, i.e.,

$$P(X|E) = P(X) \quad \Rightarrow \quad P(H|E) = P(H). \tag{12.36}$$

This means that in the case of "vacuous information" the update process does not lead to any change of probabilities – attached to direct or indirect hypotheses.

To fulfill the second requirement associated with the use of *subjective probabilities* means leaving the probabilistic basis and extending the model in an intuitive way. Following the arguments of Duda et al. [Duda, Hart, and Nilsson 1976], this way to proceed is necessary since the expert must also provide the a priori probabilities $P(H)$ and $P(X)$ and all these values – including λ and ξ – must be consistent. Already in simple inference networks, this requirement seems to be unsolvable.

A (heuristic) possibility for achieving consistency is to adjust the linear interpolation function (12.35) shown in Fig. 12.1 by breaking it into a piecewise linear function. The differences between the two figures illustrate the discrepancies between a formal and a *subjective Bayesian updating.* Other possible interpolations are given in [Duda, Hart, and Nilsson 1976]. The algebraic representation of Fig. 12.1 is given by

$$P(H|E) = \begin{cases} s & \text{if } P(X|E) \leq P(X), \\ t & \text{if } P(X) \leq P(X|E), \end{cases} \tag{12.37}$$

where

$$s \stackrel{\text{d}}{=} P(H|\overline{X}) + \frac{P(X|E)}{P(X)}(P(H) - P(H|\overline{X})) \,,$$

$$t \stackrel{\text{d}}{=} \frac{P(H) - P(H|X)P(X)}{1 - P(X)} + P(X|E)\frac{P(H|X) - P(H)}{1 - P(X)}.$$

The probabilities $P(H|X)$ and $P(H|\overline{X})$ contained in equation (12.37) are obtained by computing (12.29) and (12.31) and then applying (12.30).

To evaluate the expert rule in a real situation, the a priori probabilities $P(X)$ and $P(H)$, the likelihood ratios λ and ξ, and $P(X|E)$ as the actual "certainty" of X are needed. The first four quantities have to be given by the experts who model the domain of application. On the other hand, $P(X|E)$ has

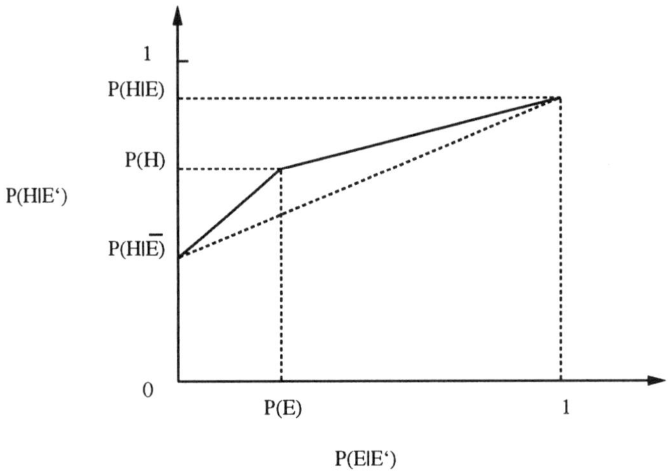

Fig. 12.1. Linear and consistent interpolation [Duda, Hart, and Nilsson 1976]

to be entered by the user while running the expert system. If there is an expert rule confirming X, the certainty $P(X|E)$ can be processed by the system.

Beyond the evaluation of the single rules looked at in the preceding paragraph, Duda et al. examine also the problem in which several expert rules concern the same hypothesis H, i.e. a node in an inference network has more than one incoming arc:

$$\begin{array}{c} \underline{\text{if}}\ X_1\ \underline{\text{then}}\ H\ \underline{\text{with}}\ (\lambda_1/\xi_1) \\ \underline{\text{if}}\ X_2\ \underline{\text{then}}\ H\ \underline{\text{with}}\ (\lambda_2/\xi_2) \\ \vdots \\ \underline{\text{if}}\ X_n\ \underline{\text{then}}\ H\ \underline{\text{with}}\ (\lambda_n/\xi_n) \end{array} \tag{12.38}$$

As a necessary condition for the *parallel combination* the following conditional independencies of multiple antecedents from both the hypothesis and its negation are assumed to be true:

$$P(X_1, \ldots, X_n|H) = \prod_{i=1}^{n} P(X_i|H) \tag{12.39}$$

$$P(X_1, \ldots, X_n|\overline{H}) = \prod_{i=1}^{n} P(X_i|\overline{H}) \tag{12.40}$$

In the case of multiple antecedents as in (12.38), given independencies (12.39) and (12.40), and if all X_i are *known to be certainly true* (i.e. $P(X_i|E_i) = 1$), then similar to equation (12.29) the odds on H are updated by

$$O(H|X_1,\ldots,X_n) = (\prod_{i=1}^{n} \lambda_i) \cdot O(H), \qquad (12.41)$$

$$\lambda_i = \frac{P(X_i|H)}{P(X_i|\overline{H})}.$$

The case of antecedents which are *known to be certainly false* is given by (compare equation (12.31))

$$O(H|\overline{X_1},\ldots,\overline{X_n}) = (\prod_{i=1}^{n} \xi_i) \cdot O(H), \qquad (12.42)$$

$$\xi_i = \frac{P(\overline{X_i}|H)}{P(\overline{X_i}|\overline{H})}.$$

Let us now take *uncertain* evidence and *inconsistent prior probabilities* into account and, therefore, consider just those arguments that led to the use of equations (12.35) and (12.37). Following Duda et al. the posterior odds $O(H|E_1,\ldots,E_n)$ are computed by multiplying the prior odds $O(H)$ and the product of the so-called *effective likelihood ratios* Δ_i:

$$O(H|E_1,\ldots,E_n) = (\prod_{i=1}^{n} \Delta_i) \cdot O(H) \qquad (12.43)$$

with the *effective likelihood ratio*

$$\Delta_i = \frac{O(H|E_i)}{O(H)}$$

If there is new evidence given by updating, e.g., $P(X_i|E_i)$, the new posterior odds $O(H|E_i)$ are computed by making use of the interpolation (12.37). Then a new effective likelihood ratio is obtained and equation (12.43) applies to compute the new posterior odds on H.

For a really simple case, the relation to probability theory and the meaning of the parallel combination formula (12.43) can be simply visualized in the special case of non-uncertain evidence. Assuming that two premises X_1 and X_2 are known to be true, the formal computation of the posterior odds $O(H|E_1,E_2)$ is shown below:

$$
\begin{aligned}
O(H|E_1,E_2) = O(H|X_1,X_2) &= \frac{P(H|X_1,X_2)}{P(\overline{H}|X_1,X_2)} \\
&= \frac{P(H,X_1,X_2)/P(X_1,X_2)}{P(\overline{H},X_1,X_2)/P(X_1,X_2)} \\
&= \frac{P(X_1,X_2|H)\cdot P(H)}{P(X_1,X_2|\overline{H})\cdot P(\overline{H})} \\
&= \frac{P(X_1|H)\cdot P(X_2|H)\cdot P(H)}{P(X_1|\overline{H})\cdot P(X_2|\overline{H})\cdot P(\overline{H})} \\
&= \lambda_1\cdot\lambda_2\cdot O(H)
\end{aligned}
$$

Just this result is obtained when using (12.37) and computing the effective likelihood ratio: applying interpolation (12.37) to the information $P(X_i|E_i) = 1$ leads to

$$
P(H|E_i) = \frac{P(H|X_i)\cdot[1-P(X_i)]}{1-P(X_i)} = P(H|X_i)
$$

Computing the effective likelihood ratio we obtain

$$
\begin{aligned}
\Delta_i = \frac{O(H|E_i)}{O(H)} &= \frac{P(H|X_i)\cdot P(\overline{H})}{P(\overline{H}|X_i)\cdot P(H)} \\
&= \frac{P(X_i|H)\cdot P(H)\cdot P(\overline{H})}{P(X_i|\overline{H})\cdot P(\overline{H}\cdot P(H)} = \lambda_i
\end{aligned}
$$

So, if the truth of evidence X_i is uncertain, a consistent interpolation function (the one shown in Fig. 12.1(b), for instance) can be used to compute the odds $O(H|E_i)$ or the probabilities $P(H|E_i)$. In each case, the final result (i.e. the final posterior probabilities) is independent of the order in which evidence is obtained and in which the leaf-nodes of the network are evaluated.

Before starting with an examination and discussion of the "subjective Bayesian method for rule-based inference systems" of Duda et al. we will look at the real expert system PROSPECTOR based on these ideas.

12.3.2 PROSPECTOR

As an example for how to use the formal model of inference networks introduced in the preceding section, we describe the way in which the "computer consultant system for mineral exploration", PROSPECTOR, performs the representation and processing of uncertain knowledge.

Beyond general knowledge concerning ore deposits, PROSPECTOR contains knowledge about geological facts. Such facts may be significant for certain ores or may disconfirm the existence of ore deposits. Corresponding to the characterizations of different classes of ores the system can be attached to different knowledge bases, the so-called *models*. So, the term "model" is used to refer to a body of knowledge about a particular domain of expertise. Therefore, the knowledge contained in PROSPECTOR consists of a number of models of certain classes of ore deposits. For instance, there is a model for nickel, another for uranium, and so on. These models may be further extended or altered, and can be used as a knowledge base for PROSPECTOR.

Consequently, the performance of the system depends also on the number of models, the type of deposits modeled, and the quality and completeness of each model. On the basis of these models the PROSPECTOR *program* aims at performing the judgmental reasoning process.

The geological characteristics of a deposit of a certain ore (i.e. a model) are represented by means of an inference network. The leaf nodes of such a network represent simple elementary assertions corresponding to "field evidence" while its internal nodes refer to geological (sub-)hypotheses. An example of a small portion of a PROSPECTOR model is shown in Fig. 12.2.

At the beginning of a session, the user enters a list of rocks and minerals observed – the so-called domain data – and other observations. The PROSPECTOR program matches these data against its models and may ask for additional information during the inference process. Finally, it provides a summary of the findings. Questions to the user are asked at the leaf nodes of an inference network. Since the user may not be able to state that something is either definitely present or definitely absent, uncertain input is taken into account. The actual certainties corresponding to the leaf-node questions are expressed on an arbitrary -5 to $+5$ scale (-5: evidence is definitely present, ..., 0: no information, ..., $+5$: evidence is definitely absent). For internal use these numbers are transformed into either probabilities or odds.

For more information concerning some geological background, details of models, implementation and test, etc., the reader is referred, for instance, to [Duda, Gaschnig, and Hart 1981]. For the sake of completeness, it should be mentioned that for PROSPECTOR the mathematical model of inference networks looked at in the previous section is extended: *three* different kinds of relations are used to specify how a change of a probability affects other probabilities, i.e. *logical relations*, *plausible relations*, and *contextual relations*.

- *logical relations* are composed out of the primitive operations of conjunction (and), disjunction (or), and negation (not). To determine the probability of a hypothesis from the probabilities of its component assertions, PROSPECTOR makes use of the fuzzy-set operations (minimum, maximum, difference from 1) as proposed by Zadeh.
- *plausible relations* correspond to "expert rules" of the form

  if premise then hypothesis with degree (λ, ξ)

  How to evaluate these rules was the main topic of the preceding section.
- *contextual relations* formulate necessary conditions which have to be established before a certain part of the network can be used in the reasoning process. For instance, assertions must sometimes be considered in a particular sequence and cannot be considered in arbitrary order.

Figure 12.2 shows a small portion of a PROSPECTOR model which characterizes the deposit of a certain kind of copper. The nodes (boxes) of this network-part represent assertions which may confirm or disconfirm the existence of the ore deposit. Above each box the a priori probability for the truth of that assertion is shown. The arcs "connecting" the nodes represent the relations mentioned above. They determine how a change in the probability of one assertion will affect those of other assertions. So they characterize how the update process is performed on the basis of the network.

The special inference rules represented by "plausible relations" are marked by the two likelihood ratios λ and ξ. They measure the degree to which a change in the evidence influences the posterior odds of (sub-)hypotheses. Both ratios have to be given by the experts (perhaps on the basis of linguistic terms) when constructing the knowledge base.

(12.2) Example. The expert rule

if "stocks"then "suggestive morphology of igneous rocks"
with degree (300, 1)

represents the knowledge

- that the existence of "stocks" (`RCS`) is a highly sufficient indicator of "suggestive morphology of igneous rocks" (`SMIR`) because $\lambda = 300$, and
- that the non-existence of `RCS` does not affect the hypothesis since $\xi = 1$.

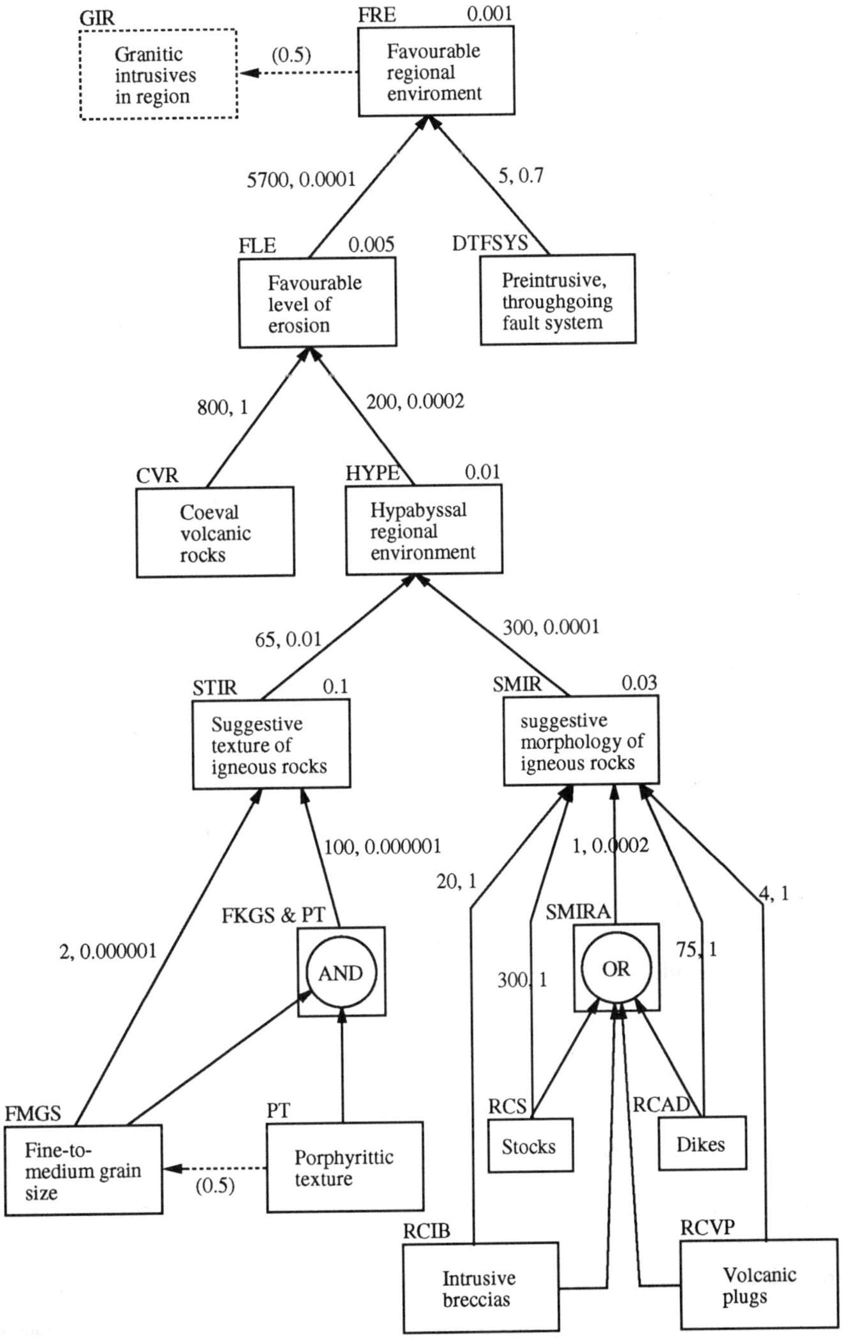

Fig. 12.2. Portion of a PROSPECTOR model for porphyry copper deposits ([Duda, Gaschnig, and Hart 1981], [Gaschnig 1982]).

For the update process, i.e. the computation of the posterior odds $O(H|E)$, one further needs the prior odds $O(H)$ of hypothesis H. Following definitions (12.28), for example, $O(\mathtt{SMIR})$ can be obtained from the a priori probability $P(\mathtt{SMIR})$ shown in Fig. 12.2:

$$O(\mathtt{SMIR}) = \frac{P(\mathtt{SMIR})}{1 - P(\mathtt{SMIR})} = \frac{0.03}{0.97}$$

Furthermore, Fig. 12.2 visualizes in which way *logical* and *plausible relations* can be combined: "stocks", "intrusive breccias", "dikes", and "volcanic plugs" are four positive indicators for SMIR because all $\lambda > 1$. On the other hand, the absence of any one of the four indicators individually is unimportant and therefore has no influence on the certainty of the hypothesis SMIR since all $\xi = 1$.

If none of the four factors is known to be present, then the node SMIRA which is the disjunction of the factors will be false. This will increase the odds on SMIR (SMIRA $\rightarrow$ SMIR, $\xi = 0.0002$).

As an example for the computation and propagation of probabilities, assume a user who indicates the existence of "intrusive breccias" (RCIB) without giving more information. Using (12.29) we get

$$O(\mathtt{SMIR}|\mathtt{RCIB}) = \lambda \cdot O(\mathtt{SMIR}) = 20 \cdot \frac{0.03}{0.97} = 0.6186$$

Because we know that $P(\mathtt{RCIB}|E) = 1$ holds for "intrusive breccias", we further get

$$O(\mathtt{SMIR}|E) = O(\mathtt{SMIR}|\mathtt{RCIB}) = 0.6186$$

by using the consistent interpolation shown in Fig. 12.1. These odds are equivalent to the a posteriori probability $P(\mathtt{SMIR}|E) = 0.382$.

The increase of the probability of SMIR from 0.03 to 0.382 now influences the probability of HYPE: from the a priori probability $P(\mathtt{HYPE}) = 0.01$ we compute the prior odds $O(\mathtt{HYPE}) = \frac{1}{99}$. Again applying (12.29) we get

$$O(\mathtt{HYPE}|\mathtt{SMIR}) = \lambda \cdot O(\mathtt{SMIR}) = 300 \cdot \frac{1}{99} = 3.03,$$

$$P(\mathtt{HYPE}|\mathtt{SMIR}) = \frac{3.03}{4.03} = 0.752.$$

Applying the interpolation (12.37) we compute the a posteriori probability $P(\mathtt{HYPE}|E) = 0.279$.[1] This change of the probability of HYPE now influences the probability of FLE, and so on. ◇

[1] This result is different from that given in [Duda, Gaschnig, and Hart 1981] because there an interpolation is chosen which is different from the one given by (12.37).

As the third type, the *contextual relations* express a condition that must be established before an assertion can be used in the reasoning process. They are also used when an assertion is geologically significant only if another assertion has already been established. For example, the whole network-part "favorable regional environment" shown in Fig. 12.2 is not considered unless there are "granitic intrusives in region". In the figure the two boxes are depicted by a dashed arrow.

12.3.3 Discussion and Related Work

We have presented the theoretical foundations of the model of inference networks. As shown, the approach is mainly based on probability theory. For practical reasons the probabilistic model is further extended by making use of a consistent interpolation function. The preceding section further shows how to draw inferences on the basis of uncertain knowledge in the system PROSPECTOR. When examining this real world application, it becomes visible that, on the basis of inference networks, the construction of knowledge bases requires great expenditure of time and data. For instance, a large amount of a priori knowledge is taken for granted: apart from two likelihood ratios for each rule the expert has to know the prior probabilities of subhypotheses.

A first attribute of the model of inference networks is that the probabilities of related (sub-)hypotheses need not sum up to 1. The reason is that, in complex knowledge bases, a priori probabilities as well as likelihood ratios may be inconsistent (which is possible because of the approximation (12.37)). Consequently, the truth of

$$\sum_{i=1}^{n} P(H_i) = 1$$

cannot be guaranteed. If, for instance, the completeness of sets of hypotheses and the disjointness of hypotheses are required, the probabilities have to be normalized.

An important problem of the approach of Duda et al. is that various kinds of *independencies* are assumed which may not conform to the real world. For instance, the second assumption (12.40) describes the conditional independence of pieces of evidence under the condition of the truth of the complementary hypotheses. Pearl argues that just this assumption cannot be met in the real world [Pearl 1982]. Also Heckerman [Heckerman 1986] considers this problem and examines the importance of the two assumptions (12.39) and (12.40) on the basis of examples.

In inference networks, only *binary variables* and characteristics are allowed. As mentioned by Pednault et al. [Pednault, Zucker, and Muresam 1981] and Glymour [Glymour 1985], and shown by Johnson [Johnson 1986] and Steve [Steve 1986], the extension of the model to multivalued variables will lead to inconsistencies. In the following we examine this important criticism that concerns the correctness and consistency of the mathematical (i.e. probabilistic) basic model, in particular the equations (12.41) and (12.42) together with the independence assumptions (12.39) and (12.40).

As argued in [Pednault, Zucker, and Muresam 1981] the model of inference networks fails if the completeness and pairwise disjointness of *more than two* elements of a hypotheses set is assumed. However, the proof by Pednault et al. was incorrect because it was based on an older theorem which itself was incorrectly proved. This was shown by Glymour [Glymour 1985] with the help of a very simple and special numerical example.

Complete proofs of the non-correctness of the mathematical theory of inference networks, which was already conjectured by Pednault, are given in [Johnson 1986] and [Steve 1986]. For instance, Johnson proves the following theorem:

(12.3) Theorem. *Assume the jointly exhaustive and mutually exclusive hypotheses $H_1, \ldots, H_n$. Further assume the evidence propositions $X_1, \ldots, X_m$ and that*

(i) $n > 2$,

(ii) $\sum_{i=1}^{n} P(H_i) = 1\ ,\ \forall i:\ P(H_i) > 0$,

(iii) $H_i \cap H_j = \emptyset\ ,\ i \neq j$,

(iv) *for all H_i the independencies (12.39) and (12.40) hold.*

Then for every hypothesis H_i there is at most one X_j that produces updating for H_i.

Proof. We assume $0 < P(H_i) < 1$, since otherwise no update for H_i is possible. We follow Johnson [Johnson 1986] in assuming $m = 2$ so that the antecedents of interest are X_1 and X_2. Following [Pednault, Zucker, and Muresam 1981], we derive

$$P(X_1, H_i) \cdot P(X_2, H_i) = P(X_1, X_2, H_i) \cdot P(H_i)$$
$$P(X_1, \overline{H_i}) \cdot P(X_2, \overline{H_i}) = P(X_1, X_2, \overline{H_i}) \cdot P(\overline{H_i})$$

from the independence assumptions (12.39) and (12.40). Since

$$P(A, \overline{B}) = P(A) - P(A, B)$$

holds, we get

$$[P(X_1) - P(X_1, H_i)] \cdot [P(X_2) - P(X_2, H_i)] \\ = [P(X_1, X_2) - P(X_1, X_2, H_i)] \cdot P(\overline{H_i})$$

from which

$$P(X_1) \cdot P(X_2) - P(X_1) \cdot P(X_2, H_i) - P(X_1, H_i) \cdot P(X_2) \\ = P(X_1, X_2) \cdot [1 - P(H_i)] - P(X_1, X_2, H_i) \tag{12.44}$$

follows. As shown by Johnson, by summing over i we obtain the equation

$$n \cdot P(X_1) \cdot P(X_2) - P(X_1) \cdot P(X_2) - P(X_1) \cdot P(X_2) \\ = P(X_1, X_2) \cdot (n-1) - P(X_1, X_2)$$

or

$$(n-2) \cdot P(X_1) \cdot P(X_2) = P(X_1, X_2) \cdot (n-2)$$

Because $n \geq 3$,

$$P(X_1) \cdot P(X_2) = P(X_1, X_2) \tag{12.45}$$

holds. Now using (12.44) and (12.45) we obtain

$$P(X_1) \cdot P(X_2, H_i) + P(X_1, H_i) \cdot P(X_2) \\ = P(X_1) \cdot P(X_2) \cdot P(H_i) + P(X_1, X_2, H_i)$$

and with (12.39)

$$P(X_1) \cdot P(X_2|H_i) \cdot P(H_i) + P(X_1|H_i) \cdot P(H_i) \cdot P(X_2) \\ = P(X_1) \cdot P(X_2) \cdot P(H_i) + P(X_1|H_i) \cdot P(X_2|H_i) \cdot P(H_i)$$

which implies

$$0 = [P(X_1) - P(X_1|H_i)] \cdot [P(X_2) - P(X_2|H_i)]$$

Because one of the factors must be 0, the equation

$$P(X_j|H_i) = P(X_j|\overline{H_i}) = P(X_j)$$

holds for either $j = 1$ or $j = 2$. Since then also

$$P(H_i|X_j) = \frac{P(H_i, X_j)}{P(X_j)} = P(H_i)$$

is true, X_1 and X_2 do not both produce updating for H_i. The more general case $m > 2$ can be proven analogously. □

This theorem has the following consequence: if an inference process is to be performed for more than two disjoint and exhaustive hypotheses which are confirmed by evidence, then this has to be done without one of the independence assumptions (12.39) and (12.40). Under this restriction the above theorem cannot be proved. Updating schemes based on both (12.39) and (12.40) are too restrictive.

Based on Johnson's theorem, Steve [Steve 1986] further shows that *if E is a piece of evidence, i.e. it is informative on the proper partition $H_1, \ldots, H_n$ ($n > 2, P(H_i) \neq 0$), its informativeness cannot be expressed by any constant odds multiplier.*

This means that under the mentioned conditions the update process cannot be performed with constant λ and ξ. For the more formal version of this assertion and the related proof the reader is referred to [Steve 1986].

Also concerning the real world use of inference networks, Lesmo [Lesmo, Saitta, and Torasso 1985] refers to the fact that already small changes of the quantities can lead to large changes of the final probabilities. However, quantities which are based on the experience of an expert, and are therefore subjective, of course may differ slightly.

The success of the real system PROSPECTOR, which indeed is based on the model of inference networks, may be traced back to the fact that at least one of the preconditions made for Johnson's theorem is not met: in relation to PROSPECTOR, Duda et al. formulate the independence assumptions (12.39) and (12.40) for dichotomous frames, i.e. for two hypotheses H and $\overline{H}$ only.

12.4 Decomposable Graphical Models

The approach recently undertaken by Pearl and Spiegelhalter for reasoning under uncertain knowledge is based on so-called *Bayesian networks* [Kim and Pearl 1983], [Pearl 1986a,b,c], [Pearl 1987]. The fundamental concept of using networks for managing uncertainty is to allow the explicit representation of *dependencies* as well as *independencies*. The model of Bayesian networks introduced in Sect. 12.4.1 represents direct application or realization of probability theory and can also be viewed as an extended and consistent reformulation of the older ideas of Duda, Hart, and Nilsson [Duda, Hart, and Nilsson 1976] who proposed the *inference networks* looked at in the preceding section. Similarities but also differences between these two approaches are pointed out in the sequel.

One of the first applications based on Pearl's ideas was presented by Andreassen et al. [Andreassen, Woldbye, Falck, and Andersen 1987]. They introduced the expert system MUNIN (MUscle and Nerve Inference Network) for interpretation of electromyographic findings. The system is a non-trivial prototype implementation for handling uncertain knowledge on the basis of networks. An overview is given in Sect. 12.4.2. In Sect. 12.4.3 we introduce the shell HUGIN (Handling Uncertainty in General Inference Networks) [Andersen, Olesen, Jensen, and Jensen 1989].

12.4.1 The Model of Pearl

A *Bayesian network* is a directed acyclic graph in which the nodes represent propositions and arcs stand for the existence of direct causal influences between nodes. The strengths of these dependencies are quantified by *conditional probabilities*. For instance, the (conditional) dependence of the two characteristics x_i and x_j may be given by $P(x_i|x_j)$. Using the chain-rule representation of joint probability distributions, then, choosing an arbitrary order on the characteristics $x_1,\ldots,x_n$, we can write

$$P(x_1,x_2,\ldots,x_n) = P(x_n|x_{n-1},\ldots,x_1)\cdot\ldots\cdot P(x_3|x_2,x_1)\cdot P(x_2|x_1)\cdot P(x_1) \tag{12.46}$$

Suppose a directed acyclic graph G in which the arrows pointing at each node x_i emanate from a set S_i of parent nodes judged to be *directly* influencing x_i. Therefore, S_i is a set of characteristics which, fixing their values, would shield x_i from the influence of all other predecessors of x_i, i.e.

$$P(x_i|S_i) = P(x_i|x_{i-1},\ldots,x_1) \tag{12.47}$$

Taking into account the meaning of $P(x_i|S_i)$ given by equation (12.47) and the fact that a separate assessment of each child-parents relationship should suffice (i.e., $S_i \subseteq \{x_1,\ldots,x_{i-1}\}$), then the chain-rule representation (12.46) can be simplified. The joint probability $P(x_1,\ldots,x_n)$ supporting the overall assessment is given by

$$P(x_1,\ldots,x_n) = \prod_i P(x_i|S_i). \tag{12.48}$$

Figure 12.3 presents an example of a Bayesian network. Assuming that the belief in the "proposition x_i" is influenced by its *direct* parent nodes only, the distribution corresponding to the graph of Fig. 12.3 can be read as

$$\begin{aligned}P(x_1,\ldots,x_7) =& P(x_7|x_6)\cdot P(x_6|x_4,x_5)\cdot P(x_5|x_2)\cdot P(x_4|x_3)\cdot\\ & P(x_3|x_1,x_2)\cdot P(x_1)\cdot P(x_2)\end{aligned}$$

Dependency relations between nodes can be detected by looking at the "blocking" of paths between the nodes (compare Sect. 12.5.2). The following

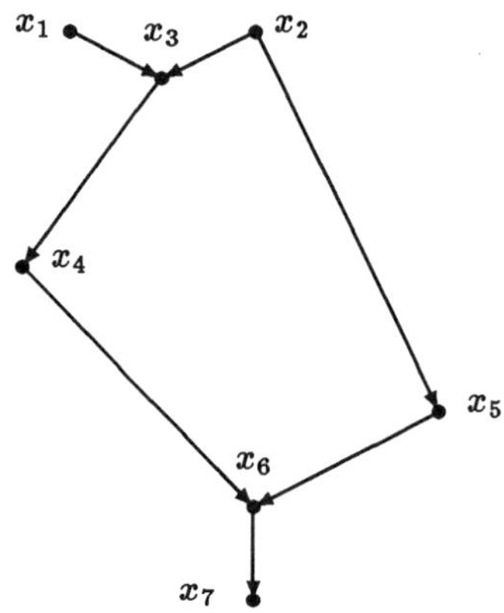

Fig. 12.3. A Bayesian network

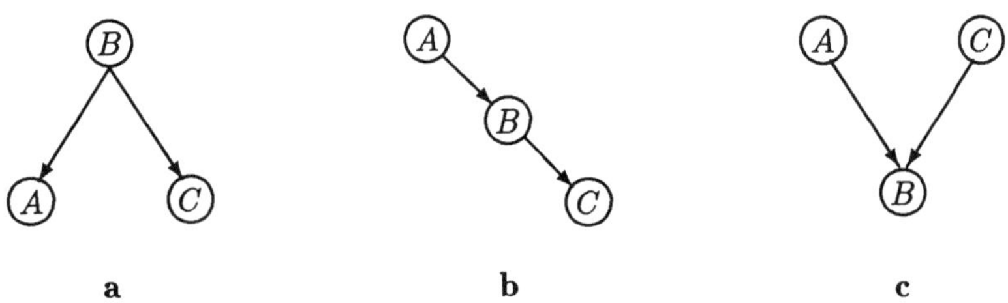

Fig. 12.4. Types of blocking

definitions are applicable to paths with directed links and sensitive to all the characteristics for which values are known at a given time. These characteristics with known values are said to be *instantiated.*

A path between nodes is said to be *blocked* if one of the following two conditions is fulfilled:

- **Diverging or cascading arrows:**
  A path is said to be blocked if an instantiated characteristic is present between two diverging or cascading arrows. In the cases (a) and (b) shown in Fig. 12.4 the two links meeting head-to-tail or tail-to-tail at node B are

blocked by B. A and C are conditionally independent, given B, so, e.g.,

$$P(C|A,B) = P(C|B)$$

and therefore

$$P(A,C|B) = P(A|B) \cdot P(C|B). \tag{12.49}$$

- **Converging arrows:**
  A path down from a node N is said to be blocked if none of its descendants is instantiated. Thus normally such a path is "blocked", unless N or any of its descendants have been instantiated. In the case (c) of Fig. 12.4 the two links meeting head-to-head at node B are *blocked by B* if neither B nor any of its descendants are instantiated. A and C are marginally independent, so, e.g.,

  $$P(C|A) = P(C)$$

  but may become dependent, given the value of B or the value of one of B's descendants.

Returning to Fig. 12.3, suppose we want to know whether there is a dependency relation between the nodes x_4 and x_5. Let S_j be sets of nodes that are instantiated. Given the set $S_1 = \{x_2\}$, the nodes x_4 and x_5 are independent because node x_2 blocks a path and no descendants of the nodes x_4 and x_5 are instantiated. But if we take another set, $S_2 = \{x_2, x_7\}$, the nodes x_4 and x_5 may not have to be independent because the node x_7 opens a path by "opening" the head-to-head connection at x_6.

To describe the process of *fusing and propagating* new information and belief through a Bayesian network, we restrict ourselves to simple tree-structured networks where each node has only one parent and – as a consequence – the converging arrows mentioned above cannot appear.

Departing from Fig. 12.3, characteristics are labelled by the capital letters A, B, C now, and their possible, mutually exclusive values subscripted: $A_1, A_2, \ldots$, $B_1, B_2, \ldots$, and $C_1, C_2, \ldots$ In order to be able to propagate belief values through such a network, the following three entities are maintained with each node (below, with the node B), and also these entities are shown in Fig. 12.5(a):

- a *fixed* conditional probability matrix $M(B|A)$ relating the node B to its immediate parent A. The matrix

  $$M(B|A)_{ij} = P(B_j|A_i)$$

  quantifies the strength of the directed link $A \to B$.

- the *current* strength of the causal or prospective support attributed to the possible values B_j of node B by the set of ancestors of B:

$$\pi(B_1), \pi(B_2), \ldots, \pi(B_n) \tag{12.50}$$

- the *current* strength of the diagnostic or retrospective support node B receives from B's descendants (i.e. the set of outgoing links):

$$\lambda(B_1), \lambda(B_2), \ldots, \lambda(B_n) \tag{12.51}$$

The meaning of the entities $\lambda(B_j)$ and $\pi(B_j)$ can be exemplified by looking at the amount of belief that is attached to a node B. Following Pearl [Pearl 1986c], the characteristics D_B^- – the data contained in network rooted at B – and D_B^+ – the data contained in the rest of the network – are introduced (Fig. 12.5(b)). Note that the problem of graph decomposition is considered to some more detail in Sect. 7.4. The amount of belief attached to B induced by the data available in $D_B^+ \cup D_B^-$ is given by the conditional probabilities

$$\mathrm{Bel}(B_j) \stackrel{\mathrm{d}}{=} P(B_j | D_B^+, D_B^-) = \gamma \cdot P(D_B^- | B_j) \cdot P(B_j | D_B^+) \tag{12.52}$$

This is a result from Bayes' theorem, where γ is a renormalization factor. Notice that in this context Bel has a meaning different from that used in the framework of the Dempster-Shafer theory. Looking at the measures $\lambda(B_j)$ and $\pi(B_j)$ introduced above, they can now be identified as

$$\begin{aligned} \lambda(B_j) &= P(D_B^- | B_j), \\ \pi(B_j) &= P(B_j | D_B^+). \end{aligned} \tag{12.53}$$

The belief vector $(\mathrm{Bel}(B_1), \mathrm{Bel}(B_2), \ldots, \mathrm{Bel}(B_n))$ reflects the overall belief currently accorded to the propositions "$B = B_j$", $j=1, \ldots, n$. This total belief can be computed using the locally available information only:

$$\mathrm{Bel}(B_j) = \gamma \cdot \lambda(B_j) \cdot \pi(B_j). \tag{12.54}$$

Before describing the way in which the task of propagating information through the network can be solved, an important difference compared to the models of the preceding sections should be pointed out: whereas in other models and systems (e.g. MYCIN) expert rules point from evidence to hypothesis (e.g. "if symptom, then disease"), the arrows in Bayesian networks point from causes (e.g. disease) to effects (e.g. symptom) (compare the remarks to equations (12.50) and (12.51)).

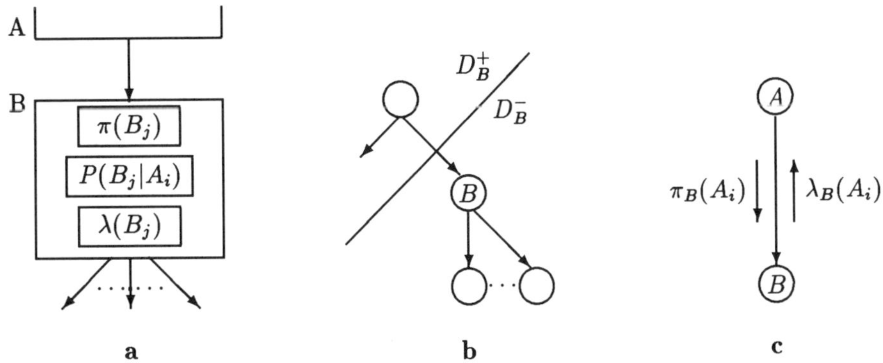

Fig. 12.5. (a) locally stored data, (b) data partitioning, (c) flow of information (from [Heinsohn and Van Loon 1988])

As reason for his choice Pearl mentions that "people often prefer to encode experiential knowledge in causal schemata and, as a consequence, rules expressed in causal format are assessed more reliably" [Pearl 1986c] (compare also the remarks of Sect. 12.1 concerning the different possibilities for making use of Bayes' formula).

However, the direction of the existing arrows does not influence the flow of information that can be executed in both directions and is triggered by two kinds of incoming information:

- *specific evidence* that arises with direct observations, and validates – with certainty – the values of a characteristic stored within a node of the network,
- *virtual evidence* that affects the belief stored in certain characteristics but corresponds to judgments only, based on undisclosed observations.

Belief propagation through a simple binary tree is triggered by changes in the belief parameters of neighboring nodes. The π-messages are sent top-down to "sensory nodes" while λ-messages are propagated bottom-up to the parent nodes (Fig. 12.5(c)). When a node is activated by receiving a message communicated by its parent or by one of its children, it first updates its parameters λ

and π. After that, this node computes messages to be posted to the neighbaring nodes. The values of the vectors are calculated by using the following formulas:

Local update triggered by the incoming data:

- $\lambda(B_j) = P(D_B^-|B_j) = \prod_{k \text{ children of } B} \lambda_k(B_j)$ [2]
- $\pi(B_j) = \sum_i P(B_j|A_i) \cdot \pi_B(A_i) \;=\; M(B|A) \times (\pi_B(A_1), \ldots, \pi_B(A_m))$ [3]
- the current belief $\text{Bel}(B_j)$ as described with (12.54)

Messages computed and posted on the message boards:

- Bottom-up:

$$\lambda_B(A_i) = \sum_j P(B_j|A_i) \cdot \lambda(B_j) \;=\; M^\top(B|A) \times (\lambda(B_1), \ldots, \lambda(B_n))$$

- Top-down: $\pi_E(B_j) = \alpha \cdot \dfrac{\text{Bel}(B_j)}{\lambda_E(B_j)}$

For proofs and examples, the reader is referred to Pearl's original papers. Applying the formulas given above, evidence can now be propagated through the network. The nodes that are affected by evidence recompute their λ and π vectors and send messages to other connected nodes. These nodes may find themselves in a similar situation.

Associated with the use of tree structures as treated above is the fact that the state of *exactly one* characteristic (the one parent node) can serve as a cause of the state of any other characteristic. A representation formalism which does not have this restriction is given by ***singly connected networks*** which allow multiple parents. A fragment of such a network is illustrated in Fig. 12.6. It is important to note that a perturbation of the causal parameter π does not affect the diagnostic parameter λ on the same link, and vice versa. As a result, the evidence is propagated through the network until the boundaries are reached. A new state of equilibrium will be reached after a finite number of updates which, in the worst case, would be equal to the diameter of the network.

---

[2] In the inference algorithms presented in the preceding sections, rules with this intent are designated as "(parallel) combination rules". The reader is also referred to the equation (12.49) which is a simplified ($k = 2$) form of the formula.

[3] To use a normalizing constant as in [Pearl 1986c] is not necessary. Both, matrix and vector, are probability distributions.

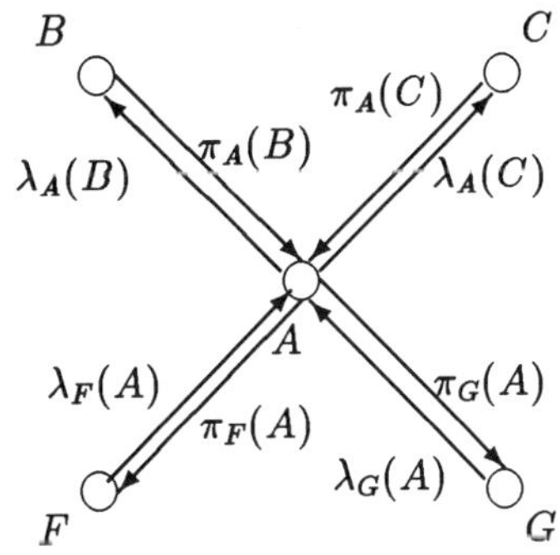

Fig. 12.6. Updating a singly connected network

12.4.2. MUNIN – An Application

One of the first applications based on Pearl's ideas was presented by Andreassen et al. [Andreassen, Woldbye, Falck, and Andersen 1987]. They introduced the expert system MUNIN[4] (MUscle and Nerve Inference Network) for interpretation of electromyographic findings. Electromyography is the diagnosis of muscle and nerve diseases through analysis of bioelectrical signals from muscle and nerve tissue.

The system is a non-trivial prototype implementation for handling uncertain knowledge on the basis of networks. Generally, this short overview aims to give the basic aspects of MUNIN, and more applicational aspects such as medical background information will be incompletely reflected.

Figure 12.7 shows the internal nodes of the network as well as the a priori probability distribution.

The left layer of nodes represents the different disease states which are taken into account by the system. The number of diseases is restricted to three, each with two to four states. In addition, the patient may be either in the state `other` if he has a disease not contained in the set of possible disease states, or in the state `normal` if he has no disease with neuro-muscular disorder.

The middle layer represents eight pathophysiological nodes, each of them with from two to nine states, and the right layer a set of fifteen finding nodes.

[4] "According to Norse mythology Munin and Hugin are the two ravens whispering intelligence into the ear of the god Odin" [Andreassen, Woldbye, Falck, and Andersen 1987].

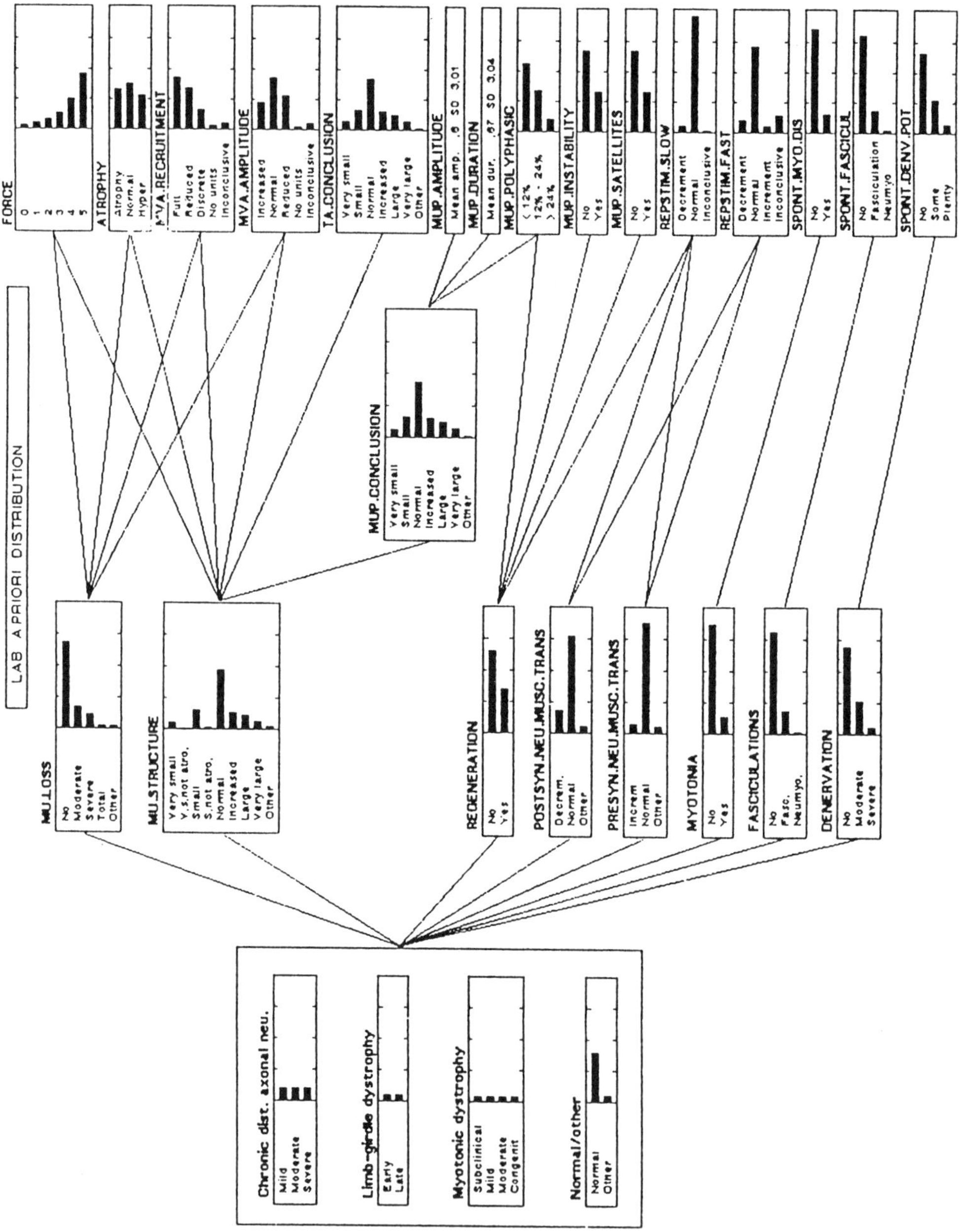

Fig. 12.7. The a priori probabilities
[Andreassen, Woldbye, Falck, and Andersen 1987]

The intermediate node MUP.CONCLUSION does not have a natural pathophysiological interpretation. For each node the probability of its states is indicated by the length of the horizontal bars.

		MU.LOSS				
		no	mod	sev	total	other
DISORDER	⋮					
	Chronic axonal neuropathy:					
	mild	0.60	0.33	0.04	0.01	0.02
	moderate	0.16	0.64	0.16	0.02	0.02
	severe	0.02	0.25	0.61	0.10	0.02
	⋮					

Fig. 12.8. The probabilities $P(\texttt{MU.LOSS}|\texttt{DISORDER})$ given 3 different disorders

Every causal link between nodes is associated with a conditional probability matrix. For instance, Fig. 12.8 shows the conditional probabilities $P(\texttt{MU.LOSS}|\texttt{DISORDER})$ for the states of `MU.LOSS` (i.e. "loss of motor units") given three of the possible eleven different disorders.

The *a priori distribution* concerning the network concepts can be determined by initializing the disease node with a priori probabilities corresponding to the observed frequencies of the diseases in the patients.

Doing so we obtain the expectations for the pathophysiological changes. Finally, the pathophysiological nodes generate expectations for the fifteen finding nodes through their causal links. Additionally, this way to generate expectations corresponding to a single disorder can be used to verify the network of medical concepts and causal links. The expected findings may be presented to a medical expert who has to identify discrepancies between his expectations and those computed by the system.

The other way of verifying the network is to enter findings typical for a disease and to compare the diagnosis of the medical expert and the probabilities generated by MUNIN. For example, in Fig. 12.9 findings corresponding to a case of "chronic axonal neuropathy" are entered. At the beginning of a diagnostic session the findings have to be entered into the finding nodes. As shown in Fig. 12.9, entered findings are indicated by broken horizontal 100% bars.

The diagnostic task then consists of updating the probabilities of all other nodes. Consequently, not only the probabilities of the disease node but also the probabilities of the remaining findings are adjusted. So the updated network probability distribution delivers diagnosis as well as predictions for remaining findings. The predicted findings still have to be verified.

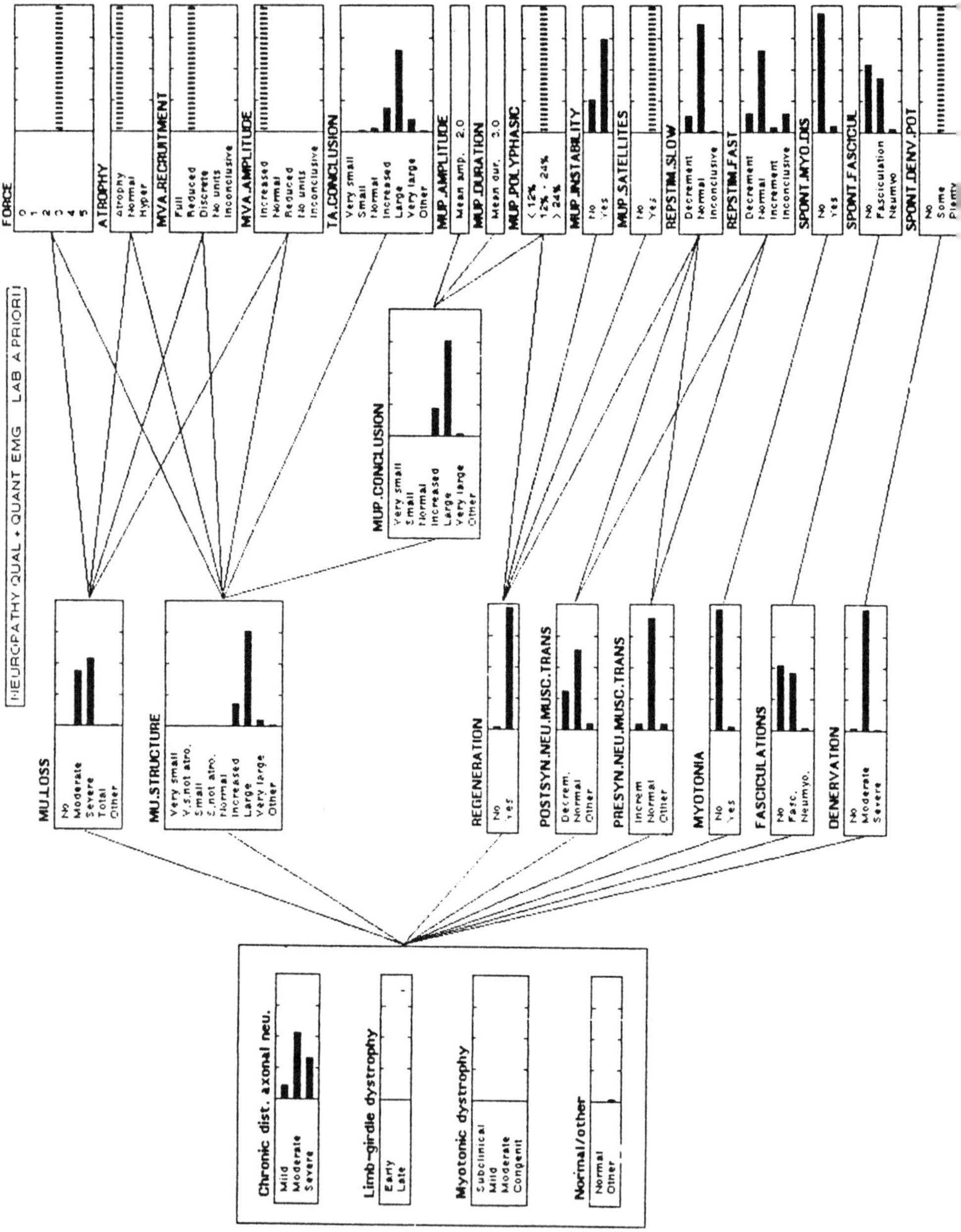

Fig. 12.9. Entering findings for a typical case of "moderate chronic axonal neuropathy" [Andreassen, Woldbye, Falck, and Andersen 1987]

The MUNIN network of Fig. 12.9 covers also the case where a node has more than one parent. An example is given with the finding node FORCE that has MU.LOSS and MU.STRUCTURE as parents. Consequently, the conditional probability matrix $P(\texttt{FORCE}|\texttt{MU.LOSS},\texttt{MU.STRUCTURE})$ contains $6 \times 5 \times 9$ probability numbers for all combinations of FORCE, MU.LOSS, and and MU.STRUCTURE, i.e. the cycle was "broken" by the joining of nodes.

However, estimating the 270 conditional probabilities on the basis of only empirical knowledge seems unpractical. For the development of MUNIN, the non-trivial task of knowledge acquisition was solved by making use of "deep knowledge" as much as possible: deep knowledge as expounded in textbooks and papers is used to reduce the problem of estimating thousands of conditional probabilities to a manageable size. A procedure for taking such commonsense knowledge into account is given in [Andreassen, Woldbye, Falck, and Andersen 1987].

An important demand regarding diagnostic systems, for example, is that they must be able to know their own limits. For this purpose the state other is attached to some of the nodes: if MUNIN is confronted with a situation outside the limits of its knowledge this will be signaled by giving other a high probability. The situation that other is supported instead of a known disease, for instance, leads to two possible interpretations: on one hand the high probability may point to an *incompleteness* of medical knowledge which was not taken into account during knowledge acquisition. On the other hand, some of the entered findings may be erroneous so that there is conflicting evidence.

12.4.3 HUGIN – A Professional Tool

The implementation of MUNIN showed that, based on the local computation paradigm, efficient and theoretically sound inference algorithms can be developed. As a consequence of this, HUGIN, an extended shell of MUNIN, was built. HUGIN is a software package for "Handling Uncertainty in General Inference Networks" and consists of the following components.

- The *editor*, which is used to specify a directed acyclic graph. It is used to specify nodes (random variables), the possible values (states), and links with the respective conditional probability tables. For simplicity we assume here that $V = \{1, \ldots, n\}$ represents the nodes and that $\Omega^{(i)}$ contains the possible values with respect to node i.
- The *compiler* that transforms this graph (which possibly contains cycles) into a tree of cliques. This is the main theoretical improvement in comparison to MUNIN and relies on improved results presented in [Lauritzen and Spiegelhalter 1988], therefore the compiler is considered in more detail later on.

- The *inference engine* that performs the reasoning process. In addition there is an interface that allows access to the inference engine from C-programs (API $\equiv$ application program interface) and a graphical runtime system.

The *qualitative* knowledge, i.e., the knowledge about dependencies, is represented by a directed acyclic graph $G = (\{1, \ldots, n\}, \mathfrak{D})$. From this graph the compiler of HUGIN constructs a tree of cliques (compare Theorem 7.38) in several steps:

(*i*) It forms the corresponding moral graph (Definition 7.41), i.e., the parents of each node are "married" and the directions in the graph are dropped. The result is an undirected graph.

(*ii*) It triangulates the moral graph by creating an order of the vertices using maximum cardinality search and fill-in algorithms with respect to that order (compare Theorem 7.37).

(*iii*) It identifies the (maximal) cliques in the graph emerging from steps (*i*) and (*ii*) and orders them according to the highest labeled vertex according to the order obtained in (*ii*). As a result we obtain a hypertree with hyperedges $C_1, \ldots, C_p$ where we can assume that $C_1 \underset{\alpha}{<} \ldots \underset{\alpha}{<} C_p$ holds. This tree has the property that if a node v is contained in any two cliques C_i and C_j, then v is also contained in all cliques on the clique path between C_i and C_j in the tree. Here the relation to Markov trees become evident, compare Def. 7.21.

(*iv*) It calculates the sets

$$S_i = C_j \cap (C_1 \cup \ldots \cup C_{i-1}) \qquad \text{and}$$
$$R_i = C_i \cap \overline{S_i} \tag{12.55}$$

for $i = 1, \ldots, p$, where $S_1 = \emptyset$ and $R_1 = C_1$. If $i > 1$ then it identifies a $j < i$ such that $S_i \subseteq C_j$. If there are more than one such j, the choice is arbitrary. The hyperedge C_i is then a parent clique of C_j. S_i is often called a *separator set.* The compiler then creates a tree of cliques according to the parent relationship obtained. A separator set is in essence the intersection of a Markov tree twig with its branch, see Def. 7.9. Note that we clarify some notations before we continue in describing step (*v*).

(12.4) Example. Consider the directed acyclic graph in Fig. 12.10a. We obtain $C_1 = \{1,2,3\}$, $R_1 = \{1,2,3\}$, $S_1 = \emptyset$, $C_2 = \{2,3,4\}$, $R_2 = \{4\}$, $S_2 = \{2,3\}$, $C_3 = \{3,5\}$, $R_3 = \{5\}$, $S_3 = \{3\}$. The tree of maximal cliques is depicted in Fig. 12.10b. In Fig. 12.10c we give a representation by explicitly drawing the separator sets. ◇

The *quantitative* knowledge of the expert has to be specified by a list of conditional probability tables. The tables express for each combination of states

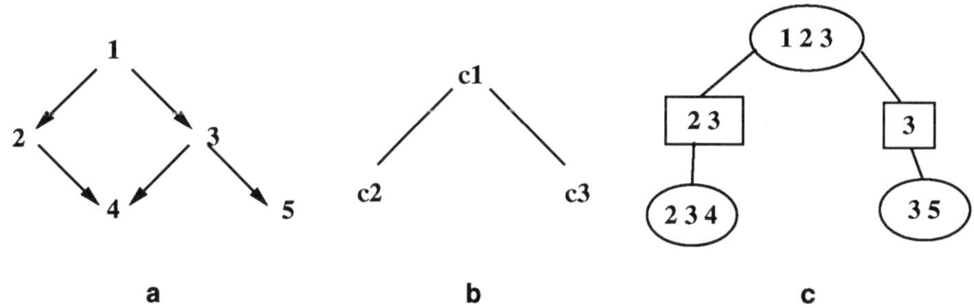

Fig. 12.10. Transformation of a directed, acyclic graph into a tree of cliques.

$\prod\limits_{w \in \mathrm{Par}(v)} \Omega^{\{w\}}$ at the parent nodes

$$\mathrm{Par}(v) \stackrel{\mathrm{d}}{=} \{w \,|\, (w,v) \in \mathcal{D}\} \tag{12.56}$$

with states $\Omega^{\{v\}}$ at node v conditional probabilities

$$K_v\big(\{\omega^{\{v\}}\} \,|\, \{\omega^{\mathrm{Par}(v)}\}\big) \geq 0,$$

for vertices without parents the (a priori) probabilities

$$K_v\big(\{\omega^{\{v\}}\}\big) \geq 0.$$

In the notation concerning dependency networks (see Sect. 7.4) the body T_h of a directed hyperedge T contains the parents of h_T, i.e. $\mathrm{Par}(h_T) = T_h$, see Def. 7.28. The probabilities must satisfy the properties

$$\sum_{\omega^{\{v\}} : \omega^{\{v\}} \in \Omega^{\{v\}}} K_v\big(\{\omega^{\{v\}}\}\big) = 1, \quad \text{and}$$

$$\sum_{\omega^{\{v\}} \in \Omega^{\{v\}}} K_v\big(\{\omega^{\{v\}}\} \,|\, \{\omega^{\mathrm{Par}(v)}\}\big) = 1 \quad \text{for all } \omega^{\mathrm{Par}(v)} \in \Omega^{\mathrm{Par}(v)}.$$

The entirety of functions K_v is referred to as a directed Markov random field [Isham 1981] defined on the moral graph.

Since the graph $G = (V, \mathfrak{D})$ is assumed to be acyclic, there is exactly one probability P on $\Omega^V \stackrel{\mathrm{d}}{=} \prod\limits_{v \in V} \Omega^{\{v\}}$ with K_v as corresponding (conditional) probabilities. Let P_A denote the marginal probability of P with respect to Ω^A for each $A \subseteq V$.

In order to obtain a representation that allows local propagation we use the decomposition

$$P_V(\{\omega^V\}) = \prod_{i=1}^{p} P_{C_i}\Big(\{\Pi_{C_i}^V(\omega^V)\}\Big) \Big/ \prod_{i=2}^{p} P_{S_i}\Big(\{\Pi_{S_i}^V(\omega^V)\}\Big), \tag{12.57}$$

where each S_i is a separator as defined in equation (12.55).

From this representation the marginal distribution on any single node may be easily obtained from the distribution of any clique which contains v.

(12.4) Example. (Continuation)

Consider the graph of Fig. 12.10a, and define $\Omega^{\{i\}} = \{\omega_i, \overline{\omega_i}\}$ for $i = 1, \ldots, 5$. Then the transition probabilities defined by

$$\begin{array}{ll}
K_1\big(\{\omega_1\}\big) = 0.2, & \\
K_2\big(\{\omega_2\} \mid \{\omega_1\}\big) = 0.8, & K_2\big(\{\omega_2\} \mid \{\overline{\omega_1}\}\big) = 0.2, \\
K_3\big(\{\omega_3\} \mid \{\omega_1\}\big) = 0.2, & K_3\big(\{\omega_3\} \mid \{\overline{\omega_1}\}\big) = 0.05, \\
K_4\big(\{\omega_4\} \mid \{(\omega_2, \omega_3)\}\big) = 0.8, & K_4\big(\{\omega_4\} \mid \{(\overline{\omega_2}, \omega_3)\}\big) = 0.7, \\
K_4\big(\{\omega_4\} \mid \{(\omega_2, \overline{\omega_3})\}\big) = 0.9, & K_4\big(\{\omega_4\} \mid \{(\overline{\omega_2}, \overline{\omega_3})\}\big) = 0.05, \\
K_4\big(\{\omega_5\} \mid \{\omega_3\}\big) = 0.8, &
\end{array}$$

determine a unique probability P on $\Omega^{\{1\}} \times \Omega^{\{2\}} \times \ldots \times \Omega^{\{5\}}$. In order to obtain the representation (12.57) we have to calculate the marginal probabilities of the "clique spaces" $\Omega^{\{1,2,3\}}$, $\Omega^{\{2,3,4\}}$, $\Omega^{\{3,5\}}$ as well as of the "separator spaces" $\Omega^{\{2,3\}}$ and $\Omega^{\{3\}}$ from the functions K_i. In this example it is an easy exercise left to the reader, but in general it is of course rather time consuming. In [Lauritzen and Spiegelhalter 1988] it is shown how to perform this task efficiently by using potential representation. ◇

Let P be a probability on Ω^V. A potential representation of P consists of a list Δ of nonempty subsets A of V and corresponding potentials ψ_A, $A \in \Delta$, which are real valued non-negative functions depending only on the states in Ω^A such that

$$P_V(\{\omega^V\}) \;=\; K \cdot \prod_{A:A\in\Delta} \psi_A(\Pi_A^V(\omega^V)), \quad \omega^V \in \Omega^V, \tag{12.58}$$

where K is a normalizing constant.

If the list of subsets is chosen to be the cliques $C_1, \ldots, C_p$ and the separators $S_2, \ldots, S_p$ constructed above, then a potential representation for P can be obtained very easily by using the conditional probability tables:

For each $v \in V$ assign a clique C_i with the property

$$\{v\} \cup \mathrm{Par}(v) \subseteq C_i.$$

This is always possible: in case that several cliques have that property, the choice is arbitrary (but then fixed). The clique assigned to v is denoted by cli(v). We obtain a potential representation of P by using the potentials

$\psi_C : \Omega^C \to [0, \infty),\ C \in \{C_1, \ldots, C_p\}$,

$$\psi_C(\omega^C) = \prod_{\substack{v:v \text{ root and} \\ \text{cli}(v)=C}} K_v\Big(\{\Pi^C_{\{v\}}(\omega^C)\}\Big) \cdot \prod_{\substack{v:v \text{ no root and} \\ \text{cli}(v)=C}} K_v\Big(\{\Pi^C_{\{v\}}(\omega^C)\} \mid \{\Pi^C_{\text{Par}(v)}(\omega^C)\}\Big),$$

$\omega^C \in \Omega^C$. If no v is assigned to a clique C we set $\psi_C(\omega^C) = 1$ for all $\omega^C \in \Omega^C$. Furthermore we set $\psi_{S_i} \equiv 1$ for $i = 2, \ldots, p$. In this special case we obtain $K = 1$.

Now we can continue to describe the methods used in by HUGIN to handle *quantitative* knowledge, which also involve several steps.

(v) Initialization
HUGIN creates the tree of cliques, where in each vertex the clique C_i, the separator S_i as well as the potentials ψ_{C_i} and ψ_{S_i} are stored. At initialization time the potentials ψ_{S_i} are identical to 1. We have the representation

$$P_V(\{\omega^V\}) = \prod_{i=1}^{p} \psi_{C_i}(\Pi^V_{C_i}(\omega^V)) \cdot \prod_{i=2}^{p} \psi_{S_i}(\Pi^V_{S_i}(\omega^V)), \quad \omega^V \in \Omega. \quad (12.59)$$

The point is that the form of this representation is *retained* under conditioning of P_V.

(vi) Instantiation of nodes
If $\omega^{\{i\}} \in \Omega^{\{(i)\}}$ is observed, then the node i is "instantiated", a clique containing i is selected and any potential not defined on the observed value is set to zero. This may occur at various parts of the network before the propagation takes place. Then the new evidence is distributed through the networks. The representation (12.59) allows an efficient local propagation scheme.

(vii) Propagation of evidence
A root clique is chosen arbitrarily in the clique tree. It sends out the message "collect evidence" to its neighbors, who communicate with *their* neighbors until it reaches the ends of the tree. Each clique then collects evidence from those neighbors *further* from the root clique, using the operations shown below for passing a message from clique A to clique B through separator S:

- calculate the marginal of ψ_A on Ω^S, denote it by ψ'_S,
- multiply ψ_B pointwise by $\frac{\psi'_S}{\psi_S}$,
- replace ψ_S by ψ'_S.

In collecting evidence each clique must await the messages from its more distant neighbors before passing on toward the root. When the root clique has collected evidence from its neighbors it normalizes its new potential to add to 1. It then distributes the evidence, which involves the same message-passing operations but working back through the tree. Afterwards all cliques and separators will hold their correct marginal distributions.

(12.5) Example. Consider the tree of cliques in Fig. 12.11a. If the node with the dot is instantiated then the flow of messages is indicated by the arrows. ◇

(12.4) Example. (Continuation)

We obtain

$$\psi_{\{2,3,4\}}(a,b,c) = K_3\big(\{c\} \mid \{b\}\big) \cdot K_2\big(\{b\} \mid \{a\}\big) \cdot K_1\big(\{a\}\big),$$

$$\psi_{\{1,2,3\}}(b,c,d) = K_4\big(\{d\} \mid \{b,c\}\big),$$

$$\psi_{\{3,5\}}(c,e) = K_5\big(\{e\} \mid \{c\}\big),$$

$$\psi_{\{2,3\}}(b,c) = 1,$$

$$\psi_{\{3\}}(c) = 1,$$

for each $(a,b,c,d,e) \in \Omega^{\{1,\dots,5\}}$ at the time of the initialization. Having observed the value ω_4, the marginal distributions of the cliques after the propagation process are as follows:

(b,c,d)	$P_{\{1,2,3\}}\big(\{(b,c,d)\}\big)$
$(\omega_2,\omega_3,\omega_4)$	0.093
$(\omega_2,\omega_3,\overline{\omega_4})$	0
$(\omega_2,\overline{\omega_3},\omega_4)$	0.733
$(\omega_2,\overline{\omega_3},\overline{\omega_4})$	0
$(\overline{\omega_2},\omega_3,\omega_4)$	0.081
$(\overline{\omega_2},\omega_3,\overline{\omega_4})$	0
$(\overline{\omega_2},\overline{\omega_3},\omega_4)$	0.093
$(\overline{\omega_2},\overline{\omega_3},\overline{\omega_4})$	0

(a,b,c)	$P_{\{1,2,3\}}\big(\{(a,b,c)\}\big)$
$(\omega_1,\omega_2,\omega_3)$	0.074
$(\omega_1,\omega_2,\overline{\omega_3})$	0.335
$(\omega_1,\overline{\omega_2},\omega_3)$	0.016
$(\omega_1,\overline{\omega_2},\overline{\omega_3})$	0.005
$(\overline{\omega_1},\omega_2,\omega_3)$	0.019
$(\overline{\omega_1},\omega_2,\overline{\omega_3})$	0.398
$(\overline{\omega_1},\overline{\omega_2},\omega_3)$	0.065
$(\overline{\omega_1},\overline{\omega_2},\overline{\omega_3})$	0.088

(c,e)	$P_{\{3,5\}}\Big(\{(c,e)\}\Big)$
(ω_3, ω_5)	0.139
$(\omega_3, \overline{\omega_5})$	0.035
$(\overline{\omega_3}, \omega_5)$	0.496
$(\overline{\omega_3}, \overline{\omega_5})$	0.330

Note that after knowing $P(\{\overline{d}\}) = 0$ we can use the reduced clique $\{2,3\}$ instead of $\{2,3,4\}$ which is done in HUGIN. $\diamond$

The complexity of these algorithms has been analysed in detail [Lauritzen and Spiegelhalter 1988]. It turns out that the algorithm has a computing time of order $O(2^m)$, where m denotes the maximal number of variables in a clique. In MUNIN, m is equal to 4, the tree of cliques has 25 vertices, the maximal number of combinations of values of variables in the cliques is 495, and the upper bound for the number of computations in the updating algorithm is 61875, a reasonably small number.

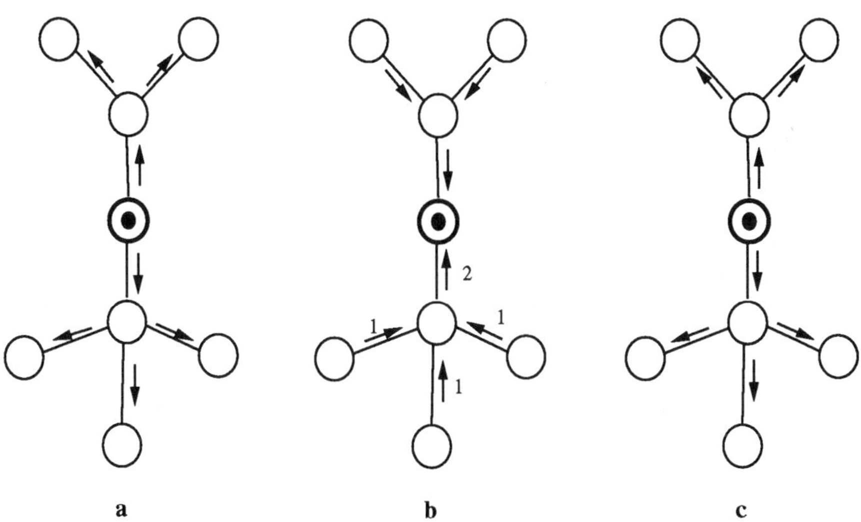

Fig. 12.11. (a) Send out message "collect evidence"; (b) collection of evidence, the 1-messages must be passed before the 2-messages can be carried out; (c) distribution of evidence

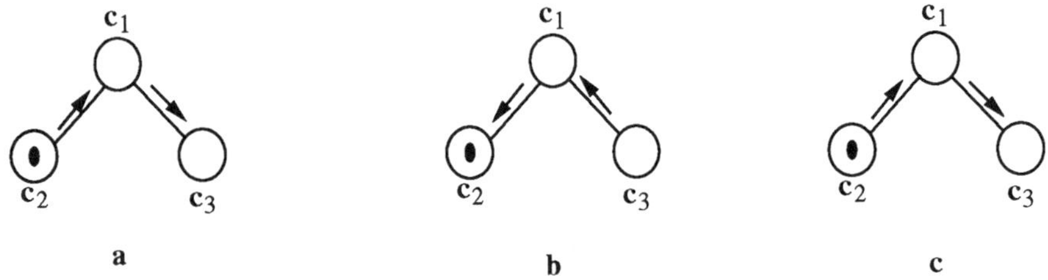

Fig. 12.12. Flow of messages: (a) Send out message "collect evidence"; (b) Collection of evidence; (c) Distribution of evidence

12.5 Propagation Based on Dependency Networks

In this section we describe in detail a *non-Bayesian model* based on dependency networks as considered in Sect. 7.4. The knowledge of each expert is encoded by sets of probabilities; in essence we consider the case (b) discussed in Sect. 6.7.

The model has been implemented by the authors, and it turns out to be very similar to Pearl's propagation scheme based on singly connected networks. HUGIN and the underlying model based on clique trees is of course much more powerful, but it is interesting to see, that starting with a constraint interpretation of expert statements one obtains the same formalisms as with a Bayesian point of view.

12.5.1 Knowledge Representation

First of all we assume that as a first step of the knowledge acquistion process the qualitative structure of the domain is determined in the form of a dependency network $C = (M, \mathfrak{T})$. This network induces a consistent numbering of the hyperedges in $\mathfrak{T}$. In order to simplify the notation we suppose the elements $T_1, \ldots, T_r$ to be consistently ordered. Each hyperedge $T_j \in \mathfrak{T}$ represents some subspace Ω^{T_j} which is subject to the knowledge we obtain from some expert space $\Theta_j = \{\theta_j\}$, i.e., we obtain a probability distribution on Ω^{T_j}. For example, θ_1 provides a marginal distribution $p_1 : \Omega^{h_{T_1}} \longrightarrow [0,1]$ (recall that T_1 is necessarily a root of the causal network) whereas in the general case θ_j provides a conditional probability distribution $p_j : \Omega^{(T_j)_h} \times \Omega^{h_{T_j}} \longrightarrow [0,1]$. So each expert θ_j determines a set $\mathcal{P}_j$ of probability distributions on Ω^M.

Let us start with expert θ_1. He considers those probability distributions to be admissible which marginally coincide with p_1, i.e.,

$$\mathcal{P}_1 = \{q : \Omega^M \longrightarrow [0,1] \mid \Pi^M_{h_{T_1}}(q)(\omega^{h_{T_1}}) = p_1(\omega^{h_{T_1}}), \quad \omega^{h_{T_1}} \in \Omega^{h_{T_1}}\}.$$

In contrast to the L-set approach we considered in Chap. 11. θ_2's information, the marginal distribution $p_2 : \Omega^{(T_2)_h} \times \Omega^{h_{T_2}} \longrightarrow [0,1]$, has to be interpreted *relative* to the state of knowledge reached so far. Thus we have

$$\mathcal{P}_2 = \{q \in \mathcal{P}_1 \mid \frac{\Pi^M_{h_{T_1} \cup h_{T_2}}(q)(\omega^{h_{T_1}}, \omega^{h_{T_2}})}{\Pi^M_{h_{T_1}}(q)(\omega^{h_{T_2}})} = p_2(\omega^{h_{T_2}} \mid \omega^{T_{j_h}}), \quad \omega^{h_{T_1} \cup h_{T_2}} \in \Omega^{h_{T_1} \cup h_{T_2}}\}$$

if T_2 is not a root, and in the general case

$$\mathcal{P}_{j+1} = \{q \in \mathcal{P}_j \mid \frac{\Pi^M_{h_{T_1} \cup \ldots \cup h_{T_{j+1}}}(q)(\omega^{h_{T_1}}, \ldots, \omega^{h_{T_{j+1}}})}{\Pi^M_{h_{T_1} \cup \ldots \cup h_{T_j}}(q)(\omega^{h_{T_2}}, \ldots, \omega^{h_{T_j}})} = p_{j+1}(\omega^{h_{T_{j+1}}} \mid \omega^{T_{j_h}}), \quad \omega^{h_{T_1} \cup \ldots \cup h_{T_{j+1}}} \in \Omega^{h_{T_1} \cup \ldots \cup h_{T_{j+1}}}\}.$$

We finally obtain

$$\mathcal{P}_r = \{q : \Omega^M \longrightarrow [0,1] \mid q(\omega^{h_{T_1} \cup \ldots \cup h_{T_r}}) = p_r(\omega^{h_{T_r}} \mid \omega^{T_{r_h}}) \cdot \ldots \cdot p_1(\omega^{h_{T_1}}), \quad \omega^M \in \Omega^M\}.$$

To verify this result, the set $\mathcal{P}_1$ may serve as the starting point of an induction over $j = 1, \ldots, r$. Assume

$$\mathcal{P}_j = \{q : \Omega^M \longrightarrow [0,1] \mid \Pi^M_{h_{T_1} \cup \ldots \cup h_{T_j}}(q)(\omega^{h_{T_1}}, \ldots, \omega^{h_{T_j}}) = p_j(\omega^{h_{T_j}} \mid \omega^{T_{j_h}}) \cdot \ldots \cdot P_1(\omega^{h_{T_1}}), \omega^{h_{T_1} \cup \ldots \cup h_{T_j}} \in \Omega^{h_{T_1} \cup \ldots \cup h_{T_j}}\}.$$

to be proven. Thus from the original definition of $\mathcal{P}_{j+1}$ we obtain

$$\begin{aligned}
\Pi^M_{h_{T_1} \cup \ldots \cup h_{T_{j+1}}}(q)(\omega^{h_{T_1}}, \ldots, \omega^{h_{T_j}}) &= p_{j+1}(\omega^{h_{T_{j+1}}} \mid \omega^{T_{j+1_h}}) \cdot \Pi^M_{h_{T_1} \cup \ldots \cup h_{T_j}}(q)(\omega^{h_{T_1}}, \ldots, \omega^{h_{T_j}}) \\
&= p_{j+1}(\omega^{h_{T_{j+1}}} \mid \omega^{T_{j+1_h}}) \cdot p_j(\omega^{h_{T_j}} \mid \omega^{T_{j_h}}) \cdot \ldots \cdot p_1(\omega^{h_{T_1}}).
\end{aligned}$$

This completes the induction.

To require the *equality* of the marginal conditional probabilities of some $q : \Omega^M \rightarrow [0,1]$ with the conditional probabilities, to be consistent means to assume that the spaces $\Omega^{(i)}$, $i \in M$, are of sufficient fineness to be the formal basis for the representation of all information concerning the respective characteristic, i.e., the sets $\Omega^{(k)}$ need not be refined. The only simplifying coarsening lies in the renunciation of additional charcteristics $X^{(m+1)}$, $X^{(m+2)}$, ... Letting $M' = M \cup \{m+1, m+2, \ldots\}$, each "continuation" $q' : \Omega^{M'} \rightarrow [0,1]$

of some probability $q : \Omega^M \to [0,1]$ yields by definition $\Pi_M^{M'}(q') = q$. So this simplification has no effect on the probabilities on Ω^M, and the use of the equality relation is reasonable (in the case of mass distributions the situation is different, compare Chap. 14).

Obviously the sets $\mathcal{P}_j$, $j = 1, \ldots, r$ are in a subset relationship, i.e., we have

$$\mathcal{P}_r \subseteq P_{r-1} \subseteq \ldots \subseteq \mathcal{P}_1$$

This property devolves to the sets of possible updates, which means that with $U_j \stackrel{\mathrm{d}}{=} \mathcal{U}_{prob}(\mathcal{P}_j)$, $j = 1, \ldots, r$, we obtain

$$\mathcal{U}_r \subseteq \mathcal{U}_{r-1} \subseteq \ldots \subseteq \mathcal{U}_1,$$

thus the integration with respect to the logical expression $T = 1$ <u>and</u>... <u>and</u> r yields

$$\mathcal{U}(T) = \bigcap_{j=1}^{r} \mathcal{U}_j = \mathcal{U}_r.$$

Considering only conditioning but not arbitrary specializations of the elements of the sets $\mathcal{P}_j$, $j = 1, \ldots, r$, is justified by our assumption that the sets $\Omega^{(k)}$, $k \in M$, are sufficient with respect to their fineness as we pointed out above.

The probabilistic formulation of an imprecise observation $E \in \Omega^M$ is the set

$$\mathcal{U}_E = \{q : \Omega^M \to [0,1] \mid q(E) = 1\}.$$

The integration of $\mathcal{U}(T)$ and $\mathcal{U}_E$ yields

$$\mathcal{U} = \mathcal{U}(T) \cap \mathcal{U}_E = \mathcal{U}_{prob}(q_0(\,.\, |E))$$

where q_0 denotes the unique element of $\mathcal{P}_r$, i.e.,

$$q_0(\omega^M) = p_r(\omega^{h_{T_r}} \mid \omega^{T_{rh}}) \cdot \ldots \cdot p_1(\omega^{h_{T_1}}). \tag{12.60}$$

Because it is its least specific element, $q_0(\,.\, | E)$ can be used as a representative of the set $\mathcal{U}$.

The transition from the so-called *apriori distribution* q_0 to the conditional (or *aposteriori*) distribution $q_0(\,.\, | E)$ can be described by the multiplication of q with some vector λ. We obtain

$$q_0(\omega^M \mid E) = \frac{1}{c}\, \lambda(\omega^M) \cdot q_0(\omega_M)\ ,\ \omega^M \in \Omega^M, \tag{12.61}$$

where $c = q_0(E)$ and $\lambda(\omega^M) = q_0(E \mid \omega^M)$. Since $E \cap \{\omega^M\}$ yields either $\{\omega^M\}$ or $\emptyset$ for all $\omega^M \in \Omega^M$ we have $\lambda(\omega^M) \in \{0,1\}$.

Since E results from several observations of single characteristics we have

$$E = \bigcap_{k:k\in M} \hat{\Pi}^M_{\{k\}}(E^{(k)}) = \bigcap_{j=1}^{r} \hat{\Pi}^M_{h_{T_j}}(E^{h_{T_j}}).$$

Therefore

$$\begin{aligned} q_0(E \mid \omega^M) &= q_0(E^{h_{T_1}} \times \ldots \times E^{h_{T_r}} \mid \omega^M) \\ &= \begin{cases} 1 & \text{if } \omega^M \in E^{h_{T_1}} \times \ldots \times E^{h_{T_r}} \\ 0 & \text{otherwise,} \end{cases} \\ &= \begin{cases} 1 & \text{if } \omega^{h_{T_1}} \in E^{h_{T_1}} \text{ and } \ldots \text{ and } \omega^{h_{T_r}} \in E^{h_{T_r}} \\ 0 & \text{otherwise,} \end{cases} \\ &- q_0(E^{h_{T_j}} \mid \omega^{h_{T_j}}) \cdot \ldots \cdot q_0(E^{h_{T_r}} \mid \omega^{h_{T_r}}) \,. \end{aligned}$$

With $\lambda_j(\omega^{h_{T_j}}) \stackrel{\mathrm{d}}{=} q_0(E^{h_{T_j}} \mid \omega^{h_{T_j}})$, $j = 1, \ldots, r$, we obtain

$$q_0(\omega^M \mid E) = \frac{1}{c} \prod_{j=1}^{r} \lambda_j(\omega^{h_{T_j}}) p_j(\omega^{h_{T_j}} \mid \omega^{T_{jh}}) \tag{12.62}$$

The conditional probability $q_0(\;.\mid E)$ is the *aposteriori* knowledge structure which encodes the total knowledge. In the sequel we have to investigate how to compute efficiently the marginals of $q_0(\;.\mid E)$ based on local computations.

12.5.2 Graph Structure and Conditional Independence

First we focus our interest on the question, how to determine from the (qualitative) structure of some dependency network $C = (M, \mathfrak{T})$ whether a probability distribution

$$q(\omega^M) = q(\omega^{h_{T_r}} \mid \omega^{T_{rh}}) \cdot \ldots \cdot q(\omega^{h_{T_1}}), \qquad \omega^M \in \Omega^M \;,$$

where $\mathfrak{T}$ is assumed to be causally consistently numbered, satisfies certain conditional independence conditions. Given a set $S_c \subseteq M$ we want to decide whether

$$q(\omega^{h_{T_i}}, \omega^{h_{T_j}} \mid \omega^{S_c}) = q(\omega^{h_{T_i}} \mid \omega^{S_c}) \cdot q(\omega^{h_{T_j}} \mid \omega^{S_c}) \;, \quad \omega^{h_{T_i} \cup h_{T_j} \cup S_c} \in \Omega^{h_{T_i} \cup h_{T_j} \cup S_c}$$

is valid or not. For this purpose we define for some set $N \subseteq M$ of vertices

$$\begin{aligned} v(N) &\stackrel{\mathrm{d}}{=} \max\{j \mid h_{T_j} \in N\}, \\ c(N) &\stackrel{\mathrm{d}}{=} \{h_{T_j} \in M \mid j \leq v(N)\}, \\ \mathfrak{T}_{c(N)} &\stackrel{\mathrm{d}}{=} \{T_j \in \mathfrak{T} \mid j \leq v(N)\} = \{T_j \in \mathfrak{T} \mid T_j \subseteq c(N)\}, \end{aligned}$$

where $\mathfrak{T}_{c(N)}$ is the set of those hyperedges required for the construction of the minimal partial causal network covering the relevant vertices (which is not neccessarily connected).

(12.6) Theorem. *Let $H = (M, \mathfrak{T})$ be a causal network. For the corresponding probability distribution $q : \Omega^M \to [0,1]$ and $M_c \subseteq M$ the condition*

$$q(\omega^{h_{T_{j1}}}, \omega^{h_{T_{j2}}} \mid \omega^{M_c}) = q(\omega^{h_{T_{j1}}} \mid \omega^{M_c}) \cdot q(\omega^{h_{T_{j2}}} \mid \omega^{M_c}),$$

$$\omega^{h_{T_{j1}} \cup h_{T_{j2}} \cup M_c} \in \Omega^{h_{T_{j1}} \cup h_{T_{j2}} \cup M_c}$$

holds, if $h_{T_{j_1}}$ and $h_{T_{j_2}}$ are unconnected with respect to $\mathfrak{T}_{c(\{h_{T_{j_1}}, h_{T_{j_2}}\} \cup M_c)}$ or otherwise, if M_c separates $h_{T_{j_1}}$ and $h_{T_{j_2}}$ (compare Def. 7.34).

Proof.

(1) If $h_{T_{j_1}}$ and $h_{T_{j_2}}$ are unconnected with respect to $\mathfrak{T}_{c\{h_{T_{j_1}}, h_{T_{j_2}}\} \cup M_c}$ then the corresponding subhypergraph divides into at least two disjoint subhypertrees, i.e., M_c divides into three disjoint subsets

$$M_c^{(1)} \stackrel{\text{d}}{=} M_c \cap \left(\{h_{T_{j_1}}\} \cup T_{j_1}\uparrow\right) \text{ containing } h_{T_{j_1}},$$

$$M_c^{(2)} \stackrel{\text{d}}{=} M_c \cap \left(\{h_{T_{j_2}}\} \cup T_{j_2}\uparrow\right) \text{ containing } h_{T_{j_2}},$$

and $M_c^{(3)} \stackrel{\text{d}}{=} M_c - (M_c^{(1)} \cup M_c^{(2)})$, the last of which one may be empty (compare Example 7.32). For each of these subhypertrees we have disjoint product forms, therefore

$$q(\omega^{h_{T_{j1}}} \omega^{h_{T_{j2}}}, \omega^{M_c}) = q(\omega^{h_{T_{j1}}} \mid \omega^{M_c^{(1)}}) \cdot q(\omega^{h_{T_{j2}}} \mid \omega^{M_c^{(2)}}) \cdot q(\omega^{M_c^{(3)}})$$

and we obtain

$$\begin{aligned}
&q(\omega^{h_{T_{j1}}}, \omega^{h_{T_{j2}}} \mid \omega^{M_c}) \\
&= \frac{q(\omega^{h_{T_{j1}}}, \omega^{M_c^{(1)}}) \cdot q(\omega^{h_{T_{j2}}}, \omega^{M_c^{(2)}}) \cdot q(\omega^{M_c^{(3)}})}{q(\omega^{M_c^{(1)}}) \cdot q(\omega^{M_c^{(2)}}) \cdot q(\omega^{M_c^{(3)}})} \\
&= \frac{q(\omega^{h_{T_{j1}}}, \omega^{M_c^{(1)}}) \cdot q(\omega^{M_c^{(2)}}) \cdot q(\omega^{M_c^{(3)}}) \cdot q(\omega^{h_{T_{j2}}}, \omega^{M_c^{(2)}}) \cdot q(\omega^{M_c^{(1)}}) \cdot q(\omega^{M_c^{(3)}})}{q(\omega^{M_c^{(1)}}) \cdot q(\omega^{M_c^{(2)}}) \cdot q(\omega^{M_c^{(3)}}) \cdot q(\omega^{M_c^{(1)}}) \cdot q(\omega^{M_c^{(2)}}) \cdot q(\omega^{M_c^{(3)}})} \\
&= q(\omega^{h_{T_{j1}}} \mid \omega^{M_c}) \cdot q(\omega^{h_{T_{j2}}} \mid \omega^{M_c}).
\end{aligned}$$

(2) Consider an arbitrary hyperedge $T_j \in \mathfrak{T}$. Its removal splits the dependency network into at least two parts as pointed out in Sect. 7.4, and we have

$$M = h_{T_j}\downarrow \cup \{h_{T_j}\} \cup T_j\uparrow .$$

$T_j\!\uparrow$ consists of the disjoint subgraphs $T_j h_{T_i}\!\downarrow$, $h_{T_i} \in T_{j_h}$, therefore we have

$$q(\omega^{T_j\Uparrow}) = \prod_{h_{T_i} \in T_{j_h}} q(\omega^{T_j h_{T_i}\Downarrow}) \ , \ \omega^{T_j} \in \Omega^{T_j} .$$

The remaining subgraph could be completed by a single root $(\emptyset, h_{T_j})$, so the distribution on $\Omega^{h_{T_j}\downarrow}$ depends only on $q(\omega^{h_{T_j}})$. Also $h_{T_j}\!\downarrow$ divides into several subgraphs $h_{T_j} T_i\!\uparrow$, $h_{T_j} \in T_{i_h}$, thus we obtain

$$q(\omega^{h_{T_j}\Downarrow} \mid \omega^{h_{T_j}}) = \prod_{T_i : h_{T_j} \in T_{i\,h}} q(\omega^{h_{T_j} T_i\Uparrow} \mid \omega^{h_{T_j}}) \ .$$

In total we have

$$q(\omega^M) = q(\omega^{h_{T_j}\Downarrow} \mid \omega^{h_{T_j}}) \cdot q(\omega^{h_{T_j}} \mid \omega^{T_j\Uparrow}) \cdot q(\omega^{T_j\Uparrow}) \ .$$

After these preliminary exertions let us consider some node x blocking (see Def. 7.34) the path between $h_{T_{j_1}}$ and $h_{T_{j_2}}$.

(2.1)Assume we have a head-to-tail blocking. Let T_j be the hyperedge with $x \in T_{j\,h}$ in the path. We obtain

$$\begin{aligned}
&q(\omega^M)\\
&= q(\omega^{h_{T_j}\Downarrow} \mid \omega^{h_{T_j}}) \cdot q(\omega^{h_{T_j}} \mid \omega^{T_{j_h}}) \cdot q(\omega^{T_j\Uparrow})\\
&= q(\omega^{h_{T_j}\Downarrow} \mid \omega^{h_{T_j}}) \cdot q(\omega^{h_{T_j}} \mid \omega^{T_{j_h}}) \cdot \Big(\prod_{\substack{i : h_{T_i} \in T_{j\,h} \\ h_{T_i} \neq x}} q(\omega^{h_{T_i} T_j\Downarrow}) \Big) \cdot q(\omega^{x T_j\Downarrow})\\
&= q(\omega^{M - x T_j\Downarrow} \mid \omega^x) \cdot q(\omega^{x T_j\Downarrow}).
\end{aligned}$$

Since $h_{T_{j_1}} \in xT_j\!\downarrow$ and $h_{T_{j_2}} \in M - xT_j\!\downarrow$ we derive

$$q(\omega^{h_{T_{j1}}}, \omega^{h_{T_{j2}}}, \omega^{M_c}) = q(\omega^{h_{T_{j1}}}, \omega^{M_c^{(1)}} \mid \omega^x) \cdot q(\omega^{h_{T_{j2}}}, \omega^{M_c^{(2)}}),$$

where $M_c^{(1)} = M_c \cap xT_j\!\uparrow$ and $M_c^{(2)} = M_c - M_c^{(1)}$. This leads to

$$\begin{aligned}
&q(\omega^{h_{T_{j1}}}, \omega^{h_{T_{j2}}} \mid \omega^{M_c})\\
&= \frac{q(\omega^{h_{T_{j1}}}, \omega^{M_c^{(1)}} \mid \omega^x) \cdot q(\omega^{h_{T_{j2}}}, \omega^{M_c^{(2)}})}{q(\omega^{M_c^{(1)}} \mid \omega^x) \cdot q(\omega^{M_c^{(2)}})}\\
&= \frac{q(\omega^{h_{T_{j1}}}, \omega^{M_c^{(1)}} \mid \omega^x) \cdot q(\omega^{M_c^{(2)}})}{q(\omega^{M_c^{(1)}} \mid \omega^x) \cdot q(\omega^{M_c^{(2)}})} \cdot \frac{q(\omega^{M_c^{(1)}} \mid \omega^x) \cdot q(\omega^{h_{T_{j2}}}, \omega^{M_c^{(2)}})}{q(\omega^{M_c^{(1)}} \mid \omega^x) \cdot q(\omega^{M_c^{(2)}})}\\
&= q(\omega^{h_{T_{j1}}} \mid \omega^{M_c}) \cdot q(\omega^{h_{T_{j2}}} \mid \omega^{M_c}).
\end{aligned}$$

(2.2)If the connection is tail-to-head, then it is handled analogously.

(2.3)In the case of a tail-to-tail blocking we have $x = h_{T_j}$. We obtain

$$q(\omega^M) = \prod_{i:h_{T_j} \in T_{ih}} q(\omega^{h_{T_j} T_i\uparrow} \mid \omega^{h_{T_j}}) \cdot q(\omega^{h_{T_j}} \mid \omega^{T_{jh}}) \cdot q(\omega^{T_j\uparrow})$$

Since $h_{T_{j_1}}$ and $h_{T_{j_2}}$ are contained in two different subgraphs $h_{T_j} T_{i_1}\uparrow$ and $h_{T_j} T_{i_2}\uparrow$ we derive

$$q(\omega^M) = q(\omega^{h_{T_j} T_{i_1}\uparrow} \mid \omega^{h_{T_j}}) \cdot q(\omega^{h_{T_j} T_{i_2}\uparrow} \mid \omega^{h_{T_j}}) \cdot q(\omega^{M-(h_{T_j} T_{i_1}\uparrow \cup h_{T_j} T_{i_2}\uparrow)})$$

Also M_c is split into three parts, and we get

$$q(\omega^{h_{T_{j1}}}, \omega^{h_{T_{j2}}}, \omega^{M_c}) = q(\omega^{h_{T_{j1}}}, \omega^{M_c^{(1)}} \mid \omega^x) \cdot q(\omega^{h_{T_{j2}}}, \omega^{M_c^{(2)}} \mid \omega^x) \cdot q(\omega^{M_c^{(3)}})$$

This yields

$$\begin{aligned} &q(\omega^{h_{T_{j1}}}, \omega^{h_{T_{j2}}} \mid \omega^{M_c}) \\ &= \frac{q(\omega^{h_{T_{j1}}}, \omega^{M_c^{(1)}} \mid \omega^x) \cdot q(\omega^{h_{T_{j2}}}, \omega^{M_c^{(2)}} \mid \omega^x) \cdot q(\omega^{M_c^{(3)}})}{q(\omega^{M_c^{(1)}} \mid \omega^{h_{T_j}}) \cdot q(\omega^{M_c^{(2)}} \mid \omega^{h_{T_j}}) \cdot q(\omega^{M_c^{(3)}})} \\ &= \frac{q(\omega^{h_{T_{j1}}}, \omega^{M_c^{(1)}} \mid \omega^x) \cdot q(\omega^{M_c^{(2)}} \mid \omega^x) \cdot q(\omega^{M_c^{(3)}})}{q(\omega^{M_c^{(1)}} \mid \omega^x) \cdot q(\omega^{M_c^{(2)}} \mid \omega^x) \cdot q(\omega^{M_c^{(3)}})} \cdot \\ &\quad \frac{q(\omega^{M_c^{(1)}} \mid \omega^x) \cdot q(\omega^{h_{T_{j2}}} \omega^{M_c^{(2)}} \mid \omega^x) \cdot q(x^{M_c^{(3)}})}{q(\omega^{M_c^{(1)}} \mid \omega^x) \cdot q(\omega^{M_c^{(2)}} \mid \omega^x) \cdot q(\omega^{M_c^{(3)}})} \\ &= \frac{q(\omega^{h_{T_{j1}}} \omega^{M_c})}{q(\omega^{M_c})} \cdot \frac{q(\omega^{h_{T_{j2}}} \omega^{M_c})}{q(\omega^{M_c})} \\ &= q(\omega^{h_{T_{j1}}} \mid \omega^{M_c}) \cdot q(\omega^{h_{T_{j2}}} \mid \omega^{M_c}) \end{aligned}$$

□

(12.7) *Remark.* This theorem directly corresponds to the *separation criterion*, considered in Sect. 12.4.1. The problem of relations between graph structures and conditional independence is studied in more detail in [Geiger and Pearl 1988, 1990]. ○

(12.8) *Remark.* The conditional independence condition

$$q(\omega^{h_{T_{j1}}}, \omega^{h_{T_{j2}}} \mid \omega^{M_c}) = q(\omega^{h_{T_{j1}}} \mid \omega^{M_c}) \cdot q(\omega^{h_{T_{j2}}} \mid \omega^{M_c})$$

is equivalent with the condition

$$q(\omega^{h_{T_{j1}}} \mid \omega^{h_{T_{j2}}}, \omega^{M_c}) = q(\omega^{h_{T_{j1}}} \mid \omega^{M_c})$$

(comp. Sect. 4.4.1) ○

12.5.3 Local Computation of Marginal Probability Distributions

Let us consider a simple dependency network $C = (M, \mathfrak{T})$. Later on we will conceive this network to be a cutout of some extended one, but for the moment assume

$$M = \{a, b, d, g, h\}$$
$$\mathfrak{T} = \{P, Q, U, V, W\}$$

where $P = (\emptyset, a)$, $Q = (\emptyset, b)$, $U = (\{a,b\}, d)$, $V = (\{d\}, g)$, $W = (\{d\}, h)$. Therefore we have

$$q(\omega^M) = q(\omega^g \mid \omega^d) \cdot q(\omega^h \mid \omega^d) \cdot q(\omega^d \mid \omega^{a,b}) \cdot q(\omega^a) \cdot q(\omega^b) \ .$$

Splitting the causal network with respect to the edge $U \in \mathfrak{T}$ we obtain

$$q(\omega^M) = q(\omega^{d\downarrow} \mid \omega^d) \cdot q(\omega^d \mid \omega^{U_h}) \cdot q(\omega^{U\uparrow}),$$

where

$$q(\omega^{d\downarrow} \mid \omega^d) = q(\omega^{g,h} \mid \omega^d) = q(\omega^g \mid \omega^d) \cdot q(\omega^h \mid \omega^d) \text{ and}$$
$$q(\omega^{U\uparrow}) = q(\omega^{a,b}) = q(\omega^a) \cdot q(\omega^b).$$

Now recall that we obtained $q : \Omega^M \to [0,1]$ by conditioning $q_0 : \Omega^M \to [0,1]$ with the actual observations $E \subseteq \Omega^M$:

$$q(\omega^M) = q_0(\omega^M \mid E) = \frac{1}{c}\, \lambda(\omega^M) \cdot q_0(\omega^M),$$

where

$$\lambda(\omega^M) = \prod_{x \in M} \lambda_j(\omega^x) \ .$$

So we have to deal with

$$q_0(\omega^M \mid E) = \frac{1}{c} \cdot \lambda_{E^{d\downarrow}}(\omega^{d\downarrow}) \cdot q_0(\omega^{d\downarrow} \mid \omega^d) \cdot \lambda_{E^d}(\omega^d) \cdot q_0(\omega^d \mid \omega^a, \omega^b) \cdot \lambda_{E^{U\uparrow}}(\omega^{U\uparrow}) \cdot q_0(\omega^{U\uparrow}).$$

For the marginal distribution for the node $d \in M$ we obtain

$$\begin{aligned} q_0(\omega^d \mid E) &= \sum_{\omega^{d\downarrow}} \sum_{\omega^{U\uparrow}} q_0(\omega^M \mid E) \\ &= \sum_{\omega^{d\downarrow}} \sum_{\omega^{U\uparrow}} \lambda_{E^{d\downarrow}}(x^{d\downarrow}) \cdot q_0(\omega^{d\downarrow} \mid \omega^d) \cdot \lambda_{E^d}(\omega^d) \cdot \\ &\quad q_0(\omega^d \mid \omega^a, \omega^b) \cdot \lambda_{E^{U\uparrow}}(\omega^{U\uparrow}) \cdot q_0(\omega^{U\uparrow}). \end{aligned}$$

Defining

$$\begin{aligned}
\lambda_{d\downarrow}(\omega^d) &\stackrel{d}{=} \sum_{\omega^{d\downarrow}} \lambda_{E^{d\downarrow}}(\omega^d) \cdot q_0(\omega^{d\downarrow} \mid \omega^d) \\
&= \sum_{\omega^g, \omega^h} \lambda_{E^g}(\omega^g) \cdot q_0(\omega^g \mid \omega^d) \cdot \lambda_{E^h}(\omega^h) \cdot q_0(\omega^h \mid \omega^d) \\
&= \left(\sum_{\omega^g} \lambda_{E^g}(\omega^g) \cdot q_0(\omega^g \mid \omega^d) \right) \cdot \\
&\quad \left(\sum_{\omega^h} \lambda_{E^h}(\omega^h) \cdot q_0(\omega^h \mid \omega^d) \right)
\end{aligned}$$

and

$$\begin{aligned}
\mu_{U\uparrow}(\omega^a, \omega^b) &= \frac{1}{c} \cdot \lambda_{E^{U\uparrow}}(\omega^{U\uparrow}) \cdot q_0(\omega^{U\uparrow}) \\
&= \frac{1}{c} \cdot \lambda_{E^a}(\omega^a) \cdot \lambda_{E^b}(\omega^b) \cdot q_0(\omega^a) \cdot q_0(\omega^b) \\
&= \frac{1}{c} \cdot \lambda_{E^a}(\omega^a) \cdot q_0(\omega^a) \cdot \lambda_{E^b}(\omega^b) \cdot q_0(\omega^b)
\end{aligned}$$

yields

$$q_0(\omega^d \mid E) = \frac{1}{c} \sum_{\omega^a, \omega^b} \lambda_{d\downarrow}(\omega^d) \cdot \lambda_d(\omega^d) \cdot q_0(\omega^d \mid \omega^a, \omega^b) \cdot \mu_{U\uparrow}(\omega^a, \omega^b).$$

So as we see in the simple dependency network of our example we can read the quantities which we need for the calculation of $q_0(\omega^d \mid E)$ easily from the product form of the probability distribution.

In general we will have to deal with more complex dependency networks. Let us consider the edge U of such a dependency network. Without loss of generality we can assume

$$U = (U_h, d),$$

i.e. $d = h_U$. Splitting the network with respect to this edge U we obtain

$$M = d\!\downarrow \cup \{d\} \cup U\!\uparrow$$

and thus

$$q(\omega^M) = q(x^{d\downarrow} \mid \omega^d) \cdot q(\omega^d \mid \omega^{U_h}) \cdot q(\omega^{U\uparrow}) .$$

(Compare the proof of Theorem 12.6, especially part (2) of this proof).

(12.9) *Remark.* The formal argument is that $d \in M$ *separates* all nodes in $d\!\downarrow$ from those in $U\!\uparrow$, therefore we have

$$q(\omega^{d\downarrow} \mid \omega^d, \omega^{U\uparrow}) = q(\omega^{d\downarrow} \mid \omega^d).$$

Moreover U_h separates d from all nodes within $U\uparrow - U_h$, so

$$q(\omega^d \mid \omega^{U\uparrow}) = q(\omega^d \mid \omega^{U_h}, \omega^{U\uparrow - U_h}) = q(\omega^d \mid \omega^{U_h}).$$

For this reason

$$q(\omega^M) = q(\omega^{d\downarrow} \mid \omega^d, \omega^{U\uparrow}) \cdot q(\omega^d \mid \omega^{U\uparrow}) \cdot q(\omega^{U\uparrow})$$

reduces to

$$q(\omega^M) = q(\omega^{d\downarrow} \mid \omega^d) \cdot q(\omega^d \mid \omega^{U_h}) \cdot q(\omega^{U\uparrow}).$$

○

Since

$$d\downarrow = \bigcup_{T:d\in T_h} dT\uparrow \text{ and } U\uparrow = \bigcup_{x:x\in U_h} xU\downarrow$$

we obtain

$$q(\omega^{d\downarrow} \mid \omega^d) = \prod_{T:d\in T_h} q(\omega^{dT\uparrow} \mid \omega^d)$$

and

$$q(\omega^{U\uparrow}) = \prod_{x:x\in U_h} q(\omega^{xU\downarrow}).$$

(12.10) *Remark.* Again the formal argument relies on the "separation criterion". Since d separates the subgraphs $dT\uparrow$, $d \in T_h$, we get the first decomposition, and since the subgraphs $xU\downarrow$, $x \in U_h$, are unconnected in $U\uparrow$ we obtain the second one. ○

So in total we have

$$q(\omega^M) = \prod_{T:d\in T_h} q(\omega^{dT\uparrow} \mid \omega^d) \cdot q(\omega^d \mid \omega^{U_h}) \cdot \prod_{x:x\in U_h} q(\omega^{xU\downarrow})$$

Note that

$$yU\uparrow = M - yU\downarrow \text{ for all } y \in U_h,$$

therefore we derive

$$\begin{aligned} q(\omega^{yU\uparrow} \mid \omega^y) &= q(\omega^{yU\uparrow} \mid \omega^{yU\downarrow}) \\ &= \frac{q(\omega^M)}{q(\omega^{yU\downarrow})} \\ &= q(\omega^{d\downarrow} \mid \omega^d) \cdot q(\omega^d \mid \omega^{U_h}) \cdot \prod_{\substack{x:x\in U_h \\ x\neq y}} q(\omega^{xU\downarrow}) \end{aligned}$$

(recall that x separates $yU\downarrow$ and $yU\uparrow$). In a similar way from

$$dT'\downarrow = M - dT'\uparrow \text{ for all } T' \text{ with } d \in T'_h$$

we obtain

$$\begin{aligned} q(\omega^{dT'\downarrow}) &= \frac{q(\omega^{dT'\downarrow}) \cdot q(\omega^{dT'\uparrow} \mid \omega^d)}{q(\omega^{dT'\uparrow} \mid \omega^d)} \\ &= \frac{q(\omega^{dT'\uparrow} \mid \omega^{dT'\downarrow}) \cdot q(\omega^{dT'\downarrow})}{q(\omega^{dT'\uparrow} \mid \omega^d)} \\ &= \frac{q(\omega^M)}{q(\omega^{dT'\uparrow} \mid \omega^d)} \\ &= \prod_{\substack{T:d\in T_h \\ T\neq T'}} q(\omega^{dT\uparrow} \mid \omega^d) \cdot q(\omega^d \mid \omega^{U_h}) \cdot q(\omega^{U\uparrow}). \end{aligned}$$

As in the discussion of the example, we have to take into account that we derived $q : \Omega^M \to [0,1]$ by conditioning $q_0 : \Omega^M \to [0,1]$ with the actual observations E. Once more we have

$$q(\omega^M) = q_0(\omega^M \mid E) = \frac{1}{c}\lambda_0(\omega^M) \cdot q_0(\omega^M)$$

where

$$\lambda_0(\omega^M) = \prod_{x\in M} \lambda(\omega^{h_x}) \, .$$

So we obtain

$$\begin{aligned} q_0(\omega^M \mid E) &= \frac{1}{c} \cdot \lambda_{E^{d\downarrow}}(\omega^{d\downarrow}) \cdot q_0(\omega^{d\downarrow} \mid \omega^d) \cdot \lambda_{E^d}(\omega^d) \\ &\quad \cdot q_0(\omega^d \mid \omega^{U_h}) \cdot \lambda_{E^{U\uparrow}}(\omega^{U\uparrow}) \cdot q_0(\omega^{U\uparrow}) \\ &= \frac{1}{c} \cdot \left(\prod_{T:d\in T_h} \lambda_{E^{dT\uparrow}}(\omega^{dT\uparrow}) \cdot q_0(\omega^{dT\uparrow} \mid \omega^d) \right) \cdot \lambda_{E^d}(\omega^d) \cdot \\ &\quad q_0(\omega^d \mid \omega^{U_h}) \cdot \prod_{x:x\in U_h} \lambda_{xU\downarrow}(\omega^{xU\downarrow}) \cdot q_0(\omega^{xU\downarrow}). \end{aligned}$$

Now let us consider the node $d \in M$. For the marginal distribution we obtain

$$\begin{aligned} q_0(\omega^d \mid E) &= \sum_{\omega^{d\downarrow}} \sum_{\omega^{U\uparrow}} q_0(\omega^M \mid E) \\ &= \frac{1}{c} \sum_{\omega^{d\downarrow}} \sum_{\omega^{U_h}} \sum_{\omega^{U\uparrow - U_h}} \lambda_{E^{d\downarrow}}(x^{d\downarrow}) \cdot q_0(\omega^{d\downarrow} \mid \omega^d) \cdot \lambda_{E^d}(\omega^d) \\ &\quad \cdot q_0(\omega^d \mid \omega^{U_h}) \cdot \lambda_{E^{U\uparrow}}(\omega^{U\uparrow}) \cdot q_0(\omega^{U\uparrow}). \end{aligned}$$

Defining

$$\begin{aligned}
\lambda_{d\downarrow}(\omega^d) &\stackrel{d}{=} \sum_{\omega^{d\downarrow}} \lambda_{E^{d\downarrow}}(\omega^d) \cdot q_0(\omega^{d\downarrow} \mid \omega^d) \\
&= \sum_{\omega^{d\downarrow}} q_0(E^{d\downarrow} \mid \omega^d) \cdot q_0(\omega^{d\downarrow} \mid \omega^d) \\
&= \sum_{\omega^{d\downarrow}} q_0(E^{d\downarrow}, \omega^{d\downarrow} \mid \omega^d) \\
&= q_0(\omega^{d\downarrow} \mid \omega^d)
\end{aligned}$$

and

$$\begin{aligned}
\mu^{U\uparrow}(\omega^{U_h}) &\stackrel{d}{=} \frac{1}{c} \cdot \sum_{\omega^{U\uparrow - U_h}} \lambda_{E^{U\uparrow}}(\omega^{U\uparrow}) \cdot q_0(\omega^{U\uparrow}) \\
&= \frac{1}{q_0(E^{U\uparrow})} \cdot \sum_{\omega^{U\uparrow - U_h}} q_0(E^{U\uparrow} \mid \omega^{U\uparrow}) \cdot q_0(\omega^{U|}) \\
&= \frac{1}{q_0(E^{U\uparrow})} \cdot \sum_{\omega^{U\uparrow - U_h}} q_0(E^{U\uparrow}, \omega^{U\uparrow}) \\
&= \frac{1}{q_0(E^{U\uparrow})} \cdot q_0(E^{U\uparrow}, \omega^{U_h}) \\
&= q_0(\omega^{U_h} \mid E^{U\uparrow})
\end{aligned}$$

yields

$$q_0(\omega^M \mid E) = \frac{1}{c} \cdot \sum_{\omega^{U_h}} \lambda_{d\downarrow}(\omega^d) \cdot \lambda_{E^d}(\omega^d) \cdot q_0(\omega^d \mid \omega^{U_h}) \cdot \mu_{U\uparrow}(\omega^{U_h}),$$

where

$$\begin{aligned}
\lambda_{d\downarrow}(\omega^d) &= \sum_{\omega^{d\downarrow}} \lambda_{E^{d\downarrow}}(\omega^d) \cdot q_0(\omega^{d\downarrow} \mid \omega^d) \\
&= \sum_{\omega^{d\downarrow}} \prod_{T:d\in T_h} \lambda_{E^{dT\uparrow}}(\omega^{dT\uparrow} \cdot q_0(\omega^{dT\uparrow} \mid \omega^d) \\
&= \prod_{T:d\in T_h} \sum_{\omega^{dT\uparrow}} \lambda_{E^{dT\uparrow}}(\omega^{dT\uparrow} \cdot q_0(\omega^{dT\uparrow} \mid \omega^d).
\end{aligned}$$

With

$$\lambda_{dT\uparrow}(\omega^d) = \sum_{\omega^{dT\uparrow}} \lambda_{E^{dT\uparrow}}(\omega^{dT\uparrow}) \cdot q_0(\omega^{dT\uparrow} \mid \omega^d) = q_0(E^{dT\uparrow} \mid \omega^d).$$

This yields

$$\lambda_{d\downarrow}(\omega^d) = \prod_{T:d\in T_h} \lambda_{E^{dT\uparrow}}(\omega^d).$$

Since

$$\begin{aligned}\mu_{U\uparrow}(\omega^{U_h}) &= \frac{1}{c}\cdot \sum_{\omega^{U\uparrow - U_h}} \lambda_{E^{U\uparrow}}(\omega^{U\uparrow})\cdot q_0(\omega^{U\uparrow}) \\ &= \frac{1}{c}\cdot \sum_{\omega^{U\uparrow - U_h}} \prod_{x:x\in U_h} \lambda_{E^{xU\downarrow}}(\omega^{xU\downarrow})\cdot q_0(\omega^{xU\downarrow}) \\ &= \frac{1}{c}\cdot \prod_{x:x\in U_h} \sum_{\omega^{xU\downarrow - U_h}} \lambda_{E^{xU\downarrow}}(\omega^{xU\downarrow})\cdot q_0(\omega^{xU\downarrow})\end{aligned}$$

with

$$\mu_{xU\uparrow}(\omega^x) = \frac{1}{c}\cdot \sum_{\omega^{xU\downarrow - x}} \lambda_{E^{xU\downarrow}}(\omega^{xU\downarrow})\cdot q_0(\omega^{xU\downarrow}) = q_0(\omega^x \mid E^{xU\downarrow})$$

we obtain

$$\mu_{U\uparrow}(\omega^{U_h}) = \prod_{x:x\in U_h} \mu_{xU\downarrow}(\omega^x)\ .$$

Since in general we cannot draw the vectors $\lambda_{dT\uparrow}(\omega^d)$ and $\mu_{xU\downarrow}(\omega^x)$ directly from the network, each node processor X^d depends on its neighbors in the network sending appropriate *messages*. Therefore the node processor ω^d for his part has to provide for some predecessor X^y, $y \in U_h$, the message

$$\lambda_{yU\uparrow}(\omega^y) = \sum_{\omega^{yU\uparrow}} q(\omega^{yU\uparrow} \mid \omega^y)$$

and for some successor ω^d linked by the hyperedge T', $d \in T'_h$, the message

$$\mu_{dT'\downarrow}(\omega^d) = \sum_{\omega^{dT'\downarrow - d}} q(\omega^{dT'\downarrow}).$$

Since from our previous exertions we know

$$q(\omega^{yU\uparrow} \mid \omega^y) = \frac{q(\omega^M)}{q(x^{yU\downarrow})} \qquad \text{and}$$

$$q(\omega^{dT'\downarrow}) = \frac{q(\omega^M)}{q(\omega^{dT'\uparrow} \mid \omega^d)}$$

we obtain easily

$$\begin{aligned}\lambda_{yU\uparrow}(\omega^y) = \frac{1}{c}\cdot \sum_{\omega^{U_h - y}} & \lambda_{d\downarrow}(\omega^d)\cdot \lambda_d(\omega^d)\cdot \\ & q_0(\omega^d \mid \omega^{U_h})\cdot \prod_{\substack{x:x\in U_h\\ x\neq y}} \mu_{xU\downarrow}(\omega^x) \qquad \text{and}\end{aligned}$$

$$\begin{aligned}\mu_{dT'\downarrow}(\omega^d) = \frac{1}{c}\cdot \sum_{\omega^{U_h}} \prod_{\substack{T:d\in T_h\\ T\neq T'}} & \lambda_{dT\uparrow}(\omega^d)\cdot \lambda_d(\omega^d)\cdot \\ & q_0(\omega^d \mid \omega^{U_h})\mu_{U\uparrow}(\omega^{U_h}),\end{aligned}$$

where

$$\lambda_{yU\uparrow}(\omega^y) = q_0(E^{yU\uparrow} \mid \omega^y) \quad \text{and}$$

$$\mu_{dT'\downarrow}(\omega^d) = q_0(\omega^d \mid E^{dT'\downarrow}).$$

(12.11) *Remark.* Note that $\lambda_{yU\uparrow}(\omega^y)$ does not depend on $\mu_{yU\downarrow}(\omega^y)$ as $\mu_{dT'\downarrow}(\omega^d)$ does not depend on $\lambda_{dT'\uparrow}(\omega^d)$, so it is guaranteed that there is no feedback. ○

In the beginning of the propagation process, when there are no observations, i.e., $\lambda_j(\omega^{h_{T_j}}) = 1$ $(E^x = \Omega^x)$, for the messages $\lambda_{dT\uparrow}(\omega^d)$ we obtain

$$\lambda_{dT\uparrow}(\omega^d) = q_0(E^{dT\uparrow} \mid \omega^d) = q_0(\Omega^{dT\uparrow} \mid \omega^d) = 1.$$

So for the initialization only a propagation from the roots to the leaves is required. Then each observation of some characteristic ω^d yields a modification of $\lambda_{E^d}(\omega^d)$ and causes a flow of messages through the causal network. Recall that

$$\lambda_{E^d}(\omega^d) = q_0(E^d \mid \omega^d) = \begin{cases} 1 & \text{if } \omega^d \in E^d \\ 0 & \text{otherwise,} \end{cases}$$

so setting $\lambda_{E^d}(\omega^d) = 0$ means considering ω^d to be impossible. Besides 0 and 1, also values $q \in [0,1]$ are reasonable, which have to be interpreted as messages from some "dummy" successor of ω^d representing the "uncertain" observation and sending λ_{E^d} as its message.

12.5.4 The Propagation Algorithm

The algorithm is based on independent node processors. Each directed hyperedge $T = (T_h, x)$ is represented by one such processor wich stores the conditional probability matrix $q_0(\omega^x \mid \omega^{T_h})$ and is able to exchange messages with its neighbors in the causal network. Figure 12.13 illustrates the structure of such a node processor. We have

- one pair of input/output ports for each predecessor X^x in the causal network to receive/send the messages

  $$\lambda_{xU\uparrow}(\omega^x) \text{ and } \mu_{xU\downarrow}(\omega^x), \qquad \text{resp.}$$

  (if ω^d is a root node, then there are of course no predecessors).

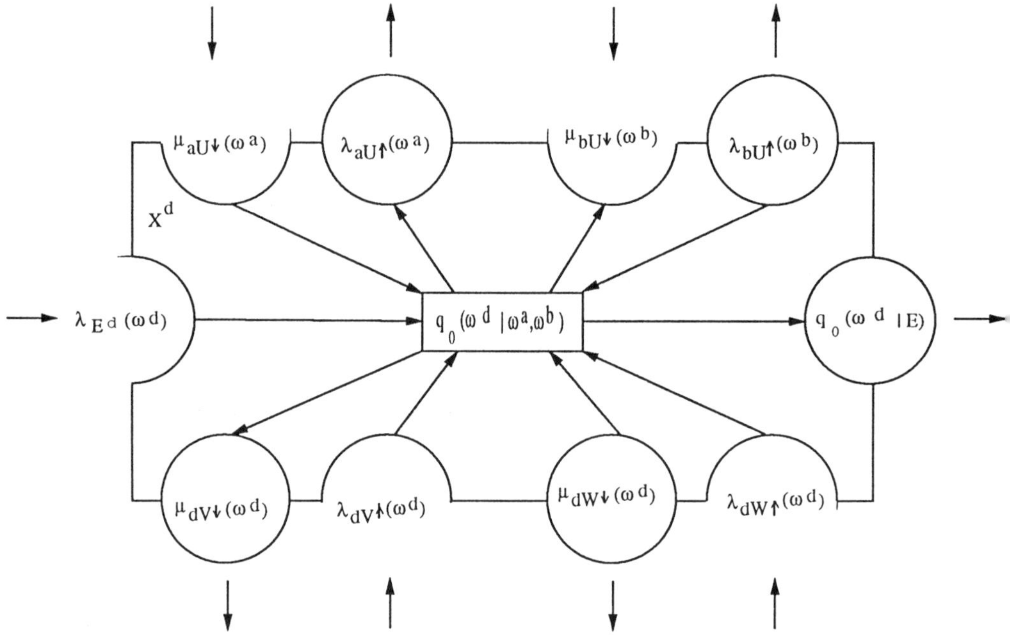

Fig. 12.13. Structure of a node processor

- one pair of input/output ports for each successor linked by the hyperedge T to receive/send the messages

$$\mu_{dT\downarrow}(\omega^d) \text{ and } \lambda_{dT\uparrow}(\omega^d), \qquad \text{resp.}$$

  (if ω^d is a leaf node, then there are of course no successors).

- an input port to receive the observation λ_{E^d} from the user and an output port to provide the result $q_0(\omega^d \mid E)$ for the user.

The structure of root and leaf nodes is somewhat different. Since root nodes do not have predecessors and since leaf nodes do not have successors in the network the respective communication facilities are not required. Moreover a root node stores instead of a conditional probability matrix a marginal probability vector.

The local algorithm of the node processor ω^d is as follows:

- If a message from the user is received then
  - calculate new messages $\lambda_{xU\uparrow}$ for all predecessors $x \in U_h$
  - calculate new messages $\mu_{dT\downarrow}$ for all successors linked by some edge T where $d \in T_h$
  - calculate new result $q_0(\omega^d \mid E)$.
- If a message from predecessor X^y is received then
  - calculate new messages for all predecessors execpt ω^y
  - calculate new messages for all successors
  - calculate new result.
- If a message from the successor linked by the hyperedge T' is received
  - calculate new messages for all predecessors
  - calculate new messages for all successors except the one connected by T'
  - calculate new result.

The respective formulas are

$$q_0(\omega^d \mid E) = \frac{1}{c} \sum_{\omega^{U_h}} \prod_{T:d\in T_h} \lambda_{dT\uparrow}(\omega^d) \cdot \lambda_{E^d}(\omega^d) \cdot q_0(\omega^d \mid \omega^{U_h}) \cdot \prod_{x:x\in U_h} \mu_{xU\downarrow}(\omega^x)$$

$$\lambda_{yU\uparrow}(\omega^d) = \frac{1}{c} \sum_{\omega^{U_h - y}} \prod_{T:d\in T_h} \lambda_{dT\uparrow}(\omega^d) \cdot \lambda_{E^d}(\omega^d) \cdot q_0(\omega^d \mid \omega^{U_h}) \cdot \prod_{\substack{x:x\in U_h \\ x\neq y}} \mu_{xU\downarrow}(\omega^x)$$

$$\mu_{dT'\downarrow}(x^d) = \frac{1}{c} \sum_{\omega^{U_h}} \prod_{\substack{T:d\in T_h \\ T\neq T'}} \lambda_{dT\uparrow}(\omega^d) \cdot \lambda_{E^d}(\omega^d) \cdot q_0(\omega^d \mid \omega^{U_h}) \cdot \prod_{x:x\in U_h} \mu_{xU\downarrow}(\omega^x)$$

As pointed out in Remark 12.11 there is no relaxation process, but after a finite and well defined number of steps the flow of messages comes to its end. Initially the causal network is ready to work for all messages $\lambda_{dT\uparrow} = \mathbb{1}$. Starting on an initializing flow of messages from the root nodes brings the network into a consistent state, where the $\mu_{dT\downarrow}$ vectors are equal to the marginal *apriori* probabilities.

Note that although there may be several messages on their way in the network at the same time, each processor can only handle one message at a time. Since the flow of messages determines the activity of the different node processors, no meta level to control the calculations is required, if an object oriented environment is used.

12.5.5 Aspects of Implementation

For the practical use of any software tool, besides the correctness and efficiency of the underlying algorithm, further aspects are of particular importance. The user has to be supported by the system whenever he is forced to provide information, and any output has to be presented adequately with respect to his requirements.

The first topic of interest in this context is the initial phase of knowledge acquisition, where the expert determines the qualitative structure of the domain. The construction of the corresponding causal network has to be graphically supported. Since a causal network is required, the expert should either be forced to add only suitable directed hyperedges such that from the beginning the system is in a consistent state or in a second step the generated hypergraph has to be transformed into a dependency network (e.g. by joining nodes).

Then the links have to be quantified by conditional probability matrices. The system should also provide support, for this purpose. Minimally the consistence of the matrices has to be assured, but a graphical presentation of the probabilities is also very useful.

The support of the consulting user is the key problem. He should be able to provide his observations simply by mouse-clicking on the screen. Moreover the result, i.e., the conditional marginal probabilities, should be graphically presented (see Fig. 12.14).

12.5.6 Numerical Example

To give a numerical example of the propagation algorithm let us return to the example from Sect. 2.2 concerning the oil contamination of waters. We consider four characteristics

$$\begin{array}{ll} \omega^{(1)} \text{ Location }, & \Omega^{(1)} = \{\text{z3,z2,z1,ca,rd,ld}\}, \\ \omega^{(2)} \text{ Authority }, & \Omega^{(2)} = \{\text{np,wp,fb,hc}\}, \\ \omega^{(3)} \text{ Measure }, & \Omega^{(3)} = \{\text{ad,fi,im,bi}\}, \\ \omega^{(4)} \text{ Decontamination }, & \Omega^{(4)} = \{\text{no,sf,pc}\}, \end{array}$$

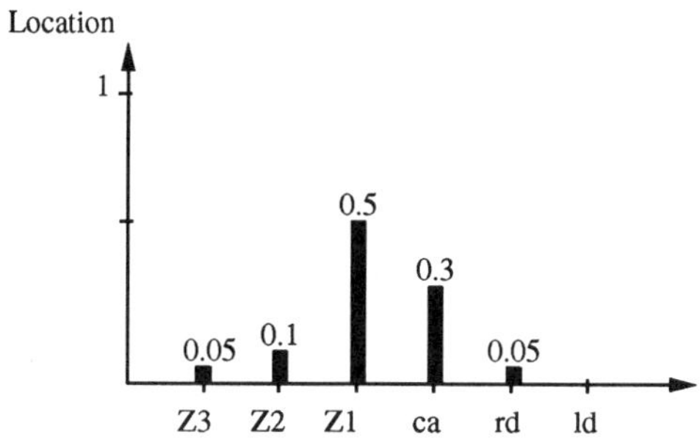

Fig. 12.14. Graphical presentation of marginal probabilities

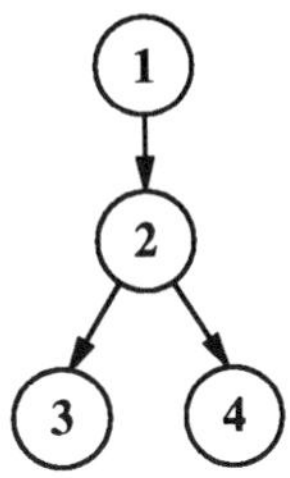

Fig. 12.15. Dependency network

where the (additional) characteristic $\omega^{(4)}$ has the possible values "none" (no), "special forces" (sf) and "private company" (pc). We know that the competent authority is determined by the location, whereas the authority decides who is charged with decontamination and what measures are taken. So the dependency network has three hyperedges, $A = (\{1\}, 2)$, $B = (\{2\}, 3)$ and $C = (\{2\}, 4)$. Figure 12.15 illustrates the structure of the network.

The node processor $X^{(1)}$ stores the marginal probability vector $p_1(\omega^{(1)})$, $\omega^{(1)} \in \Omega^{(1)}$ (recall that $X^{(1)}$ is a root node which has no predecessors). At node $X^{(2)}$ the expert knowledge concerning the authorities is stored in the form of the conditional probability matrix $p_2(\omega^{(2)} \mid \omega_i^{(1)})$, $\omega^{(1)} \in \Omega^{(1)}$, $\omega^{(2)} \in \Omega^{(2)}$. The conditional probability matrix $p_3(\omega^{(3)} \mid \omega^{(2)})$, $\omega^{(2)} \in \Omega^{(2)}$, $\omega^{(3)} \in \Omega^{(3)}$, concerns the measures the different authorities tend to take and $p_4(\omega^{(4)} \mid \omega^{(2)})$, $\omega^{(4)} \in \Omega^{(4)}$, $\omega_j^{(2)} \in \Omega^{(2)}$ describes what facilities they prefer for decontamination. The different probability matrices are listed below.

	$p_1(\cdot)$
z3	0.3
z2	0.2
z1	0.1
ca	0.1
rd	0.2
ld	0.1

$p_2(\cdot,\cdot)$	z3	z2	z1	ca	rd	ld
wp	0.9	0.8	0.6	0	0	0
qp	0	0.1	0.3	0.5	0.2	0.2
fb	0.1	0.1	0.1	0.4	0.2	0.4
hc	0	0	0	0.1	0.6	0.8

$p_3(\cdot,\cdot)$	np	wp	fb	hc
ad	0.8	0.1	0.2	0.1
fi	0.1	0.5	0.3	0.3
im	0	0.1	0.1	0
bi	0.1	0.3	0.4	0.6

$p_4(\cdot,\cdot)$	np	wp	fb	hc
ha	0.8	0.5	0.3	0.3
sf	0.2	0.3	0.5	0.1
jc	0	0.2	0.2	0.6

Figure 12.16a shows the state of the causal network after the initialization. As pointed out before, all λ–messages equal the $\mathbb{1}$–vector, whereas the μ–messages equal the marginal *apriori* probabilities. Figure 12.16b shows the impact of the observation that the decontamination is definitely performed by a private company, i.e. $E^{(4)} = \{\text{pc}\}$. Obviously the probability that the accident happend at sea decreases considerably whereas the probability that the harbor control was involved increases. Knowing that the measure taken by the competent authority was not an admonition changes the situation once more to the state illustrated by Fig. 12.16c. As we see this observation has no impact on the probabilities regarding the accident's location, but it does affect those regarding the authority involved. Finally, let us bring the observation into the system that the accident did not take place within the harbor (see Fig. 12.16d). The main impact of this information is that now the canal is the most probable location of the accident. This is not surprising, since we knew before that with probability 0.87 it did not take place at sea, so if the harbor is excluded the canal is the only remaining location which is not at sea.

This example illustrates how to handle a 4–dimensional product space with 288 elements in a simple and effective manner. The calculation of the marginal distribution and messages relies on the formulas provided in Sect. 12.5.4. Note that the numbers are calculating using more than two digits, so verification of the example may yield small divergences if the message vectors are used as they appear in the figures.

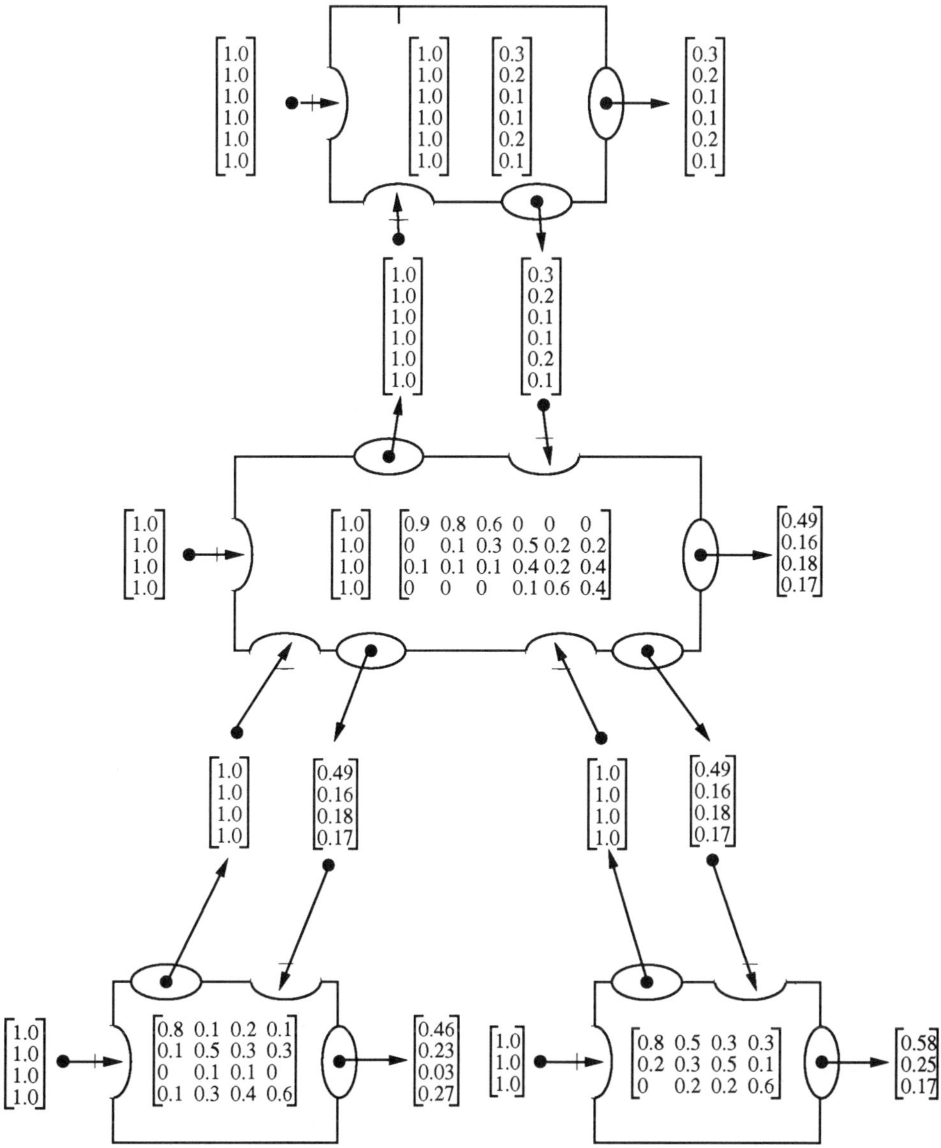

[1.0 1.0 1.0 1.0 1.0 1.0]
[1.0 1.0 1.0 1.0 1.0 1.0]
[0.3 0.2 0.1 0.1 0.2 0.1]
[0.3 0.2 0.1 0.1 0.2 0.1]
[1.0 1.0 1.0 1.0 1.0 1.0]
[0.3 0.2 0.1 0.1 0.2 0.1]
[1.0 1.0 1.0 1.0]
[1.0 1.0 1.0 1.0]
[0.9 0.8 0.6 0 0 0; 0 0.1 0.3 0.5 0.2 0.2; 0.1 0.1 0.1 0.4 0.2 0.4; 0 0 0 0.1 0.6 0.4]
[0.49 0.16 0.18 0.17]
[1.0 1.0 1.0 1.0]
[0.49 0.16 0.18 0.17]
[1.0 1.0 1.0 1.0]
[0.49 0.16 0.18 0.17]
[1.0 1.0 1.0 1.0]
[0.8 0.1 0.2 0.1; 0.1 0.5 0.3 0.3; 0 0.1 0.1 0; 0.1 0.3 0.4 0.6]
[0.46 0.23 0.03 0.27]
[1.0 1.0 1.0]
[0.8 0.5 0.3 0.3; 0.2 0.3 0.5 0.1; 0 0.2 0.2 0.6]
[0.58 0.25 0.17]

Fig. 12.16(a).

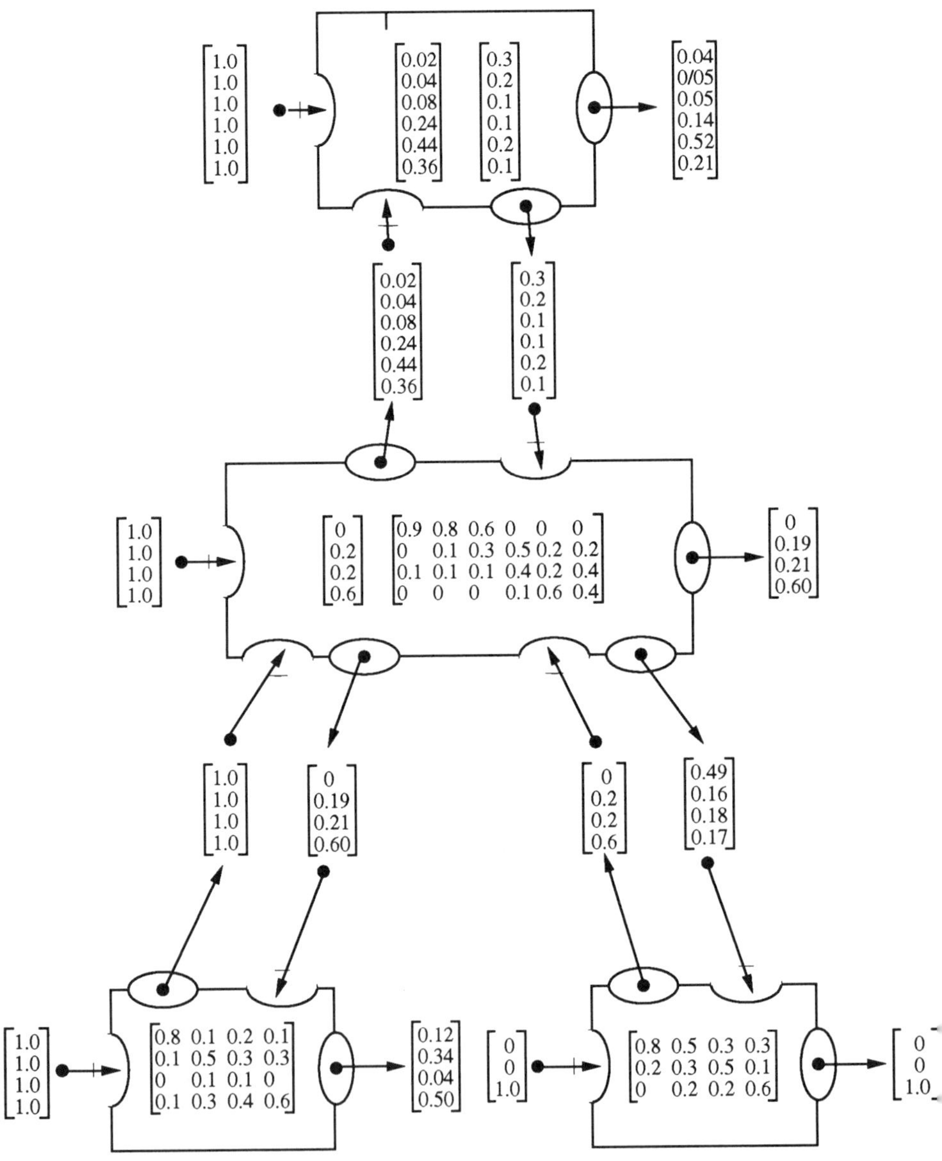

1.0 1.0 1.0 1.0 1.0 1.0
0.02 0.04 0.08 0.24 0.44 0.36
0.3 0.2 0.1 0.1 0.2 0.1
0.04 0/05 0.05 0.14 0.52 0.21
0.02 0.04 0.08 0.24 0.44 0.36
0.3 0.2 0.1 0.1 0.2 0.1
1.0 1.0 1.0 1.0
0 0.2 0.2 0.6
0.9 0.8 0.6 0 0 0
0 0.1 0.3 0.5 0.2 0.2
0.1 0.1 0.1 0.4 0.2 0.4
0 0 0 0.1 0.6 0.4
0 0.19 0.21 0.60
1.0 1.0 1.0 1.0
0 0.19 0.21 0.60
0 0.2 0.2 0.6
0.49 0.16 0.18 0.17
1.0 1.0 1.0 1.0
0.8 0.1 0.2 0.1
0.1 0.5 0.3 0.3
0 0.1 0.1 0
0.1 0.3 0.4 0.6
0.12 0.34 0.04 0.50
0 0 1.0
0.8 0.5 0.3 0.3
0.2 0.3 0.5 0.1
0 0.2 0.2 0.6
0 0 1.0

Fig. 12.16(b).

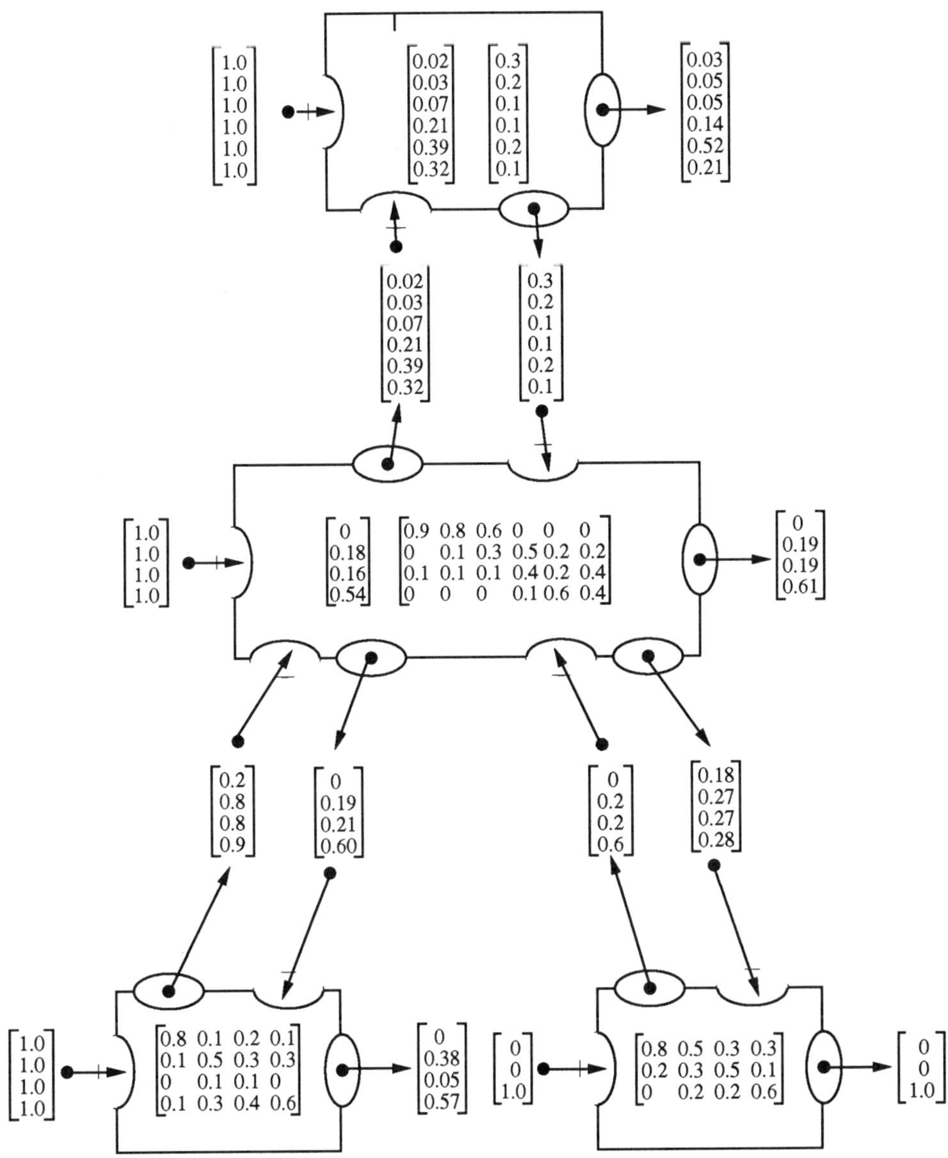

[1.0 1.0 1.0 1.0 1.0 1.0]
[0.02 0.03 0.07 0.21 0.39 0.32]
[0.3 0.2 0.1 0.1 0.2 0.1]
[0.03 0.05 0.05 0.14 0.52 0.21]
[0.02 0.03 0.07 0.21 0.39 0.32]
[0.3 0.2 0.1 0.1 0.2 0.1]
[1.0 1.0 1.0 1.0]
[0 0.18 0.16 0.54]
[0.9 0.8 0.6 0 0 0; 0 0.1 0.3 0.5 0.2 0.2; 0.1 0.1 0.1 0.4 0.2 0.4; 0 0 0 0.1 0.6 0.4]
[0 0.19 0.19 0.61]
[0.2 0.8 0.8 0.9]
[0 0.19 0.21 0.60]
[0 0.2 0.2 0.6]
[0.18 0.27 0.27 0.28]
[1.0 1.0 1.0 1.0]
[0.8 0.1 0.2 0.1; 0.1 0.5 0.3 0.3; 0 0.1 0.1 0; 0.1 0.3 0.4 0.6]
[0 0.38 0.05 0.57]
[0 0 1.0]
[0.8 0.5 0.3 0.3; 0.2 0.3 0.5 0.1; 0 0.2 0.2 0.6]
[0 0 1.0]

Fig. 12.16(c).

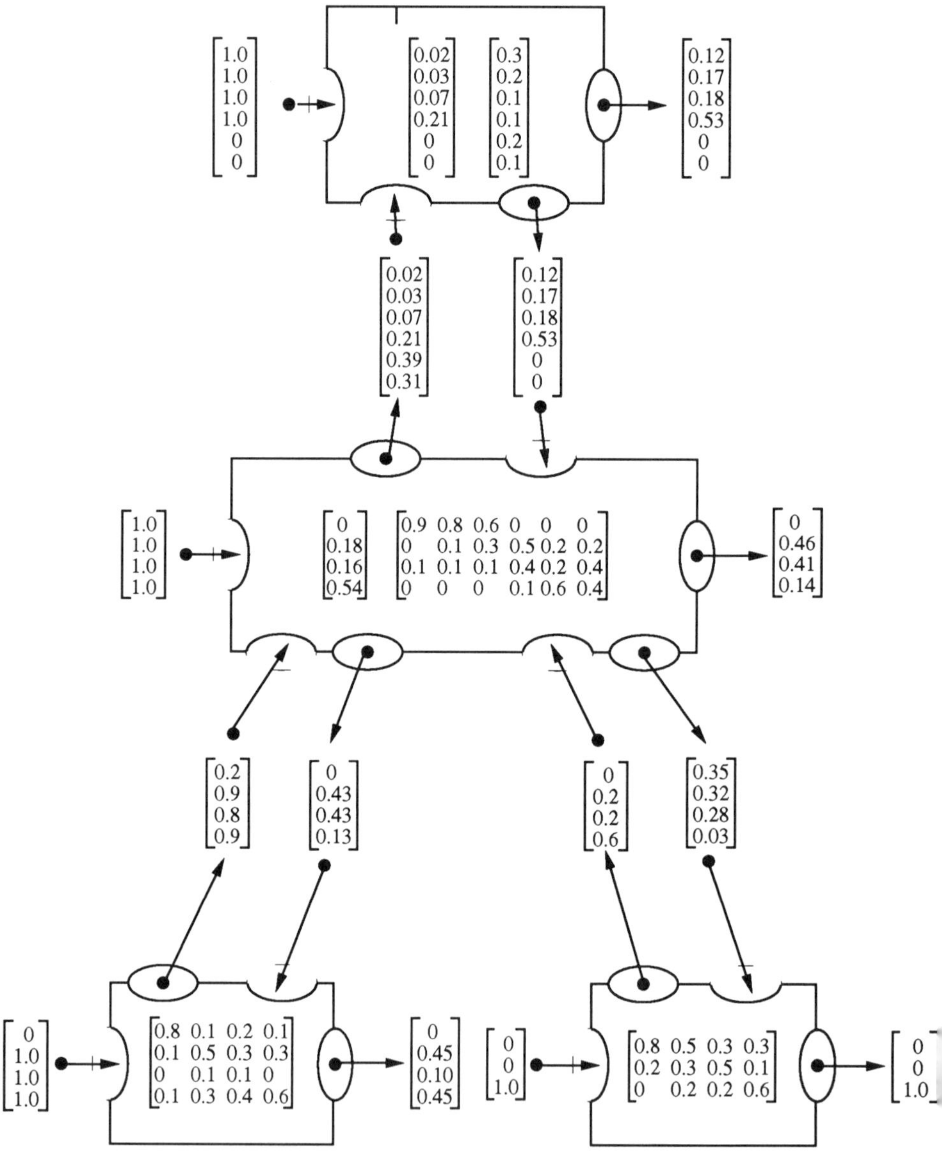

Fig. 12.16(d).

12.6. Concluding Remarks

The results of Sect. 12.2 and 12.3 are included in our monograph although these approaches are mainly of historical interest:

In inference networks only binary characteristics are allowed. As already mentioned in Chap. 12.3.3 (by referring to results presented in [Pednauld, Zucker, and Muresam 1981], [Glymour 1985], [Johnson 1986], and [Steve 1986]), the extension of the model to multivalued characteristics can lead to inconsistencies. An obvious problem of the approach of Duda et al. is the number of assumed independencies. Especially the conditional independence with respect to the negations of the hypotheses is often violated, an example of which is given in [Heckerman 1985]. In contrast, the analysis of decomposable models in the context of Bayesian probability theory turned out to be consistent. For a more detailed analysis of these models we recommend the books [Pearl 1988] and [Neapolitan 1990] as well as the excellent paper [Lauritzen and Spiegelhalter 1988, with discussion].

An improvement of decomposable models, from a knowledge representation point of view, is provided in [Heckerman 1990a]. Heckerman used an extension of the influence diagram representation called a similarity network [Heckerman 1990c]. This representation allows us to construct independent influence diagrams for subsets of a given domain. This method together with a probability assessment scheme based on partitions is used in the expert system PATHFINDER [Heckerman et al 1989].

In [Spiegelhalter and Lauritzen 1990] methods for sequential updating of conditional probabilities on directed graphical structures are considered. These methods allow us to handle imprecision in the sense that the probabilities can be updated and improved in the light of new data [Spiegelhalter 1989]. Another extension consists in using influence diagrams. Influence diagrams distinguish between those variables which are under control of a decision maker, those which are not (chances) as well as explicitely denoting a goal for solution ([Shachter 1988], [Heckerman 1990]). They allow a coherent probabilistic representation of uncertain knowledge in [Henrion 1988] some practical issues in constructing such networks are considered. Pearl [Xu and Pearl 1989] show that it is possible to extend these methods to continuous (normally distributed) variables. A very comprehensive overview to graphical models in applied statistics is containted in [Whittaker 1990].

In general, the models of Pearl and Spiegelhalter examined in Sect. 12.4 can be used to represent and to manipulate the knowledge of a domain expert in such a way that the handling of "belief" is consistent with the axioms of probability theory. With regard to the computational aspects, the model obviously provides a simple and convenient framework. In contrast to simple *rule based* inference models, on the basis of Bayesian networks a *belief maintenance system* can be derived: drawing inferences in such networks can be performed in both directions, from "symptoms" to "disease" as well as the other way around.

Cooper [Cooper 1990] has shown that for nontrivial non-singly connected Bayesian networks the problem of determining the probabilities of remaining variables, given that certain variables have been instantiated, is NP-hard and P-complete in terms of the number of nodes in the network. Probability propagation in singly connected networks on the other hand can be performed in linear time relative to the number of variables in the network [Pearl 1988]. Alternatives to avoid the complexity problems are the use of Monte Carlo techniques [Chin and Cooper 1987], the use of approximation algorithms which produce inexact bounded solution [Horvitz et al. 1989] or the creation of special case algorithms [Heckerman 1989].

Once more it has to be emphazised that the theoretical foundations underlying MUNIN and HUGIN are extensively considered in this book because they represent the present state of the art in the field of probabilistic uncertainty handling. Spiegelhalters improvements are closely related to the work of Shafer and Shenoy, which is presented in Chap. 13. We have adopted the notation of Markov-trees and deletion sequences introduced in [Shafer and Shenoy 1988] instead of junction tree's and the running intersection property, as they are more vivid, although the concepts are equivalent. Note that the approach in Sect. 12.5 is considered mainly for didactical reasons as it provides vivid insight into the formal background of the algorithms.

However, as has been made clear in [Lee 1988], Pearl's and Spiegelhalter's original model ignores two motives which have led researchers to advocate the use of the Dempster-Shafer model, for instance. One motive is related to adequately distinguishing *lack of knowledge* from *equal degrees of belief*. The other motive is that evidence partially in favor of a hypothesis should not be construed as evidence partially against the same hypothesis (i.e., in favor of its negation).

Chapter 13

Models Based on the Dempster-Shafer Theory of Evidence

Like Bayesian approaches, the Dempster-Shafer theory of evidence aims to model and quantify uncertainty by *degrees of belief*. But in contrast to Bayesian approaches it permits assignment of degrees of belief to *sets* of hypotheses rather than to hypotheses in isolation. The underlying idea is that the process of narrowing the hypothesis set with the collection of evidence is better represented in terms of this theory than in terms of Bayesian approaches. For this reason the theory can be viewed as an alternative to Bayesian modeling in probability theory.

In this chapter we mainly consider models and implementations that are based on the classical approach to evidence theory as proposed by Shafer [Shafer 1976]. His mathematical model is essentially based on the notion of *belief functions* and *Dempster's rule of combination*. This view, however, does *not* rely on probability theory: in this context it is also called the transferable belief model (see e.g. [Smets 1988]). In contrast to our treatment of belief functions there are approaches that use underlying probability spaces. Some of these concepts have already been considered earlier. Other semantics as well as their relations will be discussed in the concluding remarks of this chapter.

In Sect. 13.1 the original Dempster-Shafer theory of evidence is introduced. Section 13.2 then discusses knowledge representation aspects that arise with the use of this mathematical model. This set of approaches will be split up into three sections, on

- straightforward use of belief functions,
- belief functions in hierarchical hypothesis spaces, and
- belief propagation in Markov trees.

Section 13.3 examines proposals for performing uncertain reasoning where the knowledge of experts is assumed mainly to be represented in the form of explicit *expert rules*. In particular, the proposals of Ginsberg and Ishizuka et al. are examined.

A characteristic feature of the models looked at in Sect. 13.4 is that the algorithms for uncertain reasoning explicitly take the fact into account that the hypothesis space may be *hierarchically ordered*: the reason is that Shafer's "mathematical theory of evidence" is based on power sets. The elements of a power set can be related together by making use of subset and superset operators. In particular, Gordon and Shortliffe's extension to MYCIN's certainty factor approach and Yen's extension to the Dempster-Shafer theory are examined.

Section 13.5 mainly considers the proposals of Shafer, Shenoy, Logan, and Mellouli. A characteristic feature of these approaches is that explicit expert rules do not appear. Instead, the knowledge on dependencies is represented by the *links of a network* – the underlying idea is similar to Pearl's "Bayesian networks". As a concrete implementation the system MacEvidence is introduced.

13.1 The Mathematical Theory of Evidence

Let Ω be a *finite* "frame of discernment". Then each proposition concerning the identity of the "true value" ω can be represented by a subset of Ω. Sets including only one element correspond to elementary propositions. If we further assume that at most one elementary proposition is true, our task is to find just this true proposition (e.g. a hypothesis on a ship's location) by means of evidence. We generally say that a proposition is true if it contains the true elementary proposition. The Dempster-Shafer theory of evidence is based on the assumption that the partial (personal) belief in a proposition corresponding to a subset A of Ω is quantified by a single number $\mathrm{Bel}(A)$ in the unit interval. It is further assumed that these quantities satisfy the following properties:

$$(i)\ \mathrm{Bel}(\emptyset) = 0, \tag{13.1}$$

$$(ii)\ \mathrm{Bel}(\Omega) = 1, \tag{13.2}$$

$$(iii)\ \mathrm{Bel}(A_1 \cup \ldots \cup A_n) \geq \sum_{i=1}^{n} \mathrm{Bel}(A_i) - \sum_{i,j:1 \leq i < j \leq n} \mathrm{Bel}(A_i \cap A_j) + - \ldots + (-1)^{n+1}\mathrm{Bel}(A_1 \cap \ldots \cap A_n) \tag{13.3}$$

for every positive integer n and every collection $A_1, \ldots, A_n$ of subsets of Ω.

In evidence theory it is argued that degrees of belief should satisfy the above three properties. So, for instance, the belief in the disjunction of two propositions should at least contain the sum of the belief quantities allocated to each of it and reduced by the belief allocated to both. This motivates the above condition (13.3). However, the three requirements are of course not suitable for modeling purposes since, for example, the condition (13.3) is difficult to test. So one usually starts by introducing the concept of a *basic belief mass* (which is also called a basic probability assignment in [Shafer 1976]). Notice the formal relation to the notion of "mass distribution" in Definition 6.5.

A *basic belief mass* m is a mapping from 2^{Ω} to [0,1] such that

$$m(\emptyset) = 0 , \tag{13.4}$$

$$\sum_{A:A\subseteq\Omega} m(A) = 1 . \tag{13.5}$$

$m(A)$ defined in this way is the part of our belief that supports A, i.e. that A contains the true value, and due to lack of information does not support any strict subset of A. The sets A with $m(A) > 0$ are called *focal elements*. The function Bel_m (or simply Bel in shorthand notation) defined by

$$\mathrm{Bel}(A) = \sum_{B:B\subseteq A} m(B) \tag{13.6}$$

for all $A \subseteq \Omega$ is called the *belief function* or *credibility function*. Unlike m it measures the *total* belief that is credited to A. This definition is validated by Theorem 6.6. The corresponding plausibility function Pl_m, or Pl in short, defined by

$$\mathrm{Pl}(A) = 1 - \mathrm{Bel}(\overline{A}) = \sum_{B:A\cap B\neq\emptyset} m(B) \tag{13.7}$$

for all $A \subseteq \Omega$ measures the maximal amount of belief that can be attributed to A. The three functions m, Bel and Pl carry the same amount of information, but each has its unique interpretation. For instance, m can be expressed in terms of Bel by using the *Mõbius transformation*

$$m(A) = \sum_{B:B\subseteq A} (-1)^{|A\cap\overline{B}|} \cdot \mathrm{Bel}(B)$$

for $A \subseteq \Omega$.

Recall that

$$\mathrm{Bel}(A) + \mathrm{Bel}(\overline{A}) \leq 1 \tag{13.8}$$

holds which is Shafer's proposal concerning the representation of ignorance.

On of the main characteristics of the Dempster-Shafer approach is that it presents a formalism for combining distinct pieces of evidence, called *Dempster's rule*. Given two basic belief masses m_1 and m_2 pertaining to two sources of evidence, then the result of the combination is a new basic belief mass $m_1 \oplus m_2$ defined by

$$(m_1 \oplus m_2)(C) = \frac{\sum\limits_{A,B:A\cap B=C} m_1(A)\cdot m_2(B)}{\sum\limits_{A,B:A\cap B\neq\emptyset} m_1(A)\cdot m_2(B)} \quad \text{for } C \neq \emptyset,\ C\subseteq \Omega, \tag{13.9}$$

$$(m_1 \oplus m_2)(\emptyset) = 0.$$

$m_1 \oplus m_2$ exists if and only if

$$K \stackrel{\mathrm{d}}{=} \sum_{A,B:A\cap B=\emptyset} m_1(A)\cdot m_2(B) < 1. \tag{13.10}$$

K reflects the partially conflicting situation where two sets A and B have an empty intersection but $m_1(A)$ and $m_2(B)$ are different from zero. This is indeed in conflict with requirement (13.4) that a basic belief mass must assign zero to the empty set. Dempster's rule solves this problem by forcing $(m_1 \oplus m_2)(\emptyset) = 0$ and normalizing the remaining values of $m_1 \oplus m_2$. Notice that this implies a closed world assumption underlying the mathematical model.

The operator $\oplus$ is *commutative* and *associative*. Consequently, the final result does not depend on the order in which information is introduced and combined. We also use the notion $\mathrm{Bel}_1 \oplus \mathrm{Bel}_2$ for $\mathrm{Bel}_{m_1\oplus m_2}$ and for the plausibility analogously.

Another important feature of each reasoning model is the method used for *updating* on the basis of a true fact. In the case of the surely true proposition $A \subseteq \Omega$, in evidence theory one uses the updated belief function

$$\begin{aligned} &\mathrm{Bel}_A : 2^{\Omega} \to [0,1], \\ &\mathrm{Bel}_A(B) \stackrel{\mathrm{d}}{=} \frac{\mathrm{Bel}(B\cup\overline{A}) - \mathrm{Bel}(\overline{A})}{1-\mathrm{Bel}(\overline{A})} \end{aligned} \tag{13.11}$$

for all $B \subseteq A$. This belief function exists if $\mathrm{Pl}(A) \neq 0$.

One very problem concerns the inherent computational complexity of Shafer's theory of evidence: computing, for example, the basic belief assignment (or degree of belief Bel, or plausibility Pl, respectively) requires time exponential in the cardinality of the frame of discernment Ω. This property is obvious when looking at Dempster's rule, say. To be exact the claim of exponential complexity of Dempster's rule is slightly inaccurate. In [Orponen 1990] is demonstrated that the function computed by the rule is $\#P$-complete [Garey 1979] thus verifying its exponential complexity under the hypothesis $P \neq \#P$

A first attempt at reducing the complexity is offered in [Barnett 1981]. He shows that the computational time complexity can be reduced from exponential to linear by combining the belief functions in a simplifying order. However, to obtain this result, Barnett makes various assumptions concerning type and structure of "belief" and hypotheses to be combined. In particular, the following constraint has to be met: each piece of evidence is assumed to support only an elementary proposition (i.e. a subset with cardinality 1) or its complement. More complex subsets related to disjunctions of propositions are not allowed (see also [Gordon and Shortliffe 1984]). So, Barnett's proposal does not answer the question how "evidential reasoning" can be performed on the basis of large *hypothesis sets* in complex domains and with acceptable computational time complexity.

A solution in this direction is proposed in [Gordon and Shortliffe 1985]: again assuming Ω as the basic set of elementary hypotheses Gordon and Shortliffe consider only those subsets of Ω that are of (e.g. medical) interest for the expert. In general, this is only a small subset T of 2^{Ω}. The algorithm consists of three parts: first of all, for every element $A \in T$ all the basic beliefs confirming A and those disconfirming A are combined. Consequently, for every A this leads to two quantities $m(A)$ and $m(\overline{A})$. Secondly, the basic belief masses $m(A)$, $A \in T$, are combined. In part three of the algorithm, the basic beliefs $m(\overline{A})$, $A \in T$, are combined. Since also in this step use of the intersection operator is made, subsets $B \notin T$ may arise. Because these subsets are of less interest for the expert, in Gordon and Shortliffe's algorithm the basic belief $m(B), B \subseteq \Omega, B \notin T$, is assigned to the "next superset" of B that is an element of T.

This algorithm for performing the combination of basic belief assignments is of polynominal computational time complexity. However, one consequence of the approximation of part three is that the belief in $A, A \in T$, no longer can be characterized by an interval $[\mathrm{Bel}(A), 1 - \mathrm{Bel}(\overline{A})]$, since in general $\overline{A}$ will be not in T. Further, the result may depend on the order, the basic belief assignments are combined. So, in spite of an acceptable computational time complexity, also certain disadvantages of the algorithm are obvious. In [Kennes and Smets 1990a,b] fast algorithms for computing Dempster's rule are considered that are based on so called "fast Möbius transforms".

The case of continuous frames of discernment is examined in [Strat 1984]. In particular, he proves that the complexity can be reduced to $O(n^2)$ with n as the number of atomic propositions. But also in Strat's approach several constraints must be fulfilled.

13.2 Knowledge Representation Aspects

13.2.1 Representing Pieces of Knowledge

In order to model a piece of information in the context of belief functions first the user has to specify a suitable finite frame of discernment $\Omega = \{\omega_1, \ldots, \omega_n\}$, i.e. there must be an exhaustive set of mutually exclusive alternatives. The second step consists of choosing a basic belief mass for representing the uncertainty of the knowledge. Depending on the domain to be modeled here one has a variety of possibilities:

- If ω is definitely the "true" value then one uses the basic mass assignment

$$\begin{aligned} &m(\{\omega\}) = 1, \\ &m(A) = 0 \text{ otherwise.} \end{aligned} \tag{13.12}$$

- If there are preferences available on the singletons $\{\omega_i\}$ then one may use a *Bayesian belief function.* Here the corresponding basic mass assignment has the properties

$$\begin{aligned} &\sum_{i=1}^{n} m(\{\omega_i\}) = 1, \\ &m(A) = 0 \text{ if } |A| > 1. \end{aligned} \tag{13.13}$$

- If the only information is that the true value is definitely not in $A \subseteq \Omega$ then one should use a set function m satisfying $\mathrm{Bel}_m(\overline{A}) = 1$.

- An *explicit representation of ignorance* is possible. If we want to model "total ignorance", the so-called *vacuous mass assignment* defined by the basic belief assignment m_v is used:

$$\begin{aligned} &m_v(\Omega) = 1, \\ &m_v(A) = 0 \quad \text{for } A \subseteq \Omega,\ A \neq \Omega. \end{aligned} \tag{13.14}$$

  The combination of m_v which may be interpreted as "don't know" and a basic mass assignment m yields m again.

The ability to express total ignorance is one of the main features that is mentioned as an advantage of belief functions against the use of a single probability. Notice that also necessity measures are special belief functions: if the focal elements are nested, i.e. if they can be arranged in order such that each is contained in the following one, then we obtain the so-called *consonant belief function* as result [Shafer 1976].

The choice of a suitable frame of discernment as well as the assessment of the numbers is crucial since a non-probabilistic framework is intended. Especially the assignment of the value zero to a proposition should be handled carefully. An example may illustrate such situations.

(13.1) Example. Assume that the possible locations of a ship are given by $\Omega = \{z1, z2, z3\}$. Suppose that we obtain the evidence that "the ship is at_sea" ($\approx \{z3\}$) is believed to degree 0.9. Then we attach the mass 0.9 to $\{z3\}$ and the remaining mass 0.1 to the whole set Ω. This of course differs from a Bayesian approach where the masses have to be assigned explicitly to the elements of Ω.

We obtain $m_1(\{z3\}) = 0.9$, $m_1(\Omega) = 0.1$, and $m_1(A) = 0$, otherwise. This however leads to $\mathrm{Bel}_1(at\_sea) = 0.9$, $\mathrm{Pl}_1(at\_sea) = 1$, but $\mathrm{Bel}_1(\{z2\}) = 0$, and $\mathrm{Pl}_1(\{z2\}) - 0.1$, for instance. This is a reasonable and adequate representation of the available information. ◇

Often, pieces of information are also represented by equalities concerning belief functions such as

$$\mathrm{Bel}(A) = q$$

or inequalities such as

$$\mathrm{Bel}(A) \geq q,$$
$$\mathrm{Pl}(A) \leq p,$$

for instance. Those cases are most interesting which are related to the representation and interpretation of expert rules. As in the general case we have two semantically different approaches: the expert rule

$$\underline{\text{if}}\ A\ \underline{\text{then}}\ B\ \underline{\text{with belief}}\ p$$

can be interpreted as

$$(i)\ \underline{\text{if}}\ A\ \underline{\text{then}}\ (\ \underline{\text{belief}}\ (B)\ =\ p\ ),\ \text{or}$$
$$(ii)\ \underline{\text{belief}}\ (\ \underline{\text{if}}\ A\ \underline{\text{then}}\ B\ ) = p.$$

In the first case one could use the set of mass distributions

$$\{m | (\mathrm{Bel}_m)_A(B) = \frac{\mathrm{Bel}_m(B \cup \overline{A}) - \mathrm{Bel}_m(\overline{A})}{1 - \mathrm{Bel}_m(\overline{A})} = p\} \tag{13.15}$$

for the representation of the rule since this set of mass distributions is the only one that is compatible with our first interpretation.

In the second interpretation, the "<u>if</u> A <u>then</u> B" rule is represented by the *material implication formula* $\overline{A} \cup B$, thus converting the rule into the constraint

$$\{m | \mathrm{Bel}_m(\overline{A} \cup B) = p\}. \tag{13.16}$$

Using meta-principles such as the minimum specificity principle or maximum entropy principle one often chooses just *one* belief function that is compatible with the constraints in the interpretations (in this context Bel_1 is called less specific than Bel_2 if $\text{Bel}_1 \leq \text{Bel}_2$ holds for all $A \subseteq \Omega$).

If only two spaces, say $E = \{e_1, \ldots, e_n\}$ and $H = \{h_1, \ldots, h_m\}$, are considered then a set of expert rules is also described as a set of *conditional basic mass assignments*, i.e.

$$\underline{\text{if}}\ \{e_i\}\ \underline{\text{then}}\ \text{Bel}(B) = P(B|\{e_i\})$$

for all $B \subseteq H$ and $i = 1, \ldots, n$ (see [Smets 1988] and [Yen 1986], for instance). In this case the evaluation is of course facilitated.

13.2.2 Integration of Pieces of Evidence

If one expert describes *one* piece of evidence from different points of view, we obtain a typical constraint problem.

(13.2) Example. Assume that one expert gives the information

$$(i)\ \underline{\text{if}}\ A\ \underline{\text{then}}\ (\ \underline{\text{belief}}\ (B) = p\ ),\ \text{and}$$
$$(ii)\ \underline{\text{belief}}\ (A) = 1.$$

Then we could use the following two sets for the representation:

$$(i)\{m|(\text{Bel}_m)_A(B) = p\}$$
$$(ii)\{m|\text{Bel}_m(A) = 1\}$$

The intersection of these two sets leads to

$$\begin{aligned} &\{m|\frac{\text{Bel}_m(B \cup \overline{A}) - \text{Bel}_m(\overline{A})}{1 - \text{Bel}_m(\overline{A})} = p\ ,\ \text{Bel}_m(A) = 1\} \\ =&\{m|\text{Bel}_m(B \cup \overline{A}) = p\ ,\ \text{Bel}_m(A) = 1\} \end{aligned} \tag{13.17}$$

The calculations are facilitated since we have

$$\text{Bel}_m(A) = 1 \Rightarrow \text{Bel}_m(B \cup \overline{A}) = (\text{Bel}_m)_A(B) \tag{13.18}$$

So, in this case the two different methods for interpreting an expert rule coincide.

◇

Let us return to the interpretation in which an expert rule is viewed in the context of *material implication*. In this context, we will represent an expert rule "<u>if</u> A <u>then</u> B" also in the abbreviated form "$A \to B$". If the expert gives inequalities such as $\text{Bel}(A \to B) \geq r$, $\text{Bel}(A) \geq s$, $\text{Bel}(B \to A) \geq r'$, and $\text{Pl}(A) \leq t$, then we can at least derive inequalities to calculate the belief in B by using the properties of belief functions: from inequality (13.3) we get for general subsets $X, Y \subseteq \Omega$:

$$\begin{aligned} \text{Bel}(X \cap Y) &\geq \text{Bel}(X) + \text{Bel}(Y) - \text{Bel}(X \cup Y) \\ &\geq \max(0,\ \text{Bel}(X) + \text{Bel}(Y) - 1) \end{aligned} \tag{13.19}$$

Further, making use of the simple equation $B = (\overline{A} \cup B) \cap (A \cup B)$ leads to

$$\begin{aligned} \text{Bel}(B) &= \text{Bel}((\overline{A} \cup B) \cap (A \cup B)) \\ &\geq \max(0,\ \text{Bel}(\overline{A} \cup B) + \text{Bel}(A \cup B) - 1) \\ &\geq \max(0,\ \text{Bel}(\overline{A} \cup B) + \text{Bel}(A) - 1) \end{aligned}$$

If we use the real numbers r and s for which $r \leq \text{Bel}(A \to B)$ and $s \leq \text{Bel}(A)$ holds, we obtain

$$\text{Bel}(B) \geq \max(0,\ r + s - 1) \tag{13.20}$$

Similarly, substituting the equation $\overline{B} = (\overline{A} \cup \overline{B}) \cap (A \cup \overline{B})$ into (13.19) leads to

$$\begin{aligned} \text{Bel}(\overline{B}) &\geq \max(0,\ \text{Bel}(\overline{A} \cup \overline{B}) + \text{Bel}(A \cup \overline{B}) - 1) \\ &\geq \max(0,\ \text{Bel}(\overline{A}) + \text{Bel}(A \cup \overline{B}) - 1) \end{aligned}$$

and, with equality $\text{Pl}(X) = 1 - \text{Bel}(\overline{X})$

$$\begin{aligned} \text{Pl}(B) &\leq \min(1,\ 1 - \text{Bel}(\overline{A}) - \text{Bel}(A \cup \overline{B}) + 1) \\ &= \min(1,\ \text{Pl}(A) + 1 - \text{Bel}(A \cup \overline{B})) \\ &\leq \min(1,\ t + 1 - r') \end{aligned} \tag{13.21}$$

where t and r' are such that $t \geq \text{Pl}(A)$ and $r' \leq \text{Bel}(B \to A)$ hold. The combination of the results given by (13.20) and (13.21) together with the information given is shown below:

$$\begin{array}{rllcll} & \text{Bel} & (A & \to & B) & \geq & r \\ & \text{Bel} & (A & &) & \geq & s \\ \Longrightarrow & \text{Bel} & (& & B) & \geq & \max(0,\ r + s - 1) \end{array}$$

and

$$\begin{array}{rllcll} & \text{Bel} & (B & \to & A) & \geq & r' \\ & \text{Pl} & (& & A) & \leq & t \\ \Longrightarrow & \text{Pl} & (B & &) & \leq & \min(1,\ t + 1 - r') \end{array}$$

So, on the basis of the lower bound of belief in A and an upper bound of the plausibility of A, together with lower bounds for the two quantities $\mathrm{Bel}(A \to B)$ and $\mathrm{Bel}(B \to A)$, one can derive the two bounds for belief and plausibility concerning B – making use of the basic properties of belief functions only. The result of the sequential combination is given by the range

$$[\max(0,\ r+s-1) \leq \mathrm{Bel}(B)\ ,\ \ \mathrm{Pl}(B) \leq \min(1,\ t+1-r')\,] \qquad (13.22)$$

The related procedure however is a constraint propagation approach.

If we have distinct pieces of evidence, a constraint propagation approach is not appropriate. Instead one should use a formalism to pool the evidence or to aggregate the experts' opinions.

(13.2) Example. (continued) Let a second piece of evidence "distinct" from the first be described by a mass distribution m_2 with

$$m_2(\{z1\}) = 0.9 \text{ and } m_2(\Omega) = 0.1$$

Then the result of the combination is $m = m_1 \oplus m_2$ where

$$\begin{aligned} m(\{z1\}) &= 0.09/0.19 \approx 0.47 \\ m(\{z3\}) &= 0.09/0.19 \approx 0.47 \\ m(\Omega) &= 0.01/0.19 \approx 0.06 \end{aligned}$$

In this case the credibility is – as expected – more explicit. ◇

Another example is due to Zadeh [Zadeh 1984], [Zadeh and Ralescu 1986] – the so-called "Zadeh paradox":

(13.3) Example. Assume again the frame of discernment $\Omega = \{z1, z2, z3\}$ consisting of the possible locations "at_sea". Further, assume the basic belief masses m_1 and m_2 given by

$$\begin{array}{lclclcl} m_1(\{z1\}) &=& 0 & \quad & m_2(\{z1\}) &=& 1-k \\ m_1(\{z2\}) &=& k & & m_2(\{z2\}) &=& k \\ m_1(\{z3\}) &=& 1-k & & m_2(\{z3\}) &=& 0 \quad ,\ 0 < k << 1 \\ \multicolumn{7}{l}{m_i(A) = 0 \text{ for } A \in 2^{\Omega}\ ,\ |A| > 1\ ,\ i = 1,2} \end{array} \qquad (13.23)$$

Since both m_1 and m_2 are defined on the basis of the elementary sets $\{z1\}$, $\{z2\}$, and $\{z3\}$ of 2^{Ω}, the associated belief functions are Bayesian belief functions.

The two basic belief assignments m_1 and m_2 can be viewed as representing two information sources with the following knowledge: in the case of small k, the two sources coincide in the assertion that the outcome $z2$ is less probable. However, the two sources are different in the way that source 1 attaches a high basic belief mass to $z3$ and states the impossibility of $z1$, while, the other way around, for source 2 the outcome $z3$ is impossible and $z1$ is highly probable. Combining this widely differing information by applying formula (13.9) leads to the belief mass $m = m_1 \oplus m_2$ with

$$m(\{z2\}) = 1 \quad \text{and} \quad m(A) = 0, \text{ otherwise.}$$

Interpreting this "combined information" simply leads to the "truth of $z2$", although just $z2$ was assigned a very low basic belief by both sources, m_1 and m_2. By removing the conflicting parts of the evidence and normalizing the remaining parts, important information is discarded.

On a formal level, the above result is due to the "closed world assumption" underlying Shafer's theory: since the frame Ω is assumed to be completely defined by the set of values $\{z1, z2, z3\}$, the explicit existence of the empty set $\emptyset \in 2^{\Omega}$ – which may be interpreted by "none of the frame's values is true" – is not considered.

So, if conflicting information has to be combined, the denominator of the combination rule (13.9) becomes small or equal to zero. In the case of a zero denominator the combination cannot be performed. Otherwise the denominator serves as a normalization factor – similar to the denominator in Bayes' formula. ◇

The resolution of the Zadeh paradox requires additional knowledge on the inconsistent nature of the evidence involved. We discuss the problem again in Sect. 13.6.

Consequently, keeping the "closed world assumption" underlying the belief functions in mind the result is correct. If a hypothesis is *known to be false* on the basis of one knowledge source, this hypothesis cannot become more probable when adding other sources. However, the case of conflicting information should be at least indicated to the user because it may refer to a possible incompleteness of the modeled domain on one hand, or to some incorrectness in the data entered into the system on the other.

Shafer [Shafer 1976] has introduced a quantity which he calls the "weight of conflict" between two belief functions. This quantity is defined by

$$\text{conflict}(\text{Bel}_1, \text{Bel}_2) \stackrel{\text{d}}{=} \log \frac{1}{1-K} = -\log(1-K) \tag{13.24}$$

and can be used to indicate the problematic combinations. K is defined by equation (13.10). In the case of "absolutely conflicting information" the quantity is infinite; the case "no conflict" is indicated by "conflict$(\mathrm{Bel}_1, \mathrm{Bel}_2) = 0$".

For the sake of completeness, it should be mentioned that for handling the above situation, a theoretical framework completely different to that proposed in this book may also be considered: for instance, in [Heinsohn 1986] the theories of Dempster and Shafer have been modified by explicitly allowing the empty set as a possible focal element. The new value $\emptyset$ aims at considering the "open world assumption" and is interpreted as "none of the values given in the frame of discernment". This approach aims at completely avoiding the normalization process which – in the above case of conflicting knowledge – is said to lead to an artificial and counterintuitive increase of some basic beliefs. Instead, to maintain an explicit measure of the amount of conflict, the empty set is used and the conflict is indicated. However, implementations as well as a theoretical examination have shown that, if complex combinations of only slightly deviating knowledge sources have to be performed, the basic belief $m(\emptyset)$ will be large and is never decreased (compare the definition of K in equation (13.10)).

Additionally, as shown in Chap. 6, we have demonstrated that the "open world assumption" cannot be sufficiently modeled by simply adding the empty set to the mathematical model. An approach similar to the above was also introduced in [Smets 1988].

Concerning the *updating of belief*, suppose a belief function *bel* on Ω and let $A \subseteq \Omega$ be a completely reliable piece of evidence, i.e. the true value is certainly in the set A. Then in the Dempster-Shafer theory the updating process is performed according to formula (6.8). If $\mathrm{Bel}(\overline{A}) = 1$ holds, the knowledge and the evidence are contradictory.

Fortunately, the result Bel_A can also be formulated in terms of Dempster's rule, although integration and updating have different semantics. Let m_A the basic belief assignment defined by

$$m_{ev}(A) = 1,$$
$$m_{ev}(B) = 0 \text{ for } B \neq A.$$

Consequently,

$$\mathrm{Bel}_{ev}(B) = \begin{cases} 1 & \text{if } A \subseteq B, \\ 0 & \text{otherwise.} \end{cases}$$

Then, it is easy to show that in the case of existence

$$\mathrm{Bel}_A = \mathrm{Bel} \oplus \mathrm{Bel}_{ev} \tag{13.25}$$

holds, i.e. Dempster's rule is the essential formalism in this theory.

13.3 The Straightforward Use of Belief Functions

Concerning the question how uncertain knowledge can be propagated on the basis of a set of *explicitly given expert rules*, a multitude of more or less simple methods have been proposed in the literature. This section introduces and examines two of these inference models for rule-based diagnostic reasoning. A characteristic feature of these early models is that they were still introduced in terms of "sequential" and "parallel combinations" – similar to heuristic systems such as MYCIN.

In particular, we first introduce the model of Ishizuka et al. [Ishizuka, Fu, and Yao 1982a], which can be viewed as a straightforward extension of the probabilistic rule-based version already discussed in Sect. 12.2.1. In Sect. 13.3.2, the simple inference model of [Ginsberg 1984] is considered, who proposes a belief propagation based on the Dempster-Shafer theory. The simplicity of his approach mainly results from his restriction to dichotomous (i.e. two-valued) frames of discernment. The conclusions as well as pointers to the literature can be found in Sect. 13.3.3.

13.3.1 The Model of Ishizuka et al.

In the framework of early rule-based approaches on the basis of the Dempster-Shafer theory we start by introducing and discussing the inference model of Ishizuka et al. The mathematical foundations are mainly taken from [Ishizuka, Fu, and Yao 1982a]. Assume the characteristics X, Y, and H with the *frames of discernment*

$$\begin{aligned} X_0 &\stackrel{\mathrm{d}}{=} \Omega_X = \{x_1, \ldots, x_I\} , \\ Y_0 &\stackrel{\mathrm{d}}{=} \Omega_Y = \{y_1, \ldots, y_J\} , \\ H_0 &\stackrel{\mathrm{d}}{=} \Omega_H = \{h_1, \ldots, h_K\} , \end{aligned} \tag{13.26}$$

each consisting of mutually exclusive elements. As before, the three ranges are assumed to be exhaustive. Further, assume the states of the characteristics X and Y to be expressed in terms of the "conditional" basic belief masses[1].

$$\begin{array}{lll} m_X(X_i|E_X) , & i = 0, \ldots, 2^I - 1 , & X_i \in 2^{X_0} \\ m_Y(Y_j|E_Y) , & j = 0, \ldots, 2^J - 1 , & Y_j \in 2^{Y_0} \end{array} \tag{13.27}$$

These quantities may be processed by the system if X or Y are sub-goals in the inference process, i.e., if they are confirmed by other rules. If, on the

[1] Notice that they do not represent the same "conditional beliefs" as used in Sect. 13.2. Actually, the notion of "conditionals" is incorrect here in the context of Shafer's evidence theory.

other hand, the information is entered by a user it may be given in the form of basic belief masses, belief functions, or plausibility measures, respectively. However, especially in the case of a direct user input, only a few of the 2^I (2^J) basic beliefs will be entered and, therefore, different from zero.

On the basis of the above knowledge, the rules R1 and R2 are applied, the structure of these expert rules may look as follows:

rule R1:

$$\begin{array}{llllll} \underline{\text{if}} & X \text{ is } X_1 & \underline{\text{then}} & H \text{ is } H_{X_1} & \underline{\text{with}} & m_{H|X}(H_{X_1}|X_1) \\ \underline{\text{else if}} & X \text{ is } X_2 & \underline{\text{then}} & H \text{ is } H_{X_2} & \underline{\text{with}} & m_{H|X}(H_{X_2}|X_2) \\ & & & \vdots & & \\ \underline{\text{else}} & & & H \text{ is } H_0 & \underline{\text{with}} & 1 \end{array}$$

rule R2:

$$\begin{array}{llllll} \underline{\text{if}} & Y \text{ is } Y_1 & \underline{\text{then}} & H \text{ is } H_{Y_1} & \underline{\text{with}} & m_{H|Y}(H_{Y_1}|Y_1) \\ \underline{\text{else if}} & Y \text{ is } Y_2 & \underline{\text{then}} & H \text{ is } H_{Y_2} & \underline{\text{with}} & m_{H|Y}(H_{Y_2}|Y_2) \\ & & & \vdots & & \\ \underline{\text{else}} & & & H \text{ is } H_0 & \underline{\text{with}} & 1 \end{array}$$

The sets $X_1, X_2, \dots$ ($Y_1, Y_2, \dots$) included in the premise of rule R1 (R2) must be mutually exclusive. However, their union need not cover the complete universe of discourse X_0 (Y_0). In contrast to the sets used in the premises, some of the conclusions H_{X_l} (or H_{Y_l}, respectivly) may have non-empty intersections.

The objective of the *sequential combination* is to get information on the state of characteristic H, i.e. to infer the basic belief

$$m_H(H_k|E)\,, \quad H_k \subseteq H_0\,, \quad k = 0, \dots, 2^K - 1\,.$$

For example, to draw the inference on the basis of rule R1, the information concerning characteristic X has to be combined with the "conditional" basic belief mass attached to the rule R1. To determine the "degree of satisfaction" of the rule's premise, the basic belief assignment (13.27) must be converted to the associated belief function by applying formula (13.6) in a straightforward manner, i.e.

$$\mathrm{Bel}_X(X_l|E_X) = \sum_{X_i \subseteq X_l} m_X(X_i|E_X) \tag{13.28}$$

Consequently, the belief function Bel_X is used as input parameter for the combination formula (13.29), together with the values $m_{H|X}$ attached to rule R1.

$$\begin{aligned} m_H(H_{X_l}|E_X) &= m_{H|X}(H_{X_l}|X_l \cdot \mathrm{Bel}_X(X_l|E_X)\,, \qquad l = 1,2,\ldots \\ m_H(H_0|E_X) &= 1 - \sum_{X_l} m_{H|X}(H_{X_l}|X_l) \cdot \mathrm{Bel}_X(X_l|E_X) \end{aligned} \tag{13.29}$$

As mentioned earlier when examining other models for uncertain reasoning, also the above procedure is (implicitly) based on the *assumption of independence*: incoming evidence influences only the state of the characteristic X (as one "leaf-node" of the inference process). Unlike this leaf-node, the state of H is assumed not to be influenced (directly) by the evidence but only by the state of X.

Notice that the basic belief assignment to the subsets of H_0 is associated with a belief function of X – instead of a basic belief assignment of X.

Compared to formula (12.11) which was constructed by Ishizuka et al. for the probabilistic case, the simplicity of formula (13.29) is conspicuous. The main reason can be led back to special features of Shafer's theory: the independence of "missing belief" and "doubt", and the ability to represent "ignorance" explicitly. The first feature leads to the fact that an implicit rule-extension as given by (12.9) does not arise since support of H_{X_l} does not influence the support of $\overline{H_{X_l}}$. Consequently, an additional term like the second one in formula (12.11) is not considered. The degree of ignorance is made explicit by the quantity $m_H(H_0|E_X)$. Therefore, something like a final distribution of belief mass (compare the third term of formula (12.11)) is unnecessary.

In the same way and under the same independence assumption the second rule can be applied to compute the basic beliefs $m_H(H_{Y_l}|E_Y)$. Subsequently, the (a posteriori) basic belief assignments $m_H(H_{X_l}|E_X)$ and $m_H(H_{Y_l}|E_Y)$ are combined by applying Dempster's combination rule given by formulas (6.41) and (13.9). As already mentioned, this *parallel combination formula* assumes the independence of the characteristics X and Y with regard to the support of H.

To continue the inference process, the assignments $m_H(H_{Y_l}|E_X, E_Y)$ may be used as an input parameter for drawing inferences on the basis of another expert rule, or – if H is already the final hypothesis – can be transformed into the equivalent belief function or belief ranges, respectively. To finally visualize the uncertainty concerning the state of H, the belief ranges

$$\forall H_k \subseteq H_0 \;:\; [\,\mathrm{Bel}_H(H_k|E_X, E_Y)\,,\; \mathrm{Pl}_H(H_k|E_X, E_Y)\,]$$

may be derived from the basic belief values $m_H(H_k|E_X, E_Y)$. Before discussing the model we first show how the procedure of Ishizuka et al. will work by means of a simple example.

(13.4) Example. Assume the two frames $\Omega_X = X_0 = \{x_1, x_2, x_3\}$ and $\Omega_H = H_0 = \{h_1, h_2\}$, and the expert rule

rule R1':	<u>if</u>	X *is* $\{x_1, x_2\}$	<u>then</u>	H *is* $\{h_1\}$	<u>with</u>	0.5
			<u>and</u>	H *is* $\{h_2\}$	<u>with</u>	0.4
	<u>else</u>			H *is* H_0	<u>with</u>	1

Assume the state of X defined by the basic belief assignment

$$m_X(\{x_1\}) = 0.2\ ,$$
$$m_X(\{x_2\}) = 0.3\ ,$$
$$m_X(\{x_3\}) = 0.4\ ,$$
$$m_X(\{x_1, x_2\}) = 0.1\ .$$

The following table summarizes the information given, i.e. the belief function Bel_X that serves as an input parameter for the sequential combination (13.29), and the conditional basic beliefs given explicitly or implicitly by the expert when creating rule R1':

A	B	$\mathrm{Bel}_X(A)$	$m_{H\|X}(B\|A)$	
$\{x_1, x_2\}$	$\{h_1\}$	0.6	0.5	1)
$\{x_3\}$	$\{h_1\}$	0.4	0	3)
$\{x_1, x_2\}$	$\{h_2\}$	0.6	0.4	1)
$\{x_3\}$	$\{h_2\}$	0.4	0	3)
$\{x_1, x_2\}$	$\{h_1, h_2\}$	0.6	0.1	2)
$\{x_3\}$	$\{h_1, h_2\}$	0.4	1	3)

The index numbers in the table refer to the origin from which the conditional belief masses $m_{H|X}(B|A)$ are derived: the numbers marked by $^{1)}$ are directly given by the expert rule. The quantity $^{2)}$ is derived from

$$m_{H|X}(\{h_1, h_2\}|\{x_1, x_2\}) = 1 - m_{H|X}(\{h_1\}|\{x_1, x_2\}) - m_{H|X}(\{h_2\}|\{x_1, x_2\})$$

and measures the degree of partial ignorance. Finally, the numbers $^{3)}$ are derived from the <u>else</u>-part of the rule.

The application of sequential combination (13.29) leads to a state of H given by the basic belief assignment

$$\begin{aligned} m_H(\{h_1\}) &= 0.6 \cdot 0.5 &&= 0.30 \\ m_H(\{h_2\}) &= 0.6 \cdot 0.4 &&= 0.24 \\ m_H(\{h_1, h_2\}) &= 1 - 0.54 &&= 0.46\ . \end{aligned}$$

◇

The inference model of Ishizuka et al. can be viewed as an early straightforward example for how to propagate uncertainty in a rule-based system. As the basis for uncertainties attached to expert rules they use "conditional" basic belief measures. Before examining the model in relation to the theoretical framework of this book, let us start with a discussion of some practical aspects. Concerning these "practical aspects" we have already referred to the fact that in formula (13.29) Ishizuka et al. propose to use the belief function Bel_X instead of the corresponding basic belief mass m_X. The reason for this decision may be led back to the following fact: because the sets X_l represented in the premise of rule R1 must be mutually exclusive, the maximum number of premise-sets X_l is given by the number $I = |X_0|$. Therefore, the whole set of *subsets* of X_0 cannot be explicitly considered by the premise of a rule since $|X_0| = I << 2^I = |2^{X_0}|$.

So, for the above example-rule R1' the sets $\{x_1\}$ and $\{x_2\}$ need not appear in other parts of the premise since the premises sets X_l have to be mutually exclusive. However, but for just these two sets the basic belief values are known. Therefore, for additionally taking the values $m_X(\{x_1\})$ and $m_X(\{x_2\})$ into account when performing the sequential combination, formula (13.29) considers the *total* belief $\mathrm{Bel}_X(\{x_1, x_2\})$.

However, the inference model of Ishizuka et al. leaves some questions open. For the above example, it is not clear how to proceed

- in the case of an expert rule

if	X *is* $\{x_1\}$	then	H *is* $\{h_1\}$	with	1
else if	X *is* $\{x_2\}$	then	H *is* $\{h_1\}$	with	0.9
...					

  since formula (13.29) does not consider the basic belief

  $$m_X(\{x_1, x_2\}) = 0.1$$

  assigning the proposition "x_1 or x_2" a non-zero mass.
  Further, formula (13.29) does not take the fact into account that conclusion "H *is* $\{h_1\}$" is confirmed by both "X *is* $\{x_1\}$" and "X *is* $\{x_2\}$".

- if the state of X is defined by the basic belief assignment $m_X(\{x_2, x_3\})$ with the one focal element $\{x_2, x_3\}$, for instance. In this case the premise of rule R1' does not match, and the result of the sequential combination will be "complete ignorance" concerning the state of H.

Further information – especially concerning the use of extensions of the formalism in the expert systems SPERIL I and SPERIL II – can be found in, e.g., [Ishizuka, Fu, and Yao 1981], [Ishizuka, Fu, and Yao 1982b], [Ishizuka 1983], [Ogawa, Fu, and Yao 1985].

13.3.2 The Model of Ginsberg

In the framework of straightforward methods on the basis of the Dempster-Shafer theory, we continue with another simple proposal for uncertain reasoning taken from [Ginsberg 1984]: Ginsberg provides a procedure for propagating belief and plausibility degrees through a set of explicit "expert rules"[2]. In his approach he interprets the belief and plausibility degrees as lower and upper probability bounds of "belief", in the sense that the knowledge is insufficient for completely specifying one existing but unknown probability distribution. However, because of his restrictive assumption of *two-valued* frames of discernment, this section may also be viewed as an example of how to use belief functions with the above interpretation and how to perform sequential and parallel combinations in the simplest case.

The most important assumption made by Ginsberg is that his model applies to *dichotomous frames* only, i.e. those which account for two propositions. Consequently, the "belief" in a proposition "A" can be represented by a tuple (a, b) where a measures the extent to which one believes the proposition "A" and b measures the disbelief, i.e. belief in the complementary proposition "not A".

In terms of Dempster-Shafer theory and for $\Omega_A = A \cup \overline{A}$ the quantities are therefore defined as follows: $a = \text{Bel}(A)$ and $b = \text{Bel}(\overline{A})$. Because of the simple structure of the frames the belief Bel and mass distribution m are identical for the two focal elements A and $\overline{A}$. Consequently, the "belief interval" is given by $[a, 1-b]$. Of course, the inequality $a + b \leq 1$ has to be true, and the degree of ignorance is measured by $m(\Omega_A) = 1 - (a + b)$.

In [Ginsberg 1984] the procedure for performing the *sequential combination* in this simple framework is derived in an intuitive way instead of being formally based: in addition to the knowledge about A given above assume the frame of discernment $\Omega_C = C \cup \overline{C}$, and the expert rule

$$\underline{\text{if}}\ A\ \underline{\text{then}}\ C\ \underline{\text{with}}\ (c, d)$$

with $c = \text{Bel}(C|A)$ and $d = \text{Bel}(\overline{C}|A)$. Since, on the basis of certain evidence E, the probability of A being true is at least a, the minimum probability of C being true is at least $a \cdot c$. Similarly, the probability of $\overline{C}$ is at least $a \cdot d$.

[2] Notice that Ginsberg's original paper [Ginsberg 1984] is related to uncertainty handling in *semantic networks* instead of in explicit expert rules: similar to the idea of Rich ("default reasoning as likelihood reasoning", [Rich 1983]) which is based on "certainty factors", Ginsberg considers the "arcs" of semantic nets to be labelled with ranges of probabilities for reflecting confidence. So, for instance, the (classical) terminological assertion "bird isa flyer" would be substituted by the assertion "bird isa flyer with [0.90, 0.98]", since the first assertion is true only *usually* or for *most* birds (remember penguins, ostriches, etc.).

Notice that at this point Ginsberg leaves the original belief calculus and uses the interpretation of lower and upper probabilities. So, on the basis of available evidence E and the expert rule, the belief concerning the truth of C is given by the tuple (ac, ad), or by the interval $[ac, 1-ad]$, respectively.

For *parallel combinations* as considered in the case of two or more expert rules confirming the same conclusion, Ginsberg makes use of Dempster's original rule. For example, assume in the case of the two expert rules

$$\begin{array}{l} \underline{\text{if}}\ A\ \underline{\text{then}}\ C\ \underline{\text{with}}\ (c_1, d_1) \\ \underline{\text{if}}\ B\ \underline{\text{then}}\ C\ \underline{\text{with}}\ (c_2, d_2) \end{array}$$

the results of the two sequential combinations given by the tuples (a, b) and (c, d), with

$$\begin{aligned} a &= \mathrm{Bel}(C|A, E), \\ b &= \mathrm{Bel}(\overline{C}|A, E), \\ c &= \mathrm{Bel}(C|B, E), \\ d &= \mathrm{Bel}(\overline{C}|B, E). \end{aligned}$$

Dempster's rule instantiated on the basis of the dichotomous frame $\Omega_C = C \cup \overline{C}$ leads to three focal elements C, $\overline{C}$, and Ω_C. Defining $\overline{a} = 1-a$, $\overline{b} = 1-b$, $\overline{c} = 1-c$, and $\overline{d} = 1-d$, the mass distribution m and belief function Bel appear as follows:

$$\begin{aligned} m(C|A,B,E) &= \mathrm{Bel}(C|A,B,E) = \frac{a\overline{d} + \overline{b}c - ac}{1-(ad+bc)} = 1 - \frac{\overline{a}\,\overline{c}}{1-(ad+bc)} \\ m(\overline{C}|A,B,E) &= \mathrm{Bel}(\overline{C}|A,B,E) = \frac{b\overline{c} + \overline{a}d - bd}{1-(ad+bc)} = 1 - \frac{\overline{b}\,\overline{d}}{1-(ad+bc)} \end{aligned}$$

$$\begin{aligned} m(\Omega_C|A,B,E) &= \frac{(1-a-b)(1-c-d)}{1-(ad+bc)} = \frac{\overline{a}\,\overline{c} + \overline{b}\,\overline{d}}{1-(ad+bc)} - 1 \\ \mathrm{Bel}(\Omega_C|A,B,E) &= 1 \end{aligned}$$

Denoting by $\oplus$ the combination to be applied to tuples of belief quantities then leads to

$$(a,b) \oplus (c,d) =$$

$$\begin{cases} \left(1 - \dfrac{\overline{a}\,\overline{c}}{1-(ad+bc)}\ ,\ 1 - \dfrac{\overline{b}\,\overline{d}}{1-(ad+bc)}\right) & \text{if } ad+bc \neq 1, \\ \text{undefined} & \text{otherwise.} \end{cases} \tag{13.30}$$

The combination is undefined in the case $(0,1) \oplus (1,0)$, since the fact that a proposition has been proven both true and false indicates a *conflict* in the knowledge base.

Although the properties of Dempster's original rule have already been discussed in depth, below we will show how the most important properties appear in the special case of two-valued frames of discernment:

(a) commutativity:

$$(a_1, b_1) \oplus (a_2, b_2) = (a_2, b_2) \oplus (a_1, b_1)$$

(b) associativity:

$$((a_1, b_1) \oplus (a_2, b_2)) \oplus (a_3, b_3) = (a_1, b_1) \oplus ((a_2, b_2) \oplus (a_3, b_3))$$

(c) total ignorance:

$$(a_1, b_1) \oplus (0,0) = (a_1, b_1)$$

(d) missing disbelief:

$$(a_1, 0) \oplus (a_2, 0) = (a_1 + a_2 - a_1 a_2 \ , \ 0)$$

(e) logical certainty:

$$(a_1, b_1) \oplus (1,0) = (1,0) \text{ for } (a_1, b_1) \neq (0,1)$$

Ginsberg's model for drawing inferences on the basis of the Dempster-Shafer theory is based on the assumption of dichotomous frames of discernment. Consequently, the formulas for sequential and parallel combination get a simple structure – compared, for example, to Dempster's original rule and its computational complexity. Ginsberg's procedures, however, are only intuitively based: one main reason is that Dempster's combination rule cannot be applied in this case of lower and upper probabilities. In this case, the type-1 conditioning applies (compare Sect. 4.4).

Though simple, the approach of Ginsberg is important for quite other reasons: his model can also be viewed as a proposal for how to make use of the Dempster-Shafer theory in relation to *non-monotonic logics*. This, on the other hand, is related to the question of how to *retract* inferences and conclusions already drawn: Ginsberg demonstrated that the considered version of Dempster's rule becomes *invertible* – an aspect important in the framework of non-monotonic reasoning. In [Ginsberg 1985a] he presents arguments in favor of the answer "yes" to the question "Does probability have a place in non-monotonic reasoning?". In particular, he extends "resolution" to deal with probabilistic information. However, the general topic of non-monotonicity as well as this aspect of Ginsberg's work will be briefly considered in Chap. 15.

13.3.3 Discussion, Related Work

A characteristic feature of the two models examined above and other early models and systems is that the term *belief* is interpreted in the context of probabilities or sets of probabilities. This interpretation, however, does not fit the original ideas of Shafer who developed with his "mathematical theory of evidence" a complete mathematical framework for modeling one's *personal belief.* This framework covers algorithms for belief propagation as well as for the pooling of evidence.

One consequence of the frequent misinterpretation and application of Shafer's belief theory is that in early models and systems which propose their own algorithms for sequential and parallel combinations, often a mixture of theories (e.g. of probabilities and belief) is proposed. Typically this mixture is inherently inconsistent. "Allowed mixtures" and the relationships of the different mathematical theories were examined in detail in Chap. 6 of this book.

Another model proposed for performing evidential reasoning on the basis of expert rules is proposed by Liu. In [Liu 1986] he presents an extension of the Dempster-Shafer theory that composes the belief in a conclusion from the belief in rules and the belief in conditions – similar to the models considered above. A further goal of Liu's model is a combination of plausible reasoning with *causal reasoning* in a coherent way: to consider the causation aspects he identifies a primitive set of causal categories, the so-called *role system.* This role system aims at accounting for the *qualitative* difference of relationships between conditions and conclusions of rules [Liu 1985], [Liu 1986]. Liu considers also the inclusion of exception roles which work in the direction of "non-monotonic reasoning". This aspect has already been mentioned in the framework of Ginsberg's proposal. However, although the topic of non-monotonicity is of great importance also for *numerical* mechanisms for uncertain reasoning, the fundamental theory is still under development, also in the *non-numerical* case (compare Chap. 15).

However, there are many other models that are proposed in the literature and that may be assigned to this section on the straightforward use of belief functions.

13.4 Belief Functions in Hierarchical Hypothesis Spaces

In the preceding section we have examined different proposals for how to perform uncertain reasoning on the basis of the Dempster-Shafer theory. A characteristic feature common to all these approaches is that the knowledge of one or more "domain experts" has to be coded as *expert rules.* The expert rules have the general form "if X then H with m" and are *explicitly* contained in the system's knowledge base. Inferences are then drawn on the basis of chains of expert rules, applied to evidential knowledge entered by a user of the system.

However, when modeling a closed domain of the real world, often the frames – final hypotheses as well as intermediate values – can be hierarchically ordered. For instance, in a medical expert system certain hypotheses (i.e. the diseases) may be mutually exclusive *subtypes* of a *more general* disease. In such a *hierarchical hypothesis space* forming a taxonomy, leaf nodes represent the most specific hypotheses while internal nodes represent more general superclasses. In very early rule-based systems such as MYCIN, such hierarchical relationships had to be *explicitly* modeled by means of expert rules.

In the Dempster-Shafer theory, however, belief can be *directly* assigned to *sets of elements* without giving specific information on the subclasses. Evidence may support hypotheses in a higher level of the hierarchy, but may give no information on the evidence supporting the classes subsumed. So, on the basis of the Dempster-Shafer theory, taxonomic class relationships can be separated from those expert rules representing evidential knowledge only.

In this section we examine two approaches to diagnosis in such taxonomies. In the model considered first, Gordon and Shortliffe propose an extension to the certainty factor approach of MYCIN that simply associates certainty factors and basic beliefs and uses Dempster's rule for parallel combinations. In the framework of this book their model should be viewed as an example that makes visible the advantages of using taxonomies on the basis of the Dempster-Shafer theory. However, the formal basis of their extension remains questionable.

The second model considered is proposed by Yen and is based on an extension to the Dempster-Shafer theory that allows the uncertainties of rules to be taken into account. An implementation is Yen's prototype expert system GERTIS.

These two approaches are examined in this separate section for two reasons: first, compared to the straightforward models of Sect. 13.3 here the main emphasis is laid on "parallel combinations of hierarchically related hypotheses" rather than on sequential combinations over chains of expert rules. Second, both the model of Gordon and Shortliffe and the proposal of Yen are conceptually different from the approaches of Shafer, Shenoy, and Logan considered in Sect. 13.5: they take belief networks into account that indeed are different to taxonomies based on simple super- and sub-class relationships. Explicit expert rules do not appear.

13.4.1 Gordon and Shortliffe's Extension to MYCIN

In 1984, Gordon and Shortliffe considered the importance of Shafer's "mathematical theory of evidence" [Shafer 1976] in relation to the certainty factor model. Further, they gave ideas for integrating it in the expert system MYCIN. As motivation for the "integration" of Shafer's theoretical framework into the certainty factor model of MYCIN, in [Gordon and Shortliffe 1984], [Gordon and Shortliffe 1985] Gordon and Shortliffe emphasize the known advantages of Shafer's theory, e.g.

- Not only elementary hypotheses but also *sets* of hypotheses of a basic set may be considered and combined. By hierarchically ordering the subsets of the hypothesis space, just those of MYCIN's expert rules which describe the relation of subsuming to subsumed hypothesis sets can be disclaimed. For instance, medical interest may attach not only the elementary hypotheses {`HEP`} and {`CIRR`} but also their union {`HEP,CIRR`}, which has the medical interpretation "intrahepatic cholestasis".
  As a consequence, each confirmation and each disconfirmation of the certainty of a hypothesis (set) will influence the certainty of the hypothesis sets subsuming it. This is an implicit property of the belief function model, which in MYCIN had to be explicitly modeled with the help of expert rules.
- The generality of the Dempster-Shafer combination rule, at least from a *syntactical* viewpoint, "covers" the combination of Bayesian probabilities as well as Shortliffe's combination of identically signed certainty factors (details of this are given by Dubois et al. [Dubois and Prade 1985], for example).
- The *super-additivity* property of belief functions allows to distinguish between absence of belief in a hypothesis (set) on one hand and the belief in the complementary hypothesis (set) on the other. However, also the original certainty factor model claims to allow this, but because of the ad hoc nature of the certainty factor model a formal and consistent basis is lacking.

To give an impression of Gordon and Shortliffe's ideas we first elaborate on how to determine the "frames of discernment" on the basis of the medical expert system MYCIN. Further, we describe Gordon and Shortliffe's procedure for matching certainty factors and belief functions, and examine the way in which medical assertions will be combined.

The disjointness and completeness of the elements of hypothesis sets which have to be satisfied when applying Shafer's belief model, are given with MYCIN's "single-valued parameter" concept that was already looked at in

Sect. 9.1.2. The conceptual structures of MYCIN's knowledge representation scheme are *associative triples* of the form (OBJECT ATTRIBUTE VALUE). A hypothesis set Ω therefore consists of all triples with same object and attribute. For example, one "frame of discernment" may be the set

$$\Omega = \{ \quad (\texttt{ORGANISM1 IDENTITY } x) \mid \\ x \in \{\texttt{KLEBSIELLA}, \texttt{E.COLI}, \texttt{PSEUDOMAS}, \texttt{STREPTOCOCCUS}, \ldots\}\}$$

However, in the case of "multi-valued parameters" the above disjointness requirement is not fulfilled. But, as mentioned by Gordon and Shortliffe, the case of multiple values can be simply excluded from the integration without any restriction by handling it separately.

The next aspect to be considered is how certainty factors and belief functions are related together. Expert rules in MYCIN have the general form "<u>if</u> e <u>then</u> h <u>with</u> s" with the elementary hypothesis $h \in \Omega$ and the certainty factor $s \in [-1, +1]$.

If s is *positive* (i.e., if the truth of the premise implies a positive change of belief), then the certainty factor s is directly considered as a basic belief quantity. The mass $1 - s$ is attached to Ω, the whole set of hypotheses, and aims at measuring ignorance:

$$\begin{aligned} s \geq 0 : \quad m(\{h\}) &\stackrel{\mathrm{d}}{=} s \qquad , h \in \Omega, \\ m(\Omega) &\stackrel{\mathrm{d}}{=} 1 - s \end{aligned} \tag{13.31}$$

If, on the other hand, s is *negative*, certainty factors and basic belief values are assigned as follows

$$\begin{aligned} s < 0 : \quad m(\Omega - \{h\}) &\stackrel{\mathrm{d}}{=} -s \qquad , h \in \Omega, \\ m(\Omega) &\stackrel{\mathrm{d}}{=} 1 + s \end{aligned} \tag{13.32}$$

For the *parallel combination* discussed now, assume Ω as the exhaustive set of organism identities. For each of the three cases examined below it is visible that Dempster's rule (13.9) becomes very simple: each of the basic belief assignments m_1 and m_2 has only two focal elements, i.e., two subsets $A \subseteq \Omega$ with a basic belief different from 0. This has the consequence that the combination has at most four entries with values different from 0.

Case 1: Assume two certainty factors with the same sign. This means that two information sources confirm (disconfirm) the same associative triple. For example, if the two basic belief assignments are given by

$$\begin{array}{llllll} m_1 & (\{\texttt{PSEU}\}) & = 0.4 & \qquad m_2 & (\{\texttt{PSEU}\}) & = 0.7 \\ m_1 & (\Omega) & = 0.6 & \qquad m_2 & (\Omega) & = 0.3 \end{array}$$

we get the following combination table:

		m_2: {PSEU} 0.7	m_2: Ω 0.3
m_1	{PSEU} 0.4	{PSEU} 0.28	{PSEU} 0.12
m_1	Ω 0.6	{PSEU} 0.42	Ω 0.18

This leads to the combined basic belief $m_{12} \stackrel{\mathrm{d}}{=} m_1 \oplus m_2$ with

$$m_{12}(\{\mathtt{PSEU}\}) = 0.82 \qquad \text{and} \qquad m_{12}(\Omega) = 0.18$$

If the evaluation of each of the two expert rules leads to a confirmation of the hypothesis (set $A = \{h\}$) or to a disconfirmation of the hypothesis (set $A = \Omega - \{h\}$), then combination formula (13.9) reduces to

$$m_{12}(A) = m_1(A) + m_2(A) - m_1(A) \cdot m_2(A) \tag{13.33}$$

which is identical to the certainty factor combination formula (9.7).

Case 2: Assume two certainty factors with different signs. This means that one information source confirms the associative triple while the other source disconfirms it. For example, if the two basic belief assignments are given by

$$\begin{array}{lll} m_1 \; (\{\mathtt{PSEU}\}) = 0.4 & \qquad & m_2 \; (\overline{\{\mathtt{PSEU}\}}) = 0.8 \\ m_1 \; (\Omega) = 0.6 & & m_2 \; (\Omega) = 0.2 \end{array}$$

we obtain the combination table

		m_2: $\overline{\{\mathtt{PSEU}\}}$ 0.8	m_2: Ω 0.2
m_1	{PSEU} 0.4	$\emptyset$ 0.32	{PSEU} 0.08
m_1	Ω 0.6	$\overline{\{\mathtt{PSEU}\}}$ 0.48	Ω 0.12

In this case the application of Dempster's rule leads to a normalization with factor $1 - K = 0.68$ (for the definition of K compare equation (13.10) and Sect. 6.6) and to the basic belief assignment $m_{12} \stackrel{\mathrm{d}}{=} m_1 \oplus m_2$

$$\begin{array}{lllll} m_{12} \; (\{\mathtt{PSEU}\}) & = & 0.08/0.68 & = & 0.118 \\ m_{12} \; (\overline{\{\mathtt{PSEU}\}}) & = & 0.48/0.68 & = & 0.706 \\ m_{12} \; (\Omega) & = & 0.12/0.68 & = & 0.176 \end{array}$$

Case 3: Assume two rules confirming or disconfirming *different* hypotheses (i.e. different values of the same frame of discernment Ω). The original certainty factor approach does not cover this case and is therefore different from Dempster's combination. For instance, on the basis of the basic belief assignments

$$\begin{array}{llll} m_1 & (\{\texttt{PSEU}\}) & = 0.4 \qquad & m_2 & (\overline{\{\texttt{STREP}\}}) & = 0.7 \\ m_1 & (\Omega) & = 0.6 \qquad & m_2 & (\Omega) & = 0.3 \end{array}$$

we obtain the following combination table:

		m_2: $\overline{\{\texttt{STREP}\}}$	0.7	Ω	0.3
m_1	{PSEU} 0.4	{PSEU}	0.28	{PSEU}	0.12
m_1	Ω 0.6	$\overline{\{\texttt{STREP}\}}$	0.42	Ω	0.18

The resulting basic belief assignment m_{12} is given by

$$\begin{array}{lll} m_{12} & (\{\texttt{PSEU}\}) & = 0.40 \\ m_{12} & (\overline{\{\texttt{STREP}\}}) & = 0.42 \\ m_{12} & (\Omega) & = 0.18 \end{array}$$

In the above three cases we have not considered the final conversion of the basic belief masses into the associated belief functions. In spite of being of the same entropy, an uncertainty representation on the basis of belief functions (instead of basic beliefs) may be of more interest for a user because the "whole belief" in a hypothesis is visible. Furthermore, uncertainty can be characterized by ranges, and a space between the two bounds of a range stands for the existence of partial ignorance.

A basic belief assignment, on the other hand, may be of less interest because it measures the belief in exactly the given hypothesis—neglecting the belief in all those hypotheses that are subsumed by the given one. For instance, in case 3 examined above {PSEU} is known to be a subset of $\overline{\{\texttt{STREP}\}}$.

Applying the well-known formula (6.41) to our example by defining $m_{12} \stackrel{\text{d}}{=} m_1 \oplus m_2$ and

$$\text{Bel}_{12}(A) \stackrel{\text{d}}{=} (\text{Bel}_1 \oplus \text{Bel}_2)(A) = \sum_{B:B \subseteq A} m_{12}(B)$$

$$\text{Pl}_{12}(A) = 1 - \text{Bel}_{12}(\overline{A})$$

the belief and plausibility functions can be simply derived:

$$
\begin{array}{ll}
\mathrm{Bel}_{12}(\{\texttt{PSEU}\}) & = m_{12}(\{\texttt{PSEU}\}) = 0.40 \\
\mathrm{Bel}_{12}(\{\texttt{STREP}\}) & = 0 \\
\mathrm{Bel}_{12}(\overline{\{\texttt{PSEU}\}}) & = 0 \\
\mathrm{Bel}_{12}(\overline{\{\texttt{STREP}\}}) & = m_{12}(\overline{\{\texttt{STREP}\}}) + m_{12}(\{\texttt{PSEU}\}) = 0.82
\end{array}
$$

$$
\begin{array}{ll}
\mathrm{Pl}_{12}(\{\texttt{PSEU}\}) & = 1 - \mathrm{Bel}_{12}(\overline{\{\texttt{PSEU}\}}) = 1 \\
\mathrm{Pl}_{12}(\{\texttt{STREP}\}) & = 1 - \mathrm{Bel}_{12}(\overline{\{\texttt{STREP}\}}) = 0.18 \\
\mathrm{Pl}_{12}(\overline{\{\texttt{PSEU}\}}) & = 1 - \mathrm{Bel}_{12}(\{\texttt{PSEU}\}) = 0.60 \\
\mathrm{Pl}_{12}(\overline{\{\texttt{STREP}\}}) & = 1 - \mathrm{Bel}_{12}(\{\texttt{STREP}\}) = 1
\end{array}
$$

By joining belief and plausibility the belief ranges can be created. The following table shows the "change of belief" when combining the two knowledge sources looked at in case 3. It can be seen that the range of belief concerning $\{\texttt{PSEU}\}$ does not change because of the "vacuous character" of the range $[0,1]$. However, the belief in $\{\texttt{PSEU}\}$ leads to an increase of the belief in the subsuming hypothesis $\overline{\{\texttt{STREP}\}}$:

	before :		after :
	source 1 :	source 2 :	
$\{\texttt{PSEU}\}$	[0.4 , 1]	[0 , 1]	[0.40 , 1]
$\overline{\{\texttt{STREP}\}}$	[0.4 , 1]	[0.7 , 1]	[0.82 , 1]

We have examined the proposal of Gordon and Shortliffe for applying the Dempster-Shafer theory to MYCIN's reasoning model based on certainty factors. They have made clear that even the Dempster-Shafer theory presents a nice framework for uncertain reasoning on the basis of *hierarchical hypotheses spaces*. The integration of the Dempster-Shafer theory into the certainty factor approach leads to a simplified structure for Dempster's combination rule (13.9). This fact is due to the simplicity of the premises and conclusions of MYCIN's expert rules and to the way the integration is performed. If a hypothesis is either confirmed or disconfirmed by each of a set of rules, Dempster's combination "syntactically" reduces to the combination used in MYCIN.

Concerning their integration Gordon and Shortliffe conclude that "... this is a clear advantage over the ad hoc nature of certainty factors ..." and "... the Dempster function seems to us a more realistic reflection of the competing effects of conflicting pieces of evidence" [Gordon and Shortliffe 1984], [Gordon and Shortliffe 1985].

However, it should be noticed that their approach puts a *relative* measure (i.e. the certainty factor) on the same level with an *absolute* measure (i.e. the basic belief assignment). So, the integrated model seems to us less formally justified.

13.4.2 The Model of Yen – A Quasi-Probabilistic Approach

Below we introduce and discuss the "Dempster-Shafer based approach" to *diagnosis in hierarchically related hypotheses* that is developed by Yen [Yen 1986], [Yen 1989]. In particular, Yen extends the model of multi-valued mappings so that the uncertainty of expert rules can be represented by means of conditional probabilities. For uncertainty propagation a "probability-like" algorithm is given. Combination of evidence in this extended framework is then performed by making use of a modification of Dempster's rule. How to use this Dempster-Shafer based reasoning model for practical applications is demonstrated by Yen's implemented prototype expert system GERTIS. However, concerning the description of this system we will confine ourselves to uncertain reasoning aspects only.

In its basics the model of Yen is based on the original idea of Dempster of inducing lower and upper probabilities by a multi-valued mapping $\Gamma : E \to 2^{\Omega}$ (compare [Dempster 1967]). However, in order to take the certainty degrees of expert rules into account Yen extends Dempster's original multi-valued mapping to a probabilistic one. So, certainty degrees of expert rules can be measured by *conditional probabilities.*

For that purpose Yen [Yen 1989] introduces the notion of a *probabilistic multi-set mapping* from a space E to a space Ω as a function

$$\Gamma^* : E \to 2^{2^{\Omega} \times [0,1]}. \tag{13.34}$$

The image of an element in E, denoted by $\Gamma^*(e_i)$, is a collection of subset-probability pairs, i.e.

$$\Gamma^*(e_i) = \{(H_{i1}, P(H_{i1}|e_i)), \ldots, (H_{im}, P(H_{im}|e_i))\}$$

holds[3]. Let $E = \{e_1, \ldots, e_n\}$ be the evidence space and H_{ij} subsets of Ω such that

$$\begin{array}{lll} \text{(a)} & H_{ij} \neq \emptyset\,, & j = 1, \ldots, m \\ \text{(b)} & H_{ij} \cap H_{ik} = \emptyset\,, & j \neq k \\ \text{(c)} & P(H_{ij}|e_i) > 0\,, & j = 1, \ldots, m \\ \text{(d)} & \sum_{j=1}^{m} P(H_{ij}|e_i) = 1 & \end{array} \tag{13.35}$$

holds, where $e_i \in E$, and $H_{ij} \subseteq \Omega$ for $j = 1, \ldots, m$. In the framework of this book, for each e_i the quantity $P(H_{ij}|e_i)$ can also be viewed as a conditional basic belief assignment (i.e. $m(H_{ij}|e_i)$) to H_{ij} under the condition e_i. Consequently, $m(\cdot|e_i)$ are basic belief masses on Ω.

[3] Notice that in the context and terminology of this book the mapping induces nothing but a mass distribution on Ω.

As assumed in the conditions (13.35), each single element $e_i \in E$ is mapped to a set of (a) non-empty, (b) pairwise disjoint hypothesis groups accompanied by their conditional probabilites. Notice that an explicitly considered hypothesis group has to have a non-zero probability (c) and that the probabilities attached to the disjoint hypothesis groups must sum to 1 (d).

While in Dempster's original multi-valued mapping each expert gives exactly one hypothesis set which in his opinion is true with certainty (i.e. with probability 1), in Yen's approach an expert gives a collection of hypothesis groups which may be uncertain, i.e. a mass distribution.

On the basis of this formalism *expert rules* will have the following general structure

$$
\begin{array}{llllll}
\underline{\text{if}} & e_1 & \underline{\text{then}} & H_{11} & \underline{\text{with}} & P(H_{11}|e_1) \\
 & & & H_{12} & \underline{\text{with}} & P(H_{12}|e_1) \\
 & & & \cdots & & \\
\underline{\text{else if}} & e_2 & \underline{\text{then}} & H_{21} & \underline{\text{with}} & P(H_{21}|e_2) \\
 & & & H_{22} & \underline{\text{with}} & P(H_{22}|e_2) \\
 & & & \cdots & & \\
\cdots & & & & & \\
\underline{\text{else if}} & e_n & \underline{\text{then}} & H_{n1} & \underline{\text{with}} & P(H_{n1}|e_n) \\
 & & & H_{n2} & \underline{\text{with}} & P(H_{n2}|e_n) \\
 & & & \cdots & &
\end{array}
$$

Performing the *sequential combination* in Yen's model means combining the probability distribution on space E with the conditional probabilities given by an associated expert rule. More precisely, assuming E' to be the evidence that determines the probability distribution of space E, then the basic belief $m(H|E')$ is the total mass assigned to H by all the elements e_i associated with H under the probabilistic multi-set mapping (13.14). The sequential combination is then performed by [4]

[4] However, for some e_i the quantities $P(H|e_i)$ may be zero: this is the case if the pair is not explicitly contained in the expert rule.

Notice that in [Smets 1978] the propagation formula

$$\text{Bel}(H|E') = \sum_{B:B\subseteq E} [m(B|E') \cdot \prod_{e_i:e_i\in B} \text{Bel}(H|e_i)]$$

is proposed. If now the mass distribution m is a probability distribution then one obtains the formula

$$\text{Bel}(H|E') = \sum_{i=1}^{n} \text{Bel}(H|e_i) \cdot P(e_i|E')$$

which coincides with formula (13.36).

$$m(H|E') = \sum_{i=1}^{n} P(H|e_i) \cdot P(e_i|E') \tag{13.36}$$

Similar to most of the published approaches, Yen also assumes the conditional independence $P(H|e_i, E') = P(H|e_i)$ for all $H \subseteq \Omega$, $e_i \in E$, and $E' \subseteq E$.

The mass distribution m is a probability distribution, if all hypothesis groups H_{ij} of assumptions (13.35) (and so explicitly contained in the expert rule) are singletons. Under this condition, formula (13.36) represents a formula of probability theory for computing a posteriori probability distributions.

(13.5) Example. Assume the frames $\Omega_A = \{a, \overline{a}\}$, $\Omega_B = \{b, \overline{b}\}$, and $\Omega_H = \{h_1, h_2, h_3\}$. Further, assume the posterior probabilities given by $P(a|E_1') = P(\overline{a}|E_1') = 0.5$ and $P(b|E_2') = 0.4$, $P(\overline{b}|E_2') = 0.6$. Applying the two expert rules

rule 1:	<u>if</u>	a	<u>then</u>	$\{h_1\}$	<u>with</u>	0.4
				$\{h_3\}$	<u>with</u>	0.6
			<u>else</u>	$\{h_3\}$	<u>with</u>	1
rule 2:	<u>if</u>	b	<u>then</u>	$\{h_1\}$	<u>with</u>	0.2
				$\{h_2, h_3\}$	<u>with</u>	0.8
			<u>else</u>	$\{h_1, h_2, h_3\}$	<u>with</u>	1

to propagation formula (13.36) leads to the basic mass assignments

$$\begin{aligned} m(\{h_1\}|E_1') &= 0.5 \cdot 0.4 &&= 0.2 \\ m(\{h_3\}|E_1') &= 0.5 \cdot 0.6 + 0.5 \cdot 1 &&= 0.8 \end{aligned}$$

$$\begin{aligned} m(\{h_1\}|E_2') &= 0.4 \cdot 0.2 &&= 0.08 \\ m(\{h_2, h_3\}|E_2') &= 0.4 \cdot 0.8 &&= 0.32 \\ m(\{h_1, h_2, h_3\}|E_2') &= 0.6 \cdot 1 &&= 0.60 \end{aligned}$$

◇

Coming now to the *parallel combination*, in the original Dempster-Shafer theory basic masses are combined using Dempster's rule. However, for his extended model Yen shows that the direct application of Dempster's rule overweights the prior probabilities of the hypothesis groups. This is shown by a simple example:

(13.6) Example. Assume that for two independent evidence sources E_1' and E_2' and for the dichotomous frame $\Omega_H = H \cup \overline{H}$ the posterior basic beliefs are given by $m(H|E_1') = P(H|a)$ and $m(\overline{H}|E_1') = P(\overline{H}|a)$, since a may have been observed with certainty (i.e. $P(a|E_1') = 1$). Further, assume $m(H|E_2') = P(H|b)$ and $m(\overline{H}|E_2') = P(\overline{H}|b)$ for similar reasons. Then the application of Dempster's rule leads to

$$m(H|E_1', E_2') = \frac{P(H|a) \cdot P(H|b)}{P(H|a) \cdot P(H|b) + P(\overline{H}|a) \cdot P(\overline{H}|b)}$$

and, by using Bayes' formula and the fact that also the resulting basic belief assignment represents a probability distribution, to

$$P(H|a, b) = \frac{P(a|H) \cdot P(b|H) \cdot P(H)^2}{P(a|H) \cdot P(b|H) \cdot P(H)^2 + P(a|\overline{H}) \cdot P(b|\overline{H}) \cdot P(\overline{H})^2}$$

On the other hand, under the assumption of *conditional independence* the direct application of Bayes' rule leads to

$$P(H|a, b) = \frac{P(a|H) \cdot P(b|H) \cdot P(H)}{P(a|H) \cdot P(b|H) \cdot P(H) + P(a|\overline{H}) \cdot P(b|\overline{H}) \cdot P(\overline{H})}$$

Obviously, the two equations are different in the exponents of the prior probabilities[5]. The conclusion drawn by Yen is that Dempster's rule therefore overweights the prior probabilities. ◇

To overcome this problem, Yen introduces the so-called *basic certainty assignment* (bca) which discounts prior belief from mass distribution. The bca of a hypothesis group H_i is defined as the normalized ratio of its mass $m(H_i|E')$ to its prior probability $P(H_i)$ and "intuitively measures the belief update" [Yen 1989]:

$$C(H_i|E') = \frac{m(H_i|E') \;/\; P(H_i)}{\sum_{H_j \subseteq \Omega} m(H_j|E') \;/\; P(H_j)} \tag{13.37}$$

To combine a set of independent basic certainty assignments Dempster's rule is applied: in particular, Yen proves that for the evidential sources E_1' (for space Ω_A) and E_2' (for space Ω_B), for elements $a_i \in \Omega_A$ and $b_j \in \Omega_B$, for $P(H_k|a_i) > 0$ and $P(H_l|b_j) > 0$, and under the independence assumptions

$$\begin{aligned} P(a_i|H_k) \cdot P(b_j|H_l) &= P(a_i, b_j|H_k \cap H_l) \qquad \text{for } H_k \cap H_l \neq \emptyset \\ P(E_1'|a_i) \cdot P(E_2'|b_j) &= P(E_1', E_2'|a_i, b_j) \end{aligned}$$

the following equation (corresponding to Dempster's rule) holds:

[5] This difference has also been clarified in [Shastri and Feldman 1985], for instance.

$$C(H|E'_1, E'_2) = \frac{\sum\limits_{H_k \cap H_l = H} C(H_k|E'_1) \cdot C(H_l|E'_2)}{\sum\limits_{H_k \cap H_l \neq \emptyset} C(H_k|E'_1) \cdot C(H_l|E'_2)} \tag{13.38}$$

Finally, to obtain the combined basic belief assignment the following transformation is used:

$$m(H_i|E') = \frac{C(H_i|E') \cdot P(H_i)}{\sum\limits_{H_j \subseteq \Omega} C(H_j|E') \cdot P(H_j)} \tag{13.39}$$

More details on the above reasoning model are given in [Yen 1986], [Yen 1989]. In [Yen 1986] Yen shows that combination rule (13.38) is consistent with Bayes' theorem under conditional independence assumption.

In addition to the mathematical formalism examined above, Yen describes the implemented *prototype expert system* GERTIS (General Evidential Reasoning Tool for Intelligent Systems) which is based on the above formal evidential reasoning model. Below, we give a short overview of how knowledge is represented in GERTIS. Since we will restrict ourselves to general concepts only, the reader is referred to [Yen 1989] for more details.

GERTIS aims at diagnosing unspecified polyarthritis which can be represented in the form of a disease hierarchy consisting of mutually exclusive rheumatic diseases. The main representation concepts of GERTIS are *frames*, *hypotheses*, and *rules*.

A *frame object* directly corresponds to a frame of discernment in the notation of this book, and therefore represents an exhaustive set of mutually exclusive values for a variable. For instance, the frame which refers to all the system's final diagnoses is the "unspecified polyarthritis frame". The corresponding disease hierarchy is shown in Fig. 13.1. Leaf nodes correspond to single diagnoses while internal nodes represent unions of their children nodes. Hypotheses of interest to the physician are represented by filled squares. Because of the "closed world assumption" underlying also Yen's model, before applying GERTIS it must be known that the patient under consideration has "unspecified polyarthritis". Unlike the non-binary "unspecified polyarthritis frame", all other frames in GERTIS are binary frames.

Beyond other data, a frame object stores *belief data*, *rule data*, and *structural data*. Examples of belief data are the frame's *prior probability distribution* (needed for the transformations (13.37) and (13.39)), the *basic belief assignment* and the *basic certainty assignment* of the frame. *Rule data* include a list of rules the frame can trigger (i.e. GERTIS is a forward chaining system). A

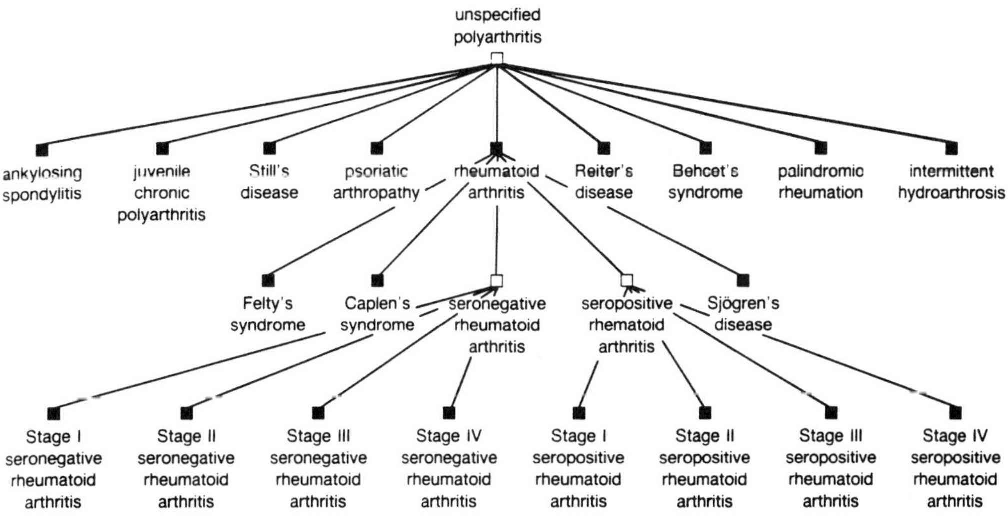

Fig. 13.1. The disease hierarchy of unspecified polyarthritis [Yen 1989]

frame object also includes *structural data* which refer to the hypothesis objects forming a taxonomic class hierarchy as in Fig. 13.1.

A *rule object* of GERTIS represents a probabilistic multi-set mapping as introduced in definition (13.35), and links a group of antecedent frames to a consequent frame. So, frame objects are linked only indirectly by expert rules. Because of the explicit representation of taxonomies, class relationships are separated from expert rules. This, however, is a major difference from conventional rule-based systems and the models considered in Sect. 13.3.

While global knowledge (e.g. the prior probability distribution) of a frame is stored in the frame object, local knowledge concerning a single hypothesis is stored in a so-called *hypothesis object.* A single hypothesis object represents a focal element, i.e., a frame's subset that is of interest to the expert. Among others, the object stores hierarchical relationships to other hypotheses represented through superclass and subclass slots (i.e. the links shown in Fig. 13.1).

An important aspect taken into account by GERTIS concerns the generation of informative explanations, especially in the considered case of hierarchically related hypotheses. For the problem domain, Yen describes an appropriate method that makes use of the effects of expert rules as well as of the knowledge on the hypothesis space structure.

Let us conclude this section with some final remarks concerning the theoretical model of Yen, where we start with some "practical aspects".

In spite of the fact that Yen's proposal is a straightforward extension of Dempster's original multi-valued mapping (i.e., for every multi-valued mapping a special probabilistic multi-set mapping can be constructed representing the same information), the direct projection of probabilistic multi-set mappings to expert rules may lead to the following problem: as is obvious from definition (13.35) the expert enters conditional probabilities when formulating rules representing relationships valid in the modeled domain. This, however, requires that the hypothesis space is constructed as an exhaustive set of *mutually exclusive* hypotheses and that the expert knows the complete (conditional) *probability distribution.* One consequence of this requirement is that some of the advantages of the Dempster-Shafer theory are not given when creating expert rules. For instance, cases of partial ignorance cannot be taken into account since the construction of an expert rule like

$$\begin{array}{llll} \underline{\text{if}} \;\; e_1 \;\; \underline{\text{then}} & \{h_1, h_2\} & \underline{\text{with}} & \cdots \\ & \{h_2, h_3\} & \underline{\text{with}} & \cdots \\ & \cdots & & \end{array}$$

is not allowed in Yen's framework because of non-exclusive hypotheses.

A question that remains open in Yen's papers concerns belief propagation over *chains of expert rules*: for instance, for two rules $a_i \rightarrow C_j$ and $c_k \rightarrow H_l$, propagation formula (13.36) does not apply, since, in general, probabilities $P(c_k)$ are not exactly known on the basis of masses $m(C_j)$.

Notice that in Yen's evidential reasoning model, complete knowledge about prior probability distributions is required. Because of partially missing domain knowledge, in GERTIS the prior probabilities of the polyarthritis frame are assumed to be uniformly distributed. Further, the formal description of the term "evidence" and its (mathematical) integration into Yen's reasoning model is missing.

For more background information concerning the application domain of GERTIS, the generation of informative explanations, or implementational aspects, for instance, the reader is referred to [Yen 1989]. Related information can also be found in [Yen 1987a], [Yen 1987b], and [Yen 1988].

13.5 MacEvidence – Belief Propagation in Markov Trees

Most of the early approaches to handling uncertainty within automated reasoning processes by means of belief functions assume the respective frame of discernment to have a hierarchical structure. This reflects that at the time when these inference procedures were developed their proponents bore in mind medical diagnosis as a typical field of application, and, as pointed out in Sect. 13.4, there we encounter manifold hierarchical structures.

In [Pearl 1986], where the "local computation paradigma" was presented, incidentially it is substantiated, that in many cases it is very promising to consider product spaces instead of hierarchies. Mellouli [Mellouli 1987] adapted the local computation paradigma and also the idea of a graphical representation of qualitative relations to belief functions. But nevertheless he sticked to hierarchical relations within the domain. Later on, in [Shafer and Shenoy 1988], where a general axiomatic framework for local computations was presented covering besides belief functions also probabilities as well as constraints (in the form of indicator functions), Shafer et.al switched to product spaces.

13.5.1 Belief Propagation in Markov Trees

The starting point of the approach is a family of *valuations* defined on subspaces of a product space Ω^M. The valuation functions attach non-negative real numbers to non-empty subsets. We start, then, with a family

$$\mathcal{V} = \{v^N : 2^{\Omega^N} \to I\!R_0^+ | N \subseteq M\} \tag{13.40}$$

of valuations. $\mathcal{V}_N$ may denote the valuations on Ω^N. Depending on the type of valuation that is actually considered (e.g. probabilities, indicator functions, or basic belief assignments which will be considered in detail later) only a subclass $\mathcal{A}_N \subseteq \mathcal{V}_N$ of valuations allow the application of the local computation scheme. The elements of $\mathcal{A}_N$ are called *admissible valuations.*

The final aim of the approach is to calculate "marginals" of the overall valuation v^M which arises from the "aggregation" of partial valuations $v^N, N \subseteq M$. For this purpose mainly two operations, *aggregation* and *projection*, are applied. Aggregation is a mapping $\otimes : \mathcal{V} \times \mathcal{V} \to \mathcal{V}$, where for $v^S \in \mathcal{V}_S$, $v^T \in \mathcal{V}_T$, $S, T \subseteq M$ we have

$$\begin{aligned} &(i) \quad v^S \otimes v^T \in \mathcal{V}_{S\cup T} \\ &(ii) \quad v^S \notin \mathcal{A}_S \vee v^T \notin \mathcal{A}_T \Rightarrow v^S \otimes v^T \notin \mathcal{A}_{S\cup T} \\ &(iii) \quad v^S \in \mathcal{A}_S \wedge v^T \in \mathcal{A}_T \Rightarrow v^S \otimes v^T \in \mathcal{A}_{S\cup T} \vee v^S \otimes v^T \notin \mathcal{A}_{S\cup T} \end{aligned} \tag{13.41}$$

If $v^S \otimes v^T \in \mathcal{A}_{S\cup T}$ then v^S and v^T are said to be *combinable.* A projection is a mapping $\Pi_S^T : \mathcal{V}_T \to \mathcal{V}_S, S \subseteq T \subseteq M$ where

$$\begin{aligned} &(i) \quad v^T \in \mathcal{A}_T \Rightarrow \Pi_S^T(v^T) \in \mathcal{A}_S \\ &(ii) \quad v^T \notin \mathcal{A}_T \Rightarrow \Pi_S^T(v^T) \notin \mathcal{A}_S \end{aligned} \tag{13.42}$$

Note that in the case of basic belief assignments, Π corresponds to the usual projection and $\otimes$ essentially represents Dempster's rule of combination.

Given a hypertree $H = (M, \mathfrak{T})$ where for each hyperedge $T \in \mathfrak{T}$ a valuation $v^T \in \mathcal{A}_T$ is specified, our goal is

- to check whether the valuations v^T , $T \in \mathfrak{T}$ are combinable,
- to carry out the aggregation, and finally
- to project the aggregated valuation to the subspace of interest.

That is, we want to calculate

$$v^S \stackrel{\text{d}}{=} \Pi_S^T(\bigotimes_{T \in \mathfrak{T}} v^T) \tag{13.43}$$

The result of the Shafer-Shenoy approach is that the local computation is possible if the operators $\otimes$ and Π obey the following axioms:

(A1) Associativity and commutativity of the aggregation

$$\begin{aligned} v^S \otimes (v^T \otimes v^U) &= (v^S \otimes v^T) \otimes v^U \\ v^S \otimes v^T &= v^T \otimes v^S \end{aligned} \tag{13.44}$$

for $S, T, U \subseteq M$.

(A2) Consonance of the projection

$$\Pi_S^T(\Pi_T^U(v^U)) = \Pi_S^U(v^U)$$

for $S, T, U \subseteq M$.

(A3) Distributivity

$$\Pi_S^{S \cup T}(v^S \otimes v^T) = v^S \otimes \Pi_{S \cap T}^T(v^T) \tag{13.45}$$

for $S, T \subseteq M$.

A simple consequence of axiom (A1) is that when several valuations are subsequently aggregated as in equation (13.43), the result does not depend on the order of succession or on the structure of parantheses. This justifies talking of

$$v^M \stackrel{\text{d}}{=} \bigotimes_{T \in \mathfrak{T}} v^T \tag{13.46}$$

as a *factorization* of v^M .

If we consider projection as the reduction of a valuation due to the deletion of variables, then because of axiom (A2) the order of deletion does not affect the result. Axiom (A3) is of importance for the local computations. It ensures that the computation of $\Pi_S^{S\cup T}(v^S \otimes v^T)$ is possible without an explicit representation of $v^S \otimes v^T$.

Let us now once more consider a hypertree $H = (M, \mathfrak{T})$. From the deletion of a twig T' we obtain a reduced hypertree $H' = (M', \mathfrak{T}')$. Let S' be the branch of T'. We have $M' = M - (T' - S')$ and $\mathfrak{T}' = \mathfrak{T} - \{T'\}$. The next theorem shows how the projection $\Pi_{M'}^M(v^M)$ can be determined.

(13.7) Theorem. *Let $H = (M, \mathfrak{T})$ be a hypertree, let $H' = (M, \mathfrak{T}')$ be the hypertree arising from H by the removal of a twig T', and let Π and $\otimes$ be the projection and aggregation operators, respectively, obeying the axioms (A1) to (A3). In this case we have*

$$\Pi_{M'}^M(v^M) = (v^{T'} \otimes \Pi_{T'\cap S'}^{S'}(v^{S'})) \otimes \bigotimes_{T\in\mathfrak{T}-\{T',S'\}} v^T \qquad (13.47)$$

Proof. The theorem follows directly from axiom (A3) where

$$v^S = \bigotimes_{T\in\mathfrak{T}'} v^T$$

and $v^T = v^{T'}$. □

The result is essential for the local computation of marginal valuations in hypertrees. If we are interested in the marginal valuation in $\Omega^{T'}$ where T' is a twig of the hypertree we have to choose a construction sequence which starts with T' (note that such a construction sequence always exists, compare Chap. 7) and use it in reverse order as a deletion sequence. By the subsequent application of Theorem 13.7 we obtain the marginal valuation.

It turns out that a representation of hypertrees by Markov trees is very useful for local computation purposes. So let $M = (\mathfrak{T}, \xi)$ be a Markov tree representation of $H = (M, \mathfrak{T})$. Consider a mapping

$$\beta : \mathfrak{T} \to \mathfrak{T}$$

which attaches to each $T \in \mathfrak{T}$ its branch according to a construction sequence of the hypertree H. β is called a *branching* for M. Assume we want once more to determine the marginal valuation on $\Omega^{T'}$ where $T' \in \mathfrak{T}$. In this case we have to choose a construction sequence starting with T' which determines a branching. If a twig is deleted as pointed out above, the valuation on $\Omega^{\beta(T)}$ is modified according to axiom (A3), i.e. $v^{\beta(T)}$ is replaced by $v^{\beta(T)} \otimes \Pi_{\beta(T)\cap T}^T(v^T)$.

In practical applications, an explicit deletion of twigs is inexpedient as it destroys the representation of the domain. Thus in the sequel, instead of hypertrees which have to be reduced we consider Markov trees where the nodes communicate, i.e. exchange messages. The nodes of Markov trees may be imagined as *processors*, able to carry out local computations. In this context a twig T' is not deleted, but the respective processor sends the marginal valuation $\Pi^T_{T\cap\beta(T)}(v^T)$ to its neighbor $\beta(T)$ to withdraw from further communication.

Each node processor administers a *current* valuation c^T defined on the space Ω^T if $T \in \mathfrak{T}$ is the respective hyperedge. c^T is initialized with v^T . A processor has to perform the following activities:

(1) Marginalization: the message for its branch is given by

$$\Pi^T_{T\cap\beta(T)}(c^T) \tag{13.48}$$

(2) Aggregation: the updated current valuation is given by

$$c^T \otimes \bigotimes_{S:\beta(S)=T} \Pi^S_{S\cap T}(v^S) \tag{13.49}$$

Note that the flow of messages does not require a meta-level control since each processor works locally and independently of the others. As soon as all messages $\Pi^S_{S\cap T}(c^T)$, $\beta(S) = T$, are available, the calculation is triggered, though obviously it is not necessary to keep the order induced by the branching.

In general our aim will be to obtain the marginal valuations for all $T \in \mathfrak{T}$. This causes no problems as the message of node S to node T is the same whenever $\beta(S) = T$ is valid for some branching β. So the local computation for a node $S \in \mathfrak{T}$ is as follows:

(1) Send to node T the message

$$v^{S\to T} \stackrel{\mathrm{d}}{=} \Pi^S_{S\cap T}(c^S \otimes \bigotimes_{T':T' \text{ neighbor of } S, T'\neq T} v^{T'\to S}) \tag{13.50}$$

Note that if S is a leaf node of the Markov tree we have

$$\{T' : T' \text{ neighbor of } S, T' \neq T\} = \emptyset.$$

(2) Calculate the marginal valuation

$$(c^S)_{\text{new}} \stackrel{\mathrm{d}}{=} c^S \otimes \bigotimes_{T':T' \text{ neighbor of } S} v^{T\to S} \tag{13.51}$$

The flow of messages starts at the leaf nodes of the Markov tree, which do not depend on other nodes. They send their messages to their branches.

The other processors aggregate incoming messages and calculate, as soon as all the information they need is available, their messages and marginal valuations (note that the leaf nodes absorb messages).

So far we have considered the general case of valuations; let us now turn to the case of belief functions. In this context the valuations are *superpotentials* of the form $h : 2^{\Omega} \to \mathbb{R}_0^+$ where

$$\sum_{A:A\neq\emptyset} h(A) > 0 \tag{13.52}$$

Superpotentials are nothing but *unnormalized mass distributions.* Let Ω^M be a product space. In the sequel h^N , $N \subseteq M$, denotes a superpotential defined on Ω^N. Projection and aggregation for superpotentials are defined as follows:

$$\begin{aligned} \Pi_S^T(h^T) &: 2^{\Omega^S} \to \mathbb{R}_0^+ \\ \Pi_S^T(h^T)(B) &= \sum_{A:\Pi_S^T(A)=B} h^T(A) \end{aligned} \tag{13.53}$$

where $S \subseteq T \subseteq M$ and

$$\begin{aligned} h^S \otimes h^T &: 2^{\Omega^{S\cup T}} \to \mathbb{R}_0^+ \\ (h^S \otimes h^T)(C) &= \sum_{A,B:\widehat{\Pi}_S^{S\cup T}(A)\cap\widehat{\Pi}_T^{S\cup T}(B)=C} h^S(A) \otimes h^T(B) \end{aligned} \tag{13.54}$$

where $S \subseteq T \subseteq M$. If

$$\begin{aligned} \widehat{\Pi}_S^T(h^S) &: 2^{\Omega^T} \to \mathbb{R}_0^+ \\ \widehat{\Pi}_S^T(h^S)(A) &= \begin{cases} h^S(B) & \text{if } A = \widehat{\Pi}_S^T(B), \\ 0 & \text{otherwise,} \end{cases} \end{aligned} \tag{13.55}$$

denotes the vacuous extension, then aggregation is (up to normalization) the Dempster-Shafer combination, i.e.

$$h^S \otimes h^T = \widehat{\Pi}_S^{S\cup T}(h^S) \oplus \widehat{\Pi}_T^{S\cup T}(h^T) \tag{13.56}$$

holds. As pointed out in Chap. 6, Dempster's rule of combination is commutative as well as associative and therefore obeys axiom (A1). The consonance of the projection (axiom (A2)) is shown in Chap. 6. Distributivity according to axiom (A3) is proved in [Shafer and Shenoy 1988].

13.5.2 MacEvidence

The system MacEvidence (designed for Apple Macintosh computers) is an implementation of the local computation scheme we presented above. It is meant to be a programming system that allows the user to specify knowledge in the so-called evidential language (i.e. in terms of basic belief assignments) and to perform reasoning processes (i.e. to apply aggregation and marginalization operators). MacEvidence provides interfaces for the graphical construction of evidential systems and for an interactive evaluation of these systems. The essential activities of a user are

- defining variables,
- defining relation nodes in order to specify structural knowledge (the structural knowledge is encoded in terms of a basic belief assignment), and
- entering evidence in order to bring in actual observations (also evidence is encoded in terms of a basic belief assignment).

The construction of a Markov tree is done by the system based on a heuristic algorithm (recall that finding an optimal Markov tree is NP-complete). Therefore the result may be far from optimal. But, as stated by Shafer and Hsia, "on the other hand, humans can often find a good, if not optimal, Markov tree arrangement just by looking at a suitable graphical depiction of the evidential system".

13.5.3 Discussion

The local computation scheme presented by Shafer and Shenoy is a charming and formally correct approach. As described in Chap. 8, we have to distinguish between structural and observational knowledge and use graphical representations of qualitative structures of the domain. The problem of finding a more or less optimal hypertree is delegated to the commonsense capabilities of the human user by providing a graphical surface for user interaction and a heuristic algorithm.

But note that the local propagation scheme strongly depends on the acceptance of Dempster's rule of combination, a topic that was extensively discussed in Chap. 6. From our point of view it is applicable only in very few special cases, and therefore the system MacEvidence has a very restricted scope of application – due to the deficiencies of the aggregation operation. If the local computation framework is is used for constraint satisfaction purposes it has valuable merits.

13.6 Conclusions

In this chapter we introduced and discussed the Dempster-Shafer theory of evidence which has received wide attention since the "seventh international conference on AI" in 1981 where three papers considering aspects on the Dempster-Shafer theory [Barnett 1981], [Friedmann 1981], [Garvey, Lowrance, and Fischler 1981] were presented. Recently, various theoretical extensions and practical applications have been proposed, some of them examined in the sections above. Belief functions were developed in [Dempster 1967], [Dempster 1968a,b] in the context of statistical studies. [Shafer 1976], [Smets 1988], [Ruspini 1986], and others considered this approach as a non-Bayesian alternative for the representation and manipulation of degrees of belief.

The basic model of evidential reasoning is intended to be a general reasoning scheme that may be used as an alternative to the classical Bayesian schemes. However, since "numbers" are attached to the subsets of a frame of discernment corresponding to propositions, the first question concerns the *semantics of these numbers*. In the classical approach due to Shafer the numbers Bel(A) represent a measure for "subjective personal credibility induced in us by evidence" [Smets 1986a,b]. So, it is claimed that belief can be handled outside a betting behavior context and do not necessarily have to be modeled on the basis of classical Bayesian approaches. In this interpretation, Dempster's rule of combination is of course central to evidence theory as it serves as a method for pooling and updating evidence.

However, there have been several objections against the use of the Dempster-Shafer theory in knowledge based systems in the above interpretation. We first consider objections against the rationality of the proposed reasoning schemes and then against the tractability of the theory. If we model credibility by means of a single number between 0 and 1, then a reasonable property obviously satisfied by each credibility measure, say cred, is given by

$$A \subseteq B \;\Rightarrow\; \mathrm{cred}(A) \leq \mathrm{cred}(B)$$

Of course, from the fact "if the true value is in A then it is necessarily in B" follows that "the credibility of A should not be larger that that of B". Also, the property

$$\mathrm{cred}(A) + \mathrm{cred}(\overline{A}) \leq 1$$

that is directly related to the modelling of ignorance, is desirable. Other properties such as

$$\min(\mathrm{cred}(A|B), \mathrm{cred}(A|\overline{B})) \leq \mathrm{cred}(A) \leq \max(\mathrm{cred}(A|B), \mathrm{cred}(A|\overline{B})),$$

which means that "if two diametrally opposed assumptions yield two different degrees of credibility in a proposition Q then the unconditional degree of credibility merited by Q should be somewhere between the two", seem to be reasonable for certain applications. This property however is satisfied neither in the belief function approach with Dempster's rule nor in the model of lower and upper probabilities with conditioning.

Beside these meta-principles, the comparision with results obtained by other theories handling uncertainty is useful: for instance, from the results of Chap. 4 it is clear that belief functions should not be used as a method for handling sets of probabilities, since belief functions are only able to represent sets of type 2. In practical applications, the uncertainty handling often seems to be unnatural. Especially, performing the updating by using Dempster's conditioning procedure gives strange results: the interval bounds turn out to be far too narrow [Weichselberger 1990], [Pearl 1990]. This indeed is not surprising, as the conditioning scheme in [Fagin and Halpern 1990] should be used as the appropriate updating scheme.

However, in this book we use the notion of mass distributions as a condensed representation of a random set, as done by Dempster in his original proposal [Dempster 1967]. When a mass distribution is given in applications, the user should assume that the underlying probability space as well as the multivalued mapping exist but are unknown. Consequently, there exist several updating mechanisms such as changing or extending the multivalued mapping, changing the underlying probabilities, or combinations of both. This discussion leads us directly to the consideration of the concept of specialization. Updating schemes are noticed by a user as mass flows, so the transferable belief model is related to our idea.

If we have the qualitative information that the underlying sources are independent, that the sensors' outcomes are conjunctively combined, and that the closed world assumption is correct, then it is reasonable to use Dempster's rule of combination for pooling evidence. But of course there are applications and situations where the truth of all these assumptions is not desirable. We conclude that under these additional qualitative requirements the result of Example 13.2 is correct, although it is of course necessary to have such meta-level information to resolve the Zadeh paradox.

Finally, one should stress the argument that the application of Shafer's theory is not "easier" because the explicit use of prior probabilities, for instance, is not necessary. This however refutes the popular misunderstanding that belief functions are the same as probability ranges – with the only difference that the theory of evidence *does not need* prior probabilities. This aspect is also discussed in [Grosof 1985b] and [Yen 1986], for instance. They propose an extension to Shafer's evidence theory which allows consideration of a priori information within the framework of an update process—similar to that of inference networks examined in Sect. 12.3.1.

On the other hand, several errors are made due to a straightforward use of belief functions and Dempster's rule, especially in the case of conditional information.

We think that a gain concerning knowledge representation facilities (e.g. in handling imprecise data) has to be paid for by recognizing that the consistent modeling of domains of the real world is even more difficult than with probability theory – which, however, is "complicated enough": the assessment of mass distributions as well as the choice of an adequate model in general (which frames of discernment, whether a product space or not, etc.).

Especially the decision making facilities have to be improved: so it is often not clear how to perform "*interval-based decision making*" (see e.g. [Loui 1985], [Yager 1985b]). For instance, if a decision has to be made between A with "belief-range" $[s_1, t_1]$ and B with "belief-range" $[s_2, t_2]$ $(A, B \subseteq \Omega)$, a problem may arise in the case of $s_1 < s_2$, $t_2 < t_1$. Also Smets considers the problem of decision making: the transferable belief model distinguishes between a *credal level* where the beliefs are quantified by belief functions, and a *pignistic level* where beliefs are quantified by probabilities. When a decision has to be made, the belief values at the credal level have to be mapped to beliefs at the pignistic level by using a suitable transformation.

Because of difficulties in eliciting precise numerical values that also have to be consistent from experts and users, the use of so-called *term sets* has been proposed. Following the proposal of Bonissone [Bonissone 1987c], a term set determines the "finest level of specifity" of the uncertainty measure users and experts can provide. However, this aspect of avoiding exact real numbers at the man-machine interface is closely related to the determination of possible *interpretations of "interval-valued likelihoods"*. For instance, the approach of Garvey, Lowrance, and Fischler [Garvey, Lowrance, and Fischler 1981] defines the "likelihood" of a proposition A as a subinterval of the unit interval [0,1]. The two bounds of the interval are the "degree of support" Bel(A) and the "degree of plausibility" Pl(A). The following table shows the interpretation of some interval-valued likelihoods as proposed by Garvey et al.

Range	*Interpretation*
$[0, 1]$	No knowledge at all about A
$[0, 0]$	A is false
$[1, 1]$	A is true
$[0.3, 1]$	The evidence partially supports A
$[0, 0.7]$	The evidence partially supports $\overline{A}$
$[0.3, 0.7]$	The evidence provides partial support for both A and $\overline{A}$
$[0.3, 0.3]$	The probability of A is exactly 0.3

While for the combination of distinct bodies of evidence Dempster's rule or an extension of it is used, very often the question how to determine belief and plausibility of complex *rule-premises* remains unanswered. A proposal for a possible procedure is given in [Garvey, Lowrance, and Fischler 1981] corresponding to the logical operations AND, OR, and NOT on two propositions A and B. Garvey et al. propose the following procedures for combining belief-ranges:

Logical operation	*Belief-range*
Intersection AND(A, B)	$[\max\{0, \mathrm{Bel}(A) + \mathrm{Bel}(B) - 1\},\ \min\{\mathrm{Pl}(A), \mathrm{Pl}(B)\}]$
Union OR(A, B)	$[\max\{\mathrm{Bel}(A), \mathrm{Bel}(B)\},\ \min\{1, \mathrm{Pl}(A) + \mathrm{Pl}(B)\}]$
Negation NOT(A)	$[1 - \mathrm{Pl}(A),\ 1 - \mathrm{Bel}(A)]$

Pearl [Pearl 1988] analyzed an analogy to the language of logic to find out that belief functions can be *formally* viewed as *probabilities of provability*. Here each theory corresponds to one focal element B_i where the elements of each focal element are considered as axioms. To each theory a probability $p_i = m(B_i)$ is assigned. Then the belief in a formula corresponding to A is the sum of all the probabilities of those theories from which A follows as logical consequence, i.e.

$$\mathrm{Bel}(A) = \sum_{B: B \rightarrow A} m(B).$$

It is stated that the set of axioms for which the proof is to be assembled is not known with certainty. Also Dempster's rule of combination can be explained by this notion.

Discussions on the relations of the Dempster-Shafer theory and the classical probability theory are given in [Hummel and Landy 1988] and [Kyburg 1987a]. Ruspini [Ruspini 1987] describes a formal model that allows the combination of unconditional and conditional probability intervals. Other proposals and models are given in [Lu and Stephanou 1984], [Prade 1985], [Bhatnagar and Kanal 1985], [Dong and Wong 1987], [Lowrance, Garvey, and Strat 1986], [Strat 1984], and [Yager 1985a,b].

In [Shenoy and Shafer 1990] the set of axioms, under which exact local computation of marginals is possible, is given. The belief function case as well as the local computation algorithms for probabilities turn out to be special cases of the so called *valuation based systems* [Shenoy 1989].

Chapter 14

Reasoning with Mass Distributions

In this chapter we use the methods presented in Chaps. 5 and 6 to establish a reasoning scheme based on the concept of a mass distribution. The qualitative knowledge is again structured by using hypertrees.

14.1 Matrix Notation for Specializations

As we showed in Sect. 6.2, a specialization is uniquely determined by functions $h_A : 2^\Omega \to [0,1]$ as defined in condition (iii) of Theorem 6.12. $h_A(B)$ specifies that portion of "belief" committed to A that in the course of refining m to m' floats to the set B. Condition (iii.a) of Theorem 6.12 assures that no evidence mass is lost, condition (iii.b) requires that the masses flow only to subsets. Those masses floating to the empty set represent partial contradictions, thus they have to be neglected and the remaining portions have to be normalized as pointed out in condition (iii.c).

14.1.1 Specialization Matrices

In a more technical notation the specialization-relation can be characterized by a matrix $V : 2^\Omega \times 2^\Omega \to [0,1]$, such that, roughly speaking, we obtain the more specific mass distribution m' by "multiplying" the vector m with the matrix V. In the sequel we use square brackets in order to indicate that we conceive the respective functions as vectors or matrices.

(14.1) Definition. *Let Ω be the frame of discernment.*

(i) A matrix $V : 2^\Omega \times 2^\Omega \to [0,1]$ is called a specialization matrix, if and only if

(a) $\sum_{B:B\subseteq\Omega} V[A,B] = 1$ *for all* $A \subseteq \Omega$

(b) $B \not\subseteq A \implies V[A,B] = 0.$

(ii) Let V be a specialization matrix and let m be a mass distribution on Ω. If

$$c \stackrel{\mathrm{d}}{=} \sum_{A:A\subseteq\Omega} \sum_{B:B\neq\emptyset} m[A]\cdot V[A,B] > 0$$

then the mass distribution $m \odot V$ is defined by

$$(m\odot V)[B] \stackrel{\mathrm{d}}{=} \begin{cases} \frac{1}{c}\cdot \sum_{A:A\subseteq\Omega} m[A]\cdot V[A,B] & \text{if } B\neq\emptyset \\ 0 & \text{otherwise,} \end{cases}$$

for all $B \subseteq \Omega$.

The main difference between the characterization of a specialization by means of a specialization matrix and by means of mass flow functions h_A, $A \subseteq \Omega$, lies in the fact that the later ones assign absolute amounts of mass whereas the former ones fix relative portions.

(14.2) Theorem. *Let m and m' be two mass distributions defined on the frame of discernment Ω. We have*

$$m' \sqsubseteq m \iff \exists V : m' = m\odot V,$$

where V is a specialization matrix.

Proof. "$\Longrightarrow$" As $m' \sqsubseteq m$, according to condition (iii) of Theorem 6.12 there are mass flow functions $h_A : 2^\Omega \to [0,1]$. Define

$$V[A,B] \stackrel{\mathrm{d}}{=} \begin{cases} \frac{h_A(B)}{m(A)} & \text{if } m(A)>0 \\ 1 & \text{if } m(A)=0 \text{ and } A=B \\ 0 & \text{otherwise,} \end{cases}$$

for all $A, B \subseteq \Omega$. V is a specialization matrix since

(i) $$\sum_{B:B\subseteq\Omega} V[A,B] = \begin{cases} \sum_{B:B\subseteq\Omega} \frac{h_A(B)}{m(A)} = 1 & \text{if } m(A)>0 \\ 1 & \text{if } m(A)=0, \end{cases}$$

(ii) $$B \not\subseteq A \implies h_A(B)=0 \implies V[A,B]=0.$$

Moreover we obtain

$$\begin{aligned} (m\odot V)[B] &= \frac{1}{c}\cdot \sum_{A:A\subseteq\Omega} m[A]\cdot V[A,B] \\ &= \frac{\sum_{A:A\subseteq\Omega} h_A(B)}{1-\sum_{A:A\subseteq\Omega} h_A(\emptyset)} = m'(B) \end{aligned}$$

for $B \neq \emptyset$ and $(m \odot V)[\emptyset] = 0 = m'(\emptyset)$, thus $m \odot V = m'$ is valid.

"$\Longleftarrow$" In the second part of the proof we have to show that from a specialization matrix $V : 2^\Omega \times 2^\Omega \to [0,1]$ mass flow functions according to condition (iii) of Theorem 6.12 can be constructed. Define

$$h_A(B) \stackrel{\text{d}}{=} m[A] \cdot V[A,B].$$

This implies

(i) $$\sum_{B:B\subseteq\Omega} h_A(B) = \sum_{B:B\subseteq\Omega} m[A] \cdot V[A,B] = m(A)$$

(ii) $$h_A(B) > 0 \implies m(A) > 0 \wedge V[A,B] > 0 \implies B \subseteq A$$

(iii) $$\frac{\sum_{A:A\subseteq\Omega} h_A(B)}{1 - \sum_{A:A\subseteq\Omega} h_A(\emptyset)} = \frac{\sum_{A:A\subseteq\Omega} m(A) \cdot V[A,B]}{1 - \sum_{A:A\subseteq\Omega} m(A) \cdot V[A,\emptyset]}$$
$$= \frac{1}{\sum_{C:C\neq\emptyset} \sum_{A:A\subseteq\Omega} m(A) \cdot V[A,C]} \cdot \sum_{A:A\subseteq\Omega} m[A] \cdot V[A,B]$$
$$= (m \odot V)[B] = m'[B]$$

□

14.1.2 Composition of Specialization Matrices

The next theorem concerns the composition of specializations, i.e. the subsequent multiplication of a mass distribution m with several specialization matrices.

(14.3) Theorem. *Let $V, W : 2^\Omega \times 2^\Omega \to [0,1]$ be two specialization matrices.*

(i) The product of V and W, given by

$$(V \cdot W)[A,C] \stackrel{\text{d}}{=} \sum_{B:B\subseteq\Omega} V[A,B] \cdot W[B,C]$$

for all $A, C \subseteq \Omega$, is again a specialization matrix, where $V \cdot W$ denotes the usual matrix multiplication.

(ii) For each mass distribution $m : 2^\Omega \to [0,1]$ we have

$$m \odot (V \cdot W) = (m \odot V) \odot W,$$

as soon as $m \odot (V \cdot W)$ or $(m \odot V) \odot W$ exists.

Proof. We have to show that the matrix $(V \cdot W)$ obeys Definition 14.1.

(i) $$\sum_{C:C\subseteq\Omega} (V \cdot W)[A,C] = \sum_{B:B\subseteq\Omega} \sum_{C:C\subseteq\Omega} V[A,B] \cdot W[B,C] = 1$$

(ii) We assumed V and W to be specialization matrices, thus we obtain

$$(V \cdot W)[A,C] = \sum_{B:A\supseteq B\supseteq C} V[A,B] \cdot W[B,C]$$

and from $C \not\subseteq A$ follows $\{B \subseteq \Omega | A \supseteq B \supseteq C\} = \emptyset$ implying $(V \cdot W)[A,C] = 0$.

In order to verify the second proposition of Theorem 14.3 consider

$$m \odot (V \cdot W)[C] = \frac{1}{c_3} \cdot \sum_{A:A\subseteq\Omega} m[A] \cdot (V \cdot W)[A,C], \quad \text{if } C \neq \emptyset \text{ where}$$

$$c_3 = \sum_{C:C\neq\emptyset} \sum_{A:A\subseteq\Omega} m[A] \cdot (V \cdot W)[A,C].$$

On the other hand we derive

$$(m \odot V)[B] = \frac{1}{c_1} \cdot \sum_{A:A\subseteq\Omega} m[A] \cdot V[A,B], \quad \text{if } B \neq \emptyset, \text{ where}$$

$$c_1 = \sum_{B:B\neq\emptyset} \sum_{A:A\subseteq\Omega} m[A] \cdot V[A,B].$$

This yields

$$(m \odot V) \odot W[C] = \frac{1}{c_2} \cdot \sum_{B:B\subseteq\Omega} (m \odot V)[B] \cdot W[B,C], \quad \text{if } C \neq \emptyset, \text{ where}$$

$$\begin{aligned} c_2 &= \sum_{C:C\neq\emptyset} \sum_{B:B\subseteq\Omega} (m \odot V)[B] \cdot W[B,C] \\ &= \sum_{C:C\neq\emptyset} \sum_{B:B\neq\emptyset} \sum_{A:A\subseteq\Omega} \frac{1}{c_1} \cdot m[A] \cdot V[A,B] \cdot W[B,C]. \end{aligned}$$

We obtain

$$m \odot (V \cdot W)[C] = \frac{1}{c_3} \cdot \sum_{A:A\subseteq\Omega} \sum_{B:B\neq\emptyset} m[A] \cdot V[A,B] \cdot W[B,C], \quad \text{if } C \neq \emptyset,$$

$$(m \odot V) \odot W[C] = \frac{1}{c_2} \cdot \sum_{B:B\neq\emptyset} \sum_{A:A\subseteq\Omega} \frac{1}{c_1} \cdot m[A] \cdot V[A,B] \cdot W[B,C], \quad \text{if } C \neq \emptyset.$$

Since

$$c_1 \cdot c_2 = c_1 \cdot \sum_{A:A\subseteq\Omega} \sum_{B:B\neq\emptyset} \sum_{C:C\neq\emptyset} \frac{1}{c_1} \cdot m[A] \cdot V[A,B] \cdot W[B,C] = c_3,$$

it follows that $m \odot (V \cdot W)[C] = (m \odot V) \odot W[C]$.

□

(14.4) Definition. *Let $V : 2^{\Omega} \times 2^{\Omega} \to [0,1]$ be a specialization matrix. V is called a strict specialization matrix if*

$$V[A,B] > 0 \implies (B = A \text{ or } B = \emptyset),$$

for all $A, B \subseteq \Omega$.

14.1.3 Properties of Specialization Matrices

As expected, strict specialization matrices characterize the strict specialization relation (compare Theorem 6.17).

(14.5) Theorem. *Let m, m' be two mass distributions defined on Ω. We have*

$$m' \leq m \iff \exists V : m' = m \odot V,$$

where V is a strict specialization matrix.

The proof of Theorem 14.5 can be carried out in strict analogy with that of Theorem 14.2. For the subsequent multiplication of a mass distribution with different strict specialization matrices we easily obtain the following theorem.

(14.6) Theorem. *Let $V, W : 2^{\Omega} \times 2^{\Omega} \to [0,1]$ be two strict specialization matrices. In this case also the product $V \cdot W$ is a strict specialization matrix.*

(14.7) *Remark.* As in the case of strict specialization matrices, just the elements of the *main diagonal* are possibly non-zero, and are often written as vectors. If the strict specialization results from the projection of an orthogonal extension (compare Theorem 6.20) then the vectors are called influence vectors. Define

$$I[A] \stackrel{\mathrm{d}}{=} V[A,A], \quad A \subseteq \Omega.$$

For the multiplication of such vector with a mass distribution m we derive:

$$(m \odot^* I)[A] \stackrel{\mathrm{d}}{=} \frac{1}{c} \cdot m[A] \cdot I[A], \quad \text{where}$$

$$c = \sum_{B:B\subseteq\Omega} m[B] \cdot I[B].$$

Obviously we have $m \odot^* I = m \odot V$, such that it is not necessary to distinguish explicitly between the operators $\odot^*$ and $\odot$.

○

The process of *revision*, i.e. the change from a mass distribution m to the revised mass distribution m_E, is important in the context of knowledge acquisition and reasoning. As we know from Sect. 6.2, revision is a special case of specialization, therefore it can be equivalently described by special specialization matrices.

Recall that revision with respect to the set $E \subseteq \Omega$ means that those masses bound to sets $A \subseteq E$ remain where they are, while those bound to sets A with $A \cap E \neq \emptyset$ float to $E \cap A$. Masses bound to sets A with $A \cap E = \emptyset$ have to be neglected since they represent (partial) contradictions of the conditioning information E and the mass distribution m. The following definition formalizes these ideas.

(14.8) Definition. *Let Ω be the frame of discernment and let $E \subseteq \Omega$ be a non-empty set. The revision matrix $R(E) : 2^\Omega \times 2^\Omega \to [0,1]$ is defined by*

$$R(E)[A,B] = \begin{cases} 1 & \text{if } B = A \cap E \\ 0 & \text{otherwise.} \end{cases}$$

Obviously $R(E)$ is a specialization matrix since in every row A there is exactly one element $R(E)[A,B] = 1$. The set B is certainly a subset of A since it is the result of the intersection of A and E.

(14.9) Theorem. *Let m be a mass distribution. For a set $E \subseteq \Omega$ with $\mathrm{Pl}_m(E) > 0$ we have*

$$m_E = m \odot R(E).$$

Proof. For $\emptyset \neq B \subseteq \Omega$ we obtain from Definition 14.8

$$\begin{aligned} \big(m \odot R(E)\big)[B] &= \frac{1}{c} \cdot \sum_{A:A\subseteq\Omega} m[A] \cdot R(E)[A,B] \\ &= \frac{1}{c} \cdot \sum_{A:A\cap E=B} m[A]. \end{aligned}$$

For the normalizing constant c we have

$$c = \sum_{B:B\neq\emptyset} \sum_{A:A\subseteq\Omega} m[A] \cdot R(E)[A,B] = \sum_{A:A\cap E\neq\emptyset} m[A].$$

Finally we derive

$$\left(m \odot R(E)\right)[B] \;=\; \frac{\sum\limits_{A:A\cap E=B} m[A]}{\sum\limits_{A:A\cap E\neq\emptyset} m[A]} \;=\; m_E(B)$$

for $\emptyset \neq B \subseteq \Omega$. If $B = \emptyset$ then $\left(m \odot R(E)\right)[\emptyset] = 0 = m_E(\emptyset)$.

□

Now let us consider two sets E_1 and E_2. In this case the stepwise revision, firstly with respect to E_1 and secondly with respect to E_2, is equivalent to the direct revision with respect to $E_1 \cap E_2$. Obviously the result does *not* depend on whether the flow of masses from A to $A \cap E_1 \cap E_2$ goes directly or via the set $A \cap E_1$.

(14.10) Theorem. *Let $E_1, E_2 \subseteq \Omega$ such that $E_1 \cap E_2 \neq \emptyset$. We have*

$$R(E_1) \cdot R(E_2) \;=\; R(E_2) \cdot R(E_1) \;=\; R(E_1 \cap E_2),$$

where $R(\,.\,)$ denotes revision matrices.

Proof. From Definition 14.8 we obtain

$$R(E_1)[A,B] = \begin{cases} 1 & \text{if } A \cap E_1 = B \\ 0 & \text{otherwise,} \end{cases}$$

$$R(E_2)[A,B] = \begin{cases} 1 & \text{if } A \cap E_2 = B \\ 0 & \text{otherwise,} \end{cases}$$

$$R(E_2 \cap E_2)[A,B] = \begin{cases} 1 & \text{if } A \cap (E_1 \cap E_2) = B \\ 0 & \text{otherwise.} \end{cases}$$

This yields

$$\begin{aligned} R(E_1) \cdot R(E_2)[A,C] &= \sum_{B:B\subseteq\Omega} R(E_1)[A,B] \cdot R(E_2)[B,C] \\ &= R(E_2)[A \cap E_1, C] \\ &= \begin{cases} 1 & \text{if } ((A \cap E_1) \cap E_2) = C \\ 0 & \text{otherwise,} \end{cases} \\ &= R(E_1 \cap E_2)[A,C], \end{aligned}$$

$A, C \subseteq \Omega$. For $R(E_2) \cdot R(E_1)$ we obtain an analogous result.

□

The use of specialization matrices leads to an interesting new concept. Some specialization matrix V represents a piece of "structural knowledge". Multiplying a mass distribution m with V means splitting the evidence masses in the light of the knowledge encoded by V. It is preferable that the "application" of V to a more specific mass distribution m' should yield a more specific result.

(14.11) Definition. *Let $V : 2^{\Omega} \times 2^{\Omega} \to [0,1]$ be a specialization matrix. V is called monotonic, if and only if*

$$m_1 \sqsubseteq m_2 \quad \Longrightarrow \quad m_1 \odot V \sqsubseteq m_2 \odot V,$$

for all mass distributions $m_1, m_2 : 2^{\Omega} \to [0,1]$ such that $m_1 \odot V$ and $m_2 \odot V$ exist.

The next example shows that not every specialization matrix is monotonic.

(14.12) Example. Let $\Omega = \{\omega_1, \omega_2\}$. Consider the mass distributions $m_1 = m^{\{\omega_1\}}$ and $m_2 = m^{\{\omega_1,\omega_2\}}$. Obviously we have $m_1 \sqsubseteq m_2$ with

V	$\emptyset$	ω_1	ω_2	Ω
$\emptyset$	1	0	0	0
ω_1	0	1	0	0
ω_2	0	0	1	0
Ω	0	0	1	0

we obtain $(m_1 \odot V) = m^{\{\omega_1\}}$ and $(m_2 \odot V) = m^{\{\omega_2\}}$, obviously $m^{\{\omega_1\}} \not\sqsubseteq m^{\{\omega_2\}}$. ◇

(14.13) *Remark.* Let m_1, m_2 be two mass distributions such that $m_1 \sqsubseteq m_2$. This implies $m_1 = m_2 \odot V'$. For a monotonic specialization matrix V we obtain

$$m_1 \sqsubseteq m_2 \quad \Longrightarrow \quad m_1 \odot V \sqsubseteq m_2 \odot V,$$

or, equivalently, $m_1 \odot V = m_2 \odot V \odot V''$. This leads to $m_2 \odot V' \odot V = m_2 \odot V \odot V''$. Obviously,

$$\forall m, V\ \exists V', V'' : \ m \odot V' \odot V = m \odot V \odot V''$$

holds whenever the products exist. ○

Note that if we use the strict specialization relation we obtain for every specialization matrix $V : 2^{\Omega} \times 2^{\Omega} \to [0,1]$

$$m_1 \leq m_2 \quad \Longrightarrow \quad m_1 \odot V \leq m_2 \odot V \tag{14.1}$$

since $(m_1 \odot V)[B] > 0$ implies that there is $A \supseteq B$ such that $m_1(A) > 0$ and $V[A,B] > 0$. From $m_1 \leq m_2$ we obtain $m_2[A] > 0$ and therefore $(m_2 \odot V)[B] > 0$.

The next theorem provides a simple means of checking easily whether a given specialization matrix is monotonic or not. It relies on a test of monotonicity for mass distributions with only one focal element.

(14.14) Theorem. *Let $V : 2^{\Omega} \times 2^{\Omega} \to [0,1]$ be a specialization matrix. If for all $A, B \subseteq \Omega$ (as far as $m^A \odot V$ and $m^B \odot V$ exist) holds*

$$m^A \sqsubseteq m^B \implies m^A \odot V \sqsubseteq m^B \odot V,$$

then V is monotonic.

Proof. We have to show that from $s \sqsubseteq t$ follows $s \odot V \sqsubseteq t \odot V$. According to condition (ii) of Theorem 6.12 this is equivalent to

$$Q_{t \odot V}(A) = 0 \implies Q_{s \odot V}(A) = 0.$$

From $Q_{s \odot V}(A) > 0$ follows that there is a set $A' \supseteq A$ with $(s \odot V)(A') > 0$. Thus there is a set $A'' \supseteq A$ such that $s(A'') > 0$ and $V[A', A''] > 0$. Since $s \sqsubseteq t$ there is $A''' \supseteq A''$ with $t(A''') > 0$. Obviously we have $(m^{A''} \odot V)[A''] > 0$. From $A'' \subseteq A'''$ we derive $m^{A''} \sqsubseteq m^{A'''}$. This implies $m^{A''} \odot V \sqsubseteq m^{A'''} \odot V$. Since $(m^{A''} \odot V)[A'] > 0$ there is $B \supseteq A'$, such that $(m^{A''} \odot V)[B] > 0$. Therefore $V[A''', B] > 0$ holds and since $t(A''') > 0$ it follows that $(t \odot V)[B] > 0$ and $Q_{t \odot V}[B] > 0$.

□

The next theorem shows that the subsequent application of monotonic specialization matrices preserves the monotonicity property.

(14.15) Theorem. *Let $V, W : 2^{\Omega} \times 2^{\Omega} \to [0,1]$ be two monotonic specialization matrices. In this case also $V \cdot W$ is a monotonic specialization matrix.*

Proof. We have to show

$$s \sqsubseteq t \implies s \odot (V \cdot W) \sqsubseteq t \odot (V \cdot W).$$

We obtain

$$\begin{aligned} s \sqsubseteq t \implies s \odot V \sqsubseteq t \odot V &\implies (s \odot V) \odot W \sqsubseteq (t \odot V) \odot W \\ &\implies s \odot (V \cdot W) \sqsubseteq t \odot (V \cdot W) \end{aligned}$$

□

The concept of monotonicity is very important; revision matrices, for example, are monotonic.

(14.16) Theorem. *Let Ω be the frame of discernment and let E be a non-empty subset of Ω. $R(E)$ is a monotonic specialization matrix.*

Proof. We have to show

$$m^A \sqsubseteq m^B \implies m^A \odot R(E) \sqsubseteq m^B \odot R(E).$$

From

$$\mathcal{Q}_{m^A \odot R(E)}(C) > 0 \implies C \subseteq A \cap E$$

together with

$$A \subseteq B \implies A \cap E \subseteq B \cap E$$

we obtain

$$\mathcal{Q}_{m^B \odot R(E)}(C) > 0$$

since $R(E)[B, B \cap E] = 1$.

□

14.2 Specializations in Product Spaces

In the context of knowledge representation in product spaces we have to deal with mass distributions of the form $m : \mathcal{Q}(\Omega^S) \to [0,1]$, i.e. only cylinder sets are focal elements, as pointed out in Sect. 6.2. In this case a revision with respect to a set $E^S \in \mathcal{Q}(\Omega^S)$ can be described by a revision matrix of the form $R(E^S) : \mathcal{Q}(\Omega^S) \times \mathcal{Q}(\Omega^S) \to [0,1]$. So if we generally require the targets of mass flows to be cylinders, then also specialization matrices have the form $V : \mathcal{Q}(\Omega^S) \times \mathcal{Q}(\Omega^S) \to [0,1]$. Note that such a matrix V has to be conceived as a shorthand notation for a matrix $V' : 2^{\Omega^S} \times 2^{\Omega^S} \to [0,1]$, where

$$V'[A,B] \stackrel{\mathrm{d}}{=} \begin{cases} V[A,B] & \text{if } A \in \mathcal{Q}(\Omega^S) \\ 1 & \text{if } A \notin \mathcal{Q}(\Omega^S) \text{ and } B = \emptyset \\ 0 & \text{otherwise.} \end{cases}$$

Whenever in the sequel specialization matrices for mass distributions on product spaces are considered, then they have to be interpreted in this way.

In the ideal case specialization matrices and mass distributions refer to the same subspace of Ω^M, but it is also possible to extend or project specialization matrices. If a specialization matrix $V : \mathcal{Q}(\Omega^S) \times \mathcal{Q}(\Omega^S) \to [0,1]$ has to be applied to a mass distribution $m : \mathcal{Q}(\Omega^T) \to [0,1], T = S + W$, then the value $V[B, B'],\ B, B' \in \mathcal{Q}(\Omega^S)$ defines that portion of $m[B \times F],\ F \in \mathcal{Q}(\Omega^W)$ that floats to $B' \times F$. Formally this means some kind of vacuous extension of V.

(14.17) Definition. *Let Ω^M be the frame of discernment and let $S, T, W \subseteq M$ be index subsets such that $T = S + W$. The extension of a specialization matrix $V : \mathcal{Q}(\Omega^S) \times \mathcal{Q}(\Omega^S) \to [0,1]$ to Ω^T is given by*

$$\widehat{\Pi}_S^T(V) : \mathcal{Q}(\Omega^T) \times \mathcal{Q}(\Omega^T) \to [0,1];$$

$$\widehat{\Pi}_S^T(V)[B \times F, B' \times F'] \stackrel{\mathrm{d}}{=} \begin{cases} V[B, B'] & \text{if } F = F' \in \mathcal{Q}(\Omega^W) \\ 0 & \text{otherwise.} \end{cases}$$

(14.18) *Remark.* It is easy to verify that $\widehat{\Pi}_S^T(V)$ can be conceived as a shorthand notation for a specialization matrix, as we have

$$\begin{aligned} \widehat{\Pi}_S^T(V)[B \times F, B' \times F'] > 0 \implies & V[B, B'] > 0 \ \wedge \ F = F' \\ \implies & B \times F \subseteq B' \times F', \end{aligned}$$

$$\begin{aligned} & \sum_{B', F' : B' \in \mathcal{Q}(\Omega^S), F' \in \mathcal{Q}(\Omega^W)} \widehat{\Pi}_S^T(V)[B \times F, B' \times F'] \\ = & \sum_{B' : B' \in \mathcal{Q}(\Omega^S)} V[B, B'] = 1. \end{aligned}$$

○

As the next theorem shows, the extension of specialization matrices is compatible with the respective set-theoretic operations.

(14.19) Theorem. *Let Ω^M be the frame of discernment and let $S, T, W \subseteq M$ be index subsets such that $T = S + W$. For $\emptyset \neq E \in \mathcal{Q}(\Omega^S)$ we have*

$$\widehat{\Pi}_S^T(R(E)) = R(\widehat{\Pi}_S^T(E)).$$

(Note that the matrices $R(\,.\,)$ are defined on the cylinders of the respective product spaces.)

Proof. From Definition 14.17 and Definition 14.8 we obtain

$$\begin{aligned} & \widehat{\Pi}_S^T(R(E))[B \times F, B' \times F'] = 1 \\ \iff & R(E)[B, B'] = 1 \ \wedge \ F = F' \\ \iff & B' = B \cap E \ \wedge \ F = F' \\ \iff & B' \times F' = (B \times F) \cap \widehat{\Pi}_S^T(E) \\ \iff & R(\widehat{\Pi}_S^T(E))[B \times F, B' \times F'] = 1. \end{aligned}$$

For 0-entries we obtain the same equivalence, so given the fact that other values do not appear this is sufficient to verify Theorem 14.7.

In the context of encoding knowledge the concept of orthogonal extension (compare Sect. 6.2) is of importance. The orthogonal extension of a mass distribution $m : \mathcal{Q}(\Omega^S) \to [0,1]$, $S \subseteq M$, to the subspace Ω^T, $S \subseteq T \subseteq M$, can be described as a specialization of the vacuous extension, where the masses originally attached to the cylinders $\widehat{\Pi}_S^T(B)$, $B \in \mathcal{Q}(\Omega^S)$, float to the sets $B \times F$, $F \in \mathcal{Q}(\Omega^W)$, $W = T - S$. For the corresponding specialization matrix $V : \mathcal{Q}(\Omega^T) \times \mathcal{Q}(\Omega^T) \to [0,1]$ we have

$$V[B, C] > 0 \implies \widehat{\Pi}_S^T(C) = B,$$

i.e. from $V[B, C] > 0$ follows $C = B \times F$, $F \in \mathcal{Q}(\Omega^W)$. For this reason instead of V an orthogonal extension matrix $V' : \mathcal{Q}(\Omega^S) \times \mathcal{Q}(\Omega^W) \to [0,1]$ is used, where

$$V'[B, F] \stackrel{\mathrm{d}}{=} V[\widehat{\Pi}_S^T(B), B \times F].$$

□

As the orthogonal extension of $m : \mathcal{Q}(\Omega^S) \to [0,1]$ we obtain the mass distribution

$$\begin{aligned} m \odot V' : \mathcal{Q}(\Omega^T) \to [0,1]; \quad (m \odot V')[B \times F] &\stackrel{\mathrm{d}}{=} (\widehat{\Pi}_S^T(m) \odot V)[B \times F] \\ &= \frac{1}{c} \cdot s[B] \cdot V'[B, F], \end{aligned}$$

if $\emptyset \neq F \in \mathcal{Q}(\Omega^W)$, where c is a normalizing constant. According to Theorem 6.20 the projection of $m \odot V'$ to Ω^S yields a strict specialization of m, i.e. $\Pi_S^T(m \odot V') \leq m$. This strict specialization can be described by an influence vector, as pointed out in Remark 14.7. The next theorem shows how this influence vector can be determined from the orthogonal extension matrix V'.

(14.20) Theorem. *Let Ω^M be the frame of discernment and let $S, T, W \subseteq M$ be index subsets such that $T = S + W$. For a mass distribution $s : \mathcal{Q}(\Omega^S) \to [0,1]$ and an orthogonal extension matrix $V : \mathcal{Q}(\Omega^S) \times \mathcal{Q}(\Omega^W) \to [0,1]$ we have $\Pi^T_S(s \odot V) = s \odot I$, where the influence vector $I : \mathcal{Q}(\Omega^S) \to [0,1]$ is given by*

$$I[B] \stackrel{\mathrm{d}}{=} \sum_{F:\emptyset \neq F \in \mathcal{Q}(\Omega^W)} V[B,F].$$

Proof. We have

$$s \odot V[B \times F] = \frac{1}{c} \cdot s[B] \cdot V[B,F],$$

where $B \in \mathcal{Q}(\Omega^S)$, $F \in \mathcal{Q}(\Omega^W)$, $B, F \neq \emptyset$. This yields

$$\begin{aligned} \Pi^T_S(s \odot V)[B] &= \sum_{\substack{C \in \mathcal{Q}(\Omega^T): \\ \Pi^T_S(C) = B}} (s \odot V)[C] \\ &= \sum_{F:\emptyset \neq F \in \mathcal{Q}(\Omega^W)} (s \odot V)[B \times F] \\ &= \sum_{F:\emptyset \neq F \in \mathcal{Q}(\Omega^W)} \frac{1}{c} \cdot s[B] \cdot V[B,F] \\ &= \frac{1}{c} \cdot s[B] \cdot I[B] \;=\; (s \odot I)[B] \end{aligned}$$

for $B \neq \emptyset$. □

If the distribution of masses within Ω^T, $T = S + W$, in the course of an orthogonal extension relies just on a subspace Ω^U, then $S = R + U$, i.e. $V'[A \times D, F] = V'[A' \times D, F]$ for all $A, A' \in \mathcal{Q}(\Omega^R)$, where $D \in \mathcal{Q}(\Omega^U)$, $F \in \mathcal{Q}(\Omega^W)$, so instead of V' we consider the matrix $V'' : \mathcal{Q}(\Omega^U) \times \mathcal{Q}(\Omega^W) \to [0,1]$;

$$V''[D,F] \stackrel{\mathrm{d}}{=} V'[A \times D, F]. \tag{14.2}$$

As the orthogonal extension of a mass distribution $s : \mathcal{Q}(\Omega^S) \to [0,1]$ we obtain

$$\begin{aligned} s \odot V'' : \mathcal{Q}(\Omega^T) \to [0,1]; \quad (s \odot V'')[A \times D \times F] &\stackrel{\mathrm{d}}{=} (s \odot V')[A \times D \times F] \\ &= (s \odot V)[A \times D \times F] \\ &= \frac{1}{c} \cdot s[A \times D] \cdot V[D,F] \end{aligned}$$

for $\emptyset \neq F \in \mathcal{Q}(\Omega^W)$. The distribution of the masses attached to the sets $\widehat{\Pi}^T_S(B)$ to $B \times F$ is done for all B with $\Pi^S_U(B) = D$ according to the same ratio. (Note that this is related to the conditional independence of probabilities.)

(14.21) *Remark.* If $R = S$, i.e. $U = \emptyset$, then the orthogonal extension matrix has the form $V : \mathcal{Q}(\Omega^{\emptyset}) \times \mathcal{Q}(\Omega^W) \to [0,1]$, and we obtain

$$t[B \times F] = \frac{1}{c} \cdot s[B] \cdot V[\{\varepsilon\}, F].$$

The "matrix" $V : \mathcal{Q}(\Omega^{\emptyset}) \times \mathcal{Q}(\Omega^W) \to [0,1]$ can be conceived as a mass distribution $w : \mathcal{Q}(\Omega^W) \to [0,1]$; $w[F] = V[\{\varepsilon\}, F]$ (if we assume $V[\{\varepsilon\}, \emptyset] = 0$). Obviously in this case $s \odot V$ means simply the combination of "independent" mass distributions. ○

The revision of mass distributions on product spaces is generally done with respect to information concerning subspaces of Ω^M. If an observation yields, for example, the set $E \in \mathcal{Q}(\Omega^S)$ then a mass distribution $t : \mathcal{Q}(\Omega^T) \to [0,1]$, $S \subseteq T \subseteq \Omega$, is revised with respect to $\widehat{\Pi}_S^T(E)$. If t results from the orthogonal extension of a mass distribution $s : \mathcal{Q}(\Omega^S) \to [0,1]$, then it is *not* possible to perform the revision within Ω^S before the extension, as the next example shows.

(14.22) Example. Let $\Omega^{(1)} = \{\text{z1}, \text{z2}, \text{z3}\}$ and $\Omega^{(2)} = \{\text{tv}, \text{ps}\}$. Consider the mass distribution

$$m : \mathcal{Q}(\Omega^{\{(1)\}}) \to [0,1]; \;\; m(A) = \begin{cases} \frac{1}{4} & \text{if } A = \{\text{z1}\} \\ \frac{1}{4} & \text{if } A = \{\text{z2}\} \\ \frac{1}{2} & \text{if } A = \Omega^{(1)} \\ 0 & \text{otherwise.} \end{cases}$$

By applying the orthogonal extension matrix $V : \mathcal{Q}(\Omega^{(1)}) \times \mathcal{Q}(\Omega^{(2)}) \to [0,1]$ given below we obtain the mass distribution $m \odot V : \mathcal{Q}(\Omega^{\{1,2\}}) \to [0,1]$.

V	$\emptyset$	{tv}	{ps}	$\Omega^{(2)}$
$\emptyset$	1	0	0	0
{z1}	0	1	0	0
{z2}	0	$\frac{1}{2}$	$\frac{1}{2}$	0
{z3}	0	0	1	0
{z1,z2}	0	$\frac{3}{4}$	$\frac{1}{4}$	0
{z1,z3}	0	$\frac{1}{2}$	$\frac{1}{2}$	0
{z2,z3}	0	$\frac{1}{4}$	$\frac{3}{4}$	0
$\Omega^{(1)}$	0	0	0	1

The revision of $m \odot V$ with respect to $\widehat{\Pi}_{\{1\}}^{\{1,2\}}(E)$, $E = \{z1, z2\}$ yields the mass distribution $(m \odot V)_{\widehat{\Pi}_{\{1\}}^{\{1,2\}}(E)} = m \odot \big(V \cdot R(\widehat{\Pi}_{\{1\}}^{\{1,2\}}(E))\big)$:

$(m \odot V)_{E \times \Omega^{(2)}}$	$\emptyset$	{tv}	{ps}	$\Omega^{(2)}$
$\emptyset$	0	0	0	0
{z1}	0	$\frac{1}{4}$	0	0
{z2}	0	$\frac{1}{8}$	$\frac{1}{8}$	0
{z3}	0	0	0	0
{z1,z2}	0	0	0	0
{z1,z3}	0	0	0	$\frac{1}{2}$
{z2,z3}	0	0	0	0
$\Omega^{(1)}$	0	0	0	0

Starting with the revision of m with respect to E we obtain

$$m_E : \mathcal{Q}(\Omega^{(1)}) \to [0,1]; \quad m_E(A) = \begin{cases} \frac{1}{3} & \text{if } A = \{a_1\} \\ \frac{2}{3} & \text{if } A = \{a_3\} \\ 0 & \text{otherwise;} \end{cases}$$

and the orthogonal extension of m_E yields $m_E \odot V = (m \odot R(E)) \odot V$:

$m_E \odot V$	$\emptyset$	{tv}	{ps}	$\Omega^{(2)}$
$\emptyset$	0	0	0	0
{z1}	0	$\frac{1}{3}$	0	0
{z2}	0	0	0	0
{z3}	0	0	0	0
{z1,z2}	0	0	0	0
{z1,z3}	0	$\frac{1}{3}$	$\frac{1}{3}$	0
{z2,z3}	0	0	0	0
$\Omega^{(1)}$	0	0	0	0

Obviously we have $(m \odot R) \odot V \neq m \odot \big(V \cdot R(\widehat{\Pi}_{\{1\}}^{\{1,2\}}(E))\big)$. ◇

If an orthogonal extension matrix of the form $V : \mathcal{Q}(\Omega^U) \times \mathcal{Q}(\Omega^W) \to [0,1]$ is given (compare equation (14.2)), then the revision with respect to a set $E \in \mathcal{Q}(\Omega^R)$ can be carried out before the extension.

(14.23) Theorem. *Let Ω^M be the frame of discernment and let $R, S, T, U, W \subseteq M$ be index subsets such that $T = S + W$ and $S = R + U$. In this case for an orthogonal extension matrix $V : \mathcal{Q}(\Omega^U) \times \mathcal{Q}(\Omega^W) \to [0,1]$ we have*

$$R(\widehat{\Pi}_R^S(E)) \cdot V \;=\; V \cdot R(\widehat{\Pi}_R^T(E))$$

for $E \in \mathcal{Q}(\Omega^R)$.

Proof. Recall that the matrix $V : \mathcal{Q}(\Omega^U) \times \mathcal{Q}(\Omega^W) \to [0,1]$ is a shorthand notation for a matrix $\widetilde{V} : \mathcal{Q}(\Omega^S) \times \mathcal{Q}(\Omega^W) \to [0,1]$, where

$$\Pi_U^S(B) = \Pi_U^S(B') = D \implies \widetilde{V}[B,F] = \widetilde{V}[B',F] = V[D,F],$$

$B, B' \in \mathcal{Q}(\Omega^S)$, $F \in \mathcal{Q}(\Omega^W)$ (compare equation 14.1).

Let $B = A \times D$, $B' = A' \times D'$, $A, A' \in \mathcal{Q}(\Omega^R)$, $D, D' \in \mathcal{Q}(\Omega^U)$. By Definition 14.17 and Theorem 14.19 we obtain

$$\begin{aligned}
&\Big(R(\widehat{\Pi}_R^S(E)) \cdot V\Big)[B, B' \times F] \\
&\quad = R(\widehat{\Pi}_R^S(E))[B,B'] \cdot \widetilde{V}[B',F] \\
&\quad = \widehat{\Pi}_R^S(R(E))[A \times D, A' \times D'] \cdot V[D',F] \\
&\quad = \begin{cases} R(E)[A,A'] \cdot V[D',F] & \text{if } D = D' \\ 0 & \text{otherwise,} \end{cases} \\
&\quad = \begin{cases} V[D,F] & \text{if } B' = (A \cap E) \times D \\ 0 & \text{otherwise.} \end{cases}
\end{aligned}$$

$$\begin{aligned}
&V \cdot R(\widehat{\Pi}_R^T(E))[B, B' \times F] \\
&\quad = \widetilde{V}[B,F] \cdot R(\widehat{\Pi}_R^T(E))[B \times F, B' \times F'] \\
&\quad = V[D,F] \cdot R(\Pi_R^T(E))[A \times D \times F, A' \times D' \times F'] \\
&\quad = \begin{cases} V[D,F] \cdot R(E)[A,A'] & \text{if } D \times F = D' \times F' \\ 0 & \text{otherwise,} \end{cases} \\
&\quad = \begin{cases} V[D,F] & \text{if } B' = (A \cap E) \times D \\ 0 & \text{otherwise.} \end{cases}
\end{aligned}$$

□

Theorem 14.23 concerned the impact of revision in subspaces to the results of orthogonal extensions. Now let us turn to the impact of specializations performed on subspaces. Let $S, T, W \subseteq M$ such that $T = S + W$. The projection of a specialization of $t : \mathcal{Q}(\Omega^T) \to [0,1]$ described by the extension $\widehat{\Pi}_S^T(V)$ of a specialization matrix $V : \mathcal{Q}(\Omega^S) \times \mathcal{Q}(\Omega^S) \to [0,1]$ yields the same result as the direct application of V to $\Pi_S^T(t)$.

(14.24) Theorem. *Let Ω^M be the frame of discernment and let $S, T, W \subseteq M$ be index subsets, where $T = S + W$. For a mass distribution $t : \mathcal{Q}(\Omega^T) \to [0,1]$ and an orthogonal extension matrix $V : \mathcal{Q}(\Omega^S) \times \mathcal{Q}(\Omega^S) \to [0,1]$ we have*

$$\Pi_S^T(t \odot \widehat{\Pi}_S^T(V)) \;=\; \Pi_S^T(t) \odot V.$$

Proof. We have

$$\begin{aligned}
\Pi_S^T(t \odot \widehat{\Pi}_S^T(V))[B'] &= \sum_{\substack{C:\emptyset \neq C \in \mathcal{Q}(\Omega^T) \\ \Pi_S^T(C)=B'}} (t \odot \widehat{\Pi}_S^T(V))[C] \\
&= \sum_{F:\emptyset \neq F \in \mathcal{Q}(\Omega^W)} (t \odot \widehat{\Pi}_S^T(V))[B' \times F] \\
&= \sum_{F:\emptyset \neq F \in \mathcal{Q}(\Omega^W)} \frac{1}{c} \cdot \sum_{B:\emptyset \neq B \in \mathcal{Q}(\Omega^S)} t[B \times F] \cdot V[B, B'] \\
&= \frac{1}{c} \cdot \sum_{B:\emptyset \neq B \in \mathcal{Q}(\Omega^S)} V[B, B'] \cdot \sum_{F:\emptyset \neq F \in \mathcal{Q}(\Omega^W)} t[B \times F] \\
&= \frac{1}{c} \cdot \sum_{B:\emptyset \neq B \in \mathcal{Q}(\Omega^S)} \Pi_S^T(t)[B] \cdot V[B, B'] \\
&= \Pi_S^T(t) \odot V[B'],
\end{aligned}$$

for $\emptyset \neq B' \in \mathcal{Q}(\Omega^S)$.

□

(14.25) *Remark.* The most important consequence of Theorem 14.19 and Theorem 14.24 is that for a mass distribution $s : \mathcal{Q}(\Omega^S) \to [0,1]$ an orthogonal extension matrix and a set $E \in \mathcal{Q}(\Omega^S)$ we obtain

$$\begin{aligned}
&\Pi_S^T\Big((s \odot V) \odot R(\widehat{\Pi}_S^T(E))\Big) \\
&\qquad = \Pi_S^T(s \odot V) \odot R(E) \\
&\qquad = s \odot I \odot R(E),
\end{aligned}$$

where I is the influence vector derived from V according to Theorem 14.20.

14.3 Knowledge Representation with Mass Distributions

As in the case of probabilistic reasoning (compare Sect. 12.5) we assume that in a first step of the knowledge acquisition process the qualitative structure of the domain has been determined in the form of a dependency network $C = (M, \mathfrak{T})$. The second step is to specify for every link in the dependency network the *quantitative* dependencies.

For each hyperedge $T_j \in \mathfrak{T}$ (recall that we assume the elements of $\mathfrak{T}$ to be consistently numbered) an expert θ_j provides a (monotonic) orthogonal extension matrix V_j that determines how to extend mass distributions on $\Omega^{T_{jh}}$ to $\Omega^{T_{jh} \cup h_{T_j}}$, i.e.

$$V_j : \mathcal{Q}(\Omega^{T_{jh}}) \times \mathcal{Q}(\Omega^{h_{T_j}}) \to [0,1].$$

This matrix specifies the flow of evidence masses from cylinders $A \in \mathcal{Q}(\Omega^{T_{jh}})$ to sets $A \times B$, $B \in \mathcal{Q}(\Omega^{h_{T_j}})$, i.e. the expert directs the flow of evidence masses to smaller sets. So each expert formulates his knowledge *relative* to the state of knowledge reached before he makes his statement. The assumption that for each hyperedge there is one expert is just introduced for mathematical convenience. Of course, the mass flows are often specified by one human expert. From this point of view it is obvious that the matrices V_j are assumed to be monotonic, since their application should increase the specificity of knowledge.

The acquisition of the different matrices is done according to the consistent ordering of $\mathfrak{T}$. The starting point is the state of total ignorance, where nothing is known but the "closed world assumption" mentioned before, which is represented by the mass distribution $m_v : \mathcal{Q}(\Omega^{\emptyset}) \to [0,1]$; $m_v(\{\varepsilon\}) = 1$, the only mass distribution on $\Omega^{\emptyset}$. In the light of additional knowledge all possible orthogonal extensions are derivable from m_v.

So, in the state of total ignorance, i.e. before any expert has been asked, we have

$$\mathcal{K}_0 = \{m : \mathcal{Q}(\Omega^{\emptyset}) \to [0,1] \mid m \leq m_v\}.$$

Note that in this case the condition $\Pi_{\emptyset}^{M}(m) \leq m_v$ is equivalent to $\Pi_{\emptyset}^{M}(m) = m_v$, which is satisfied by each mass distribution on Ω^M, since there is only one mass distribution on $\Omega^{\emptyset} = \{\varepsilon\}$. The set $\mathcal{M}_0$ is therefore more of conceptual than of practical interest. Expert θ_1 specifies (quasi-relative to m_v) an orthogonal extension matrix of the form

$$V_1 : \mathcal{Q}(\Omega^{\emptyset}) \times \mathcal{Q}(\Omega^{h_{T_1}}) \to [0,1],$$

which has only one column (and is of course monotonic), since $\Omega^{\emptyset}$ has only one element and so $\mathcal{Q}(\Omega^{\emptyset})$ contains only one non-empty set. In practice θ_1 formulates a marginal mass distribution $m_1 : \mathcal{Q}(\Omega^{h_{T_1}}) \to [0,1]$. This information leads to the set

$$\mathcal{K}_1 = \{m : \mathcal{Q}(\Omega^{h_{T_{j1}}}) \to [0,1] \mid \exists\, m_0 \in \mathcal{M}_0 : m \le m_0 \odot V_1\}$$

which contains those mass distributions which are consistent from the point of view of θ_1. From θ_2 we obtain an orthogonal extension matrix

$$V_2 : \mathcal{Q}(\Omega^{T_{2h}}) \times \mathcal{Q}(\Omega^{h_{T_2}}) \to [0,1]$$

leading to the set

$$\mathcal{K}_2 = \{m : \mathcal{Q}(\Omega^{h_{T_1} \cup h_{T_2}}) \to [0,1] \mid \exists\, m_1 \in \mathcal{M}_1 : m \le m_1 \odot V_2\}$$

and in the general case

$$V_j : \mathcal{Q}(\Omega^{T_{jk}}) \times \mathcal{Q}(\Omega^{h_{T_j}}) \to [0,1]$$

which yields

$$\mathcal{K}_j = \{m : \mathcal{Q}(\Omega^{h_{T_1} \cup \ldots \cup h_{T_j}}) \to [0,1] \mid \exists\, m_{j-1} \in \mathcal{M}_{j-1} : m \le m_{j-1} \odot V_j\}.$$

In order to solve the recursion let us firstly consider the set $\mathcal{M}_2$. Since $m_1 \in \mathcal{M}_1$ we have $m \le m_v \odot V_1$ and from equation (14.1) we obtain

$$m \odot V_2 \le m_v \odot V_1 \odot V_2.$$

The use of this technique leads to

$$\mathcal{K}_r = \{m : \mathcal{Q}(\Omega^{h_{T_1} \cup \ldots \cup h_{T_r}}) \to [0,1] \mid m \le m_v \odot V_1 \odot \ldots \odot V_r\}.$$

Under the aspect of knowledge acquisition, it is not the sets $\mathcal{K}_j$ that are of interest but the sets of those mass distributions on Ω^M compatible with the marginal mass distributions within the sets $\mathcal{K}_j$. Thus we define

$$\mathcal{M}_j = \{m : \mathcal{Q}(\Omega^M) \to [0,1] \mid \Pi^M_{h_{T_1} \cup \ldots \cup h_{T_j}}(m) \in \mathcal{K}_j\}.$$

The sets $\mathcal{M}_j$ are the sets of those mass distributions on Ω^M compatible with the information of the experts θ_j, $j = 1, \ldots, r$. Since the amount of information grows with every additional expert taken into account (recall that the matrices V_j are assumed to be monotonic) the sequence $\mathcal{M}_1, \mathcal{M}_2, \mathcal{M}_3, \ldots$ should be a decreasing sequence of sets, i.e.

$$\mathcal{M}_r \subseteq \mathcal{M}_{r-1} \subseteq \ldots \subseteq \mathcal{M}_1.$$

To verify this property we have to show that $m \in \mathcal{M}_{i+1}$ implies $m \in \mathcal{M}_i$ for $i = 1, \ldots, r$:

$$\begin{aligned} m \in \mathcal{M}_{i+1} &\implies \Pi^M_{h_{T_1} \cup \ldots \cup h_{T_{i+1}}}(m) \in \mathcal{K}_{j+1} \\ &\implies \exists\, m_i \in \mathcal{K}_i : \Pi^M_{h_{T_1} \cup \ldots \cup h_{T_{i+1}}}(m) \le m_i \odot V_{i+1}. \end{aligned}$$

Since the strict specialization relation is invariant against projection (see equation (6.17)) we obtain

$$\Pi^{h_{T_1}\cup\ldots\cup h_{T_{i+1}}}_{h_{T_1}\cup\ldots\cup h_{T_i}}\left(\Pi^{M}_{h_{T_1}\cup\ldots\cup h_{T_{i+1}}}(m)\right) \;\leq\; \Pi^{h_{T_1}\cup\ldots\cup h_{T_{i+1}}}_{h_{T_1}\cup\ldots\cup h_{T_i}}(m_i \odot V_{i+1}).$$

As the projection of an orthogonal extension yields a strict specialization (compare Theorem 6.20) we obtain

$$\Pi^{M}_{h_{T_1}\cup\ldots\cup h_{T_i}}(m) \leq \Pi^{h_{T_1}\cup\ldots\cup h_{T_{i+1}}}_{h_{T_1}\cup\ldots\cup h_{T_i}}(m_i \odot V_{i+1}) \;\leq\; m_i\,.$$

Since $m_i \in \mathcal{K}_i$ this implies $m \in \mathcal{M}_i$. Finally we obtain

$$\mathcal{M}_r = \{m : \mathcal{Q}(\Omega^M) \to [0,1] \mid \Pi^M_M(m) \in \mathcal{K}_r\} = \mathcal{K}_r.$$

Define $m_0 \stackrel{\text{d}}{=} m_v \odot V_1 \odot \ldots \odot V_r$. The final result of the knowledge acquisition process is the set $\mathcal{M}_r = \{m' \mid m' \leq m_0\}$. Taking into account the possibility of user observation, where revision is used as the general updating scheme, we obtain the sets $\mathcal{U}_j \stackrel{\text{d}}{=} \mathcal{U}_{revis}(\mathcal{M}_j)$, $j = 1,\ldots,r$. The subset relationship of the sets $\mathcal{M}_j$, $j = 1,\ldots,r$ devolves to the sets of possible updates, thus the integration with respect to the logical expression $T = 1$ <u>and</u> ... <u>and</u> r (see Chap. 6) yields

$$\mathcal{U} \stackrel{\text{d}}{=} \mathcal{U}(T) \;=\; \bigcap_{j=1}^{r} \mathcal{U}_j \;=\; \mathcal{U}_r \;=\; \mathcal{U}_{revis}(\{m' \mid m' \leq m_0\}).$$

The total actual observation

$$E \stackrel{\text{d}}{=} \bigcap_{i=1}^{m} \widehat{\Pi}^M_{\{i\}}(E^{(i)}) \;=\; \bigcap_{i=1}^{r} \widehat{\Pi}^M_{h_{T_j}}(E^{h_{T_j}})$$

is encoded by the set $\mathcal{U}_E = \mathcal{U}^{refin}_{revis}(\{m^E\})$. The integration of $\mathcal{U}$ and $\mathcal{U}_E$ yields

$$\mathcal{U}_{revis}(\{m' \mid m' \leq m_0\}) \cap \mathcal{U}^{refin}_{revis}(\{m^E\}) \;=\; \mathcal{U}_{revis}(\{m'' \mid m'' \leq (m_0)_E\})$$

as the next theorem shows.

(14.26) Theorem. *Let Ω be the frame of discernment and let m be a mass distribution on Ω. For $E \subseteq \Omega$ we obtain*

$$\mathcal{U}_{revis}(\{m' \mid m' \leq m\}) \cap \mathcal{U}^{refin}_{revis}(\{m^E\}) \;=\; \mathcal{U}_{revis}(\{m'' \mid m'' \leq m_E\}).$$

Proof.

(i) Consider $n \in \mathcal{U}_{revis}(\{m' | \ m' \leq m\}) \cap \mathcal{U}_{revis}^{refin}(\{m^E\})$. That means there are m', E' such that $n = m'_{E'}$ where $m' \leq m$. Furthermore we know $n \sqsubseteq m^E$.

Define $E'' \stackrel{\rm d}{=} E \cap E'$ and

$$\mathcal{B}(C) \stackrel{\rm d}{=} \{A | \ C = A \cap E'' \text{ and } m_E(A) > 0\}.$$

Note that

$$\begin{aligned} n(C) > 0 &\Longrightarrow m'_{E'}(C) > 0 \ \wedge \ C \subseteq E \\ &\Longrightarrow \exists A^* : C = A^* \cap E' \ \wedge \ m'(A^*) > 0 \\ &\Longrightarrow C = A^* \cap E' \cap E \ \wedge \ m_E(A^*) > 0 \\ &\Longrightarrow C = A^* \cap E'' \ \wedge \ m_E(A^*) > 0 \\ &\Longrightarrow A^* \in \mathcal{B}(C) \\ &\Longrightarrow \mathcal{B}(C) \neq \emptyset. \end{aligned}$$

Define

$$m''(A) = \begin{cases} \frac{n(C)}{|\mathcal{B}(C)|} & \text{if } A \in \mathcal{B}(C) \\ 0 & \text{otherwise.} \end{cases}$$

Obviously we have $m'' \leq m_E$.

Furthermore we obtain

$$m''_{E''}(C) = \sum_{A: A \cap E'' = C} m''(A) = \sum_{A: A \in \mathcal{B}(C)} m''(A) = n(C).$$

Thus $n \in \mathcal{U}_{revis}(\{m'' | \ m'' \leq m_E\})$.

(ii) Now consider $n \in \mathcal{U}_{revis}(\{m'' | \ m'' \leq m_E\})$. That means there are m'', E'' such that $n = m''_{E''}$ and $m'' \leq m_E$.

Define $E' \stackrel{\rm d}{=} E \cap E'$ and

$$\mathcal{D}(C) = \{A | \ C = A \cap E' \text{ and } m(A) > 0\}.$$

Note that

$$\begin{aligned} n(C) > 0 &\Longrightarrow m''_{E''}(C) > 0 \\ &\Longrightarrow \exists A^* : C = A^* \cap E'' \ \wedge \ m''(A^*) > 0 \\ &\Longrightarrow C = A^* \cap E'' \cap E \ \wedge \ m_E(A^*) > 0 \\ &\Longrightarrow C = A^* \cap E' \ \wedge \ \exists A^{**} : (A^{**} \cap E = A^* \ \wedge \ m(A^{**}) > 0) \\ &\Longrightarrow \exists A^{**} : (A^{**} \cap E' = A^{**} \cap E \cap E'' \\ &\qquad\qquad = A^* \cap E'' = C \ \wedge \ m(A^{**}) > 0) \\ &\Longrightarrow \exists A^{**} : A^{**} \in \mathcal{D}(C) \\ &\Longrightarrow \mathcal{D}(C) \neq \emptyset. \end{aligned}$$

Define

$$m'(A) = \begin{cases} \frac{n(C)}{|\mathcal{D}(C)|} & \text{if } A \in \mathcal{D}(C) \\ 0 & \text{otherwise.} \end{cases}$$

Obviously we have $m' \leq m(A)$

$$m'_{E'}(C) = \sum_{A:A\cap E'=C} m'(A) = \sum_{A:A\in\mathcal{D}(C)} m'(A) = n(C)$$

and as $E' \subseteq E$ we obtain $n(C) \sqsubseteq m^E$. □

The mass distribution $(m_0)_E$ therefore can be used as a representative of the set $\mathcal{U} \cap \mathcal{U}_E = \{m' | \; m' \leq (m_0)_E\}$. We obtain

$$\begin{aligned} n \stackrel{\mathrm{d}}{=} (m_0)_E &= m_v \odot V_1 \odot \ldots \odot V_r \odot R(E) \\ &= m_v \odot V_1 \odot \ldots \odot V_r \odot \prod_{j=1}^{r} R(\widehat{\Pi}^M_{h_{T_j}}(E^{h_{T_j}})). \end{aligned} \quad (14.3)$$

This mass distribution encodes the combined expert and observational knowledge. The user is interested mainly in the marginal mass distributions or in the belief and plausibility functions induced by n. The change of marginal mass distributions due to the change from the a priori mass distribution m_0 to the a postiori mass distribution $(m_0)_E$ reflects the evaluation of the actual problem in the light of the expert knowledge. Although the problem of knowledge representation with belief functions is solved in principle by the above reflections, they raise the question of what an efficient algorithm for the calculation of the marginal beliefs looks like. A direct representation of $\mathcal{Q}(\Omega^M)$ is obviously impossible. In the next section we present an algorithm which is based on local computations.

14.4 Local Computations of Marginal Mass Distributions

Let us once more consider firstly a simple dependency network $C = (M, \mathfrak{T})$ in order to clarify our basic ideas. Later on we will conceive this simple dependency network to be a cutout of a more extensive one to derive more general results.

So let us start with $C = (M, \mathfrak{T})$ where

$$M = \{a, b, c, d, g, h\},$$
$$\mathfrak{T} = \{P, Q, U, V, W\},$$

$P = (\emptyset, a)$, $Q = (\emptyset, b)$, $U = (\{a, b\}, d)$, $V = (\{d\}, g)$, and $W = (\{d\}, h)$. From the knowledge acquisition process we may obtain the following quantifications of the dependencies:

$$\begin{aligned}
m_a &: \mathcal{Q}(\Omega^{\{a\}}) \to [0,1], \\
m_b &: \mathcal{Q}(\Omega^{\{b\}}) \to [0,1], \\
V_U &: \mathcal{Q}(\Omega^{\{a,b\}}) \times \mathcal{Q}(\Omega^{\{d\}}) \to [0,1], \\
V_V &: \mathcal{Q}(\Omega^{\{d\}}) \times \mathcal{Q}(\Omega^{\{g\}}) \to [0,1], \\
V_W &: \mathcal{Q}(\Omega^{\{d\}}) \times \mathcal{Q}(\Omega^{\{h\}}) \to [0,1],
\end{aligned}$$

where m_a, m_b are marginal mass distributions and V_U, V_V, V_W are orthogonal extension matrices. In addition there may be the observations $E^{\{a\}} \in \mathcal{Q}(\Omega^{\{a\}})$, $E^{\{b\}} \in \mathcal{Q}(\Omega^{\{b\}})$, $E^{\{d\}} \in \mathcal{Q}(\Omega^{\{d\}})$, $E^{\{g\}} \in \mathcal{Q}(\Omega^{\{g\}})$, and $E^{\{h\}} \in \mathcal{Q}(\Omega^{\{h\}})$. We obtain

$$\begin{aligned}
(m_0)_E = m_a \odot m_b \odot V_U \odot R\big(\widehat{\Pi}^{\{a,b,d\}}_{\{a,b\}}(E^{\{a,b\}})\big) \\
\odot V_V \odot V_W \odot R\big(\widehat{\Pi}^{\{d,g,h\}}_{\{g,h\}}(E^{\{d,g\}})\big) \\
\odot R\big(\widehat{\Pi}^{\{d,g,h\}}_{\{d\}}(E^{\{d\}})\big)
\end{aligned}$$

with

$$\begin{aligned}
m_{U\uparrow} &\stackrel{\mathrm{d}}{=} m_{Ua\downarrow} \odot m_{Ub\downarrow}, \quad \text{where } m_{Ua\downarrow} = m_a,\; m_{Ub\downarrow} = m_b, \\
V_{d\downarrow} &\stackrel{\mathrm{d}}{=} V_{dV\uparrow} \cdot V_{dW\uparrow} \cdot R\big(\widehat{\Pi}^{d\downarrow}_{\{a\}}(E^{\{d\}})\big), \quad \text{where } V_{dV\uparrow} = V_V,\; V_{dW\uparrow} = V_W,
\end{aligned}$$

and derive

$$\begin{aligned}
(m_0)_E &= m_{U\uparrow} \odot V_U \odot V_{d\downarrow} \odot R\big(\widehat{\Pi}^{d\downarrow}_{\{d\}}(E^{\{d\}})\big) \\
&= m_{Ua\downarrow} \odot m_{Ub\downarrow} \odot V_U \odot V_{dV\uparrow} \odot V_{dW\uparrow} \odot R\big(\widehat{\Pi}^{\{a,b\}\cup d\downarrow}_{\{d\}}(E^{\{d\}})\big).
\end{aligned}$$

Projection yields

$$\begin{aligned}
\Pi^M_{\{d\}}\big((m_0)_E\big) &= \Pi^{\{a,b,d\}}_{\{d\}}\Big(\Pi^M_{\{a,b,d\}}\big((m_0)_E\big)\Big) \\
&= \Pi^{\{a,b,d\}}_{\{d\}}\Big(m_{Ua\downarrow} \odot m_{Ub\downarrow} \odot V_U \odot I_{dV\uparrow} \odot I_{dW\uparrow} \\
&\qquad \odot R\big(\widehat{\Pi}^{\{a,b,d\}}_{\{d\}}(E^{\{d\}})\big)\Big)
\end{aligned}$$

where

$$I_{dV\uparrow} = \Pi^{dV\uparrow}_{\{d\}}(V_{dV\uparrow}),\; I_{dW\uparrow} = \Pi^{dW\uparrow}_{\{d\}}(V_{dW\uparrow})$$

are influence vectors.

Now let us consider the general case of a more extensive network. We obtain

$$m_0 = (m_{U\uparrow} \odot V_U) \odot V_{d\downarrow}$$

where

$$m_{U\uparrow} : \mathcal{Q}(\Omega^{U\uparrow}) \to [0,1]; \quad m_{U\uparrow}(A^{U\uparrow}) \stackrel{\mathrm{d}}{=} m_v(\{\varepsilon\}) \cdot \prod_{T:T\in\mathfrak{T}^{U\uparrow}} V_T[A^{T_h}, A^{h_T}]$$

$$V_{d\downarrow} : \mathcal{Q}(\Omega^{d}) \times \mathcal{Q}(\Omega^{d\downarrow}) \to [0,1]; \quad V_{d\downarrow}[A^d, A^{d\downarrow}] \stackrel{\mathrm{d}}{=} \prod_{T:T\in\mathfrak{T}^{d\downarrow}} V_T[A^{T_h}, A^{h_T}]$$

which mean $m_{U\uparrow}$ is a mass distribution on $\mathcal{Q}(\Omega^{U\uparrow})$ and $V_{d\downarrow}$ is an orthogonal extension matrix. We obtain

$$m_{U\uparrow}[\,.\,] = \prod_{x:x\in U_h} m^{xU\downarrow}[\Pi^{U\uparrow}_{xU\downarrow}(\,.\,)],$$

$$V_{d\downarrow}[\,.\,,.\,] = \prod_{T:d\in T_h} V_{dT\uparrow}[\,.\,,\Pi^{d\downarrow}_{dT\uparrow}(\,.\,)],$$

where

$$m^{xU\downarrow} : \mathcal{Q}(\Omega^{xU\downarrow}) \to [0,1];$$

$$m^{xU\downarrow}[A^{xU\downarrow}] \stackrel{\mathrm{d}}{=} m_v(\{\varepsilon\}) \cdot \prod_{T:T\in\mathfrak{T}^{xU\downarrow}} V_T[A^{T_h}, A^{h_T}],$$

$$V_{dT\uparrow} : \mathcal{Q}(\Omega^{d}) \times \mathcal{Q}(\Omega^{dT\uparrow}) \to [0,1];$$

$$V_{dT\uparrow}[A^d, A^{dT\uparrow}] \stackrel{\mathrm{d}}{=} \prod_{T:T\in\mathfrak{T}^{dT\uparrow}} V_T[A^{T_h}, A^{h_T}].$$

In total we obtain

$$m_0 = \left(\prod_{x:x\in U_h} m^{xU\downarrow}\right) \odot V_T \odot \left(\prod_{T:d\in T_h} V_{dT\uparrow}\right).$$

As pointed out in Sect. 14.3, the a priori mass distribution m_0 has to be revised with respect to the observation $E \in \mathcal{Q}(\Omega^M)$. The result is

$$(m_0)_E = m_0 \odot R(E)$$

(compare equation 14.3), where the revision matrix $R(E)$ can be decomposed as follows:

$$R(E) = R(\widehat{\Pi}^M_{U\uparrow - U_h}(E^{U\uparrow - U_h})) \cdot R(\widehat{\Pi}^M_{U_h}(E^{U_h})) \cdot R(\widehat{\Pi}^M_{d}(E^{d})) \cdot R(\widehat{\Pi}^M_{d\downarrow}(E^{d\downarrow})).$$

Thus we obtain

$$(m_0)_E = \left(m^{M^{U\uparrow}} \odot R(\widehat{\Pi}^M_{U\uparrow - U_h}(E^{U\uparrow - U_h}))\right)$$

$$\odot\, V_V \cdot R(\widehat{\Pi}^M_{U_h}(E^{U_h}))$$

$$\odot\, V_{d\downarrow} \cdot R(\widehat{\Pi}^M_{d\downarrow}(E^{d\downarrow})) \cdot R(\widehat{\Pi}^M_{d}(E^{d})).$$

A further decomposition yields

$$m^{U\uparrow} \odot R(\widehat{\Pi}^{U\uparrow}_{U\uparrow - U_h}(E^{U\uparrow - U_h})) = \prod_{x:x\in U_h} m^{Ux\downarrow} \odot R(\widehat{\Pi}^{Ux\downarrow}_{Ux\downarrow - x}(E^{Ux\downarrow - x}))$$

$$V_{d\downarrow} \cdot R(\widehat{\Pi}^{d\downarrow + d}_{d\downarrow}(E^{d\downarrow})) = \prod_{T:d\in T_h} V_{dT\uparrow} \cdot R(\widehat{\Pi}^{dT\uparrow + d}_{dT\uparrow}(E^{dT\uparrow})).$$

Note that

$$m^{U\uparrow} \odot R(\widehat{\Pi}^{U\uparrow}_{U\uparrow - U_h}(E^{U\uparrow - U_h})) \quad \text{and} \quad m^{Ux\downarrow} \odot R(\widehat{\Pi}^{Ux\downarrow}_{Ux\downarrow - x}(E^{Ux\downarrow - x}))$$

are mass distributions and that

$$V_{d\downarrow} \cdot R(\widehat{\Pi}^{d\downarrow + d}_{d\downarrow}(E^{d\downarrow})) \quad \text{and} \quad V_{dT\uparrow} \cdot R(\widehat{\Pi}^{dT\uparrow + d}_{dT\uparrow}(E^{dT\uparrow}))$$

are orthogonal extension matrices.

We have to calculate the projection of $(m_0)_E$ to the space $\Omega^{U_h \cup h_U}$:

$$\begin{aligned}
\Pi^M_{U_h \cup h_U}(m_0)_E &= \Pi^{M^{U\uparrow + d}}_{U_h + d}\left[\Pi^M_{M^{U\uparrow + d}}(m_0)_E\right] \\
&= \Pi^{M^{U\uparrow + d}}_{U_h + d}\Big[m_{U\uparrow} \odot R(\widehat{\Pi}^{U\uparrow}_{U\uparrow - U_h}(E^{U\uparrow - U_h})) \\
&\qquad \cdot V_U \cdot R(\widehat{\Pi}^{U\uparrow + d}_{U_h}(E^{U_h})) \cdot I_{d\downarrow} \cdot R(\widehat{\Pi}^{U\uparrow + d}_{d}(E^{d}))\Big] \\
&= p_{U\uparrow} \odot V_U \cdot R(\widehat{\Pi}^{U_h + d}_{U_h}(E^{U_h})) \cdot I_{d\downarrow} \cdot R(\widehat{\Pi}^{U\uparrow + d}_{d}(E^{d}))
\end{aligned}$$

where

$$p_{U\uparrow} \stackrel{\mathrm{d}}{=} \Pi^{U\uparrow}_{U_h}\Big(m_{U\uparrow} \odot R(\widehat{\Pi}^{U\uparrow}_{U\uparrow - U_h}(E^{U\uparrow - U_h}))\Big),$$

$$I_{d\downarrow} \stackrel{\mathrm{d}}{=} \Pi^{d\downarrow + d}_{d}\Big(V_{d\downarrow} \cdot R(\widehat{\Pi}^{d\downarrow + d}_{d\downarrow}(E^{d\downarrow}))\Big).$$

As the sets $Ux\downarrow$, $x \in U_h$, and $dT\uparrow$, $d \in T_h$, are disjoint, we obtain

$$p_{U\uparrow}[\,.\,] = \prod_{x:x\in U_h} p_{Ux\downarrow}[\Pi^{U_h}_x(\,.\,)],$$

$$I_{d\downarrow}[\,.\,] = \prod_{T:d\in T_h} I_{dT\uparrow}[\,.\,],$$

where

$$p_{Ux\downarrow} = \Pi^{Ux\downarrow}_x\Big(m_{Ux\downarrow} \odot R(\widehat{\Pi}^{Ux\downarrow}_{Ux\downarrow - x}(E^{Ux\downarrow - x}))\Big),$$

$$I_{dT\uparrow} = \Pi^{dT\uparrow}_d\Big(V_{dT\uparrow} \cdot R(\widehat{\Pi}^{dT\uparrow + d}_{dT\uparrow}(E^{dT\uparrow}))\Big).$$

In total we therefore obtain

$$\begin{aligned} \Pi^M_{\{d\}}(m_0)_E &= \Pi_d^{U_h \cup d}\Big(\prod_{x:x\in U_h} p_{Ux\downarrow} \odot V_U \odot R\big(\widehat{\Pi}_{U_h}^{U_h+d}(E^{U_h})\big)\Big) \\ &\quad \odot \prod_{T:d\in T_h} I_{dT\uparrow} \odot R\big(\widehat{\Pi}_d^{U_h+d}(E^d)\big)\Big). \end{aligned}$$

This calculation has to be done by node processor X^U. As at the node processor X^U only the orthogonal extension matrix V_U and the user observation E^d are available, the other quantities have to be submitted by its neighbors. For the same reason X^U has to provide messages.

A predecessor X^T needs the influence vector $I_{xT\uparrow}$ (let $x = h_T$) given by

$$I_{xT\uparrow} = \Pi_x^{xT\uparrow}\Big(V_{xT\uparrow} \cdot R\big(\widehat{\Pi}_{xT\uparrow}^{xT\uparrow+x}(E^{xT\uparrow})\big)\Big).$$

According to

$$xT\downarrow = d\downarrow \cup \{d\} \cup \bigcup_{\substack{h_S:h_S\in U_h,\\ h_S\neq x}} h_S S\downarrow,$$

we obtain

$$\begin{aligned} &V_{xT\uparrow} \cdot R\big(\widehat{\Pi}_{xT\uparrow}^{xT\uparrow+x}(E^{xT\uparrow})\big) \\ &= \Big[\prod_{\substack{h_S:h_S\in U_h,\\ h_S\neq x}} m_{h_S S\downarrow} \odot V_U \odot R\big(\widehat{\Pi}_{U_h-h_S}^{U_h+d}(E^{U_h-h_S})\big)\Big] \\ &\quad \odot V_{d\downarrow} \cdot R\big(\widehat{\Pi}_{d\downarrow}^{d\downarrow-d}(E^{d\downarrow})\big) \cdot R\big(\widehat{\Pi}_d^{d\downarrow+d}(E^d)\big)] \end{aligned}$$

which means

$$\begin{aligned} I_{xT\uparrow} &= \Pi_x^{U_h+d}\Big[\prod_{\substack{h_S:h_S\in U_h,\\ h_S\neq x}} p_{h_S S\downarrow} \odot V_U \odot R\big(\widehat{\Pi}_{U_h-h_S}^{U_h+d}(E^{U_h-h_S})\big) \\ &\quad \odot I_{d\downarrow} \cdot R\big(\widehat{\Pi}_d^{d\downarrow+d}(E^d)\big)\Big]. \end{aligned}$$

On the other hand the successor W needs a message $p_{dW\downarrow}$, i.e.

$$\Pi_d^{dW\downarrow}\Big(m^{dW\downarrow} \odot R\big(\widehat{\Pi}_{dW\downarrow-d}^{dW\downarrow}(E^{dW\downarrow-d})\big)\Big).$$

We know

$$dW\uparrow = U\uparrow \cup \{d\} \cup \bigcup_{\substack{T:d\in T_h,\\ T\neq W}} dT\uparrow$$

thus we obtain

$$\begin{aligned} & m_{dW\downarrow} \odot R\big(\widehat{\Pi}^{dW\downarrow}_{dW\downarrow - d}(E^{dW\downarrow - d})\big) \\ & \quad = m_{U\uparrow} \odot R\big(\widehat{\Pi}^{U\uparrow}_{U\uparrow - U_h}(E^{U\uparrow - U_h})\big) \odot V_U \cdot R\big(\widehat{\Pi}^{U\uparrow + d}_{U_h}(E^{U_h})\big) \\ & \qquad \odot \prod_{\substack{T: d \in T_h, \\ T \neq W}} V^{dT\uparrow} \cdot R\big(\widehat{\Pi}^{dT\uparrow + d}_{dT\uparrow}(E^{dT\uparrow})\big) \end{aligned}$$

i.e.

$$p_{dW\downarrow} = \widehat{\Pi}^{U_h + d}_{d}\Big[p_{U\uparrow} \odot V_U \cdot R\big(\widehat{\Pi}^{U_h + h_U}_{U_h}(E^{U_h})\big)\Big] \cdot \prod_{\substack{T: d \in T_h, \\ T \neq W}} I_{dT\uparrow}.$$

Note that also the observation E^d has to be submitted to X^W.

14.5 The Propagation Algorithm

The basis of the algorithm we present in this section is the idea of independent node processors exchanging messages. For each hyperedge $T \in \mathfrak{T}$ of the causal network there is one processor X^T which stores the respective orthogonal extension matrix, quantifying the causal dependencies of the characteristics $X^{(i)}$, $i \in T_h$ and the characteristic X^{h_T}. All node processors are of the same structure, which is depicted in Fig. 14.1.
We have

- an input port to receive messages E^{h_T} from the user, concerning his or her observation of characteristic X^{h_T}
- for each predecessor S (i.e. $h_S \in T_h$) a pair of input/output ports to receive the message $(p_{Th_S\downarrow}, E^{h_T})$ and to send the message $I_{Th_S\uparrow}$
- for each successor U (i.e. $h_T \in U_h$) a pair of input/output ports to receive the message $I_{Uh_T\uparrow}$ and to send the message $(p_{Uh_T\downarrow}, E^{h_U})$
- an output port to provide the marginal revised mass distribution for the user.

The structure of root and leaf nodes is somewhat different. To simply the structure shown in Fig. 14.1, no facilities for the communication with parents are required for root nodes, and similarly leaf nodes need not be able to communicate with children.

In the sequel we assume the knowledge acquisition process to be completed, i.e. at every node the orthogonal extension matrices (i.e. the marginal mass distribution at the root nodes) are available. In the beginning all observations E^d are assumed to be equal to the respective domain, which means there is no observational knowledge, i.e. for the initialization process we have $R(\,.\,) = \mathbb{1}$.

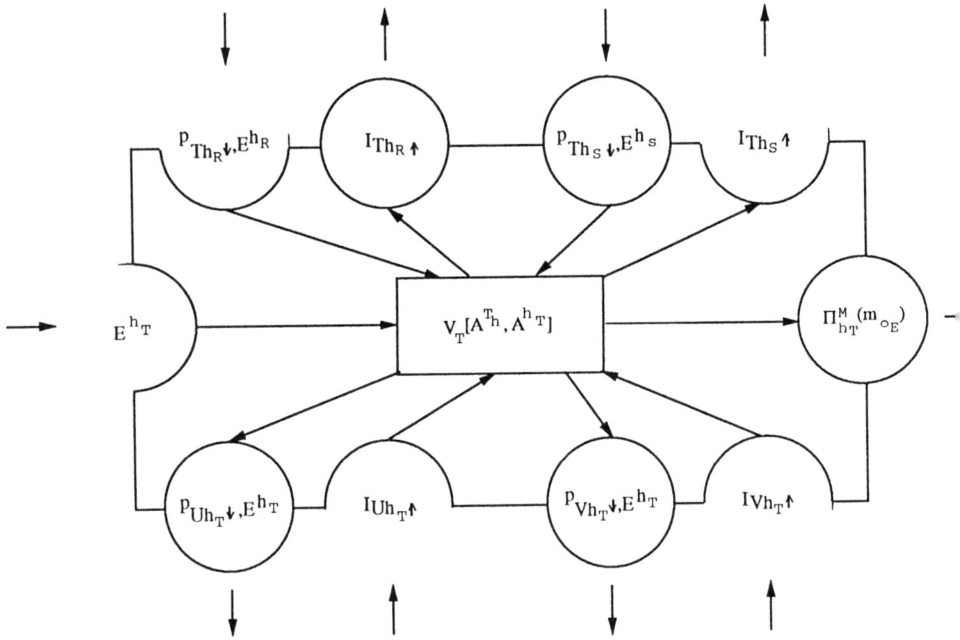

Fig. 14.1. Node processor

To bring up the system each node has to send its messages to all neighbors. Since the node processors are assumed to work independently, this can be done in parallel. The local algorithm for every node is as follows.

If the node processor X^T receives the observation E^{h_T} from the user it has to perform the following activities:

(i) The calculation of the new marginal revised mass distribution:

$$\Pi^M_{h_T}(m_{0_E})[B^{h_T}] = \sum_{A^{h_T}} \sum_{B^{T_h}} \sum_{C^{T_h}} \left(\prod_{S:h_S \in T_h} p_{Th_S\downarrow}[\Pi^{T_h}_{h_S}(A^{T_h})] \right) \cdot V_T[A^{T_h}, A^{h_T}] \cdot R(E^{T_h})[A^{T_h}, B^{T_h}] \cdot \left(\prod_{U:h_T \in U_h} I_{Uh_T\uparrow}[A^{h_T}] \right) \cdot R(E^{h_T})[A^{h_T}, B^{h_T}].$$

(ii) The calculation of the new messages for parent S:

$$I_{T h_S \uparrow}[A^{h_S}] = \sum_{B^{h_T}} \sum_{A^{h_T}} \sum_{B^{T_h - h_S}} \sum_{A^{T_h - h_S}} \left(\prod_{\substack{S' : h_{S'} \in T_h \\ S' \neq S}} p_{T h_{S'} \downarrow}[\Pi^{T_h}_{h_{S'}}(A^{T_h})] \right)$$

$$\cdot V_T[A^{T_h}, A^{h_T}] \cdot R(E^{T_h - h_S})[\Pi^{T_h}_{T_h - h_S}(A^{T_h}), B^{T_h - h_S}]$$

$$\cdot \left(\prod_{U : h_T \in U_h} I_{U h_T \uparrow}[A^{h_T}] \right) \cdot R(E^{h_T})[A^{h_T}, B^{h_T}].$$

(iii) The calculation of the new messages for child U:

$$p_{h_T U \downarrow}[A^{T_h}] = \sum_{B^{T_h}} \sum_{A^{T_h}} \left(\prod_{S : h_S \in T_h} p_{T h_S \downarrow}[\Pi^{T_h}_{h_S}(A^{T_h})] \right) \cdot V_T[A^{T_h}, A^{h_T}]$$

$$\cdot R(E^{T_h})[A^{T_h}, B^{T_h}] \cdot \left(\prod_{\substack{U' : h_T \in U'_h \\ U' \neq U}} I_{U h_T \uparrow}[A^{h_T}] \right)$$

As we see from the above equations, the message $I_{T h_S \uparrow}$ sent to parent S does not depend on the message $p_{T h_S \downarrow}$ received from this node. The same holds for the exchange of messages with child U. $p_{U h_T \downarrow}$ does not depend on $I_{U h_T \uparrow}$. So there is no feedback and the algorithm terminates after a well defined number of steps. The maximum path length a flow of messages has to cover is therefore given by the diameter of the causal network. So if a dependency network on the basis of belief functions is to be installed, then first the qualitative structure of the domain has to be determined by the construction of a dependency network, then the dependencies have to be quantified by the orthogonal extension matrices, and only after that is the network ready for operation.

14.6 Aspects of Implementation

Besides the algorithm just presented, there are several further aspects which are important for a useful software tool designed to represent uncertain knowledge on the basis of belief functions. The structure and the surface of such a system have to meet certain requirements if a successful application is to be result.

First the process of knowledge acquisition has to be supported. The expert should be able to determine first of all the qualitative structure of the domain in the form of a hypergraph. This should be graphically supported. The change from an arbitrary hypergraph to a causal network by the join of nodes should be performed either by direct user intervention or automatically. Another way to ensure that a dependency network arises from this first part of the knowledge acquisition process is to provide only those tools for the construction of the dependency graphs (on the screen) which cannot perform any action destroying the dependency network property. The possible structure of such a dependency network may be as depicted in Fig. 14.2.

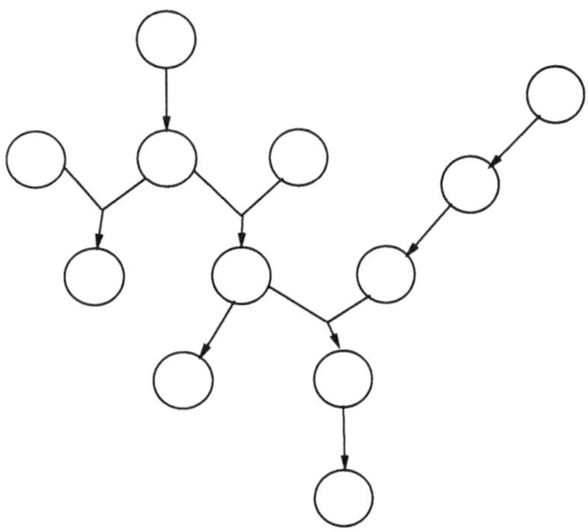

Fig. 14.2. A simple dependency network

The second step the expert has to perform is to determine the domains for the different characteristics. If a node represents a compound characteristic the single domains $\Omega^{(i)}$, $i \in h_S$ should be specified.

After that the expert has to quantify the dependencies he or she has stated to exist. That means he or she has to determine the orthogonal extension matrices. Recall that a typical orthogonal extension matrix has the form

$$V[A^{T_h}, A^{h_T}], \quad A^{T_h} \in \mathcal{Q}(\Omega^{T_h}), \; A^{h_T} \in \mathcal{Q}(\Omega^{h_T}).$$

If we let $T_h = \{i,j\}$, $h_T = \{k\}$, $\Omega^{(1)} = \{z1, \ldots, rd\}$, $\Omega^{(2)} = \{hc, \ldots, fb\}$, $\Omega^{(3)} = \{ad, \ldots, im\}$, then we have

$$|\mathcal{Q}(\Omega^{\{1,2\}})| = 2^6 \cdot 2^4 = 2^{10},$$
$$|\mathcal{Q}(\Omega^{\{3\}})| = 2^4$$

which means $2^4 \cdot 2^{10} = 16 \cdot 1024 = 16,384$ entries to specify this orthogonal extension matrix. It is obvious that no expert can determine 16,384 numbers, so he has to be powerfully supported. One possibility to do this is to start with a default assignment of V where

$$V[A^{T_h}, A^{h_T}] = \begin{cases} 1 & \text{if } A^{h_T} = \Omega^{h_T} \\ 0 & \text{otherwise,} \end{cases}$$

which represents total ignorance. Only if the expert explicitly specifies changes of these default values are they modified.

Also the user needs support. Firstly he should have a graphical representation of the causal network that enables him to bring in his observations, secondly there should be a graphical presentation of the results to facilitate the correct interpretation of the results. The system should provide not the mass distribution directly but the induced belief and plausibility function, and not of all sets but only of those the expert specifies to be of interest.

In addition he should be able to ask for arbitrary sets. Moreover it might be useful to defer the normalization to the output of the results and to display the portion of evidence mass on which the result is based (e.g. 0.3%) to give a measure of the degree of "pathology" of the actual case under consideration.

The above example shows the limits of this approach. The complexity of this matrices explodes if the size of the sets involved increases, especially if compound characteristics or nodes with multiple parents are considered. But if the domains are fairly small (e.g., $\Omega^{(1)} = \{$ friend/enemy$\}$, etc.) and the dependency network is mainly a tree then this approach can yield useful results.

An additional increase in efficiency can be obtained by omitting the graphical surface in process control applications or by efficient algorithms for the representation of orthogonal extension matrices, making use of the fact that they tend to be sparse.

Chapter 15

Related Research

In this book we restrict our attention to numerical methods that are based on measure-theoretic concepts. Logical approaches to uncertainty and vagueness are not treated, and due to space limitations we shall not present concepts such as probabilistic logic within the book. Although one could argue that probabilistic logic is also a numerical method, we regard it is more as a combination of a calculus for uncertainty handling and a logical calculus, but not as a pure measure-theoretic concept.

In this chapter we give a brief overview of several concepts not considered in the rest of the book. For a more comprehensive overview we refer to the collection of paper [Shafer and Pearl 1990] where several key papers are reprinted.

15.1 Nonstandard Logics

In contrast to numerical methods, symbolic techniques emphasize the nature of relations between propositions and use these relations to reason about uncertainty and vagueness. It seems to be natural to use logics for mathematical modeling purposes in this case. Often it is sufficient to consider the "standard logics" such as propositional logic or predicate logic. Inferences on the basis of these logics are well considered and supported by mechanical proof procedures or programming languages such as PROLOG. Although it is often possible to represent a term that contains uncertainty by a purely standard expression, in general it is not sufficient to use only standard logics in the case of knowledge based systems: learning leads us often to realize that something we took as true indeed is not.

In order to enrich the classical logical calculi we can use two different strategies (compare [Smets et al. 1988]):

- we use "deviant logics" where some of the things they say use roughly the same vocabulary as classical logics but may have different or enriched semantics with different proof procedures, or

- we extend the logics in that we talk about things that classical logic cannot, by extending the vocabulary of logic.

  A logic can be both deviant and extended.

Deviant logics such as *circumscription* [McCarthy 1980] or *default logic* [Reiter 1980] are used to model nonmonotonic and default reasoning. The aim is to treat inferences of the form "in the absence of any information to the contrary, assume X". As one might conclude from this assertion, there may exist *exceptions* to X. What we know about the world is represented as a general statement which is true except under certain specifiable conditions. Consequently, the situation is characterized by a kind of uncertainty that arises from the *incompleteness of knowledge*. The semantics of the term "non-monotonic" are directly associated with the process of *retracting knowledge*. If a domain can be modelled on the basis of assertions of this kind and if appropriate quantitative parameters are unavailable, then default or assumption-based inferences on the basis of generalizations are powerful methods for uncertain reasoning.

For basic information on non-monotonic and default logics the reader is referred to [McDermott and Doyle 1980], [Reiter 1980], and [Besnard 1989]. More information on the current state of research in this area can be found in [Ginsberg 1987], and [Reinfrank et al. 1989], for instance. It is obvious that all these approaches that make default assumptions and specify exceptions are really different from approaches that assign quantities such as conditional probabilities, for instance. Truth-maintenance systems can be regarded as the practical side of non-monotonic logic [de Kleer 1986].

Other deviant logics discussed in the context of knowledge based systems are *intuitionistic logic* [Clarke et al 1988] and *epistemic logic* [Moore 1985].

Extended logics are obtained by using additional modal operators [McDermott 1982] or by introducing degrees of truth in the case of multi-valued logics [Rosser and Turquette 1952]. The Łukasiewicz Logic Łukasiewicz and Tarski 1930] is a well known multivalued logic.

15.2 Integrating Uncertainty Calculi and Logics

Beside the purely logical approaches to capturing uncertainty, there exist several attempts to marry numerical methods with logical inference schemes.

Assume that there are two logical propositions $t \equiv$ "the ship A is at sea" and $s \equiv$ "the ship B is at the refueling dock", both of which may be either true or false. Now suppose that because of the incompleteness of knowledge the expert does not know whether the propositions s and t are true or false. But he can specify his "uncertainty" about the truth of s and the truth of t. Then it is the aim of the overall calculus to evaluate, for instance, the uncertainty that expression "s or t" is true.

The best-considered approach in this area is the integration of predicate logic with probability theory called *probabilistic logic* [Nilsson 1986]. The aim of the combination is achieved by interpreting logical formulas as subsets of elementary events referred to as sets of possible worlds. For an evaluation of the desired probabilities it is necessary to integrate all information by using a joint probability measure that is always assumed to exist. The techniques used for reasoning are often constraint satisfaction methods. It turns out that the basic pattern of inference in classical logic can be extended to probabilistic logic. For example, the resolution principle becomes [Suppes 1966]:

$$P(s \ \underline{\text{or}}\ t) \geq \max(0, P(s \ \underline{\text{or}}\ t) + P((\underline{\text{not}}\ s) \ \underline{\text{or}}\ t) - 1)$$

The marriage is of course not restricted to probability and classical predicate logic. One can also integrate possibility theory and predicate logic [Dubois and Prade 1988] in order to obtain a *possibilistic logic*. A similar approach can also be used in the context of belief functions [Smets 1988]. Another strategy would be to use a nonstandard logic and to combine it with an uncertainty calculus. Some results are already described in the literature, such as epistemic logics and probability [Ruspini 1987] or default logic and probability [Neufeld 1989]. Recently, terminological logics [Nebel 1990] have been extended by means of probability theory [Heinsohn and Owsnicki 1988], [Heinsohn 1991]. The relation between uncertainty and logical non-monotonicity is also refered to as *defeasibility* ([Grosof 1990], [Bonissone et al 1990]).

One of the first researchers who has examined the role of numerical uncertainty handling in the framework of *nonmonotonic logics* is Ginsberg. In Sect. 13.3.2 we have already examined that part of his proposal which provides a procedure for propagating belief and plausibility degrees through a set of explicit "expert rules". Moreover, the work presented in [Ginsberg 1984] is related to the question how to *retract* inferences and conclusions already drawn. The reader is also referred to [Ginsberg 1985a] where Ginsberg presents arguments in favor of the answer "yes" to the question "Does probability have a place in non-monotonic reasoning?". The aspect of non-monotonicity and numerical uncertainty handling is also considered in [Grosof 1986]. However, the problem of "belief retraction" or "belief revision" is still only incompletely solved — even if numerical measures of uncertainty are involved.

15.3 Symbolic Methods

The set of models and applications introduced and examined in this book employ *numerical* calculi. Concerning these numerical approaches it is often criticized that they represent uncertainty only on the basis of numerical relations among propositions and fail to represent information explicitly that is important in reasoning about uncertainty.

In the following, we summarize some aspects that may be viewed as *disadvantages* of numeric models. So, as argued in [Bonissone 1987a], numerical approaches

- generally tend to impose some restrictions upon the type and structure of the information, e.g.,
  - the hypotheses within a frame of discernment have to be jointly exhaustive and mutually exclusive, or/and
  - the evidence have to be absolutely or conditionally independent.
- represent uncertainty as a *precise quantity* (scalar or interval) on a given scale. If so, experts and users must provide these uncertainty quantities which have to be *precise and consistent.* The system's output is then equally precise. However, to require this level of precision may be unrealistic.
- use *one* calculus for representing uncertainty and drawing inferences on the basis of uncertain information (e.g., for the sequential and parallel combinations). This calculus is selected a priori. So, any dynamic change in aggregation operators is prevented.
- remove the conflicting part if conflicting information is detected. For instance, if Dempster's rule is used for parallel combinations the non-conflicting portions are normalized and propagated as if the conflict never existed.

Based on these or similar arguments, *symbolic* methods are proposed. In contrast to numeric models they emphasize the nature of relations between propositions and use these relations to reason about uncertainty. In [Clark 1990] the differences between quantitative and symbolic techniques are pointed out as follows (here from the viewpoint of symbolic methods, however):

- Symbolic techniques have a shorter history than quantitative techniques.
- Symbolic techniques derive inspiration more from patterns of competent human reasoning.
- Symbolic techniques make fewer and weaker assumptions about independency and exclusivity and are therefore more robust in some circumstances. However, by making stronger assumptions, quantitative approaches achieve greater precision in the combination of evidence.

- Symbolic techniques are more amenable to implementation of metalevel control.

The most important (areas of) symbolic approaches that may be discussed are

- the nonnumeric model of Fox [Fox 1986], and
- Cohen's theory of endorsements [Cohen 1985].

Since a detailed introduction and examination of these and other symbolic approaches would fill another book we confine ourselves to a very few remarks on these two nonnumeric models and give pointers into the literature.

Basic to the work of Fox is that there exist qualitatively different aspects of uncertainty which can be formalized *symbolically* and can be used for reasoning. In his work on semantic aspects of uncertainty Fox describes a set of different qualitative states of uncertainty. The central uncertainty terms are as follows [Fox 1986]:

- P *is possible* if conditions that are necessary for P are not violated.
- P *is plausible* if P is possible and there is an argument in support of P or the balance of argument is for P.
- P *is probable* if P is possible and there is at least one item of evidence in favor of P.
- P *is certain* if a sufficient condition for P is true.
- P *is believed* if it is reported by a credible source, or a reliable device or procedure, or if it is the most probable or plausible of the alternatives.
- P *is likely* if the summary of evidence and argument is in favor of P.
- P *is suspected* if it is not believed but there is evidence or argument to support it.
- P *is doubted* if not-P is suspected or a competitor of P is suspected.
- P *is assumed* if it is asserted in domain knowledge (e.g. by default) or derived by general knowledge and there is no evidence or argument to the contrary.

As argued by Fox, for large applications, the data or time necessary for deriving quantitative parameters such as conditional probabilities may be unavailable. On the basis of the above weaker *qualitative distinctions*, facts and rules can be formulated in an intelligible manner. Consequently, the original ideas of Fox are contrary to the models considered in this book which handle uncertainty by means of quantitative measures. Note that recently Fox [Fox et al. 1989] has demonstrated how quantitative measures and quantitative combination procedures can be employed in a qualitative framework.

Cohen's *theory of endorsements* [Cohen 1985], [Cohen and Grinberg 1983a,

1983b] is based on the assumption that states of uncertainty are composites of reasons for belief and disbelief. Strength of evidence then is a summary of factors that pertain to certainty. So, in Cohen's approach, it is important to know about the *type and sources of uncertainty* in order to select an appropriate strategy for resolving or discounting this uncertainty. In the theory proposed by Cohen, *endorsements* are data structures that summarize reasons for believing and disbelieving. They classify the justifications according to support or conflict between evidence and a proposition. They further refer to the actions that can be applied for resolving uncertainty. An implementation of this theory is the system SOLOMON [Cohen 1985] which was used to construct FOLIO [Cohen and Liebermann 1983], a financial portfolio management system. In SOLOMON, propositions and endorsements are represented on the basis of first-order predicate calculus. Endorsements are attached to

- data (relating to their source, type, and accuracy),
- rules (characterizing the relation between condition and conclusion, for instance),
- conclusions (denoting whether they are corroborative, or conflicting, for instance),
- tasks (characterizing the potential outcome of the task),
- resolution procedures.

When the goal driven strategy fails to produce a sufficiently well supported conclusion to satisfy a task goal, a task is generated that aims to resolve (e.g. by obtaining missing information, or by removing conflicts) or discount (e.g. by selecting a course of action that effectively covers all outcomes) the uncertainty.

The endorsement approach represents uncertainty in a declarative way. To discuss this model, however, more details would be necessary. In the framework of this book it is sufficient to conclude that Cohen's heuristic proposal is best applied in real-world domains where uncertainty (in the above interpretation) can be made explicit and where a numerical ranking based on sound algorithms is not necessary or not possible.

The characterization of the general differences between numeric and symbolic approaches to uncertainty handling as well as the pros and cons of them is given in [Bhatnagar and Kanal 1985], [Bonissone 1987b], and [Clark 1990], for example.

Since in the framework of this book we focus mainly on numerical methods for uncertain reasoning, we did not consider *causation aspects* of an inference. Following Pearl [Pearl 1988a] (p.18), "causation is a language with which one can talk efficiently about certain structures of relevance relationships, with the objective of separating the relevant from the superfluous".

So, causation aspects concern the condition-conclusion relationships and their qualitative difference. However, one really important aspect when building a concrete system is the combination of both numerical uncertain reasoning with *causal reasoning* in a coherent way that aims at accounting for the *qualitative* difference of relationships between conditions and conclusions of rules.

15.4 Conclusions

In our view, *the nonnumerical approaches alone* are inadequate to represent and summarize measures of uncertainty. It is really important to have a theoretical framework that allows us to handle numerical information on uncertainty and vagueness. Such models were developed in the framework of this book.

When trying to develop systems that aim at modeling parts of the real world and aspects of intelligent behavior, for instance, several other aspects and requirements naturally come into play. In addition to quantitative knowledge, qualitative information be exist, uncertain knowledge may have to be retracted, arguments for and against decisions may have to be reviewed, more specific user interaction in decision making may be required, and intelligible explanations may have to be produced. Consequently, the models examined in this book serve as a *basis* for developing a systems that allow us to draw correct inferences on the basis of numerical uncertain knowledge. This basis should not now be viewed as being exclusive but has to be extended by other aspects and features such as those discussed above.

However, the consideration of extensions such as integrating numeric and symbolic approaches, or numerical methods and non-monotonic theories, for instance, would involve exceeding the scope of this book and would be radically incomplete because such topics are only partially explored. One important area of current research is the use of both quantitative and symbolic handling of uncertainty in the same application.

References

The following abbreviations are used:

AAAI	(American) National Conference on Artificial Intelligence
ACC	American Control Conference
AI	Artificial Intelligence
GWAI	German Workshop on Artificial Intelligence
IJCAI	International Joint Conference on Artificial Intelligence
IPMU	International Conference on Information Processing and Management of Uncertainty in Knowledge-Based Systems
PAMI	IEEE Transactions on Pattern Analysis and Machine Intelligence

Abel, S. (1988) The Sum-and-Lattice-Points Method Based on an Evidential-Reasoning System Applied to the Real-Time Vehicle Guidance Problem. In: [Lemmer and Kanal 1988], pp. 365 - 370

Abramson, B. (1986) An Explanation of and Cure for Minimax Pathology. In: [Kanal and Lemmer 1986], pp. 495 - 504

Aczel, J. (1966) Lectures on Functional Equations and their Applications. Academic Press, New York

Adams, J.B. (1976), A Probability Model of Medical Reasoning and the MYCIN Model. Mathematical Biosciences **32**, 177 - 186

Adams, K. (1966) Probability and the Logic of Conditionals. In: Hintikka, J., Suppes, P. (eds.) Aspects of Inductive Logic. North-Holland, Amsterdam

Adams, K. (1975) The Logic of Conditionals. Reidel, Dordrecht

Agosta, J.M. (1990) The Structure of Bayes Networks for Visual Recognition. In: [Shachter, Levitt, Kanal, and Lemmer 1990], pp. 397 - 405

Akaike, H. (1974) A New Look at the Statistical Model Identification. IEEE Transactions on Automatic Control **19**, 716 - 723

Aleliunas, R.A. (1988) A New Normative Theory of Probabilistic Logics. Proc.of the 7th Biennial Conf. of the Canadian Society for Computational Studies of Intelligence (CSCSI-88) Edmonton, Alta, pp. 67 - 74

Aleliunas, R. (1990) A Summary of a New Normative Theory of Probabilistic Logic. In: [Shachter, Levitt, Kanal, and Lemmer 1990], pp. 199 - 206

Allen, A.O. (1978) Probability, Statistics, and Queuing Theory. Academic Press, New York

Andersen, S.K., Olesen, K.G., Jensen, F.V., Jensen, F. (1989) HUGIN – A Shell for Building Bayesian Belief Universes for Expert Systems. Proceedings of the 11th IJCAI, Detroit

Andreassen, S., Woldbye, M., Falck, B., Andersen, S.K. (1987) MUNIN – A Causal Probabilistic Network for Interpretation of Electromyographic Findings. Proc. of the 10th IJCAI, Milan, pp. 366 - 372

Anger, B. (1977) Representation of Capacities. Math. Annals **229**, 245 - 258

Arnborg, S., Corneil, D.G., Prokurowski, A. (1987) Complexity in Finding Embeddings in a k-Tree. SIAM J. Alg. Disc. Math. **8**: 2, 277 - 284

Aubin, J.P., Frankowska, H, (1990) Set-Valued Analysis. Birkhäuser, Boston

Aumann, J. (1965) Integral of Set Valued Functions. J. Math. Anal. Appl. **12**, 1 - 22

Bacchus, F. (1990) On Probability Distributions over Possible Worlds. In: [Shachter, Levitt, Kanal, and Lemmer 1990], pp. 217 - 226

Baldwin, J.F. (1979) A New Approach to Approximate Reasoning Using a Fuzzy Logic. Fuzzy Sets and Systems **2**, 309 - 325

Baldwin, J.F. (1987) Evidential Support Logic Programming. Fuzzy Sets and Systems **24**, 1 - 26

Barnett, J.A. (1981) Computational Methods for a Mathematical Theory of Evidence. Proc. of the AAAI. Vancouver, pp. 868 - 875

Barth, S.W., Norton, S.W. (1988) Knowledge Engineering within a Generalized Bayesian Framework. In: [Lemmer, Kanal 1988], pp. 103 - 114

Bayes, T. (1963) An Essay Towards Solving a Problem in the Doctrine of Chances. Phil. Trans **3**, 370 - 418. Reproduced in: Deming, W.E., Haffner (eds.) Two papers by Bayes. New York

Bellman, R.E., Giertz, M. (1973) On the Analytic Formalism of the Theory of Fuzzy Sets. Information Sciences **5**, 149 - 156

Beran, R.J. (1971) On Distribution-Free Statistical Inference with Upper and Lower Probabilities. Ann. Math. Statist. **42**, 157 - 168

Berenstein, C., Kanal, L.N., Lavine, D. (1986) Consensus Rules. In: [Kanal and Lemmer 1986], pp. 27 - 32

Berge, C. (1973) Graphs and Hypergraphs. Translated from French by E.Minicka. North-Holland, Amsterdam

Berger, J.O. (1980) Statistical Decision Theory. Springer, New York

Besnard, P. (1989) Default Logic: First Workshop on Uncertainty in AI. Springer-Verlag, Berlin

Bhatnagar, R.K., Kanal, L.N. (1985) Handling Uncertain Information: A Review of Numeric and Non-Numeric Methods. Also in: [Kanal and Lemmer 1986], pp. 3 - 26

Binford, L.B., Hota, N., Hemphill, G. (1989) Bayesian Inference in Model-Based Machine Vision. In: [Kanal, Levitt, and Lemmer 1989], pp. 73 - 95

Biswas, G., Anand, T.S. (1989) Using the Dempster-Shafer Scheme in a Mixed-Initiative Expert System Shell. In: [Kanal, Levitt, and Lemmer 1989], pp. 223 - 239

Black, M. (1937) Vagueness. An Exercise in Logical Analysis. Phil. Sci. **4**, 427 - 455

Black, P.K. (1987) Is Shafer General Bayes? In: Third Workshop on Uncertainty in Artificial Intelligence. Seattle, pp. 2 - 9

Black, P.K., Laskey, K.B. (1990) Hierarchical Evidence and Belief Functions. In: [Shachter, Levitt, Kanal, and Lemmer 1990], pp. 207 - 215

Bonissone, P.P. (1983) Coping with Uncertainty in Expert Systems: A Comparative Study. Proc. of the ACC, pp. 1230 - 1232

Bonissone, P.P. (1987a) Summarizing and Propagating Uncertain Information with Triangular Norms. Int. J. of Approximate Reasoning **1**: 1, 71 - 101

Bonissone, P.P. (1987b) Plausible Reasoning. In: Shapiro, S.C. (ed.), Encyclopedia of Artificial Intelligence, Vol.2, John Wiley, New York, pp. 854 - 863

Bonissone, P.P. (1987c) Using T-norm Based Uncertainty Calculi in a Naval Situation Assessment Application. In: [Kanal, Levitt, and Lemmer 1989], pp. 241 - 256

Bonissone, P.P. (1990) Now that I Have a Good Theory of Uncertainty, What Else Do I Need. In: [Henrion, Shachter, Kanal, and Lemmer 1990], pp. 237 - 253

Bonissone, P.P., Cyrluk, D.A., Goodwin, J.W., Stillman, J. (1990) Uncertainty and Incompleteness: Breaking the Symmetry off Defeasible Reasoning, In: [Henrion, Shachter, Kanal, and Lemmer 1990], pp. 67 - 85

Bonissone, P.P., Decker, K.S. (1985) Selecting Uncertainty Calculi and Granularity: An Experiment in Trading-off Precision and Complexity. In: [Kanal and Lemmer 1986], pp. 217 - 247

Bonissone, P.P., Gans, S.S., Decker, K.S. (1987) RUM: A Layered Architecture for Reasoning with Uncertainty. Proc. of the 10th IJCAI, Milan, pp. 891 - 898

Bonissone, P.P., Tong, R.M. (1985) Reasoning with Uncertainty in Expert Systems. Int. J. of Man-Machine Studies **22**: 3, 241 - 250

Bonissone, P.P., Wood, N.C. (1989) T-Norms Based Reasoning in Situation Assessment Applications. In: [Kanal, Levitt, and Lemmer 1989], pp. 241 - 256

Booker, L.B., Hota, N. (1988) Probabilistic Reasoning about Ship Images. In: [Lemmer, Kanal 1988], pp. 371 - 379

Booker, L.B., Hota, N., Hemphill, G. (1989) Computing Belief Commitments Using Tensor Products. In: [Kanal, Levitt, and Lemmer 1989], pp. 97 - 107

Breese, J.S., Fehling, M.R. (1990) Control of Problem Solving: Principles and Architecture. In: [Shachter, Levitt, Kanal, and Lemmer 1990], pp. 59 - 68

Buchanan, B.G., Duda, R.O. (1983) Principles of Rule-Based Expert Systems. In: Yovits, M.C. (ed.) Advances in Computers. Vol. 22, Academic Press, New York, pp. 164 - 216

Buchanan, B.G., Shortliffe, E.H. (1984) Rule-Based Expert Systems: The MYCIN Experiments of the Stanford Heuristic Programming Project. Addison-Wesley, Reading, MA

Buckley, J.J., Siler, W., Tucker, D.M. (1986) A Fuzzy Expert System. Fuzzy Sets and Systems **20**, 1 - 16

Buckley, J.J., Tucker, D.M. (1989) Second Generation Fuzzy Expert System. Fuzzy Sets and Systems , 271 - 284

Bundy, A. (1985) Incidence Calculus: A Mechanism for Probabilistic Reasoning. J. of Automated Reasoning **1**, 263 - 283

Buntine, W. (1989) Decision Tree Induction Systems: A Bayesian Analysis. In: [Kanal, Levitt, and Lemmer 1989], pp. 109 - 127

Campos, L.M. de, Lamata, M.T., Moral, S. (1988) Logical Connectives for Combining Fuzzy Measures. In: Ras, Z.W., Saitta, L. (eds.) Methodologies for Intelligent Systems. Elsevier Science Publishing Co., North-Holland, New York, pp. 11 - 18

Campos, L.M. de, Lamata, M.T., Moral, S. (1990) The Concept of Conditional Fuzzy Measure. Int. J. of Intelligent Systems **5**, 237 - 246

Carnap, R. (1971) Logical Foundations of Probability Theory. University of Chicago Press, Chicago

Caruana, R.A. (1989) The Automatic Training of Rule Bases that Use Numerical Uncertainty Representations. In: [Kanal, Levitt, and Lemmer 1989], pp. 347 - 356

Cayrol, M., Farreny, H., Prade, H. (1982) Fuzzy Pattern Matching. Kybernetes **11**, 103 - 116

Cecile, M., McLeish, M., Pascoe, P., Taylor, W. (1990) Induction and Uncertainty Management Techniques Applied to Veterinary Medical Diagnosis. In: [Shachter, Levitt, Kanal, and Lemmer 1990], pp. 369 - 381

Cendrowska, J., Bramer, M.A. (1984) A Rational Reconstruction of the MYCIN Consultation System. Int. J. of Man-Machine Studies **20**, 229 - 317

Chandrasekaran, B., Tanner, M.C. (1986) Uncertainty Handling in Expert Systems: Uniform vs. Task-Specific Formalisms. In: [Kanal and Lemmer 1986], pp. 35 - 67

Charniak, E., Riesbeck, C.K., McDermott, D.V. (1980) Artificial Intelligence Programming. Lawrence Erlbaum, Hillsdale, New York

Chatalic, P., Dubois, D., Prade, H. (1987a) A System for Handling Relational Dependencies in Approximate Reasoning. Proc. of the 3rd International Expert System Conference. London, pp. 495 - 502

Chatalic, P., Dubois, D., Prade, H. (1987b) An Approach to Approximate Reasoning Based on the Dempster Rule of Combinations. Int. J. Expert Systems **1**, 67 - 85

Chateauneuf, A., Jaffrey, J.Y. (1989) Some Characterizations of Lower Probabilities and other Monotone Capacities through the Use of Möbius Inversion. Mathematical Social Sciences **17**, 263 - 283

Chavez, R.M., Cooper, G.F. (1990) KNET: Integrating Hypermedia and Normative Bayesian Modeling. In: [Shachter, Levitt, Kanal, and Lemmer 1990], pp. 339 - 349

Chen, K. (1988) Learning to Predict: An Inductive Approach. In: [Lemmer and Kanal 1988], pp. 115 - 123

Cheeseman, P. (1983) A Method of Computing Generalized Bayesian Probability Values for Expert Systems. Proc. of the 8th IJCAI. Karlsruhe, pp. 198 - 202

Cheeseman, P. (1985) In Defense of Probability. Proc. of the 9th IJCAI. Los Angeles, pp. 1002 - 1009

Cheeseman, P. (1985) Probability Versus Fuzzy Reasoning. In: [Kanal and Lemmer 1986], pp. 85 - 102

Cheeseman, P. (1988) An Inquiry into Computer Understanding. Computational Intelligence **4**, 58 - 66

Chen, S.-S. (1988) Some Extensions of Probabilistic Logic. In: [Lemmer and Kanal 1988], pp. 221 - 227

Chin, H.L., Cooper, G.F. (1987) Stochastic Simulation of Bayesian Belief Networks. Proceedings of the 3rd AAAI Workshop of Uncertainty in AI, Seattle

Chin, H.L., Cooper, G.F. (1989) Bayesian Belief Network Inference Using Simulation. In: [Kanal, Levitt, and Lemmer 1989], pp. 129 - 147

Choquet, G. (1953) Theory of Capacities. Ann. Inst. Fourier **5**, 131 - 295

Clark, D.A. (1990) Numerical and Symbolic Approaches to Uncertainty Management in AI. Artificial Intelligence Review **4**: 2, 109 - 146

Clark, D.A. et al. (1988) Responses to 'An AI View of the Treatment of Uncertainty' by Alessandro Safiotti. Knowledge Engineering Review **3**, 59 - 91

Clarke, M.R.B., Gabbay, D.M. (1988) An Intuitionistic Basis for Non-Monotonic Reasoning. In: [Smets et al.1988], pp. 163-178

Cohen, P.R. (1985) Heuristic Reasoning about Uncertainty: An Artificial Intelligence Approach. Research Notes in Artificial Intelligence. Morgan Kaufman, Boston

Cohen, M.S. (1986) An Expert System Framework for Non-Monotonic Reasoning About Probabilistic Assumptions. In: [Kanal and Lemmer 1986], pp. 279 - 293

Cohen, P.R. (1985) Representativeness and Uncertainty in Classification Systems. AI Magazine **6**: 3, 136 - 149

Cohen, P.R. (1989) Steps Towards Programs that Manage Uncertainty. In: [Kanal, Levitt, and Lemmer 1989], pp. 257 - 266

Cohen, P.R., Day, D., Delisio, J., Greenberg, M., Kjieldsen, R., Suthers, D., Berman, P. (1987) Management of Uncertainty in Medicine. Int. J. of Approximate Reasoning **1**: 1, 103 - 116

Cohen P.R., Grinberg, M.R. (1983a) A Theory of Heuristic Reasoning about Uncertainty. AI Magazine, pp. 17 - 24

Cohen, P.R., Grinberg, M.R. (1983b) A Framework for Heuristic Reasoning About Uncertainty. Proc. of the 8th IJCAI. Karlsruhe, pp. 355 - 357

Cohen, P.R., Liebermann, M.D. (1983) FOLIO: An Expert Assistent for Portfolio Managers. Proc. of the 8th IJCAI. Karlsruhe, pp. 212 - 214

Cohen, P.R., Shafer, G., Shenoy, P.P. (1987) Modifiable Combining Functions. AI for Electronic Design, Analysis and Manufacture **1**: 1, 47 - 57. Also in: [Kanal, Levitt, and Lemmer 1989], pp. 357 - 373

Cooper, G.F. (1989) An Algorithm for Computing Probabilistic Propositions. In: [Kanal, Levitt, and Lemmer 1989], pp. 3 - 14

Cooper, G.F. (1990) The Computational Complexity of Probabilistic Inference Using Bayesian Belief Networks. Artificial Intelligence **42**, 393 - 405

Cox, R. (1946) Probability, Frequency and Reasonable Expectation. Am. J. of Physics **14**: 1, 1-13

Craddock, A.J., Browse, R.A. (1988) Belief as Summarization and Meta-Support. In: [Lemmer and Kanal 1988], pp. 229 - 236

Dalkey, N.C. (1986) Inductive Inference and the Representation of Uncertainty. In: [Kanal and Lemmer 1986], pp. 393 - 397

Dalkey, N.C. (1988) Models vs. Inductive Inference for Dealing with Probabilistic Inference. In: [Lemmer and Kanal 1988], pp. 3 - 9

Dalkey, N.C. (1989) The Inductive Logic of Information Systems. In: [Kanal, Levitt, and Lemmer 1989], pp. 375 - 386

D'Ambrosio, B. (1987) Truth Maintenance with Numeric Certainty Estimates. Proc 3rd IEEE Conf. on AI Applications. Orlando, FL, pp. 244 - 49

D'Ambrosio, B. (1989) A Hybrid Approach to Reasoning Under Uncertainty. In: [Kanal, Levitt, and Lemmer 1989], pp. 267 - 283

D'Ambrosio, B. (1990) Process, Structure, and Modularity in Reasoning with Uncertainty. In: [Shachter, Levitt, Kanal, and Lemmer 1990], pp. 15 - 25

Darroch, J.N., Lauritzen, S.L., Speed, T.P. (1980) Markov Fields and Log-Linear Models for Contingency Tables. Ann. Statist. **8**, 522 - 539

Dawid, A.P. (1979) Conditional Independence in Statistical Theory. J. Royal Stat. Society Series A **41**: 1, 1 - 31

Dawid, A.P. (1984) Statistical Theory. The Prequential Approach (with Discussion). J. Royal Stat. Society Series A **147**, 278 - 292

Dawid, A.P., Lauritzen, S.L. (1989) Markov Distributions, Hyper Markov Laws, and Meta Markov Models on Decomposable Graphs, with Applications to Bayesian Learning in Expert Systems. Res. Rep. R-89-31, Inst. Elec. Sys., Aalborg Univ.

Davis, R., Buchanan, B.G., Shortlife, E.H. (1977) Production Rules as a Representation for a Knowledge-Based Consultation System. Artificial Intelligence **8**, 15 - 45

Dean, T., Kanazawa, K. (1990) Probabilistic Causal Reasoning. In: [Shachter, Levitt, Kanal, and Lemmer 1990], pp. 27 - 42

Dechter, R., Pearl, J. (1988) Network-Based Heuristics for Constraint-Satisfaction Problems. Artificial Intelligence **34**: 1, 1 - 38

DeFinetti, B. (1937a) La Prévision ses Lois Logiques. Ressources Objectives Ann. Inst. Henri Poincaré **7**, 1 - 68

DeFinetti, B. (1937b) Foresight: Its Logical Laws, Its Subjective Sources. Ann. Inst. H. Poincaré **7**, 1 - 68. Translated by Kyburg, H.E. in: [Kyburg, Smokler 1980], pp. 55 - 118

DeFinetti, B. (1970) Theory of Probability, Vol. 1, Wiley, London

DeFinetti, B. (1972) Probability, Induction and Statistics, Wiley, New York

DeFinetti, B. (1975) Theory of Probability, Vol. 2, Wiley, New York

DeFinetti, B. (1977) Probabilities of Probabilities: A Real Problem or a Misunderstanding? In: Aykac, A., Brumet, C. (eds.) New Developments in the Applications of Bayesian Methods. North-Holland, Amsterdam, pp. 1 - 10

DeGroot, M. (1970) Optimal Statistical Decisions. McGraw-Hill, New York

Dempster, A.P. (1967) Upper and Lower Probabilities Induced by a Multivalued Mapping. Ann. Math. Stat. **38**, 325 - 339

Dempster, A.P. (1968a) Upper and Lower Probabilities Generated by a Random Closed Interval. Ann. Math. Stat. **39**, 957 - 966

Dempster, A.P. (1968b) A Generalization of Bayesian Inference. J. Royal Statistical Society Series B **30**, 205 - 245

Dempster, A.P. (1972) Covariance Selection. Biometrica **28**, 157 - 175

Dempster, A.P., Kong, A. (1988) Uncertainty, Evidence and Artificial Analysis. Research Report S-120. Dep Statistic **38**, 325 - 339

Dempster, A.P., Laird, N., Rubin, D.B. (1977) Maximum Likelihood from Incomplete Data via the EM Algorithm (with Discussion). J. Royal Statatitical Society Series B **39**, 1 - 38

Diaconis, P., Zabell, S.L. (1982) Updating Subjective Probability. J. American Statistical Association **77**, 822 - 830

Diaconis, P., Zabell, S.L. (1983) Some Alternatives to Bayes' Rule. Tech. Rep. **205**, Dept. Statistics, Stanford Univ. In: Grofman, B., Guillermo, O. (eds.) Information Pooling and Group Decision Making. JAI Press, Greenwich, CN, pp. 25 - 38

Dirac, G.A. (1961) On Rigid Circuit Graphs. Abh. Math. Sem. Univ. Hamburg **25**, 71 - 76

Domotor, Z. (1981) Higher Order Probabilities. Philosophical Studies **40**, 31 - 46

Dong, W.M., Wong, F.S. (1987) Propagation of Evidence in Rule Based Systems. Int. J. of Man-Machine Studies. **26**, 551 - 566

Doyle, J. A (1979) Truth Maintenance Systems. Artificial Intelligence **12**: 3, 231 - 272

Dubois, D., Prade, H. (1980a) Fuzzy Sets and Systems: Theory and Applications. Academic Press, New York

Dubois, D., Prade, H. (1982a) A Class of Fuzzy Based Triangular Norms. Int. J. General Systems **8**, 43 - 61

Dubois, D., Prade, H. (1982b) On Several Representations of an Uncertain Body of Evidence. In: Gupta, M.M., Sanchez, E. (eds.) Fuzzy Information and Decision Process. North-Holland, Amsterdam, pp. 167 - 181

Dubois, D., Prade, H. (1984) Fuzzy Logics and the Generalized Modus Ponens Revisited. Cybernetics and Systems **15**, 293 - 331

Dubois, D., Prade, H. (1985) Combination and Propagation of Uncertainty with Belief-Functions – A Reexamination. Proc. of the 9th IJCAI. Los Angeles, pp. 111 - 113

Dubois, D., Prade, H. (1986a) A Set-Theoretic View of Belief Functions. Logical Operations and Approximations by Fuzzy Sets. Int. J. General Systems **12**, 193 - 226

Dubois, D., Prade, H. (1986b) The Principle of Minimum Specificity as a Basis of Evidential Reasoning. Proc. of IPMU. Paris, pp. 40 - 43 (extended abstract)

Dubois, D., Prade, H. (1986c) On the Unicity of Dempster Rule of Combination. Int. J. of Intelligent Systems **1**, 133 - 142

Dubois, D., Prade, H. (1988a) An Introduction to Possibilistic and Fuzzy Logics. In: [Smets, Mamdani, Dubois, and Prade 1988], pp. 287 - 326

Dubois, D., Prade, H. (1988b) Default Reasoning and Possibility Theory. Artificial Intelligence **35**: 2, 243 - 257

Dubois, D., Prade, H. (1988c) Possibility Theory: An Approach to Computerized Processing of Uncertainty. Plenum Press, New York

Dubois, D., Prade, H. (1988d) The Treatment of Uncertainty in Knowledge Based Systems using Fuzzy Set and Possibility Theory. Int. J. of Intelligent Systems **3**, 141 - 165

Dubois, D., Prade, H. (1988e) On Fuzzy Syllogisms. Computational Intelligence **4(2)**, 171 - 179

Dubois, D., Prade, H. (1989) Fuzzy Sets, Probability and Measurement. European J. of Operational Research **40**, 135 - 154

Dubois, D., Prade, H. (1990a) Consonant Approximations of Belief Functions. To appear in: Int. J. of Approximate Reasoning.

Dubois, D., Prade, H. (1990b) Inference in Possibilistic Hypergraphs. 3rd International Conference on Information Processing and Management of Uncertainty. Paris, 228 - 230

Dubois, D., Prade, H. (1990c) Fuzzy Sets in Approximate Reasoning. Part 1: Inference with Possibility Distributions. Part 2: Logical Approaches. To appear in: Fuzzy Sets and Systems

Dubois, D., Prade, H. (1990d) Modeling Uncertain and Vague Knowledge in Possibility and Evidence Theories. In: [Shachter, Levitt, Kanal, and Lemmer 1990], pp. 283 - 302

Dubois, D., Prade, H., Toncas, J.M. (1990) Inference with Imprecise Numerical Quantifiers. In: Ras, Z., Zemankova, M. (eds.) Intelligent Systems: State of the Art and Future Directions. Ellis Horwood, Chichester, UK

Duda, R.O., Hart, P.E. (1973) Pattern Recognition and Scene Analysis. Wiley, New York

Duda, R.O., Hart, P.E., Barnett, P., Gasching, J., Konoloige, K., Reboh, R., Slocom, J. (1978) Development of the PROSPECTOR Consultant System for Mineral Exploration. Final Report for SRI Projects 5821 and 6915, SRI International Artificial Intelligence Center

Duda, R., Gaschnig, J., Hart, P. (1981) Model Design in the PROSPECTOR Consultant System for Mineral Exploration. In: Michie, D. (ed.) Expert Systems in the Microelectronic Age. Edinburgh University Press, pp. 153 - 167

Duda, R.O., Hart, P.E., Nilsson, N.J. (1976) Subjective Bayesian Methods for Rule-Based Inference Systems. Proc. of National Computer Conference. AFIPS, Vol.45, pp. 1075 - 1082

Edwards, D., Havránek, T. (1985) A Fast Procedure for Model Search in Multidimensional Contingency Tables. Biometrika **72**, 339 - 351

Edwards, D., Havránek, T. (1987) A Fast Model Selection Procedure for Large Families of Models. J. American Statistical Association **82**, 205 - 211

Edwards, D., Kreiner, S. (1983) The Analysis of Contingency Tables by Graphical Models. Biometrika **70**, 553 - 562

Einhorn, H.J., Hogarth, R.M. (1985) Ambiguity and Uncertainty in Probabilistic Inference. Psychological Review **92**, 4 - 25

Erman, L.D., Hayes-Roth, F., Lesser, V. R., Reddy, D. R. (1980) The Hearsay-II Speech-Understanding System: Integrating Knowledge to Resolve Uncertainty. Computing Surveys **12**, 213 - 253

Fagin, R., Halpern, J.Y. (1988) Reasoning about Knowledge and Probability. In: Asilomar, C., Vardi, M. (eds.) Proc. of the Second Conference on Theoretical Aspects of Reasoning about Knowledge. Morgan Kaufmann, Los Altos, pp. 277 - 293

Fagin, R., Halpern, J.Y. (1989) Uncertainty, Belief, and Probability. Proc. Int. Joint Conference on AI. Detroit, pp. 1161 - 1167

Falkenhainer, B. (1986) Towards a General-Purpose Belief Maintenance System. Proc. 2nd Workshop on Uncertainty in Artificial Intelligence. Univ. of Pennsylvania, Philadelphia, pp. 71 - 76. Also in: [Lemmer and Kanal 1988], pp. 125 - 131

Feller, W. (1966) An Introduction to Probability and its Applications. Vol. I - II, Wiley, New York

Ferguson, T.S. (1967) Mathematical Statistics: A Decision Theoretic Approach. Academic Press, New York

Field, H. (1978) A Note on Jeffrey Conditionalization. Philosophy of Science **45**, 361 - 367

Fine, T. (1973) Theories of Probability. Academic Press, New York

Fishburn, P.C (1986) The Axioms of Subjective Probability. Statistical Science **1**: 3, 335 - 358

Fisz, M. (1980) Wahrscheinlichkeitsrechnung und Mathematische Statistik. VEB Deutscher Verlag der Wissenschaften, Berlin

Forgy, C. (1981) OPS5 User's Manual. Technical Report: 5, CMU-CS-81-135, Carnegie-Mellon University, Pittsburgh

Fox, J. (1985) Judgement, Policy and the Harmony Machine. Proc. of the 9th IJCAI. Los Angeles, pp. 1284 - 1288

Fox, J. (1986a) Three Arguments for Extending the Framework of Probability. In: [Kanal and Lemmer 1986], pp. 447 - 458

Fox, J. (1986b) Knowledge, Decision Making, and Uncertainty. In: Gale, W.A. (ed.) Artificial Intelligence and Statistics. Readings, MA

Fox, J., Barber, D., Bardhan, K.D. (1980) Alternatives to Bayes? Methods of Information in Medicine **19**: 4, 210 - 215

Fox, J., Kempf, K.G. (1988) Planning, Scheduling, and Uncertainty in the Sequence of Future Events. In: [Lemmer and Kanal 1988], pp. 395 - 401

Fox, J., O'Neil, M., Glowinski, A., Clark, D.A. (1989) Decision Making as a Logical Process. In: Research and Development in Expert Systems V. Cambridge University Press

French, S. (1984) Fuzzy Decision Analysis, Some Criticism. Management Science **20**, 29 - 44

French, S. (1985) Group Consensus Probability Distributions: A Critical Survey. In: Bernado et al. (eds.) Bayesian Statistics. North-Holland, Amsterdam

Friedman, L. (1981) Extended Plausible Inference. Proc. of the 7th IJCAI. Vancouver, pp. 487 - 495

Frydenberg, M., Lauritzen, S.L. (1989) Decomposition of Maximum Likelihood in Mixed Interaction Models. Biometrika **76**: 3, 539 - 555

Fung, R.M., Chong, C.Y. (1986) Metaprobability and Dempster-Shafer in Evidential Reasoning. In: [Kanal and Lemmer 1986], pp. 295 - 302

Gale, W. (1986) Artificial Intelligence and Statistics. Addison Wesley, Reading, MA

Gärdenfors, P. (1979) Forecasts, Decisions and Uncertain Probabilities. Erkenntnis **14**, 159 - 181

Gärdenfors, P. (1988) Knowledge in Flux. MIT Press, Cambridge, MA

Gärdenfors, P., Sahlin, N.E. (1982) Unreliable Probabilities, Risk Tasking, and Decision Making. Synthese **53**, 361 - 386

Gärdenfors, P. Sahlin, N.E. (1988) Decision, Probability and Utility. Selected Readings. Cambridge University Press, Cambridge

Gaifman, H. (1986) A Theory of Higher Order Probabilities. In: Halpern, J.Y. (eds.) Theoretical Aspects of Reasoning about Knowledge. Morgan Kaufman. Los Altos

Garey, M.R., Johnson, D.S. (1979) Computers and Intractability: A Guide to the Theory of NP-Completeness, Freeman, New York

Garvey, T.D. (1986) Evidential Reasoning for Land-Use Classification. AI Center, SRI Int., Menlo Park, CA

Garvey, T.D., Lowrance, J.D., Fischler, M.A. (1981) An Inference Technique for Integrating Knowledge from Disparate Sources. Proceedings Seventh Intern. Joint Conf. on Artificial Intelligence. Vancouver, pp. 319 - 325

Gaschnig, J. (1982) PROSPECTOR An Expert System for Mineral Exploration. In: Michie, D. (ed.) Introductory Readings in Expert Systems. Gordon and Breach, New York, pp. 47 - 64

Gavril, T. (1972) Algorithms for Minimum Coloring Maximum Clique, Minimum Coloring by Cliques, and Maximum Independent Set of a Chordal Graph. SIAM J. Comput. **1**, 180 - 187

Gebhardt, J., Kruse, R. (1991) The Context Model — A Uniform Approach to Vagueness and Uncertainty. 4th IFSA Wold Congress, Brussels

Geiger, D., Pearl, J. (1990) On the Logic of Causal Models. In: [Shachter, Levitt, Kanal, and Lemmer 1990], pp. 3 - 14

Geiger, D., Verma, T., Pearl, J. (1990) *d*-Separation: From Theorems to Algorithmus. In: [Henrion, Shachter, Kanal, and Lemmer], pp. 139 - 148

Geman, S., Geman, D. (1984) Stochastic Relaxations. Gibbs Distribution and the Bayesian Restoration of Images. IEEE Trans. on Pattern Analysis and Machine Intelligence. PAMI-6: 6, 721 - 742

Genesereth, M.R., Nilsson, N.J. (1987) Logical Foundations of Artificial Intelligence. Morgan Kaufmann, Los Altos

Genest, C., Zidek, J.V. (1986) Combining Probability Distributions: A Critique and an Annotated Bibliography. Statistical Science. **1**: 1, 114 - 148

Ginsberg, M.L. (1984) Non-Monotonic Reasoning Using Dempster's Rule. Proc. of the AAAI-84. Austin, TX, pp. 126 - 129

Ginsberg, M.L. (1985) Implementing Probabilistic Reasoning. In: [Kanal and Lemmer 1986], pp. 331 - 338

Ginsberg, M.L. (1985a) Does Probability have a Place in Non-Monotonic Reasoning? Proc. of the 9th IJCAI. Los Angeles, pp. 107 - 110

Ginsberg. M.L. (1987) Readings in Non-Monotonic Reasoning. Morgan Kaufmann, Los Altos

Giles, R. (1982) Foundation for a Possibility Theory. In: Gupta, M.M., Sanchez, E. (eds.) Fuzzy Information and Decision Processes. North-Holland, Amsterdam, pp. 183 - 195

Glover, F. (1990) Artificial Intelligence, Heuristic Frameworks, and Tabu Search. To be published in: Managerial and Decision Economics

Glymour, C. (1985) Independence Assumptions and Bayesian Updating. Artificial Intelligence **25**, 95 - 99

Glymour, C., Scheines, R., Spirtes, P., Kelly, K. (1987) Discovering Causal Structure. Artificial Intelligence, Philosophy of Science and Statistical Modelling. Academic Press, New York

Godo, L., DeMantaras, R.L., Sierra, C., Verdaguer, A. (1988) Managing Linguistically Expressed Uncertainty in MILOKO: Application to Medical Diagnosis. Artificial Intelligence Communications **1**: 1, 14 - 31

Goguen, J.A. (1967) L–Fuzzy Sets. J. of Math. Analysis and Applications **18**: 1, 145 - 174

Goldman, S.A., Rivest, R.L. (1988) A Non-Iterative Maximum Entropy Algorithm. In: [Lemmer and Kanal 1988], pp. 133 - 148

Golmard, J.-L., Mallet, A. (1989) Learning Probabilities in Causal Trees from Incomplete Databases. Manuscript

Golumbic, M.C. (1980) Algorithmic Graph Theory and Perfect Graphs. Academic Press, London

Good, I.J. (1950) Probability and the Weighting of Evidence. Griffin, London

Good, I.J. (1960) Weights of Evidence, Corroboration, Explanating Power, Information and the Utility of Experiments. J. of the Royal Statistical Society Series B **22**, 319 - 331

Good, I.J. (1961) A Causal Calculus. Philosophy of Science **11**, 305 - 318

Good, I.J. (1962) Subjective Probability as the Measure of a Non-Measurable Set. In: Nagel, N., Suppes, P, Tarski, A. (eds.) Logic, Methodology, and the Philosophy of Science. Stanford University Press, Stanford, pp. 319 - 329

Good, I.J. (1982) Discussion of [Shafer 1982b]: Lindley's Paradox. J. of the American Statistical Association **77**: 378, 325 - 351

Good, I.J. (1983) Good Thinking: the Foundations of Probabilty and its Applications. Univ. Minnesota Press, Minneapolis

Goodman, I.R. (1970) The Multivariate Analysis of Qualitative Data: Interaction among Multiple Classifications. J. Amer. Statist. Assoc. **65**, 226 - 256

Goodman, I.R. (1987) A Measure-Free Approach to Conditioning. Proc. 3rd Workshop on Uncertainty in AI, Seattle, pp. 270 - 277

Goodman, I.R., Nguyen, H.T. (1985) Uncertainty Models for Knowledge-Based Systems, North-Holland, Amsterdam

Gordon, J., Shortliffe, E.H. (1984) The Dempster-Shafer Theory of Evidence. In: Buchanan, B.G., Shortlife, E.H. (eds.) Rule-Based Expert Systems: The MYCIN Experiments of the Stanford Heuristic Programming Project. Addison-Wesley, Reading, MA, pp. 272 - 292

Gordon, J., Shortliffe, E.H. (1985) A Method for Managing Evidential Reasoning in a Hierarchical Hypothesis Space. Artificial Intelligence **26**, 323 - 357

Graham, I., Jones, L.P. (1988) Expert Systems – Knowledge, Uncertainty, and Decision. Chapman and Hall Computing, London

Grosof, B.N. (1985a) An Inequality Paradigm for Probabilistic Knowledge: The Logic of Conditional Probability Intervals. In: [Kanal and Lemmer 1986], pp. 257 - 275

Grosof, B.N. (1985b) Evidential Confirmation as Transformed Probability: On the Duality of Priors and Updates. In: [Kanal and Lemmer 1986], pp. 153 - 166

Grosof, B.N. (1986) Non-Monotonicity in Probabilistic Reasoning. In: Proc. Workshop on Uncertainty in Artificial Intelligence. University of Pennsylvania, Philadelphia, pp. 91 - 98

Grosof, B.N. (1988) Non-Monotonicity in Probabilistic Reasoning. In: [Lemmer and Kanal 1988], pp. 237 - 249

Grosof, B.N. (1990) Defeasible Reasoning and Uncertainty: Comments. In: [Henrion, Shachter, Kanal, and Lemmer 1990], pp. 61 - 66

Haack, S. (1979) Do we Need Fuzzy Logic? Int. J. Man-Machine Studies **11**, 437 - 445

Hacking, I. (1965) Logic of Statistical Inference. Cambridge University Press, Cambridge, UK

Hacking, I. (1975a) All Kinds of Possibility. Philosophical Review **84**, 321 - 328

Hacking, I. (1975b) The Emergence of Probability. Cambridge University Press, Cambridge, UK

Haddawy, P., Frisch, A.M. (1990) Modal Logics of Higher-Order Probability. In: [Shachter, Levitt, Kanal, and Lemmer 1990], pp. 133 - 148

Hager, G., Durrant-White, H.F. (1988) Information and Multi-Sensor Coordination. In: [Lemmer and Kanal 1988], pp. 381 - 394

Hager, G., Mintz, M. (1989) Estimation Procedures for Robust Sensor Control. In: [Kanal, Levitt, and Lemmer 1989], pp. 285 - 300

Hájek, P. (1985) Combining Functions for Certainty Degrees in Consulting Systems. Int. Journal of Man-Machine Studies **22**, 59 - 76

Halmos, P.R. (1950) Measure Theory. Van Nostrand, Princeton

Hamburger, H. (1986) Representing, Combining and Using Uncertain Estimates. In: [Kanal and Lemmer 1986], pp. 399 - 414

Hampel, F.R. (1971) A General Qualitative Definition of Robustness. Ann. Math. Statist. **42**, 1887 - 1896

Hanson, S.J., Bauer, M. (1986) Machine Learning, Clustering, and Polymorphy. In: [Kanal and Lemmer 1986], pp. 415 - 428

Harper, W., Stalnaker, R., Pearce, P. (1981) IFs, Conditionals, Belief, Decision, Chance, and Time. Reidel, Dordrecht

Hawthorne, J. (1988) A Semantic Approach to Non-Monotonic Entailments. In: [Lemmer and Kanal 1988], pp. 251 - 262

Hayes, J.E. Michie, D., Mikulich, L.I. (1979) Machine Intelligence 9, Wiley, New York

Heckerman, D.E. (1986) Probabilistic Interpretations for MYCIN's Certainty Factors. In: [Kanal and Lemmer 1986], pp. 167 - 196

Heckerman, D.E. (1988) An Axiomatic Framework for Belief Updates. In: [Lemmer and Kanal 1988], pp. 11 - 22

Heckerman, D.E. (1989) A Tractable Inference Algorithm for Diagnosis Multiple Deseases. Proc. of the 5th Workshop on Uncertainty in AI, Windsor

Heckerman, D. (1990a) An Empirical Comparison of Three Inference Methods. In: [Shachter, Levitt, Kanal, and Lemmer 1990], pp. 283 - 302

Heckerman, D.E. (1990b) Probabilistic Similarity Networks. Networks **20**

Heckerman, D. (1990c) Similarity Networks for the Construction of Multiple-Fault Belief Networks. Proc. of the 6th Conference on Uncertainty in AI, Cambridge, Mass., pp 32-39

Heckerman, D., Horvitz, E.J. (1988) The Myth of Modularity in Rule-Based Systems for Reasoning with Uncertainty. In: [Lemmer and Kanal 1988], pp. 23 - 34

Heckerman, D., Horvitz, E.J., Nathwani, B.N. (1989) Update on the Pathfinder Project. Proceedings of the 13th Symposium on Computer Applications in Medical Care. Washington, pp. 203 - 207

Heckerman, D., Jimison, H. (1989) A Bayesian Perspective on Confidence. In: [Kanal, Levitt, and Lemmer 1989], pp. 149 - 160

Heines, J.M. (1983) Basic Concepts in Knowledge-based Systems. Machine-Mediated Learning **1**: 1, 65 - 95

Heinsohn, J. (1986) Weiterentwicklung und Implementierung eines Inferenzmechanismus auf der Basis von Belief-Funktionen. Diploma Thesis, Tech. Univ. Braunschweig (in German)

Heinsohn, J. (1987) Handling Uncertain Knowledge – A Review of Numeric Methods. Philips Research Laboratories Hamburg, Report MS-H 3966/86 (in German)

Heinsohn, J., Van Loon, J.J. (1988) Numerical Measures for Handling Uncertainty – Looked at from a Bayesian Perspective. Philips Research Laboratories Hamburg, Report MS-H 4795/88

Heinsohn, J., Owsnicki-Klewe, B. (1988) Probabilistic Inheritance and Reasoning in Hybrid Knowledge Representation Systems. In: Hoeppner, W. (ed.) Proc. of the 12th German Workshop on Artificial Intelligence (GWAI-88), Informatik-Fachberichte Bd.181. Springer, Berlin, pp. 51 - 60

Heinsohn, J. (1991) A Probabilistic Extension for Term Subsumption Languages. Research Report, German Research Center for Artificial Intelligence (DFKI), (forthcoming)

Henkind, S.J. (1988) Imprecise Meanings as a Cause of Uncertainty in Medical Knowledge-Based Systems. In: [Lemmer and Kanal 1988], pp. 35 - 41

Henrion, M. (1987b) Uncertainty in Artificial Intelligence: Is Probability Epistemologically and Heuristically Adequate? In: Mumpower, J. et al.(eds.) Expert Judgement and Expert Systems NATO ASI F Vol. 35. Springer, Berlin

Henrion, M. (1987) Comparing Rule Based Expert Systems with Bayes' Belief Networks for Representing Complex Uncertain Expert Knowledge. 1987 Joint Statistical Meetings. American Statistical Association, San Francisco, pp. 161 - 170

Henrion, M. (1988) Propagating Uncertainty in Bayesian Networks by Probabilistic Sampling. In: [Lemmer and Kanal 1988], pp. 149 - 163

Henrion, M. (1989) Some Practical Issues in Constructing Belief Network. In: [Kanal, Levitt, and Lemmer 1989], pp. 161 - 173

Henrion, M., Shachter, R.D., Kanal, L.N., Lemmer, J.F. (1990) Uncertainty in Artificial Intelligence 5, North Holland, Amsterdam

Holtzman, S., Breese, J. (1986) Exact Reasoning About Uncertainty: On the Design of Expert Systems for Decision Support. In: [Kanal and Lemmer 1986], pp. 339 - 345

Hörmander, L. (1954) Sur la Fonction d'Appui des Ensembles Convexes dans un Enspace Localement Convexe. Arkiv. Mat. **3**, 181 - 186

Horvitz, E.J. (1989) Reasoning about Beliefs and Actions Under Computational Resource Constraints. In: [Kanal, Levitt, and Lemmer 1989], pp. 301 - 324

Horvitz, E., Heckerman, D. (1985) The Inconsistent Use of Measures of Certainty in Artificial Intelligence Research. In: [Kanal and Lemmer 1986], pp. 137 - 152

Horvitz, E., Heckerman, D., Langlotz, C. (1986) A Framework for Comparing Alternative Formalisms for Plausible Reasoning. Proc. of the AAAI-86. Philadelphia, pp. 210 - 214

Horvitz, E.J., Suermondt, H.J., Cooper, G.F. (1989) Bounded Conditioning, Flexible Inference for Decision under Source Resources. Proc. of the 5th Workshop of Uncertainty in AI, Windsor

Howard, R.A. (1989) Knowledge Maps. Management Science **35**, 903 - 922

Howard, R.A., Matheson, J.E. (1989) Influence Diagrams. In: Howard, R.A., Matheson, J.E. (eds.) Readings on the Principles and Applications of Decision Analysis, Vol. 2, Strategic Decisions Group, pp. 721 - 762

Hsia, Y.T., Shenoy, P.P. (1989) An Evidential Language for Expert Systems. In: Ras, Z.W. (ed.) Methodologies for Intelligence Systems **4**, 9 - 16

Huber, P.J. (1968) Robust Confidence Limits. Zur Wahrscheinlichkeitstheorie und verw. Gebiete **10**, 269 - 278

Huber, P.J. (1973) The Use of Choquet Capacities in Statistics. Bull. Int. Stat. Inst. **XLV**: 4, 181 - 188

Huber, P.J. (1981) Robust Statistics. Wiley, New York

Huber, P.J., Strassen, V. (1974) Minimax tests and the Neyman-Pearson Lemma for Capacities. Ann. Statist. **1**, 251 - 263

Hummel, R., Landy, M. (1988) Evidence as Opinions of Experts. In: [Lemmer and Kanal 1988], pp. 43 - 53

Hunter, D. (1986) Uncertain Reasoning Using Maximum Entropy Inference. In: [Kanal and Lemmer 1986], pp. 203 - 209

Hunter, D. (1987) Dempster-Shafer Versus Probabilistic Logic. AAAI Third Workshop on Uncertainty in Artificial Intelligence, Seattle, pp. 22 - 29

Hunter, D. (1990) Parallel Belief Revision. In: [Shachter, Levitt, Kanal, and Lemmer 1990], pp. 241 - 251

Hutchinson, T.A. (1989) Post-Marketing Suveillance, Causality Assessment and a Bayesian Expert System Approach. Pharmaceut. Med. **4**, 149 - 153

Ichihashi, H., Tanaka, H. (1989) Jeffrey-like Rules of Conditioning for the Dempster-Shafer Theory of Evidence. Int. J. of Approximate Reasoning **3**, 119 - 238

Isham, V. (1981) An Introduction to Spatial Point Processes and Markov Random Fields. Int. Statistical Review **49**, 21 - 43

Ishizuka, M. (1983) Inference Methods Based on Extended Dempster and Shafer's Theory for Problems with Uncertainty/Fuzziness. New Generation Computing, 159 - 268

Ishizuka, M., Fu, K.S., Yao, J.T.P. (1981) Inexact Inference for Rule-Based Damage Assessment of Existing Structures. Proc. of the 7th IJCAI. Vancouver, pp. 837 - 842

Ishizuka, M., Fu, K.S., Yao, J.T.P. (1982a) Inference Procedures under Uncertainty for the Problem Reducing Method. Information Science **28**, 179 - 206

Ishizuka, M., Fu, K.S., Yao, J.T.P. (1982b) SPERIL: An Expert System for Damage Assessment of Existing Structures. Proc. 6th Int. Conference on Pattern Recognition, 932 - 937

Jaffrey, J.Y. (1988a) Application of Linear Utility Theory for Belief Functions. In: Bouchon, B., Saita, L., Yager, R.R. (eds.) Uncertainty and Intelligent Systems. Lecture Notes in Computer Science 313. Springer, Berlin, pp. 1 - 8

Jaffrey, J.Y. (1988b) Linear Utility Theory for Belief Functions. Operation Research Letters **8**, 107 - 112

Jain, P., Agogino, A.M. (1990) Stochastic Sensitivity Using Fuzzy Influence Diagrams. In: [Shachter, Levitt, Kanal, and Lemmer 1990], pp. 79 - 92

Jaynay, E. (1968) Prior Probability. IEEE Trans. System Sci. Cybernetics (SSC-4), 227 - 241

Jaynes, E.T. (1979) Where Do We Stand on Maximum Entropy. In: Levine and Tribus (eds.): The Maximum Entropy Formalism. MIT Press, Cambridge, MA

Jeffrey, R. (1965) The Logic of Decision. McGraw-Hill, New York

Jeffrey, R. (1983) The Logic of Decision. 2nd ed. Univ. Chicago Press, Chicago

Jensen, F.V. (1988) Junction Trees – A New Characterization of Decomposable Hypergraphs. JUDEX Research Report, Aalborg, Denmark

Jensen, F.V. (1990) Calculation in HUGIN of Probabilities for Specific Configurations – A Trick with Many Applications. Univ. of Aalborg, Institute for Electronic Systems, R. 90-6

Jensen, F.V., Lauritzen, S.L., Olesen, K.G. (1990) Bayesin Updating in Recursive Graphical Models by Local Computations. Computational Statistics Quarterly, to appear

Jimison, H.B. (1990) Generating Explanations of Decision Models Based on an Augmented Representation of Uncertainty. In: [Shachter, Levitt, Kanal, and Lemmer 1990], pp. 351 - 365

Johnson, R.W. (1986) Independence and Bayesian Updating Methods. Artificial Intelligence **29**, 217 - 222. Also Intervals. In: [Kanal and Lemmer 1986], pp. 197 - 201

Kadesch, R.R. (1986) Subjective Inference with Multiple Evidence. Artificial Intelligence **28**, 333 - 341

Kahneman, D., Slovic, P., Tversky, A. (1982) Judgment under Uncertainty: Heuristics and Biases. Cambridge University Press, Cambridge, UK

Kalagnanam, J., Henrion, M. (1990) A Comparison of Decision Analysis and Expert Rules for Sequential Diagnosis. In: [Shachter, Levitt, Kanal, and Lemmer 1990], pp. 271 - 281

Kanal, L.N., Lemmer, J.F. (1986) Uncertainty in Artificial Intelligence. North-Holland, Amsterdam

Kanal, L.N., Levitt, T.S., Lemmer, J. (1989) Uncertainty in Artificial Intelligence 3. North-Holland, Amsterdam

Kellerer, H.G. (1964) Verteilungsfunktionen mit gegebenen Marginalverteilungen. Wahrscheinlichkeitstheorie und verw. Gebiete **3**, 247 - 270

Kendall, D.G. (1974) Foundation of a Theory of Random Sets. In: Harding, E.F., Kendall, D.G. (eds.) Stochastic Geometry. pp. 322 - 376

Kennes, R., Smets, P. (1990a) Fast Algorithm for Dempster-Shafer Theory. 3rd International Conference on Information Processing and Management of Uncertainty in Knowledge Based Systems. Paris, pp. 99 - 101

Kennes, R., Smets, P. (1990b) Computational aspects of the Möbius Transform. Proc. of the 6th Conference on Uncertainty in AI, Cambridge, Mass., pp 344 - 351

Kenney, D.A., Raiffa, H. (1976) Decisions with Multiple Objectives: Preferences and Value Tradeoffs. Wiley, New York

Keynes (1921) A Treatise on Probability. Macmillan, London

Kim, J.H., Pearl, J. (1983) A Computational Model for Causal and Diagnosic Reasoning in Inference Systems. Proc. of the 8th IJCAI. Karlsruhe, pp. 190 - 193

De Kleer, J. (1986) An Assumption-Based TMS. Artificial Intelligence **28**, 127 - 162

Klir, G.J., Folger, T.A. (1988) Fuzzy Sets, Uncertainty and Information. Prentice Hall, New York

Kohlas, J. (1988a) Conditional Belief Structures. Probability in Engineering and Information Science **2**: 4, 415 - 433

Kohlas, J. (1988b) Propagating Belief Functions Through Constraint Systems. Institute for Automation and Operations Research. Univ. of Fribourg, No. 161

Kohlas, J. (1989) Modeling Uncertainty with Belief Functions in Numerical Models. European J. of Operations Research **40**, 377 - 388

Kohlas, J., Monney, P.A. (1988) Propagating Information and Belief in a Network of Linked Variables. Institute for Automation and Operations Research. Univ. of Fribourg, No. 151

Kohlas, J. (1990) A Mathematical Theory of Hints, Univ. of Fribourg, Working paper: 173

Kolmogorov, A.N. (1950) Foundations of the Theory of Probability. Chelsea. New York

Kong, A. (1986) Multivariate Belief Functions and Graphical Models. Doct. Diss. Dept. of Statistics, Harvard Univ., Cambridge, MA

Koopmann, B.O. (1940) The Axioms and Algebra of Intuitive Probability. Ann. Math. **41**, 269 - 292

Krantz, D.H., Duncan, L.R., Suppes, P., Tverska, A. (1971) Foundations of Measurement. Academic Press, New York

Kruse, R., Gebhardt, J. (1989) On a Dialog System for Vague Data. Proc. 3rd IFSA Congress, Seattle

Kruse, R., Meyer, K.D. (1987) Statistics with Vague Data. Reidel, Dordrecht

Kruse, R., Schwecke, E., (1989) On the Treatment of Cyclic Dependencies in Causal Networks. Proc. 3rd IFSA Congress, Seattle

Kruse, R., Schwecke, E. (1990a) On the Representation of Uncertain Knowledge in the Context of Belief Functions. 8th Intern. Congress of Cybernetics and Systems, New York

Kruse, R., Schwecke, E. (1990b) On the Combination of Information Sources. Proc. of the 3rd IPMU Conference, Paris, 440 - 442

Kruse, R., Schwecke, E. (1990c) Fuzzy Reasoning in Multidimensional Space of Hypotheses. Int. J. Approximate Reasoning **4**, 47 - 68

Kruse, R., Schwecke, E. (1990d) Specialization, A New Concept for Uncertainty Handling with Belief Functions. Int. J. General Systems, to appear

Kruse, R., Schwecke, E., Klawonn, F. (1991) On a Tool for Reasoning with Mass Distributions. Proc. of the 12th IJCAI, Sydney, Australia

Kwok, S.W., Carter, C. (1990) Multiple Decision Trees. In: [Shachter, Levitt, Kanal, and Lemmer 1990], pp. 327 - 335

Kyburg, H.E. (1961) Probability and the Logic of Rational Belief. Wesleyan University Press, Middleton

Kyburg, H.E., Jr. (1974) The Logical Foundations of Statistical Inference. Reidel, Dordrecht

Kyburg, H.E., Jr. (1987a) Bayesian and Non-Bayesian Evidential Updating. Artificial Intelligence **31**, 271 - 294

Kyburg, H.E., Jr. (1987b) Representing Knowledge and Evidence for Decision. In: Bouchon, B., Yager, R.R. (eds.) Uncertainty in Knowledge-Based Systems. Lecture Notes in Computer Science 286. Springer, Berlin, pp. 30 - 40

Kyburg, H.E., Jr. (1987c) Higher Order Probabilities, In: [Kanal,Levitt, and Lemmer 1989], pp. 15 - 22

Kyburg, H.E., Jr. (1988) Knowledge. In: [Lemmer and Kanal 1988], pp. 263 - 272

Kyburg, H.E., Jr. (1989) Higher Order Probabilities. In: [Kanal, Levitt, and Lemmer 1989], pp. 15 - 22

Kyburg, H.E., Jr. (1990a) Epistemological Relevance and Statistical Knowledge. In: [Shachter, Levitt, Kanal, and Lemmer 1990], pp. 159 - 168

Kyburg, H.E., Jr. (1990b) Probabilistic Inference and Non-Monotonic Inference. In: [Shachter, Levitt, Kanal, and Lemmer 1990], pp. 319 - 326

Kyburg, H.E., Jr., Smokler, M.E. (1964) Studies in Subjective Probabilities. Wiley, New York

Lane, D.A. (1989) Subjective Probability and Causality Assessment. Applied Stochastic Models and Data Analysis **5**, 53 - 76

Laskey, K.B., Cohen, M.S., Martin, A.W. (1989) Representing and Eliciting Knowledge about Uncertain Evidence and its Implications. IEEE Transactions on Systems, Man and Cybernetics **19**, 536 - 545

Laskey, K.B., Lehner, P. (1990) Assumptions, Belief, and Probabilities. Artificial Intelligence **41**, 65 - 77

Lauritzen, S.L. (1989) Mixed Graphical Association Models (with discussion). Scand. J. Statist. **16**, 273 - 306

Lauritzen, S.L., Dawid, A.P., Larsen, B.N., Leimer, H.-G. (1990) Independence Properties of Directed Markov Fields. Networks, to appear

Lauritzen, S.L., Speed, T.P., Vijayan, K. (1984) Decomposable Graphs and Hypergraphs. J. of the Australian Math. Soc., Series A **36**, 12 - 29

Lauritzen, S.L., Spiegelhalter, D.J. (1988) Local Computations with Probabilities on Graphical Structures and their Application to Expert Systems. J. of the Royal Stat. Soc., Series B, **50**: 2, 157 - 224

Lauritzen, S.L., Wermuth, N. (1989) Graphical Models for Associations Between Variables, some of which are Qualitative and some Quantitative. Ann. Statist. **17**, 31 - 57

Lawrence, J.D. (1982) Dependency-Graph Models of Evidential Support. COINS Technical Report. pp. 82 - 26

Lee, C.H. (1988) A Comparison of Two Evidential Reasoning Schemes. Artificial Intelligence **35**, 127 - 134

Lemmer, J.F. (1983) Generalized Bayesian Updating of Incompletely Specified Distributions. Large Scale Systems **5**, 51 - 68, In: [Kanal and Lemmer 1986], pp. 339 - 345

Lemmer, J., Kanal, L.N. (1988) Uncertainty in Artificial Intelligence 2. North-Holland, Amsterdam

Levitt, T.S. (1986) Model-Based Probabilistic Situation Inference in Hierarchical Hypothesis Spaces. In: [Kanal and Lemmer 1986], pp. 347 - 356

Levitt, T.S., Binford, T.O., Ettinger, G.J. (1990) Utility-Based Control for Computer Vision. In: [Shachter, Levitt, Kanal, and Lemmer 1990], pp. 407 - 422

Lesmo, L., Saitta, L., Torasso, P. (1985) Evidence Combination in Expert Systems. Int. J. of Man-Machine Studies **22**: 3, 307 - 326

Levi, I. (1982) Ignorance, Probability, and Rational Choice. Synthese **53**, 387 - 417

Levi, I. (1983) Consonance, Dissonance and Evidentiary Mechanisms. In: Gärdenfors, P., Hansson, B., Sahlin, N. (eds.) Evidentiary Value: Philosophical, Judicial, and Psychological Aspects of a Theory. Library of Theoria 15. C.W.K. Gleerups, Lund, Sweden, pp. 27 - 43

Levi, L. (1984) Decisions and Revisions: Philosophical Essays on Knowledge and Value. Cambridge University Press, Cambridge, UK

Levitt, T.S. (1988) Bayesian Inference for Radar Imagery Based Surveillance. In: [Lemmer and Kanal 1988], pp. 413 - 421

Lewis, D. (1976) Probabilities of Conditionals and Conditional Probabilities. Philosophical Review **85**, 297 - 315

Li, Z.-N., Uhr, L. (1988) Evidential Reasoning in a Computer Vision System. In: [Lemmer and Kanal 1988], pp. 403 - 412

Lindley, D.V. (1982) Scoring Rules and the Inevitability of Probability. Intern. Statist. Rev. **50**, 1 - 26

Lindley, D.V. (1987) The Probability Approach to the Treatment of Uncertainty in Artificial Intelligence and Expert Systems. Statistical Science **2**: 1, 3 - 44

Liu, G.S.-H. (1985) Knowledge Structures and Evidential Reasoning in Decision Analysis. In: [Kanal and Lemmer 1986], pp. 203 - 316

Liu, G.S.-H. (1986) Causal and Plausible Reasoning in Expert Systems. Proc. of the AAAI-86, Philadelphia, pp. 220 - 225

Liu, X., Gammerman, A. (1988) A Hybrid Approach to Deductive Uncertain Inference. Int. J. of Man-Machine Studies **28**, 671 - 681

Loui, R.P. (1986) Interval-Based Decisions for Reasoning Systems. In: [Kanal and Lemmer 1986], pp. 459 - 472

Loui, R.P. (1988) Computing Reference Classes. In: [Lemmer and Kanal 1988], pp. 273 - 289

Loui, R.P. (1990) Evidential Reasoning Compared in a Network Usage Prediction Testbed: Preliminary Report. In: [Shachter, Levitt, Kanal, and Lemmer 1990], pp. 253 - 269

Lowrance, J.D., Garvey, T.D., Strat, T.M. (1986) A Framework for Evidential-Reasoning Systems. Proc. of the AAAI-86. Philadelphia, pp. 896 - 903

Lukasiewicz, J., Tarski, A. (1930) Untersuchungen über den Aussagenkalkül. J. Royal Stat. Society et Lettres Varsowe III, **23**, pp. 1 - 21

Lummer, G. (1968) Algebres de Fonctions et Espaces de Hardy. Lecture Notes in Mathematics 75, Springer, Berlin

Lu, S.Y., Stephanou, H.E. (1984) A Set-Theoretic Framework for the Processing of Uncertain Knowledge. Proc. of the AAAI-84. Austin, TX, pp. 216 - 221

Magrez, P., Smets, P. (1989) Epistemic Necessity, Possibility and Truth. Tools for Dealing with Imprecision and Uncertainty in Fuzzy Knowledge-Based Systems. Int. J. of Approximate Reasoning **3**: 1, 35 - 57

Maier, D. (1983) The Theory of Relational Databases. Pitman, London

Maier, D., Ullmann, J.D. (1982) Connections in Acyclic Hypergraphs. In: Proc. of the ACM Symposium on Principles of Database Systems. ACM, New York, pp. 34 - 39

Mamdani, E.H., Gaines, B.R. (1981) Fuzzy Reasoning and its Applications. Academic Press, London

De Mántaras, R.L. (1990) Approximate Reasoning. Ellis Horwood, Chichester, UK

Martin-Clouaire, R., Prade, H. (1985) On the Problems of Representation and Propagation of Uncertainty in Expert Systems. Int. J. of Man-Machine Studies **22**: 3, 251 - 264

Matheron, G. (1975) Random Sets and Integral Geometry. Wiley, New York

Matheron, G. (1989) Estimating and Choosing. Springer, Berlin

McCarthy, J.M. (1980) Circumscription – A Form of Nonmonotonic Reasoning. AI **13**, 27 - 39

McDermott, D. (1982) Nonmonotonic Logic II: Nonmonotonic and Modal Theories, JACM **29**, 33 - 57

McDermott, D., Doyle, J. (1980) Non-Monotonic Logic 1. AI **13**, 41 - 72

McLeish, M. (1988) Probabilistic Logic: Some Comments and Possible Use for Nonmonotonic Reasoning. In: [Lemmer and Kanal 1988], pp. 55 - 62

McLeish, M. (1989) Nilsson's Probabilistic Entailment Extended to Dempster-Shafer Theory. In: [Kanal, Levitt, and Lemmer 1989], pp. 23 - 34

Van Melle, W.J. (1980) System Aids in Constructing Consultation Programs. UMI Research Press, Ann Arbor, MI

Van Melle, W.J., Shortliffe, E.H., Buchanan, B.G. (1984) EMYCIN: A Knowledge Engineer's Tool for Constructing Rule-Based Expert Systems. In: [Buchanan and Shortliffe 1984]

Mellouli, K. (1987) On the Propagation of Beliefs in Networks Using the Dempster-Shafer Theory of Evidence. Doct. Diss., School of Business, University of Kansas, Lawrence

Michie, D. (1982) Introductory Readings in Expert Systems. Gordon and Breach, New York

Mizumoto, M., Fukami, S., Tanaka, K. (1979) Several Methods for Fuzzy Conditional Inferences. Proceedings 18th Conference on Decision and Control. Fort Lauderdale, FL, pp. 777 - 782

Moore, R.E. (1966) Interval Analysis. Prentice-Hall, Englewood Cliffs, NJ

Moore R. (1979) Methods and Applications of Internal Analysis. SIAM Studies on Applied Mathematics **2**, Philadelphia

Moore, R.C. (1985) Semantical Considerations on Nonmonotonic Logic. AI **25**, 75 - 94

Nau, D., Purdom, P., Tzeng, C.-H. (1986) An Evaluation of Two Alternatives to Minimax. In: [Kanal and Lemmer 1986], pp. 505 - 509

Neapolitan, R.E. (1990) Probabilistic Reasoning in Expert Systems: Theory and Algorithms, Wiley, New York

Nebel, B. (1990) Reasoning and Revision in Hybrid Representation Systems. Lecture Notes in Computer Science **422**, Springer, Berlin

Negoita, C. (1985) Expert Systems and Fuzzy Systems. Addison-Wesley, Reading MA

Neufeld, E. (1989) Defaults and Probability; Extensions and Coherence. In: Brachman, R. J., Levesque, H. J., and Reiter, R. (eds.) Proc. of the 1st International Conference on Principles of Knowledge Representation and Reasoning (KR89). Toronto, CA, pp. 312 - 323

Neufeld, E., Poole, D. (1989) Towards Solving the Multiple Extension Problem: Combining Defaults and Probability, In: [Kanal, Levitt, and Lemmer 1989], pp. 35 - 44

Neufeld, E., Poole, D., Aleliunas, R. (1990) Probabilistic Semantics and Defaults. In: [Shachter, Levitt, Kanal, and Lemmer 1990], pp. 121 - 131

Neumann von, J., Morgenstern, O. (1974) Theory of Games and Economic Behavior. 2nd ed. Princeton University Press, Princeton,

Nguyen, H.T. (1978) On Random Sets and Belief Functions. Journal of Mathematical Analysis and Applications. **65**, 531 - 542

Nilsson, N.J. (1986) Probabilistic Logic. Artificial Intelligence **28**, 71 - 87

Norton, S.W. (1988) An Explamation Mechanism for Bayesian Inferencing Systems. In: [Lemmer and Kanal 1988], pp. 165 - 173

Ogawa, H., Fu, K.S., Yao, J.T.P. (1985) An Inexact Inference for Damage Assessment of Existing Structures. Int. J. of Man-Machine Studies **22**: 3, 295 - 306

Olesen, K.G., Kjaerulff, U., Jensen, F., Jensen, F.V., Falck, B., Andreassen, S., Andersen, S.K. (1989) A MUNIN Network for the Median Nerve - A Case Study on Loops. Applied Artificial Intelligence. Special Issue: Towards Causal AI Models in Practice

Oppel, U.G. (1990) HUGIN, A Software Package for the Construction of Expert Systems Based on Causal Probabilistic Networks. Proceedings of the FAW Workshop on Uncertainty in Knowledge-Based Systems, Ulm

Orponen, P. (1990) Dempster's Rule of Combination is #P-Complete. Artificial Intelligence **44**, 245 - 253

Paass, G. (1988) Probabilistic Logic. In: Smets, P., Mamdani, A., Dubois, D., Prade, H. (eds.) Non Standard Logics for Automated Reasoning. Academic Press, London, pp. 213 - 251

Pawlak, Z. (1982) Rough Sets. Int. J. of Computer and Information Sciences **1**: 5, 341 - 356

Pearl, J. (1982) Reverend Bayes on Inference Engines: A Distributed Hierarchical Approach. Proc. of the AAAI-82. Pittsburgh, pp. 133 - 136

Pearl, J. (1986a) On Evidential Reasoning in a Hierarchy of Hypotheses. Artificial Intelligence **28**, 9 - 15

Pearl, J. (1986b) Probabilistic Reasoning using Graphs. Proc. of the IPMU, Paris

Pearl, J. (1986c) Fusion, Propagation and Structuring in Belief Networks. Artificial Intelligence **29**, 241 - 288

Pearl, J. (1986d) A Constraint-Propagation Approach to Probabilistic Reasoning. In: [Kanal and Lemmer 1986], pp. 357 - 369

Pearl, J. (1987a) Distributed Revision of Composite Beliefs. Artificial Intelligence **33**, 173 - 215

Pearl, J. (1987b) Evidential Reasoning Using Stochastic Simulation of Causal Models. Artificial Intelligence **33**, 245 - 257

Pearl, J. (1988a) Probabilistic Reasoning in Intelligent Systems. Networks of Plausible Inference. Morgan Kaufmann, New York

Pearl, J. (1988b) On Probability Intervals. Int. J. of Approximate Reasoning **3**, 211 - 216

Pearl, J. (1988c) On Logic and Probability. Computational Intelligence **4**: 1, 99 - 103

Pearl, J. (1988d) On Probabilistic Inference in Intelligent Systems. Morgan Kaufman, San Mateo, CA

Pearl, J. (1988e) Distributed Revision of Belief Commitment in Composite Explanations. In: [Lemmer and Kanal 1988], pp. 291 - 315

Pearl, J. (1990) Reasoning with Belief Functions: An Analysis of Compatibility. Int. J. of Approximate Reasoning **4**, 363 - 389

Pearl, J., Dechter, R. (1989) Learning Structure from Data: A Survey. Proceedings COLT 89, Santa Cruz, 230 - 244

Pednauld, E.P.D., Zucker, S.W., Muresam, L.V. (1981) On the Independence Assumption Underlying Subjective Bayesian Updating. Artificial Intelligence **16**, 213 - 222

Planchet, B. (1989) Credibility and Conditioning. J. of Theoretical Probability. **2**: 3, 289 - 299

Popper, K.L. (1975) The Logic of Scientific Discovery. Scientific Edition, New York

Prade, H. (1983) A Synthetic View of Approximate Reasoning Techniques. Proc. of the 8th IJCAI. Karlsruhe, pp. 130 - 136

Prade, H. (1985) A Computational Approach to Approximate and Plausible Reasoning with Applications to Expert Systems. IEEE Transactions on Pattern Analysis and Machine Intelligence. **7**: (3), 260 - 283

Prade, H., Negoita, C.V. (1986) Fuzzy Logic in Knowledge Engineering. Verlag TÜV Rheinland, Köln

Puri, M.L., Ralescu, D. (1982) A Possibility Measure is Not a Fuzzy Measure. Fuzzy Sets and Systems **7**, 311 - 314

Quinlan, J.R. (1983a) INFERNO, A Cautious Approach to Uncertain Inference. The Computer Journal **26**: 3, 255 - 269

Quinlan, J.R. (1983b) Consistency and Plausible Reasoning. Proc. of the 8th IJCAI. Karlsruhe, pp. 137 - 144

Quinlan, J.R. (1985) Internal Consistency in Plausible Reasoning Systems. New Generation Computing **3**, 157 - 180

Raiffa, H.A. (1968) Decision Analysis: Introductory Lectures on Choices Under Uncertainty. Addison-Wesley, Reading, MA

Raiffa, H.A., Schlaifer, R. (1968) Applied Statistical Decision Theory. MIT Press, Cambridge, MA

Ramsey, F.P. (1931) Truth and Probability. In: Kyburg, H.E., Smokler, H.E. (eds.) Studies in Subjective Probability. Wiley, New York (1980) pp. 61 - 92

Rebane, G., Pearl, J. (1989) The Recovery of Causal Poly-Trees from Statistical Data. In: [Kanal, Levitt, and Lemmer 1989], pp. 175 - 182

Reinfrank, M., DeKleer, J., Ginsberg, M.L., Sandewall, E. (1989) Non-Monotonic Reasoning. Proc. of the 2nd Intern. Workshop. Lecture Notes in AI **346**. Springer-Verlag, Berlin

Reiter, R. (1980) A Logic for Default Reasoning. AI **13**, 81 - 132

Rendell, L. (1986) Induction, of and by Probability. In: [Kanal and Lemmer 1986], pp. 429 - 443

Renyi, A. (1970) Probability Theory. North-Holland, Amsterdam

Rich, E. (1983) Default Reasoning as Likelihood Reasoning. Proc. of the AAAI-83. Washington, DC, pp. 348 - 351

Richter, M. (1988) Prinzipien der künstlichen Intelligenz. Teubner, Stuttgart

Rissanen, J. (1987) Stochastic Complexity (with Discussion). J. Royal Statistical Society Series B **49**, 223 - 239 and 253 - 265

Rivett, B.H.P. (1975) Behavioural Problems of Utility Theory. In: White, D.J., Bowen, K.C. (eds.) The Role and Effectiveness of the Theory of Decision in Practice. Hodder and Stoughton, London, pp. 21 - 27

Rollinger, C.R. (1983) How to Represent Evidence-Aspects of Uncertainty Reasoning. Proc. of the 8th IJCAI. Karlsruhe, pp. 358 - 361

Rollinger, C.R. (1984) Die Repräsentation natürlichsprachlich formulierten Wissens – Behandlung der Aspekte Unsicherheit und Satzverknüpfung. KIT-Report, Technische Universität Berlin, Germany

Rosser, J.N., Turquette, A. (1952) Many-Valued Logic. North-Holland, Amsterdam

Ruspini E.H. (1986) The Logical Foundations of Evidential Reasoning. Technical Note 408, SRI International, Menlo Park, CA

Ruspini, E. (1987) Epistemic Logics, Probability and the Calculus of Evidence. Proc. 10th Int. Joint Conf. on AI (IJCAI-87), Milan, pp. 924 - 931

Ruspini, E. (1989) The Semantics of Vague Knowledge. Revue Internationale de Systemique **3**: 4, 387 - 420

Ruspini, E. (1990) Similarity Based Models for Possibilistic Logics. Information Processing and Management of Uncertainty in Knowledge Based Systems. 3rd Int. Conf., pp. 56 - 58

Saffiotti, A. (1988) An AI View of the Treatment of Uncertainty. Knowledge Engineering Review **2**, 75 - 98

Sanchez, E., Zadeh, L.A. (1987) Approximate Reasoning in Intelligent Systems, Decision, and Control. Pergamon Press, Oxford

Saunders, D. (1989) Improvements to INFERNO. In: Cohn, A.D. (ed.) Proceedings of the Seventh Conference of the Society for the Study of Artificial Intelligence and Simulation of Behaviour. Pitman and Kaufmann, Brighton, pp. 105 - 112

Savage, L.J. (1954) The Foundations of Statistics, Wiley, New York

Savage, L.J. (1962) The Foundations of Statistical Inference, Wiley, New York

Schefe, P. (1980) On Foundations of Reasoning with Uncertain Facts and Vague Concepts. Int. J. of Man-Machine Studies **12**, 35 - 62

Schocken, S. (1988) On the Rational Scope of Probabilistic Rule-Based Inference Systems. In: [Lemmer and Kanal 1988], pp. 175 - 189

Schwartz, S.M., Baron, J., Clarke, J.R. (1988) A Causal Bayesian Model for the Diagnosis of Appendictis. In: [Lemmer and Kanal 1988], pp. 423 - 434

Schwarz, G. (1978) Estimating the Dimension of a Model. Ann. Statist. **6**, 461 - 464

Schweizer, B., Sklar, A. (1961) Associative Functions and Statistical Triangle Inequalities. Publ. Math. Debrecen **8**, 169 - 186

Schweizer, B., Sklar, A. (1983) Probabilistic Metric Spaces. North-Holland, New York

Shachter, R.D. (1986a) Intelligent Probabilistic Inference. In: [Kanal and Lemmer 1986], pp. 371 - 382

Shachter, R.D. (1986b) Evaluatiny Influence Diagrams. Operation Research **33**(6)

Shachter, R.D. (1988a) Probabilistic Inference and Influence Diagrams. Operation Research **36**: 4

Shachter, R.D. (1988b) DAVID: Influence Diagram Processing System for the Macintosh. In: [Lemmer and Kanal 1988], pp. 191 - 196

Shachter, R.D. (1990) A Linear Approximation Method for Probabilistic Inference. In: [Shachter, Levitt, Kanal, and Lemmer 1990], pp. 93 - 103

Shachter, R.D., Bertrand, L.J. (1989) Efficient Inference on Generalized Fault Diagrams. In: [Kanal, Levitt, and Lemmer 1989], pp. 325 - 332

Shachter, R.D., Eddy, D.M., Hasselblad, V., Wolpert, R. (1989) A Heuristic Bayesian Approach to Knowledge Acquisition: Application to the Analysis of Tissue-Type Plasminogen Activator. In: [Kanal, Levitt, and Lemmer 1989], pp. 183 - 190

Shachter, R.D., Heckermann, D. (1988) A Backwards View for Assessment. In: [Lemmer and Kanal 1988], pp. 317 - 324

Shachter, R.D., Levitt, T.S., Kanal, L.N., Lemmer, J.F. (1990) Uncertainty in Artificial Intelligence 4. North-Holland, Amsterdam

Shafer, G. (1976) A Mathematical Theory of Evidence. Princeton University Press, Princeton

Shafer, G. (1978) Nonadditive Probabilities in the Work of Bernoulli and Lambert. Arch. History Exact Sci. **19**, 309 - 370.

Shafer, G. (1979) Allocations of Probability. Ann. of Prob. **7**, 827 - 839

Shafer, G. (1982a) Belief Functions and Parametric Models. J. Royal Statistical Society Series B, **3**, 322 - 352

Shafer, G. (1982b) Lindley's Paradox. J. of the American Statistical Association **77**: 378, 325 - 351

Shafer, G. (1985) Conditional Probability. Int. Statistical Review **53**: 3, 261 - 277

Shafer, G. (1986a) Savage Revisited. Statistical Science **1**, 463 - 501

Shafer, G. (1986b) The Combination of Evidence. Int. J. of Intelligent Systems **1**, 155- 179

Shafer, G. (1986c) Probability Judgement in Artificial Intelligence. In: [Kanal and Lemmer 1986], pp. 127 - 135

Shafer, G. (1987a) Probability Judgements. Artificial Intelligence and Expert Systems. **2**: 1, 3 - 44

Shafer, G. (1987b) Belief Functions and Possibility Measures. In: Bedzek, J. (ed.) Analysis of Fuzzy Information, Vol. 1, Mathematics and Logic. CRC Press, Boca Raton, FL, pp. 51 - 84

Shafer, G., Logan, R. (1987) Implementing Dempster's Rule for Hierarchical Evidence. Artificial Intelligence **33**, 271 - 298

Shafer, G., Pearl, J. (1990) Readings in Uncertain Reasoning. Morgan Kaufmann, San Mateo, CA

Shafer, G., Shenoy, P.P. (1988) Local Computations in Hypertrees. School of Business, Working Paper No. 201. University of Kansas, Lawrence, KA

Shafer, G., Shenoy, P., Mellouli, K. (1987) Propagating Belief Functions in Qualitative Markov Trees. Intern. J. of Approximate Reasoning **1**, 349 - 400

Shafer, G., Tversky, A. (1985) Languages and Designs for Probability Judgement. Cognitive Science **9**, 309 - 339

Shastri, L. (1985) Evidential Reasoning in Semantic Networks: A Formal Theory and its Parallel Implementation. Ph. D. Thesis, University of Rochester, Rochester, NY

Shastri, L., Feldman, J.A. (1985) Evidential Reasoning in Semantic Networks: A Formal Theory. Proc. of the 9th IJCAI. Los Angeles, pp. 465 - 473

Shenoy, P.P. (1989) A Valuation-Based Language for Expert Systems. Int. J. for Approximate Reasoning **3**: 5, 383 - 411

Shenoy, P.P. (1990a) Valuation-Based Systems for Propositional Logic. Methodologies for Intelligent Systems **5**, North-Holland, to appear

Shenoy, P.P. (1990b) Valuation-Based Systems for Discrete Optimization. Proceedings of the 6th Conference on Uncertainty in Artificial Intelligence. Boston

Shenoy, P.P., Shafer, G. (1987) Propagating Belief Functions with Local Computations. IEEE Expert **1**: 3, 43 - 52

Shenoy, P.P., Shafer, G. (1988) An Axiomatic Framework for Bayesian and Belief-Function Propagation. Proceedings of the 4th Workshop on Uncertainty in Artificial Intelligence. Minneapolis, pp. 307 - 314

Shenoy, P.P., Shafer, G. (1990) Axioms for Probability and Belief-Function Propagation. In: [Shachter, Levitt, Kanal, and Lemmer 1990], pp. 169 - 198

Shenoy, P., Shafer, G., Mellouli, K. (1988) Propagation of Belief Functions. A Distributed Approach. In: [Lemmer and Kanal 1988], pp. 325 - 335

Shoham, Y. (1988) Reasoning about Change. MIT Press. Cambridge, MA

Shore, J.E. (1986) Relative Entropy, Probabilistic Inference, and AI. In: [Kanal and Lemmer 1986], pp. 211 - 215

Shortliffe, E.H. (1976) Computer Based Medical Consultations: MYCIN. Elsevier, New York

Shortliffe, E.H., Buchanan, B.G., (1975) A Model for Inexact Reasoning in Medicine. Mathematical Biosciences **23**, 351 - 379

Slack, T.B. (1989) Advantages and a Limitation of Using LEG Nets in a Real Time Problem. In: [Kanal, Levitt, and Lemmer 1989], pp. 191 - 197

Smets, P. (1978) Un Modèle Mathématico-Statistique Simulant le Processus du Diagnostic Médical. Doctoral Dissertation at the Free University of Brussels, Presses Universitaires de Bruxelles

Smets, P. (1979) Modèle Quantitatif du Diagnostic Médical. Bulletin de l'Académie Royale de Médecine de Belgique **134**, 330 - 343

Smets, P. (1981a) Medical Diagnosis: Fuzzy Sets and Degree of Belief. Fuzzy Sets and Systems **5**, 259 - 266

Smets, P. (1981b) The Degree of Belief in a Fuzzy Event. Information Sciences **25**, 1 - 19

Smets, P. (1986a) Combining Non Distinct Pieces of Evidence. Proc. NAFIPS86, New Orleans, LA, pp. 544 - 548

Smets, P. (1986b) Bayes Theorem Generalized for Belief Functions. Proceedings European Conference on Artificial Intelligence **86** Vol. II, 164 - 170

Smets, P. (1987) Belief Functions and Bayes Theorem. In: Chorafas, D., Rowe, A. (eds.) Artificial Intelligence and Expert Systems. AMK, Berlin

Smets, P. (1988) Belief Functions. In: [Smets, Mamdani, Dubois, and Prade 1988], pp. 253 - 286

Smets, P. (1989) Constructing the Pignistic Probability Function in a Context of Uncertainty. Proc. of the 5th Workshop on Uncertainty in AI. Windsor, Canada, 319 - 326 , also in [Henrion, Shachter, Kanal, and Lemmer 1990], 29 - 40

Smets, P. (1990a) Resolving Misunderstandings About Belief Functions: A Response to J. Pearl's Criticisms in "Reasoning with Belief Functions: An Analysis of Complexity". Technical Report. IRIDIA, Université Libre de Bruxelles, Belgium

Smets, P. (1990b) Varieties of Ignorance and the Need for Well-founded Theories. To appear in: Information Sciences

Smets, P. and Magrez, P. (1987) Implication in Fuzzy Logic. Int. J. of Approximate Reasoning **1**: 4, 327 - 347

Smets, P., Mamdani, A., Dubois, D., Prade, H. (1988) Non-Standard Logics for Automated Reasoning. Academic Press, London

Smith, C.A.B. (1961) Consistency in Statistical Inference and Decision. J. Royal Statistical Society Series B **23**, 1 - 37

Smith, C.A.B. (1965) Personal Probability and Statistical Analysis. J. Royal Statistical Society Series A **128**, 469 - 499

Smith, J.Q. (1989) Influence Diagrams for Statistical Modeling. Ann. of Statistics **17**: 2, 654 - 672

Smith, R., Self, M., Cheeseman, P. (1988) Estimating Uncertain Spatial Relationships in Robotics. In: [Lemmer and Kanal 1988], pp. 435 - 461

Smithson, M. (1989) Uncertainty and Ignorance. Springer, New York

Solomonoff, R. (1986) The Application of Algorithmic Probability to Problems in Artificial Intelligence. In: [Kanal and Lemmer 1986], pp. 473 - 491

Soula, G., Vialettes, B., San Marco, J.L. (1983) PROTIS; A Fuzzy Deduction-Rule System. In: Proc. Medinfo-83. North-Holland, Amsterdam, pp. 533 - 536

Spoes, M. (1989) Syllogistic Inference under Uncertainty. Psychologie Verlagsunion, München

Spohn, W. (1990) A General Non-Probabilistic Theory of Inductive Reasoning. In: [Shachter, Levitt, Kanal, and Lemmer 1990], pp. 149 - 158

Spiegelhalter, D.J. (1985) Probabilistic Reasoning in Predictive Expert Systems. Also in: [Kanal and Lemmer 1986], pp. 47 - 67

Spiegelhalter, D.J. (1986a) A Statistical View of Uncertainty in Expert Systems. In: [Gale 1986], 17 - 55

Spiegelhalter, D.J. (1986b) Probabilistic Reasoning in Predictive Expert Systems. In: [Kanal and Lemmer 1986], pp. 47 - 67

Spiegelhalter, D.J. (1986c) Computers, Expert Systems and ADRs: Can Causality Assessment be Automated? Drug Inform. J. **20**, 543 - 550

Spiegelhalter, D.J. (1987) Probabilistic Expert Systems in Medicine. Practical Issues in Handling Uncertainty **2**: 1, 25 - 30

Spiegelhalter, D.J. (1987) A Unified Approach to Imprecision and Sensitivity of Beliefs in Expert Systems. In: [Kanal, Levitt, and Lemmer 1989], 199 - 208

Spiegelhalter, D.J. (1988) Commentary on 'An AI View of the Treatment of Uncertainty' by Alessandro Safiotti. Knowledge Engineering Review, **3**, 78 - 79

Spiegelhalter, D.J. (1989) A Unified Approach to Imprecision and Sensitivity of Beliefs in Expert Systems. In: [Kanal, Levitt, and Lemmer 1989], pp. 199 - 208

Spiegelhalter, D.J., Franklin, R.L.G., Bull, K. Assessment, Criticism and Improvement of Imprecise Subjective Probabilities for a Medical Expert System. In: [Henrion, Shachter, Kanal, and Lemmer 1990], pp 285 - 294

Spiegelhalter, D.J., Knill-Jones, R. (1984) Statistical and Knowledge-Based Approaches to Clinical Decision-Support Systems. J. Royal Statistical Society Series A **147**, 35 - 77

Spiegelhalter, D.J., Lauritzen, S.L. (1990) Sequential Updating of Conditional Probabilities on Directed Graphical Structures. Networks, **20**, 579 - 605

Star, S. (1989) Theory-Based Inductive Learning: An Integration of Symbolic and Quantitative Methods. In: [Kanal, Levitt, and Lemmer 1989], pp. 401 - 424

Star, S. (1990) Generating Decision Structures and Causal Explanations for Decision Making. In: [Shachter, Levitt, Kanal, and Lemmer 1990], pp. 43 - 57

Steve, G. (1986) Probabilistic Inferential Engines in Expert Systems: How Should the Strength of Rules be Expressed? Proc. of the IPMU. Paris, pp. 451 - 455

Stoyan, D., Kendall, W.S., Mecke, J. (1987) Stochastic Geometry and its Applications, Wiley, Chichester, UK

Strassen, V. (1964) Messfehler und Information. Zur Wahrscheinlichkeitstheorie und verw. Gebiete **2**, 273 - 305

Strat, T. (1984) Continuous Belief Functions for Evidential Reasoning. Proc. Fourth National Conf. on Artificial Intelligence. Austin, TX, pp. 308 - 313

Strat, T. (1989) Making Decisions with Belief Functions. Proc. 5th Workshop on Uncertainty in AI. Windsor, Canada, pp. 351 - 360

Strat, T., Lowrance, J.D. (1989) Explaining Evidential Analysis. Int. J. of Approximate Reasoning **3**, 299 - 353

Sugeno, M. (1974) Theory of Fuzzy Integrals and its Applications. Ph. D. Thesis, Tokyo Institute of Technology, Japan

Sugeno, M. (1985) Industrial Applications of Fuzzy Control, North-Holland, Amsterdam

Sundberg, R. (1975) Some Results about Decomposable (or Markov-Type) Models for Multidimenional Contingency Tables: Distribution of Marginals and Partitioning of Tests. Scand. J. Statist. **2**, 71 - 79

Suppes, P. (1966) Probabilistic Inference and the Concept of Total Evidence. In: Hintikka, J., Suppes, P. (eds.) Aspects of Inductive Logics. North-Holland, Amsterdam

Suppes, P., Zukotti, M. (1977) On Using Random Relations to Generate Upper and Lower Probabilities. Synthesis **36**, 427 - 440

Szeliski, R. (1989) Bayesian Modeling of Uncertainty in Low-Level Vision. Kluwer, Boston

Szolovits, P., Patil, R.S., Schwartz, W.B. (1988) Artificial Intelligence in Medical Diagnosis. Annals Internal Medicine **108**, 80 - 87

Szolovits, P., Pauker, S.G. (1987) Categorical and Probabilistic Reasoning in Medical Diagnosis. Artificial Intelligence **11**, 115 - 144

Tarjan, R.E., Yannakakis, M. (1984) Simple Linear-Time Algorithms to Test Chordality of Graphs, Test Acyclicity of Hypergraphs, and Selectively Reduce Acyclic Hypergraphs. SIAM J. Computing **13**: 3, 566 – 579

Tessen, B. (1989) Internal Probability Propagation, University of Bergen, Norway

Tong, R.M., Appelbaum, L.A. (1988) Experiments with Interval-Valued Uncertainty. In: [Lemmer and Kanal 1988], pp. 63 - 75

Torgersen, E.N. (1970) Comparison of Experiments when the Parameter Space is Finite. Wahrscheinlichkeitstheorie und verw. Gebiete **16**, 219 - 249

Tribus, M. (1969) Rational Descriptions, Decisions, and Design. Pergamon Press, NY

Shenoy, P.P., Shafer, G., Melloulli, K. (1988) Propagation of Belief Functions: A Distributed Approach. In: [Lemmer and Kanal 1988], pp. 325 - 335

Ursic, S. (1988) Generalizing Fuzzy Logic Probabilistic Inferences. In: [Lemmer and Kanal 1988], pp. 337 - 362

Vaughan, D.S., Perrin, B.M., Yadrick, R.M., Holden, P.D., Kempf, K.G. (1986) An Odds Ratio Based Inference Engine. In: [Kanal and Lemmer 1986], pp. 383 - 389

Verma, T., Pearl, J. (1990a) Causal Networks: Semantics and Expressiveness. In: [Shachter, Levitt, Kanal, and Lemmer 1990], pp. 69 - 76

Verma, T., Pearl, J. (1990b) Equivalence and Synthesis of Causal Models. Proc. of the 6th Conference of Uncertainty in AI. Cambridge, MA, pp. 220 - 227

Wald, A. (1950) Consistent Families of Measures and Their Extensions. Theory of Probability and Appl. **7**, 147 - 163

Walley, P., Fine, T.L. (1982) Towards a Frequentist Theory of Upper and Lower Probability Ann. of Statistics **10**: 3, 741 - 761

Wang, P.Z., Liu, X.H., Sanchez, E. (1986) Set Valued Statistics and its Application to Earthquake Engineering. Fuzzy Sets and Systems **18**, 347 - 356

Wang, P.Z., Sanchez, E. (1982) Treating a Fuzzy Set as a Projectable Random Set. In: Gupta, M.M., Sanchez, E. (eds.) Fuzzy Information and Decision Processes, North-Holland, Amsterdam, pp. 213 - 220

Webber, B.L., Nilsson, N.J. (1981) Readings in Artificial Intelligence. Tioga, Palo Alto, CA

Weichselberger, K., Pöhlmann, S. (1990) A Methodology for Uncertainty in Knowledge Based Systems. Lecture Notes in Artificial Intelligence **419**, Springer, Berlin

Weber, S. (1983) A General Concept of Fuzzy Connectives, Negations and Implications Based on t-Norms and t-Conorms. Fuzzy Sets and Systems **11**, 115 - 134

Weiss, S.M., Kulikowski, C.A. (1979) EXPERT: A System for Developing Consultation Models. Proc. of the 6th IJCAI. Tokyo, 942 - 947

Weiss, S.M., Kulikowski, C.A. (1983) A Practical Guide to Designing Expert Systems. Chapman and Hall, London

Weiss, S.M., (1978) A Model-Based Method for Computer Aided Medical Decision Making. Artificial Intelligence **11**, 145 - 172

Wellman, M.P. (1988) Qualitative Probabilistic Networks for Planning under Uncertainty. In: [Lemmer and Kanal 1988], pp. 197 - 208

Wen, W.X. (1990) Minimum Cross Entropy Reasoning in Recursive Causal Networks. In: [Shachter, Levitt, Kanal, and Lemmer 1990], pp. 105 - 119

Wermuth, N. (1976) Model Search Among Multiplicative Models. Biometrics **32**, 253 - 263

Wermuth, N., Lauritzen, S.L. (1983) Graphical and Recursive Models for Contingency Tables. Biometrika **70**, 537 - 552

Wermuth, N., Lauritzen, S.L. (1990) On Substantive Research Hypotheses, Conditional Independence Graphs and Graphical Chain Models (with Discussion). J. Royal Statistical Society Series B **52**, 21 - 72

Whalen, T., Schott, B. (1985) Alternative Logics for Approximate Reasoning in Expert Systems: A Comparative Study. Int. J. of Man-Machine Studies **22**: 3, 327 - 346

Whittaker, J. (1989) Graphical Models in Applied Multivariate Statistics. Wiley, Chichester, UK

Williams, P.M. (1982) Discussion. In: Shafer, G. Belief Functions and Parametric Models. J. Royal Statistical Society Series B **44**, 322 - 352

Wilson, N. (1990) Rules, Belief Functions, and Default Logic. Proc. Workshop Uncertainty in AI, Boston

Von Winterfeldt, D., Edwards, W. (1986) Decision Analysis and Behavioral Research. Cambridge University Press, New York

Wise, B.P. (1988) Experimentally Comparing Uncertain Inference Systems to Probability. In: [Lemmer and Kanal 1988], pp. 89 - 99

Wise, B.P. (1989) Satisfaction of Assumption is a Weak Predictor of Performance. In: [Kanal, Levitt, and Lemmer 1989], pp. 45 - 54

Wise, B.P., Henrion, M. (1985) A Framework for Comparing Uncertain Inference Systems to Probability. In: [Kanal and Lemmer 1986], pp. 69 - 83

Wise, B.P., Perrin, B.M., Vaughan, D.S., Yadrick, R.M. (1989) Evaluation of Uncertain Inference Models III: The Role of Tuning. In: [Kanal, Levitt, and Lemmer 1989], pp. 55 - 62

Xu, L., Pearl, J. (1989) Structuring Causal Tree Models with Continuous Variables. In: [Kanal, Levitt, and Lemmer 1989], pp. 209 - 219

Yadrick, R.M., Perrin, B.M., Vaughan, D.S., Holden, P.D., Kempf, K.G. (1988) Evaluation of Uncertain Inference Models I: PROSPECTOR. In: [Lemmer and Kanal 1988], pp. 77 - 87

Yager, R.R. (1980) Aspects of Possibilistic Uncertainty. Int. J. of Man-Machine Studies **12**: 3, 283 - 298

Yager, R.R. (1983) Some Relationships between Possibility, Truth and Certainty. Fuzzy Sets and Systems **11**, 151 - 156

Yager, R.R. (1984) Approximate Reasoning as a Basis for Rule Based Expert Systems. IEEE Transactions on Systems, Man, and Cybernetics **14**, 636 - 643

Yager, R.R. (1985a) Reasoning with Uncertainty for Expert Systems. Proc. of the 9th IJCAI. Los Angeles, pp. 1295 - 1297

Yager, R.R. (1985b) A General Approach to Decision Making with Evidential Knowledge. In: [Kanal and Lemmer 1986], pp. 317 - 327

Yager, R.R. (1985c) On the Dempster-Shafer Framework and new Combination Rules. Information Sciences **41**, 93 - 137

Yager, R.R. (1986) The Entailment Principle for Dempster Shafer Granules. Int. J. Intelligent Systems **1**, 247 - 262

Yager, R.R. (1987) Using Approximate Reasoning to Represent Default Knowledge. Artificial Intelligence **31**, 99 - 112

Yager, R.R. (1988) On Implementing Usual Values. In: [Lemmer and Kanal 1988], pp. 209 - 217

Yager, R.R. (1989) On the Representation of Commonsense Knowledge by Possibilistic Reasoning. Int. J. of Man-Machine Studies **31**, 587 - 610

Yager, R.R., Ovchinnikov, S., Tong, R.M. and Nguyen, H.T. (1987) Fuzzy Sets and Applications. (Selected papers by Zadeh, L.A.), Wiley, New York

Yannakakis, M. (1981) Computing the Maximum Fill-in is NP Complete. SIAM J. Algebraic Discrete Methods **2**, 77 - 79

Yeh, A. (1990) Predicting the Likely Behaviors of Continuous Nonlinear Systems in Equilibrium. In: [Shachter, Levitt, Kanal, and Lemmer 1990], pp. 383 - 395

Yen, J. (1986) A Reasoning Model Based on an Extended Dempster-Shafer Theory. Proc. of the AAAI-86. Philadelphia, pp. 125 - 131

Yen, J. (1987a) Can Evidence be Combined in the Dempster-Shafer Theory. In: Proc. Workshop on Uncertainty in Artificial Intelligence. University of Washington, Seattle, pp. 70 - 76

Yen, J. (1987b) Implementing Evidential Reasoning in Expert Systems. In: [Kanal, Levitt, and Lemmer 1989], pp. 333 - 344

Yen, J. (1988) Generalizing the Dempster-Shafer Theory to Fuzzy Sets. In: Proc. Workshop on Uncertainty in Artificial Intelligence. Univ. of Minnesota, MN, pp. 382 - 390

Yen, J. (1989a) GERTIS: a Dempster-Shafer Approach to Diagnosing Hierarchical Hypotheses. Communications of the ACM **32**: 5, 573 - 585

Yen, J. (1989b) Can Evidence be Combined in the Dempster-Shafer Theory. In: [Kanal, Levitt, and Lemmer 1989], pp. 63 - 69

Yen, J. (1989c) Implementing Evidential Reasoning in Expert Systems. In: [Kanal, Levitt, and Lemmer 1989], pp. 333 - 344

Yen, J. (1990) A Framework of Fuzzy Evidential Reasoning. In: [Shachter, Levitt, Kanal, and Lemmer 1990], pp. 227 - 239

Zadeh, L.A. (1965) Fuzzy Sets. Information and Control **8**, 338 - 353

Zadeh, L.A. (1968) Probability Measures of Fuzzy Events. J. Math. Anal. Appl. **10**, 421 - 427

Zadeh, L.A. (1971a) Similarity Relations and Fuzzy Orderings. Information Sciences **3**, 177 - 200

Zadeh, L.A. (1971b) Quantitative Fuzzy Semantics. Information Sciences **3**, 159 - 176

Zadeh, L.A. (1975a) Calculus of Fuzzy Restrictions. In: Zadeh. L.A., Fu, K.S., Tanaka, K., Shimura, M. (eds.) Fuzzy Sets and Their Applications to Cognitive and Decision Processes. Academic Press, New York, pp. 1 - 39

Zadeh, L.A. (1975b) The Concept of a Linguistic Variable and its Applications to Approximate Reasoning. Part I, Inform. Sci. **8**, 199 - 249, Part II, Inform. Sci. **8**, 301 - 357, Part III, Inform. Sci. **9**, 43 - 80

Zadeh, L.A. (1975c) Fuzzy Logic and Approximate Reasoning. Synthese **30**, 407 - 428

Zadeh, L.A. (1978) Fuzzy Sets as a Basis for a Theory of Possibility. Fuzzy Sets and Systems **1**, 3 - 28

Zadeh, L.A. (1979a) Approximate Reasoning Based on Fuzzy Logic. Proc. of the 6th ICJAI. Tokyo, pp. 1004 - 1010

Zadeh, L.A. (1979b) A Theory of Approximate Reasoning. In: Michie, D. Hayes, J.E., Mikulich, L.I. (eds.) Machine Intelligence **9**, pp. 149 - 194. Elsevier, Amsterdam

Zadeh, L.A. (1981) Possibility Theory and Soft Data Analysis. In: Cobb, L., Thrall, R.M. (eds.) Mathematical Frontiers of the Social and Policy Sciences. Westview, Boulder, CO, pp. 69 - 129

Zadeh, L.A. (1983) The Role of Fuzzy Logic in the Management of Uncertainty in Expert Systems. Fuzzy Sets and Systems **11**, 199 - 227

Zadeh, L.A. (1984) A Mathematical Theory of Evidence. Book Review, AI Magazine **5**: 3, 81 - 83

Zadeh, L.A. (1985a) Syllogistic Reasoning as a Basis for Combination of Evidence in Expert Systems. Proc. of the 9th IJCAI. Los Angeles, 417 - 419

Zadeh, L.A. (1985b) Syllogistic Reasoning in Fuzzy Logic and its Application to Causality and Reasoning with Dispositions. IEEE Trans. on Systems, Man, and Cybernetics **15**: 6, 745 - 763

Zadeh, L.A. (1986a) Is Probability Theory Sufficient for Dealing with Uncertainty in AI? A Negative View. In: [Kanal and Lemmer 1986], pp. 103 - 116

Zadeh, L.A. (1986b) A Simple View of the Dempster-Shafer Theory of Evidence and its Implication for the Rule of Combination. AI Magazine **7**: 2, 85 - 90

Zadeh, L.A. (1988) PRUF: A Meaning Representation Language for Natural Languages. Int. J. of Man-Machine Studies **10**, 395 - 460

Zadeh, L.A., Ralescu, A. (1986) On the Combinability of Evidence in the Dempster-Shafer Theory. In: Proc. Workshop on Uncertainty in Artificial Intelligence. Univ. of Pennsylvania, pp. 347 - 349

Zarley, D., Hsia, Y-T., Shafer, G. (1988) Evidential Reasoning Using DELIEF. Proc 7th Natl. Conf. on AI (AAAI-88), St. Paul, MN, pp. 205 - 209

Zimmer, A.C. (1986) What Uncertainty Judgements Can Tell About the Underlying Subjective Probabilities. In: [Kanal and Lemmer 1986], pp. 249 - 258

Zimmermann, H.J. (1986) Fuzzy Sets, Decision Making, and Expert Systems. Kluwer Academic Press, Dordrecht

Index

L

M

N

Springer Series
Artificial Intelligence

Former Subseries of SYMBOLIC COMPUTATION

N. J. Nilsson: Principles of Artificial Intelligence. XV, 476 pages, 139 figs., 1982

J. H. Siekmann, G. Wrightson (Eds.): Automation of Reasoning 1. Classical Papers on Computational Logic 1957-1966. XXII, 525 pages, 1983

J. H. Siekmann, G. Wrightson (Eds.): Automation of Reasoning 2. Classical Papers on Computational Logic 1967-1970. XXII, 638 pages, 1983

L. Bolc (Ed.): The Design of Interpreters, Compilers, and Editors for Augmented Transition Networks. XI, 214 pages, 72 figs., 1983

M. M. Botvinnik: Computers in Chess. Solving Inexact Search Problems. With contributions by A. I. Reznitsky, B. M. Stilman, M. A. Tsfasman, A. D. Yudin. Translated from the Russian by A. A. Brown. XIV, 158 pages, 48 figs., 1984

L. Bolc (Ed.): Natural Language Communication with Pictorial Information Systems. VII, 327 pages, 67 figs., 1984

R. S. Michalski, J. G. Carbonell, T. M. Mitchell (Eds.): Machine Learning. An Artificial Intelligence Approach. XI, 572 pages, 1984

C. Blume, W. Jakob: Programming Languages for Industrial Robots. XIII, 376 pages, 145 figs., 1986

J. W. Lloyd: Foundations of Logic Programming. Second, extended edition. XII, 212 pages, 1987

L. Bolc (Ed.): Computational Models of Learning. IX, 208 pages, 34 figs., 1987

L. Bolc (Ed.): Natural Language Parsing Systems. XVIII, 367 pages, 151 figs., 1987

N. Cercone, G. McCalla (Eds.): The Knowledge Frontier. Essays in the Representation of Knowledge. XXXV, 512 pages, 93 figs., 1987

G. Rayna: REDUCE. Software for Algebraic Computation. IX, 329 pages, 1987

D. D. McDonald, L. Bolc (Eds.): Natural Language Generation Systems. XI, 389 pages, 84 figs., 1988

L. Bolc, M. J. Coombs (Eds.): Expert System Applications. IX, 471 pages, 84 figs., 1988

C.-H. Tzeng: A Theory of Heuristic Information in Game-Tree Search. X, 107 pages, 22 figs., 1988

H. Coelho, J. C. Cotta: Prolog by Example. How to Learn, Teach and Use It. X, 382 pages, 68 figs., 1988

L. Kanal, V. Kumar (Eds.): Search in Artificial Intelligence. X, 482 pages, 67 figs., 1988

H. Abramson, V. Dahl: Logic Grammars. XIV, 234 pages, 40 figs., 1989

R. Hausser: Computation of Language. An Essay on Syntax, Semantics, and Pragmatics in Natural Man-Machine Communication. XVI, 425 pages, 1989

P. Besnard: An Introduction to Default Logic. XI, 201 pages, 1989

A. Kobsa, W. Wahlster (Eds.): User Models in Dialog Systems. XI, 471 pages, 113 figs., 1989

B. D'Ambrosio: Qualitative Process Theory Using Linguistic Variables. X, 156 pages, 22 figs., 1989

V. Kumar, P. S. Gopalakrishnan, L. N. Kanal (Eds.) Parallel Algorithms for Machine Intelligence and Vision. XI, 433 pages, 148 figs., 1990

Y. Peng, J. A. Reggia: Abductive Inference Models for Diagnostic Problem-Solving. XII, 284 pages, 25 figs., 1990

A. Bundy (Ed.): Catalogue of Artificial Intelligence Techniques. Third, revised edition. XV, 179 pages, 1990

D. Navinchandra: Exploration and Innovation in Design. XI, 196 pages, 51 figs., 1991

R. Kruse, E. Schwecke, J. Heinsohn: Uncertainty and Vagueness in Knowledge Based Systems. Numerical Methods. XI, 491 pages, 59 figs., 1991